History of Mathematics Education

Series Editors
Nerida F. Ellerton
M.A. Ken Clements

More information about this series at http://www.springer.com/series/13545

Kay Owens and Glen Lean with Patricia Paraide
and Charly Muke

History of Number

Evidence from Papua New Guinea and Oceania

Springer

Kay Owens
Charles Sturt University
Dubbo, NSW, Australia

Glen Lean

with

Patricia Paraide
Divine Word University
Madang, Papua New Guinea

Charly Muke
St. Teresa's College
Abergowrie, QLD, Australia

ISSN 2509-9736 ISSN 2509-9744 (electronic)
History of Mathematics Education
ISBN 978-3-319-45482-5 ISBN 978-3-319-45483-2 (eBook)
DOI 10.1007/978-3-319-45483-2

Library of Congress Control Number: 2016953735

Printed on acid-free paper

This Springer imprint is published by Springer Nature
The registered company is Springer International Publishing AG
The registered company address is: Gewerbestrasse 11, 6330 Cham, Switzerland

Foreword

This book is remarkable in many ways. It contributes fresh data to discussions about the history of numbers in the South Sea regions. The data sets are huge, reflecting collections of number information over some 20 years. The book was not written by the chief researcher Glendon Lean, who sadly died in 1995. It was he who collected data on the many field trips he made while living in Papua New Guinea. He was fluent in the lingua franca and a couple of the PNG languages with some understanding of several others and thus was well able to interpret the rich data sources available to a scholar like him.

I was privileged to collaborate with him and finally to help him complete his PhD, which consisted of the thesis plus four equally huge appendices. These together document not just the routine number ideas that we take for granted in our western European culture but the many and varied ways in which number and number representations can be found in diverse cultures, still to this day. In addition, he used the historical and linguistic background data to illustrate the antiquity of when these seem likely to have developed. He also illustrates how variations and similarities occurred.

Dr Kay Owens is equally to be congratulated on creating a book which helps to clarify several of the current issues about the development of number ideas in a rich cultural context like that of PNG. With Patricia Paraide and the other researchers, Kay has illustrated the significance of this history of number to mathematics education today. It redresses the lack of previous knowledge of these ancient systems. It is a formidable resource for future scholars, and hopefully this book will inspire them to continue the analyses that Glendon Lean and Kay Owens have started. The ethnographic database that Dr Owens has created at the University of Goroka has helped not just the PNG scholars but scholars from everywhere to explore what this study has started. It would be a tragedy if it were unavailable to others or left like that. Glendon Lean's commitment to his colleagues and communities in PNG deserves much more.

Monash University
Clayton, VIC, Australia

Alan Bishop

Preface

Papua New Guinea and Oceania (including West Papua, the other half of New Guinea) are unique in having the highest number of distinct languages in the smallest land mass in the world and also in having related languages spread across huge ocean spaces. This extraordinary ecology creates an incredible opportunity to explore the history of number. There are indeed 1 300 languages in this region. That means 1 300 cultures. The story that unfolds in this book reaches back to an era between 1 500 and 40 000 years ago, and yet the story is still active and part of the lives of the same genealogies of people today. This active story was very much the case during the period of first contact with Europeans which began sporadically in the late 1700s but was only significant after the 1880s. Places in this story have been among some of the last to be infiltrated by European colonialists, some as recently as 60 years ago. Some of the data drawn upon for this book were the first to be recorded for the language and culture at first contact by Europeans, but much of the data were less than 50 years old. Data from living cultures were mostly recorded in the 1970s and have been compared where possible with older records. Based on variations between existing languages that are linguistically and mathematically analysed together with evidence from other disciplines such as archaeology, the ancient story of number is told.

My research and experience in Papua New Guinea (PNG) has informed my writing of this story as it did an earlier Springer book *Visuospatial Reasoning: An Ecocultural Perspective on Space, Geometry and Measurement Education*. When that book was finished, Ken Clements, one of the editors of the new series on the history of mathematics education, emailed me. He asked if I would be willing to write a history of number from the perspective of Papua New Guinea and Oceania based on Glen Lean's work. This request came because he was well aware, unlike most historians of mathematics and mathematics education, of the incredible work of his friend and colleague Glendon Angove Lean, and he was aware that my interest in Glen's work and experience in Papua New Guinea provided a good basis for writing this book. It is a story that needs to be known by others.

In 1973, I went to live in Lae, PNG, and to work in the Mathematics Department at the PNG University of Technology beginning a life-long friendship with Glen Lean who sadly passed away in 1995, a week after a special ceremony to confer his PhD. For 20 years, Glen collected data on counting systems of Papua New Guinea and Oceania which culminated in his thesis being finished in 1993, but it remained unpublished until this book incorporated it. His academic mentor and executor, Alan Bishop, sent as much of Glen's collected materials to the University of Goroka (UoG) in PNG where the Glen Lean Ethnomathematics Centre (GLEC) was established. With my husband, we were able to catalogue Glen's photocopied papers, but unfortunately half of the material was lost through the climes and transport of PNG (some we were able to replace). The University of Goroka holds the most complete copy of his resources in one place. With funds from USA's National Science Foundation, with a small team of PNG lecturers and research assistants, we were able to work on Glen's collection of counting systems and put them into a database available from the GLEC website, but the database has not been readily available for some years. This database was developed from poor and incomplete electronic copies and a rare hard copy of his provincial summaries upon which he developed his thesis.

We prepared his thesis for the GLEC website together with copies of some journal articles and other materials that Glen had sourced. The references in this book support the arguments presented in the following chapters but also provide the data sources for around 100 languages given in the Appendices and selected as evidence for the arguments. The extent of the reference list indicates how much material Glen collated for 900 languages, that is, 75% of the region's languages. However, the data collected by Glen from students and others from 1968 to 1988 provide the really rich collection of living data that are evidentially telling the ancient story of their ancestors. I say "evidentially" as this book also presents the analytic, linguistic and archaeological evidence to support this statement.

Chapter 2 indicates the nature of Glen's data collection and reflects Glen's unique abilities that meant that he was able to collate the data from written records and thousands of questionnaires completed over 20 years by PNG and Oceania students and teachers. He searched maps, he asked questions, he visited remote places and with his linguistic, historical and mathematical skills, he was able to analyse and synthesise this data. His data from people also relied on his rapport with the many PNG people with whom he associated. Again Glen's rapport was an exceptional skill to which his many friends and colleagues can testify. My forays into collecting and checking data have indicated how extraordinary his work was and the nature of his decision-making in summarising the data (e.g. Owens, 2001). One of the main issues has been that of collecting data from people whose languages and to some extent cultures are rapidly changing. Another problem was in the multiple names given to any one language and the variations between villages who speak the same language or even the same dialect of the language.

To convert Glen's work into a book readable in APA style from the poor electronic copies of his thesis with its footnote format that referred by code to the tables in his huge appendices required considerable tedious effort and PNG knowledge. His work has been extended by more recent linguistic, anthropological and archaeological papers. While Glen did place his thesis into the cultural contexts of the languages and counting systems, this book goes further. It also gives credit to the work of Papua New Guineans and others who have pursued ethnomathematics as a significant area for PNG mathematics education. Patricia Paraide and Charly Muke are authors on chapters which specifically incorporate research on their own cultures. I would also like to acknowledge the team of linguists at the Australian National University; many SIL linguists and translators; Ralph Lawton and the team of Kilivila translators; Geoffery Saxe; my former UoG colleagues Wilfred Kaleva, Martin Imong, Rex Matang (deceased), Gairo Onagi and Samuel Kopamu who have shared their counting systems and ideas with me; my many PNG friends, colleagues and students who have accepted me into their families; and researchers Joseph Fisher, Peter Dwyer, Mark Donahue and Jadran Mimica whose works have significance for this book. Geoff Smith and Sue Holzknecht are long-term friends and were colleagues at the PNG University of Technology working in the fields of language and linguistics. They have shared much with me prior to my writing this book. I sincerely thank them.

This book portrays a history of number from an ancient time uninfluenced by events occurring in the Middle East region. It is a story that begins at an often much older time than developments in the Middle East and upon which the Indo-European centric histories of number are often based. Briefly, Lean argued that the types of counting systems within different phyla and protolanguages indicate the antiquity of the systems when combined with archaeological evidence. The system classifications also indicate how counting systems spread and changed. I have been fascinated by the alternative systems of counting portrayed in this story, and I trust the reader will also be enthralled. It is my belief that without this history being taught, there is a serious lack of richness in the history of mathematics and counting systems for any school student across the world.

Owens, K. (2001). The work of Glendon Lean on the counting systems of Papua New Guinea and Oceania. *Mathematics Education Research Journal, 13*(1), 47–71.

Dubbo, NSW, Australia Kay Owens

Acknowledgements

This book is only possible because many, many Papua New Guineans and linguists shared their languages and aspects of their mathematical culture with others. I wish to thank them sincerely. I hope this book does justice to knowledge generated by the sharing of many, many communities. I wish to thank my very patient husband who has also proof read the whole book and Sandra Stewart, Melissa McNair and Laura White for their assistance with formatting the many tables at different stages of preparing this manuscript. I also acknowledge the encouragement of Ken Clements and Alan Bishop in their respective roles as series editor and Glen's supervisor and academic executor who are both pleased that Glen's work is being more widely recognised in the world as we believe it is ground--breaking. It has been an honour to extend his work in this book.

Contents

List of Figures

List of Tables

Chapter 1
An Overview of the Studies, Papua New Guinea, Oceania, Languages and Migrations

Kay Owens and Glen Lean

Abstract This multidisciplinary study draws on archaeological information, linguistics, and an understanding of mathematics. A summary of the background archaeology for New Guinea and Oceania provides some evidence of the longevity of these cultures and the archaeological evidence for the spread of languages in New Guinea and Oceania. The diversity of language groups is a result of movement, colonisation, influence and innovation over time. The overview presented in this chapter permits the reader to link the pursuing discussion in a time and place. The chapter finishes with an overview of the book that sets out the diversity of counting system cycles, where they are established, and how they may have developed.

Keywords Lapita • migrations in the Pacific • Oceania • Papua New Guinea • prehistory • studies of counting systems • West Pacific languages

Introduction

Over the past 250 years the study of systems of numeration as they occur in natural languages has sometimes been of interest to mathematicians and others while at other times it has been of little interest. During the nineteenth century, we find that the published literature of the explorers, administrators and missionaries of the European colonising powers provides a rich source of information on the "exotic" languages and customs of the peoples of Africa, the Americas, the Far East, and Oceania. Indeed this material still serves today as a major source of data on non-European counting systems. The study of numerals and counting systems was initiated largely by scholars with an interest in linguistics and anthropology but during the twentieth century there was a marked drop in interest on the part of both linguists and anthropologists and the subject was relegated to one with only marginal status (Hurford, 1987). Numerals, it would seem, are strange animals which do not possess, for the linguist, the same intrinsic interest as other aspects of linguistic analysis, or, for the ethnographer, the same intellectual challenge as unravelling a complex kinship system or social structure. They are, perhaps, best left to the historians or philosophers of mathematics.

Philosophers of mathematics tend to take the view that the study of natural language numerals has little bearing on the understanding of the nature of number, an abstract concept independent of nominal linguistic vagaries. Those sections which deal specifically with number in standard histories of

K. Owens (✉)
Charles Sturt University, Dubbo, NSW, Australia
e-mail: kowens@csu.edu.au

G. Lean

K. Owens et al., *History of Number*, History of Mathematics Education,
DOI 10.1007/978-3-319-45483-2_1

mathematics (Boyer & Merzbach, 1989; Dantzig, 1954; Struik, 1987), usually begin with the advent of written numerals perhaps for no other reason than that historians prefer to work from written records. Also, until relatively recently, there has been a marked Eurocentric bias in the Western accounts of the development of mathematical ideas, and even the contributions of India, China and Islam have been given scant attention (Joseph, 2000). In most cases, the occurrence of mathematical ideas as they existed in "primitive" tribes, is dealt with in a speculative or dismissive paragraph or two.

There are several points which should be made concerning this lack of interest in the numerals and counting systems which existed in societies outside the usual ambit of historians. First, those societies which do not possess a tradition of historical documentation are those in which the majority of the world's languages are spoken; their exclusion from the historical record means that the numerically greater part of the counting systems used by humankind is ignored. Second, those societies which are the focus of historical scholarship possess a relatively small range of counting system types. Generally speaking, their counting systems possess a base of 10, although several systems have features or irregularities which suggest vestiges of base 12 or base 20 systems: English and German being examples of the former, and French and Dutch being examples of the latter. The inference which may be drawn from this highly selective view is that humankind's response to the enumeration of its world is largely consistent across both time and cultures. Third, it may be useful to consider whether the absence of historical documentation for these societies means that we are unable to construct a *prehistory* of number. Given the information that has accumulated in the disciplines of anthropology and archaeology, an extensive database of natural language numerals and counting systems, together with the methods developed by comparative linguists, it may be theoretically possible to construct a tentative outline of the prehistoric development of the concept of number in human societies in the millennia preceding the advent of written numerals.

In addition, many who study non-Western systems often consider them as ancient and of historical interest only. They are unaware that there are Indigenous communities today that can trace back their ancestry and may continue to use foundational practices for counting and mathematical practices today. Finally there are some that dismiss much of the old record of counting systems as irrelevant or simplistic and of need of being replaced by modern and universal systems of number. In so doing, they fail to realise the importance of identity in the learning of mathematics and its value in establishing mathematical identity. It is theorised that ecocultural identity is an important way forward in establishing a self-regulating and motivated learner who can think mathematically (Owens, 2014, 2015).

Fortunately, since the mid-1960s, there has been a renaissance of interest in documenting the mathematical ideas existing in non-Western societies. This has been partly due to the work of cross-cultural psychologists, for example Berry and Dasen (1974), Cole, Gay, Glick, and Sharp (1971), Dasen and de Ribaupierre (1987), Gay and Cole (1967), Saxe (1985, 1991, 2012; Saxe & Esmonde, 2005), and Pica, Izard and Dehaene and colleagues (e.g., Izard, Pica, Spelke, & Dehaene, 2011; Pica, Lemer, Izard, & Dehaene, 2004). All these researchers have been interested in situated cognition and the cultural context of learning and thinking, especially as it relates to the development of mathematical concepts. In particular, a literature of "ethnomathematics" has been established which has dealt, for example, with the mathematical ideas existing in African societies (Zaslavsky, 1973) and in North and Central American Indian societies (Ascher, 1994, 2002; Closs, 1986). The ethnomathematical approach has even been extended to include the study of identifiable cultural (and sub-cultural) groups within Western societies as well as non-Western societies, and is concerned, for example, with the mathematics used by gamblers (Pickles, 2013) or supermarket shoppers (D'Ambrosio, 2006; Lave, 1988). Paulus Gerdes (1999) and colleagues, in many publications, were able to link school mathematics to the clever cultural mathematics of many different cultural groups in many different practices, one of which was creating 3D objects from 2D surfaces or through different patterns and structures of weaving. More importantly Gerdes and his co-workers noted how new designs were being created and produced through apparent visual patterning. Adam (2010) indicated how advanced mathematical manipulation could be discussed with basket weavers who could discuss the relevance of certain changes. Eglash (1999) documented the use of fractals in African societies and subsequently with colleagues extended the link of ethnomathematics to technology (e.g., Eglash, 2010) while Rosa and Orey (2012) significantly illustrated the link between ethnomathematics and advanced mathematics through "ethnomodelling."

The renewed interest in counting systems of non-Western societies has made its mark in more recent histories of number and mathematics as may be found in works by Flegg (1984, 1989), Ifrah (1987), Van der Waerden and Flegg (1975a, 1975b) and Joseph (2000), but others such as Cooke (2011) recall the ancient systems with little update. Much of the material, however, on which these studies were based derives from the nineteenth century (Frobenius, 1900). Until the work of Pacific Resources in Education and Learning after 1990, there had not been an increase in the database commensurate with the increase of interest in the systems themselves. This is particularly true of the large geographical region which encompasses Melanesia, Polynesia and Micronesia and which has received little attention in the ethnomathematical literature and virtually none in the historical literature, even though this region accounts for over a quarter of the world's languages (Pawley, 1981). Goetzfridt's (2008) extensive bibliography and analysis of these references showed how significant the mathematics of these cultures were and that there was recorded material although much of this knowledge is oral (Goetzfridt, 2012). Lean's (1992) study, in part, attempted to redress the situation by, first, providing documentation of the natural language numeral systems used in the cultures of the New Guinea area and the Pacific, and, second, by investigating the implications of these for the history, or prehistory, of number. It remains the most extensive study of the region and a classic source for further exploration of the history of number or the number systems of the region. Some recent anthropological linguistic studies have critiqued earlier migration studies that were based on Lapita pottery findings across the Pacific. Lean had used the earlier theories to develop his thesis on migration of languages. For the most part we present Lean's argument and discuss recent research in Chapters 9 and 10 (Ross, 1988).

The diversity of the languages and the rapid change in languages and cultural experiences and practices means that an awareness of the early contact data together with a recognition of developmental change are important for several reasons. First, they are important for transitions from cultural mathematics to school mathematics. It provides awareness that may enhance the value of retention of culture, promote identification and rediscovery of cultural knowledge, and most importantly use their strengths to initiate an understanding of school mathematics. Second, they are important for bilingual education in a Papua New Guinea (PNG) context where there is limited knowledge of how counting systems can be incorporated well into bilingual education or majority-English language of instruction (espoused by the current re-reform of education). Finally, they are important for a greater understanding of the history of number in the world and for an understanding of school mathematics in a global context, including any one child having a greater appreciation of the composite (group) units of the base 10 place value system.

Purpose of the Study

The original aims of this study were, first, to clarify the number and nature of the counting systems used in the foundational societies of Papua New Guinea and, second, to investigate whether the accumulated data had any implications for several theoretical questions which arise in the literature of the history and development of the concept of number in human societies. For example, were the structural/cyclic features of natural language numeral systems invented independently in various societies in different geographical locations, or were they diffused from a single source? Is there any evidence for a chronological developmental sequence of counting system types; that is to say, can we construct something approaching a prehistory of number?

Much of the existing data on numerals and counting systems derive, as noted above, from information collected during the nineteenth century in the foundational societies of Africa and the Americas. Published data are also available for a number of Polynesian societies of the Pacific region, however the amount of data available for PNG appeared to be relatively small until Glen Lean began his work. Even though the population of PNG, being about 6.5 million, is a small proportion of the world's total, the linguistic situation is nevertheless very complex with a total of about 850 languages, or a sixth of the world's spoken languages within its borders (SIL, nd). To document the nature of the counting systems of even half the total number of these languages would make a significant addition to the database from which scholars have worked in the past. This was the significant 22-year study undertaken by Glendon Angove Lean (prior to his death in 1995) and this forms the basis of this book. We extend on this material and explore one question in more depth: how cultural contexts have impacted on counting and how that has influenced the development of number through different usages.

Tour of the Major Studies

Smith's Studies

We draw on several other studies of counting or number systems in Papua New Guinea that are forming significant parts of the argument for this book. These include Geoff Smith's (1984) study of Morobe's counting systems. This work informed Glen Lean and will be incorporated into the chapters of this book based on Lean's work. Smith made an exhaustive study of the Morobe counting systems. He collected his data by visiting most of the languages in situ in the Province especially if he was unable to source reliable older speakers in Lae, the capital. This meant long walks into the mountains on either side of the large Markham valley. Both Smith and Lean used a framework of cycle systems for describing the multitude of systems and both frameworks are elaborated further in Chapter 2. Lean based his database of counting systems in Morobe largely on Smith's work. A number of the Morobe languages are also spoken in neighbouring Provinces such as Madang and Eastern Highlands Province. Smith (1978) had earlier recorded the counting system of the Kiwai in Western Province and made a summary of counting in the highlands regions in terms of business encounters. In addition, he noted that in many areas the efficient formal counting of valuable objects on ceremonial occasions was necessary, and commented on how the people in the Highlands Region of PNG recorded the quantity of goods (Smith, 1981). He found complex cycles associated with the distribution of pigs, shells, and other valuables in the Western Highlands, Southern Highlands, and Enga Provinces. He noted that a variety of counting methods can be used for recording these transactions.

Lean's Study

Lean began the collection of data on PNG natural language numerals and counting systems early in 1968 and continued to collect these from both primary and secondary sources until 1988. The documentation of the accumulated data began in 1985 and a final revised version was completed at the beginning of 1991: this material can be found in the databases held on the Glen Lean Ethnomathematics Centre website (GLEC, 2008). In all, data were acquired on the numerals, counting or tally systems of 532 languages, or just over 70% of the total, then believed to be about 750. Generally speaking, the languages for which it was difficult to obtain data were those with 500 speakers or less; in the Madang Province alone some 93 languages fall into that category. In the ensuing years, SIL have determined many more languages identified by the people as distinct languages. In Lean's work much of this detail was lost when he collated the information from a number of different sources, especially those completed by students on his Counting System Questionnaire as he tried to select the best representation from his sources.

In contrast to the relatively homogeneous cultural and linguistic situation existing in Polynesia, the diversity of the cultures and languages existing in Melanesia has been commented on by many scholars (e.g., Carrier, 1981; Kettenis, 1978; Lancy, 1983; Saxe, 1979, 1981a, 1981b, 1981c, 1982; Smith, 1978). This is particularly true of PNG where the diversity appears to be greatest and in this sense is untypical of its neighbours with the exception, perhaps, of Timor Leste and of West Papua (currently controlled by Indonesia), the complementary western half of the island of New Guinea. In order to see whether this diversity extended to counting systems and whether, therefore, the PNG data already gathered were untypical of the region as a whole, it was decided to extend the geographical scope of the original study to include the remainder of Melanesia, which is termed "Island Melanesia", and the islands of Polynesia and Micronesia.

This study, then, deals with the counting systems of Papua New Guinea and Oceania. Oceania includes West Papua, the Solomon Islands, Vanuatu, New Caledonia, Fiji, and Rotuma. In addition, Lean included all of Polynesia, both Triangle and Outlier, and those islands of Micronesia on which Oceanic Austronesian languages are spoken: effectively, that excludes the western part of Micronesia which contains the majority of the Marianas, where Chamorro is spoken, the Palau (Belau) Islands, and Yap. The islands to the west and south of West Papua, which include Timor, Halmahera, and the Kai and Aru Islands, have also been excluded although we do have some data from Timor Leste. The other major area excluded in the original analysis by Lean is Australia and although some reference will be made to Australian languages these are, nevertheless, untypical of the rest of Oceania whose languages are generally classified as either Austronesian (AN) or Non-Austronesian (NAN)–also called Papuan. These classifications are discussed further below. In Oceania, a further 450 languages or so are spoken, thus giving a total for the region under consideration, if we include PNG, of about 1300 (see Figure 1.1 which indicates this region). Occasionally it has been useful to refer to New Guinea by which is meant the combined regions of West Papua and PNG. However, there are now many studies of number in Australian languages and these will be referred to in this book as comparisons and in terms of the importance of this historical study for mathematics education with Indigenous communities more widely.

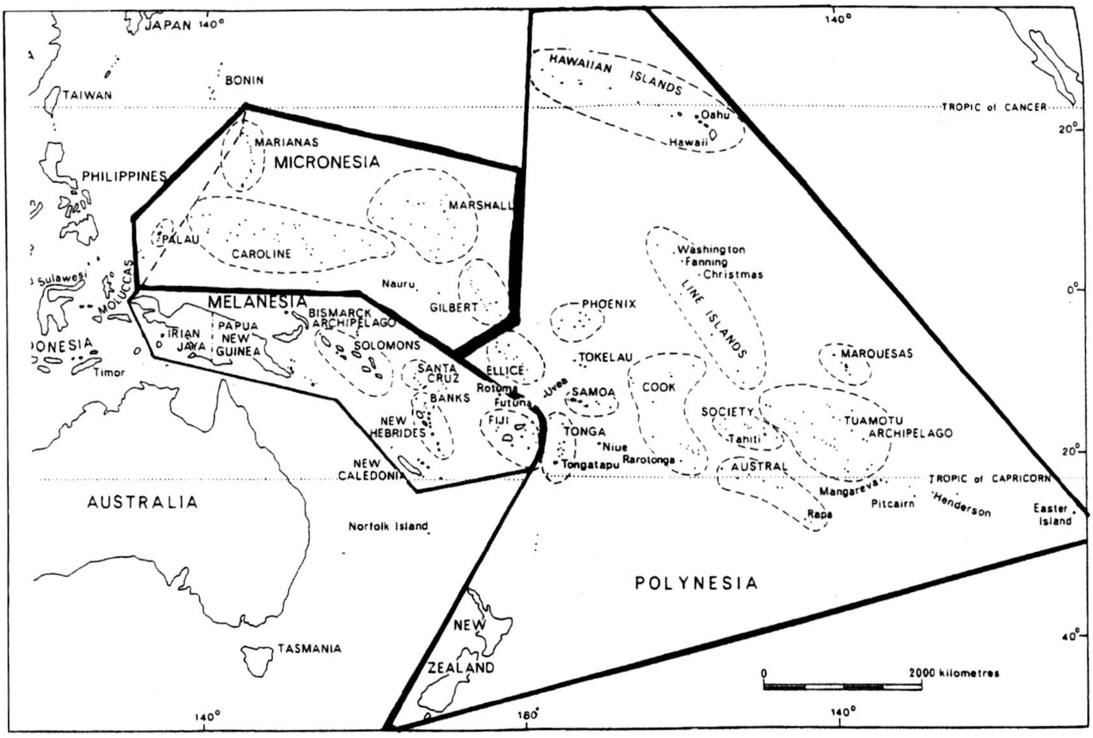

Figure 1.1. Melanesia, Micronesia, and Polynesia.

Muke's Study in Wahgi Valley, Jiwaka PNG

Charly Muke undertook a study of counting of his own language in Wahgi valley in the Jiwaka Province (formerly part of Western Highlands Province). Like Lean, Muke (2000) undertook a cultural study of counting based on the view that mathematics is not, as the Ancient Greeks and as mathematicians down the ages have suggested, as singular and definite but developed in a cultural context (Ernest, 1996, 2012). His ethnographic study of his own language addressed the following research questions:

- What are the Wahgi counting systems of both the past and present in terms of:
  - verbal communications: counting names
  - symbols: both written and on artefacts
  - practice: other practices of counting?
- What are the social relationships (e.g. conceptual, understanding, values, beliefs etc.) in the culture associated with counting practices?
- In what contexts are Wahgi counting systems found?
- How can this information be used to contribute to the learning of mathematics in the school system?

One strength of Muke's study is that he not only spoke the language dialect Yu Wooi with an understanding of the other dialects but also gathered data from 73 participants (30 teachers, 40 students and 3 villagers), 72 through questionnaire and one through lengthy discussions with his father, a villager. From this number of participants, 58 were speakers of Yu Wooi including the interviewee, living in nine tribes. The others spoke the Kumai, Yu Nimbang and North Wahgi dialects. As a result he was aware of the diversity within the counting system and its context.

Paraide's Study of Tolai Language and Cultural Practices

Patricia Paraide (2010) documented the number and measurement knowledge of her Tolai community (East New Britain Province) and their cultural activities including those related to business. One of her research questions was "How is Indigenous knowledge of number and measurement positioned and/or constructed, compared to Western number and measurement?" She particularly researched this question within the cultural context at the time of her study but cognisant of her cultural knowledge heritage. For the Tolai, number and measurement knowledge play an important part of their roles in community. Number and measurement knowledge is integrated into the people's everyday activities such as gardening, fishing, dancing, feasting, and building houses and canoes. Paraide also presented possible reasons for the diminishing status of some of the counting systems of particular items in this community. Moreover, she focused on the various power plays that occur between teachers, parents, and students which contribute to changes in attitudes and behaviour towards vernacular instruction and Indigenous knowledge.

For her case study, she used classroom and community observations, teachers, community members, parents' interviews, students' discussions, field notes, informal discussions, participation in the research participants' community activities, and her knowledge of her own Indigenous number and measurement, as sources of data which are commonly used in qualitative methods of research. Interviews with Elders were especially important. The students in this community speak

Tinatatuna (also called Tolai or Kuanua), which was also the language of instruction in the Elementary School and bilingual instruction in the Lower Primary School site. At the time of the study, all the Lower Primary teachers in the school site spoke Tinatatuna. The participation of these two school sites allowed her to pose provocative questions regarding the implementation of vernacular and bilingual instruction and the integration of Indigenous and Western mathematical knowledge at this level. She focused on ten children who were in Grade 2 in Elementary School and their transition to Primary School (Grade 3). She held focus groups with the children and also interviewed the teachers of these two schools and the ten parents, and observed four mathematics lessons in each classroom.

Owens' Studies

When Lean died, his documents were sent back to PNG to the University of Goroka. Kay Owens undertook to put the counting systems into a database and make it available on the website of the Glen Lean Ethnomathematics Centre (GLEC, 2008). At the time, Rex Matang was Director of the Centre and he assisted with this work.[1] Subsequently, he carried out a study to explore the impact of language of instruction on students' learning of mathematics concepts. Although the data are now available in Word documents, the original database is still available (it includes maps and diagrams that are not in the Word documents) and basic data are available in an EXCEL file. Owens also read and selected for digitising appropriate sections of the available documents that formed the basis of Lean's work. As a result, she became fully immersed in Lean's study, and has written a number of papers on the work (e.g., Owens, 2001). In addition, Owens and Wilfred Kaleva undertook a study on measurement in PNG.[2] There was a considerable amount of interviewing and gathering of data on counting and its use in practice during this study. They read all 350 projects written by students at the University of Goroka and analysed their contents. Many of the students included counting systems in their data within cultural contexts. Owens visited several villages to gather data and in the process collected further counting system data that expanded on that collected by Lean. She also collected data from linguistics research published since Lean's time. This work also informs the chapters of this book and the status of local counting systems in a changing society. One of her pursuits was to investigate the finding by Wassmann and Dasen on the body-part tally system of the Yupno (Wassmann & Dasen, 1994). She carried out a comparative analysis with Lean's work. The two former researchers found a counting system from one male Elder who gave details of a system which had not been previously recorded and was not known by other age groups within the community. Owens (2001) critically examined their findings in the light of Lean's work and followed up the likelihood of such a system during the measurement study (Owens, 2010; Owens & Kaleva, 2008) with villagers from the Yupno area, in particular from the village in which Wassmann and Dasen carried out their developmental psychology research work with Keck (Keck, 1998; Wassmann, 1997).

[1] PREL, Pacific Resources in Education and Learning in Hawaii included this project in a USA National Science Foundation grant.

[2] This study was funded by a Charles Sturt University Competitive Grant.

Saxe's Studies

Results of a significant study on how developments and change in number systems occurred provided rich data for Oksapmin, collected at several points in time by Geoffery Saxe (2012, Saxe & Esmonde, 2005). This study has also informed the discussion in this book. Like Wassmann and Dasen's (1994) study on Yupno counting, Saxe's study had its origins in concepts and ideas found in literature from anthropological cognitive psychology or culture-based developmental psychology. The original work by Saxe recorded the body-part tally system of the Oksapmin (Saxe, 1981a, 1982). Subsequently, he returned and recorded the impact of change on the counting systems of significant events such as people going away to work for a salary, using trade stores, and schooling. Like Paraide, he looked at power relationships and developed a model to explain the developmental changes in individuals and society on number concepts and language meaning. These studies have also informed the arguments in this book.

Matang's Studies

Rex Matang died before completing the write-up of his doctoral study, but ideas from that study were important for mathematics education. Matang first gathered data from children in elementary schools in his own Kâte area (Matang & Owens, 2006). He gave each child an individual interview on early arithmetic and collated and compared the results from schools where the language of instruction was Kâte or Kâte with the creole Tok Pisin or with English. The questions were asked in the language that the child used to respond. Subsequently he retested some children and also collected data from other areas of PNG where he had taught and learnt sufficient of the language to communicate. He compared the conceptual understanding and fluency of children who had some teaching in Tok Ples (the language of the village) with those who had none (Matang & Owens, 2014).

Other Studies

A number of studies have significantly informed this book and are recognised as providing often fairly recent materials on the counting systems of PNG. There are too many to mention them all here but I select a few to illustrate the diverse nature of these studies, often within a single language or language Family. Perhaps this highlights how extraordinary Lean's work was in attempting to collate and analyse the majority of the hundreds of languages. However, although he himself appreciated and did discuss some of the cultural contexts for some of these languages, he focused on the nature and history of the number systems.

Fisher's Study. Joseph Fisher (~2010) provided details relating to the counting system of Kuruti, a language of Manus Province, PNG. In doing so, he noted the complexities of a classificatory system in more detail than was provided in Lean's records. He also noted the importance of his study for teaching elementary children how arithmetic works both in Kuruti and in English and how they could support each other.

Mimica's Study. Jardin Mimica's (1988) study considered the development of rational thought of the Iqwaye in which number were incorporated into the kinship system and cosmology of the Iqwaye who live on the border of Eastern Highlands, Gulf and Morobe. Part of the analysis compared the system with neighbouring counting systems. This was an extraordinary insight into the complexity of the culture in which numbers were used and developed up to very large, indeed, infinity.

Dwyer and Minnegal's Study. Peter Dwyer and Monica Minnegal (2016) first collected data in 1979 for Etolo in the Southern Highlands on the border of Western Province. With the records of Lean and a number of other long-term partial residents among the Strickland and Bosavi language groups, they have recently summarised data on the body-part tally systems of these various languages.

Prehistory in New Guinea

During the period of the last ice-age, Australia and New Guinea were connected along with other islands. There is evidence, from northern Australia, of human occupation some 55 000 years ago. There is evidence from places in the Owen Stanley Range of occupation as early as 52 000 years ago, in Manus of early occupation, New Ireland of occupation around 33 000 years ago, from Buka in Bougainville from 29 000 and New Britain from 11 900 years ago. Initially the people were hunter gatherers although there were managed grasslands and forests. There is, from 10 000 years ago, evidence of agriculture in the highland valleys where drainage was used to run off excess water and pollen records suggest land clearing while along the coast and in the Sepik and Ramu basin (north coast rivers) 6 700 years ago with evidence of tree agriculture (O'Connell & Allen, 2004; White, O'Connell, Koettig, 1982). This is discussed further in Chapter 10.

In nearby regions such as the Solomon Islands and Vanuatu, most studies have focused on material records of human occupation occurring with the appearance of the Lapita culture indicated by a specific designed pottery. However, there is evidence of older occupation and the languages and influences on counting systems provides evidence of these earlier occupations. The earliest appearances of the Lapita pottery were in the Bismarck Archipelago (New Guinea Islands region) with varied and complex pots with finely executed decoration dating from 3 600 to 3 200 years ago. The second period from 3 200 years ago occurs in the Solomons, Vanuatu and New Caledonia with fewer pot designs and courser stamp designs. The designs did not occur on pieces around 2 500 years ago. The pottery in the third area and period is found in Fiji and Western Polynesia, from 3 000 years ago and continuing until 2 000 years ago, with fewer designs and course stamps. At the same time Oceanic pigs, dogs and chickens and rat appear across these regions with Island South East Asia as their source. In addition, an adze reminiscent of Island South East Asia and necklaces occur together with obsidian from Talasea in West New Britain and Lou Island (Manus area) (Clark, 2003; Green, 2003; Sheppard, Walter, & Roga, 2010; Spriggs, 2011). Pre-Lapita, this obsidian is found on the north coast of New Guinea and as far away as Nissan Island towards Bougainville but after Lapita it is found from Borneo to Fiji. There is also greater forest clearing, erosion and agricultural evidence. South East Asian and New Guinea mainland plants are found spreading out across the Lapita region. There is some evidence of continuity from Pre-Lapita to Lapita times with some nut exploitation, shell adzes, one-piece fish hooks, beads, necklaces and armlets. An earlier obsidian blade culture at Talasea finished but some of the rock shelters also seemed not to be used (Spriggs, 2006). Spriggs suggested that the Island Southeast Asian influence mixed with the existing culturals of the Bismarck and developed for over 400 years before moving out over the Pacific. Furthermore, the non-dentate-stamped pottery from the New Guinea area and obsidian spread from the Bismarck out to Fiji from 1 500 years ago suggesting some continuity of contact.

One piece of Lapita pottery has been found on the mainland north coast of New Guinea, but along the south coast there is extensive long-distance evidence of trade of Fergusson Island (Milne Bay Province) obsidian and red-slipped pottery from the coast from 200 CE. There is also evidence of the Bismarck Lapita material culture on the islands of the Torres Strait of Australia who had visitors from PNG (Butler & Dean, 2013). The Sepik-Ramu Madang pottery style occurs from 500 CE but it may have been earlier. Thus there is evidence of on-going relations between groups over many years and that this continued when the Oceanic Austronesians crossed the Pacific and returned for trade or other reasons.

Although the Mariana Islands of Micronesia seem to have been inhabited from 3 200 years ago, and there appear to be Island Southeast Asia links, there may also be some links to plain and rim-notched Lapita pottery. Hence in Micronesia, there are two distinct groups of Austronesian languages.

The breakup of the Polynesian into Tongic and Nuclear Polynesia coincides with the split in the material culture of the Lapita pottery between Samoa and Tonga. Plain ware occurred around 3 000 years ago in Samoa with some dentated ware occurring in Tonga until 2 000 years ago. The languages and pottery further east in Polynesia all developed from Samoa from 2 200 years ago. The Papuan Tip Cluster also coincided with the spread of pottery 1 800 years ago, although more recently the Mailu developed a non-Austronesian language. The North New Guinea Cluster also developed in contact with, incorporating some features of and replacing Non-Austronesian languages within the last 2 500 years. While the pig appears as early as 6 000 years ago (some say 10 000), the word for pig in many Non-Austronesian languages has been borrowed from Austronesian languages. Finally, the genetic studies where available tend to support the Lapita and linguistic records (Spriggs, 2006). The Polynesian populations show evidence of having come through the Bismarck area and perhaps later when more influence of Melanesia's New Guinea Non-Austronesians occurred through intermarrying in this area, there was a further migration (Spriggs, 2006).

The Linguistic Situation

Generally speaking, Lean tried to take into account the linguistic situation of all regions given in the *Language Atlas of the Pacific Area, Part 1: New Guinea Area, Oceania, Australia*, edited by Wurm and Hattori (1981). The languages of New Guinea and Oceania are classified into two distinct groups: (a) the Papuan languages, referred to as non-Austronesian (NAN), and (b) the Austronesian (AN) languages. Each of these groups is considered below. Although SIL Ethnologue now takes account of many more languages, especially in the Highlands region, it has not been possible to re-examine Lean's questionnaire data in relation to these languages because we do not have his original data. However, some new data are discussed in Chapters 7 and 11 as part of the discussion on change. Although Lean described the languages as Non-Austronesian (Papuan) and Austronesia, we also know that there are a number of languages that are sometimes called hybrid languages as discussed in the following paragraphs, in the Ethnologue, and in Chapters 7 and 11.

The Papuan (Non-Austronesian) Languages

Wurm (1982b) noted that

> the Papuan languages are generally regarded as radically different from the Austronesian languages and for that matter from all other known language groups in the world. There may only be a remote possibility of some link existing between the languages spoken by the aboriginal population of the Andaman Islands, to the south of Burma, and some Papuan language groups in the extreme west of the New Guinea area. (p. 225)

Wurm indicated, however, that several groups of NAN languages show strong influence by AN languages although it seems likely that such influence was from Oceanic, or Pre-Oceanic, AN, rather than a more archaic form. Wurm pointed out that, in general, the NAN languages are not thought to be related to the Australian languages. However, there are some features such as groupings of five which are similar and these are discussed in Chapter 13.

The majority of the NAN languages are located on the New Guinea mainland. Some are also found to the west, in north Halmahera and Timor Leste. They also extend to the east of the mainland and are found in Rossel Island (Milne Bay Province), in the New Britain area, New Ireland, and Bougainville Island. Seven NAN languages are scattered throughout the largely AN-speaking Solomon Islands, although that is the limit of their eastward extension (one language, Kazukuru, has become extinct). We do not find NAN languages in the remainder of Melanesia, although Wurm (1982b) indicates that some of the eastern Vanuatan languages may display some Papuan influence.

Of the two types of languages spoken in the New Guinea area, by far the majority are NAN. According to Wurm (1982a, 1982b) there were 722 NAN languages spoken in the region under consideration: 543 in Papua New Guinea, 172 in West Papua, and 7 in the Solomon Islands. Some dialects of these languages might now be considered separate languages increasing the number. Since the 1950s a great deal of work has been carried out in investigating and classifying the NAN languages which, at one time, were thought to be largely unrelated. Although it is unnecessary to provide the fine details of the classification here (an individual language's classification is given in each data table on the website), it is nevertheless useful to indicate the major classificatory phyla for the NAN languages together with the number of languages assigned to each group (arranged in descending order of magnitude). Those are given in Table 1.1.

Table 1.1
The Number of Languages in Each NAN Phylum (from Lean's Thesis)

NAN Language Phyla	Languages in Phylum***
Trans New Guinea Phylum	>500
Sepik-Ramu Phylum	98
Torricelli Phylum	48
East Papuan Phylum*	23
West Papuan Phylum**	13
Sko Phylum-level Stock	8
Arai Phylum-level Family	6
Geelvink Bay Phylum	6
Kwomtari Phylum-level Stock	5
East Vogelkop Phylum-level Stock	3
Amto-Musan Phylum-level Family	2
Isolates	8

Notes: *Wurm gave 27 languages for the East Papuan Phylum, however 4 of these were extinct by the 1980s so 23 is stated in Table 1.1.

**Wurm gave 24 languages in the West Papuan Phylum, 11 are spoken in North Halmahera which was not considered in Lean's study so only 13 are listed.

***SIL *Ethnologue* lists more languages for the different phylum as the languages are recognised as distinct by the people from the neighbouring languages.

The distribution of these phyla in New Guinea are indicated in Figure 1.2. Clearly a large section of the New Guinea mainland is grouped into one particular phylum, Trans New Guinea Phylum. Capell (1969) provided further classificatory evidence of many of these languages in terms of the importance of the verb to which infixes and suffices were added to complete the communication. However, not all these languages had similar counting systems. They varied considerably from hardly using any numbers or numerals to those that had a stronger emphasis on number. It should also be noted that these languages interweave with Austronesian languages in coastal regions. The database based on Lean's (1992) work gives details not only of the counting system but the linguistic classification as known in 1992. There may be some changes to this information as a result of further linguistic study and people's recognition of differences.

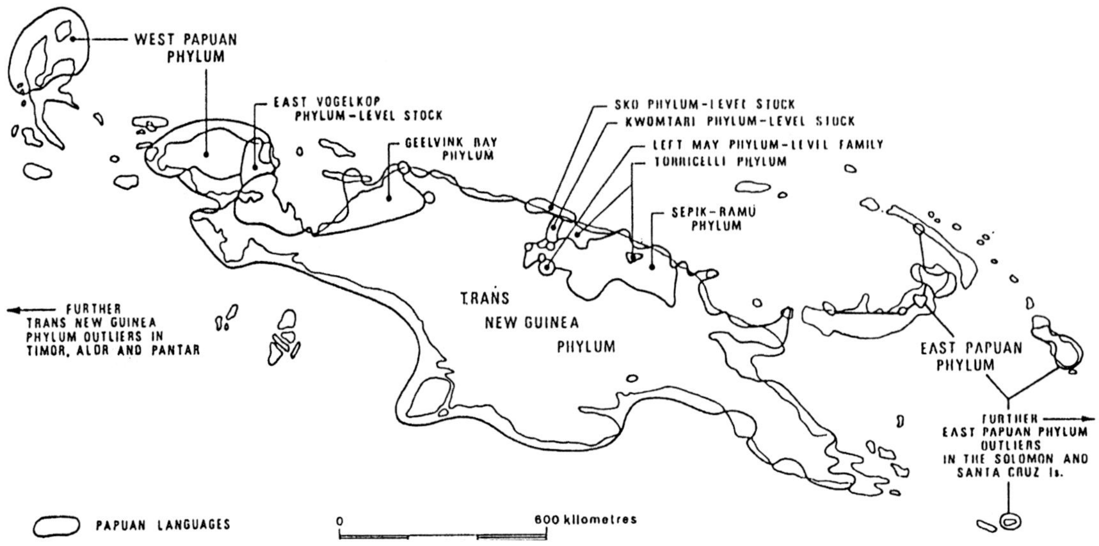

Figure 1.2. Map of distribution of NAN languages in New Guinea.

The Austronesian Languages

Although the large majority of the NAN languages are spoken on the New Guinea mainland, and especially in the interior highland regions, the AN languages are dispersed over an enormous geographical area stretching from Madagascar in the west to Easter Island in the east. They are found in Formosa, the Philippines, southern Vietnam, Indonesia and Malaysia. Eugene Chan and Richard Parker have collated extensive lists of the counting system data for Austronesian languages but this information is not readily available. In the region under consideration here, the AN languages are spoken mainly in the coastal areas of the New Guinea mainland, throughout almost all of Island Melanesia as well as the atolls and islands of Micronesia and Polynesia (Tryon, 1982). All the AN languages are regarded as descendants of Proto Austronesian (PAN), thought to be spoken in Taiwan at about 7 000 years ago (Tryon, 1982). At some time after this date the Austronesians dispersed, some eventually migrating into the Borneo and western Indonesian area. A migration into the New Guinea region, between 5 000 and 4 000 years ago (Ross, 1989; Tryon, 1982), brought the ancestors of all Austronesians of the Pacific area and who established, somewhere in the northeast of Papua New Guinea (possibly the Willaumez Peninsula of West New Britain), the homeland of Proto Oceanic (Ross, 1989).

In the region under consideration there are two major distinct groups of AN languages: the Oceanic and non-Oceanic. Ross (1988) notes that

> there is today widespread acceptance among Austronesianists of the "Oceanic hypothesis", according to which all the Austronesian languages of Oceania east of a line drawn from north to south through the western Pacific are descended from a single proto language, today named "Proto Oceanic" (POC). Its descendants are known simply as the "Oceanic languages". This line divides Chamorro (Marianas Islands) and Belau (formerly Palau) from the rest of Micronesia and bisects the north coast of the island of New Guinea at 138° longitude. (p. 1)

This essentially means that virtually all of the AN languages dealt with here are Oceanic, with some exceptions in the western part of West Papua. The distribution of the AN languages throughout the New Guinea mainland is shown in Figure 1.3. The high-order subgroups of Oceanic given in Table 1.2 seem fairly well supported.

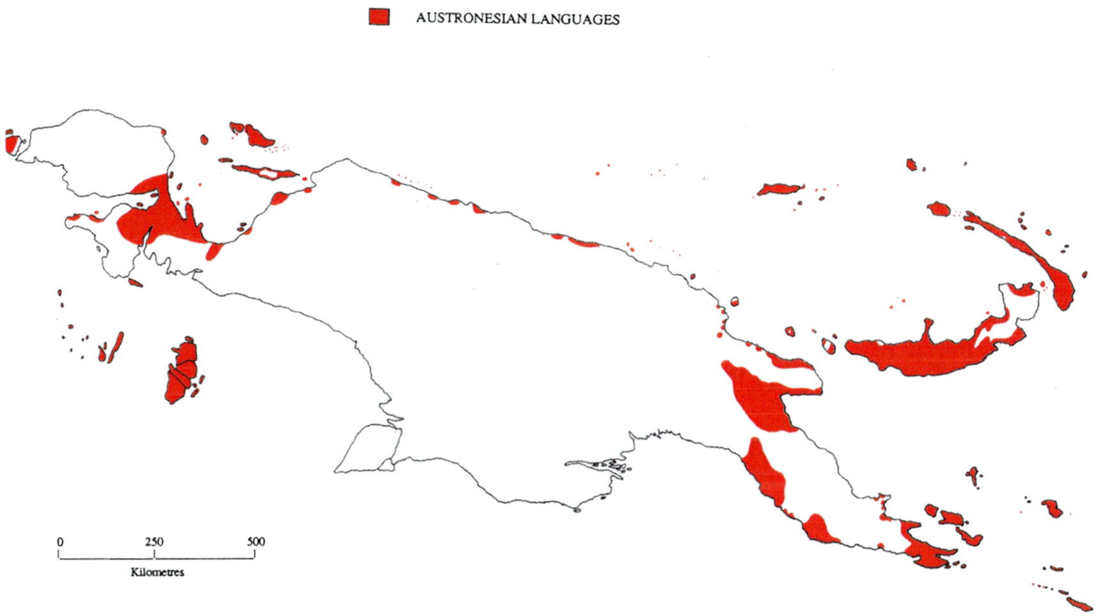

Figure 1.3. Distribution of AN languages in New Guinea.

Table 1.2
Higher Order Subgroups of Oceanic

Admiralty Islands	Southeast Solomonic	Southern Oceanic
St Matthias Islands	North/Central Vanuatu	Central Pacific
Western Oceanic	South Vanuatu	Nuclear Micronesian

Note. Subgroups of Oceanic were given in Pawley and Ross (2006).

The Oceanic AN languages are themselves classified further. Research by Ross (1988) on the AN languages of western Melanesia and by Lynch and Tryon (1985) on eastern Melanesia and the Pacific region have provided the following picture (the implications for migrations are discussed subsequently).

The Oceanic languages spoken in Papua New Guinea and the northwest Solomon Islands may be classified into just two first order sub-groups: the Admiralties Cluster (located in the Manus Province of PNG), and Western Oceanic. The latter comprises three clusters:

1. The North New Guinea Cluster, which includes the languages of the Huon Gulf-Markham Valley area, parts of New Britain and the north to northeast coast of the New Guinea mainland.
2. The Papuan Tip Cluster, consisting of the languages of the Milne Bay and Central Provinces.
3. The Meso-Melanesian Cluster, comprising the languages of the New Ireland Province (with the exception of the St Matthias group), several New Britain languages, those of Bougainville (previously called the North Solomons Province) of PNG, and the northwest Solomon Islands group.

A third sub-group of Oceanic languages has been delineated by Lynch and Tryon. This is the Central/Eastern Oceanic sub-group which comprises, according to Ross "all the languages of north and central Vanuatu, Fiji, Polynesia and Micronesia ... together with the southeast Solomons, Utupua and Vanikoro, the south Vanuatu, and possibly the Loyalties and New Caledonia groups" (Ross, 1989, p. 136)

To summarise the situation regarding the AN languages spoken in the region encompassing New Guinea and Oceania, the information given in Table 1.3 shows the total number of identified languages spoken in each country at the time of Lean's study.

Table 1.3
The Number of AN Languages in Each Country Dealt With in This Study

Country	Number of AN Languages in that Country	Country	Number of AN Languages in that Country
PNG	206	New Caledonia	128
West Papua	39	Fiji	2
Solomon Islands	56	Rotuma	1
Vanuatu	105	Polynesia	24

Theories of Migration and Prehistory

The complex linguistic situation existing in New Guinea and the remainder of Island Melanesia is paralleled by the obviously apparent cultural and genetic heterogeneity of the region. While it might seem to be an impossible task to unravel the prehistoric conditions which have led to the contemporary diversity, it is precisely in this region of the world where a marriage between archaeology and the reconstructive methods of historical linguistics is producing fruitful results. The reconstruction of the prehistory of the Oceanic languages and their speakers is rather more advanced, and probably more accurate, than that of the NAN languages. Nevertheless, a summary of some current theories regarding the prehistoric migrations of the groups ancestral to today's speakers of the NAN and Oceanic AN languages are presented in this section insofar as they are relevant to subsequent arguments.

Papuan Language Migrations

The picture of the putative migrations of NAN speakers into the New Guinea area that is given here is that first put forward by Wurm, Laycock, Voorhoeve, and Dutton (1975), reiterated by Wurm (1982a, 1982b) and by Tryon (1984), and investigated further by Voorhoeve (1987).

About 60 000 years ago the first immigrants arrived into the New Guinea region. These were probably Australoids who came from the west and entered the northern part of what was then the single New Guinea-Australian continent. They spread into New Guinea and south into Australia. Approximately 10 000 years ago, after the last ice age, the increased sea level brought what is now the Torres Strait into existence and New Guinea and Australia were separated. The theory put forward by Wurm (1982b) is concerned with the establishment of the NAN languages in New Guinea and does not deal with the origin of the Australian languages and whether in fact there were several Australoid migrations. Whatever remnants of an Australoid population there were in New Guinea have either disappeared or have been largely engulfed by subsequent Papuan migrations. Wurm (1982a) speculated that it is "possible to suggest that Sepik-Ramu Phylum speakers may constitute some remnants of the earlier Australoid population in the New Guinea area who have been strongly affected by thousands of years of contact with Papuans" (Wurm, 1982a, p. 237).

According to Wurm's early research, between about 10 000 to 15 000 years ago, i.e. shortly after, or not long before, the separation of New Guinea and Australia, the first NAN migration came from the west of New Guinea and spread in an easterly direction right across the mainland overlaying the extant Australoid population. According to Wurm (1982a),

> it seems possible that languages descended from ancestral language forms spoken by these first Papuan immigrants are still surviving today in the form of the isolates and the members

of at least some of the small phylic groups ... there is a possibility that members of the East Papuan Phylum, or at least an element in them, may also be derived from the languages of the first Papuan immigrants. (p. 243)

Other archaic language groups which may be a survival from this migration are the West Papuan Phylum and the Torricelli Phylum. Voorhoeve (1987) suggested that the West Papuan Phylum may have been the result of a westward migration by some Torricelli Phylum speakers.

Wurm (1982a) argued that "a few millennia later, a second NAN migration appears to have entered the New Guinea area and spread through much of it with the languages carried by this migration overlaying the language picture brought in by the first migration" (p. 234).

The main NAN migration is thought to have taken place about 5 000 BP when a group immediately to the west of the New Guinea mainland moved into the Vogelkop-Bomberai Peninsula and moved eastward right through the island. The languages introduced by this migration are those that eventually formed the current Trans New Guinea Phylum. Because the Trans New Guinea Phylum languages appear to have adopted some archaic AN loanwords it is assumed that they were in contact with AN groups prior to the migration across New Guinea. The Trans New Guinea Phylum migration took perhaps about 1 000 years to spread right across the mainland and eventually to the southeastern Papuan tip where East Papuan Phylum languages may have been displaced eastwards to Rossel Island, New Britain, New Ireland and to Bougainville and the Solomon Islands, reaching as far east as Santa Cruz. Tryon (1984) noted that "the existence of Papuan languages southeast of Santa Cruz has not been demonstrated, although there is some speculation that they may have once extended as far as southern Vanuatu and New Caledonia" (p. 153). Figure 1.4 shows some of the main migrations, according to Wurm, that occurred in the New Guinea mainland. Chapter 11 suggests that earlier datings are probable.

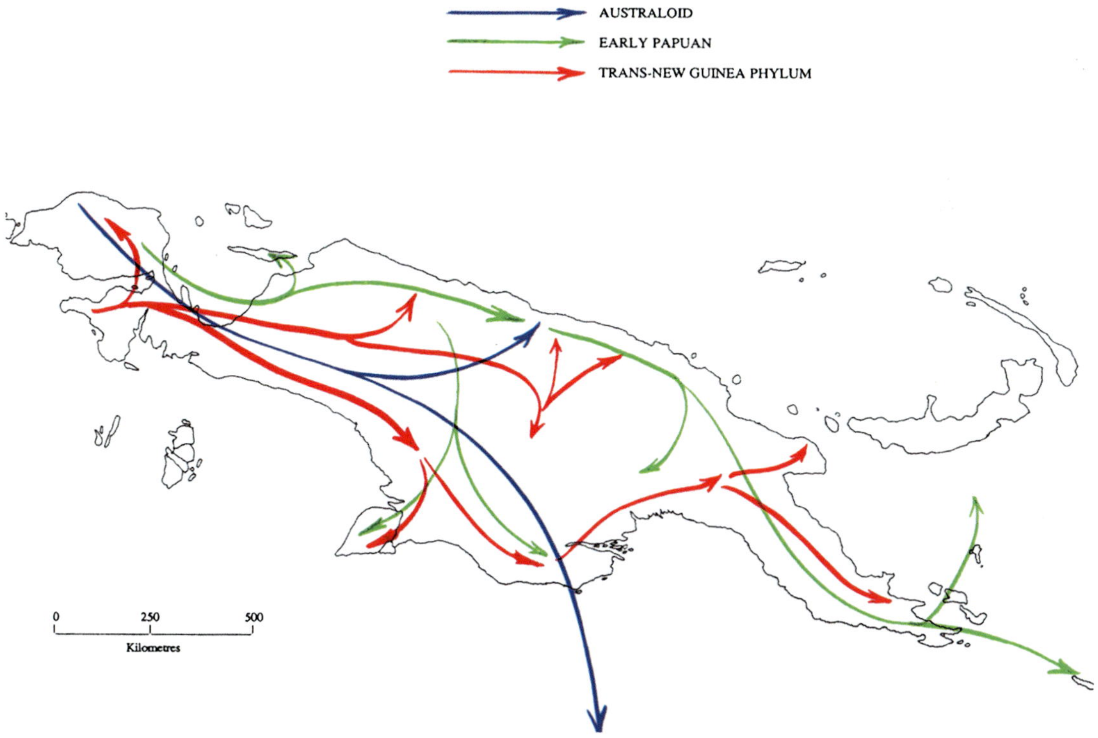

Figure 1.4. Early language migrations in New Guinea.

Oceanic Austronesian Language Migrations

The linguistic reconstruction of the prehistory of all the Oceanic AN languages was outlined above. The migrations within Papua New Guinea will be discussed first, followed by the putative situation for the remainder of Oceania. Figure 1.5 shows the Oceanic Austronesian migrations from the Bismarck Sea in Papua New Guinea.

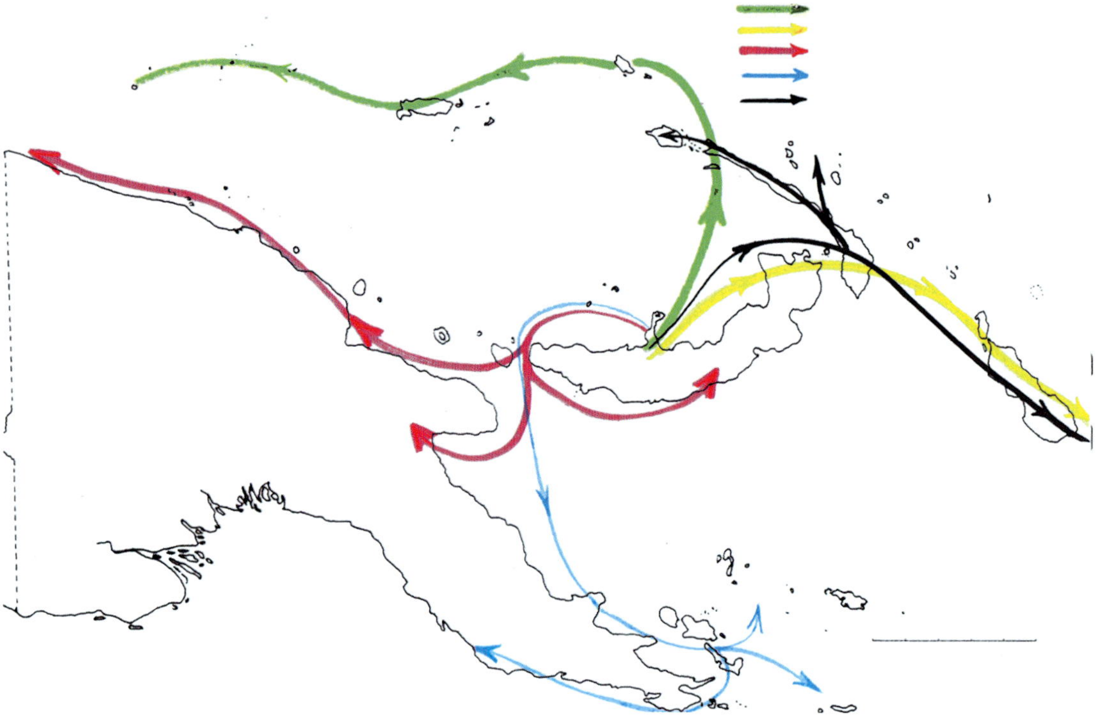

Figure 1.5. Oceanic AN migrations in Papua New Guinea.

Claims by Ross (1988, 1989) are now summarised:

1. Speakers of Pre-Oceanic moved from the eastern Indonesia area and settled in what is now New Britain, probably in or near the Willaumez Peninsula. This migration may have occurred between 4 500 and 5 000 years ago (Ross, 1989; Tryon, 1984).
2. After a period of stability in which the innovations characteristic of Proto Oceanic developed, two early migrations of Proto Oceanic speakers out of the homeland occurred: first, a movement eastward and probably southwards into the Solomon Islands resulted in the establishment of the homeland of Proto Central-Eastern Oceanic; and second, a movement via New Ireland and the St. Matthias group northwest to the Admiralty Islands, established the homeland of Proto Admiralty.

Ross (1989) had two caveats: (a) it may be that more than one movement needs to be posited to account for all the Central/Eastern Oceanic languages, and (b) the location of the Proto Central-Eastern Oceanic homeland may have been in the Bougainville or the northwest Solomon Islands area, i.e. the area now occupied by Meso-Melanesian Western Oceanic speakers. In fact, Kennedy (1983) surmised that the archaeological evidence implies "settlement in the Admiralties in the period 4 500–5 000 BP" (p. 115), and that the immigrants came "from the New Britain-New Ireland area" (p. 118).

After these two early migrations, there was a further period of stability in the homeland region where the innovations characteristic of Western Oceanic occurred. After a period when Western Oceanic developed east and west linkages, or perhaps dialect chains, the west linkage gradually dispersed giving rise to the North New Guinea Cluster of languages extending from the Huon Gulf region north and northwest to the Madang and Sepik coasts. A migration southwards by a Western Oceanic dialect established the homeland of Proto Papuan Tip somewhere in the D'Entrecasteaux Islands or the nearby mainland. The eastern linkage of Western Oceanic was ancestral to the Meso-Melanesian Cluster and one or perhaps several migrations to southern New Ireland gave rise to that region as a centre of dispersal, some groups spreading north and east to occupy the rest of New Ireland and its offshore islands, and other groups moving south to establish the Northwest Solomonic group of languages—see Figure 1.5 for the main Oceanic AN migrations in the New Guinea region.

Although it is possible to designate subgroupings of the Austronesian languages, this is made difficult due to contact effects but support is given by lexicostatistical evidence of similar word origins. There is disputed but generally agreed recognition that Austronesia as a language developed on the island of Taiwan some 5 000 years ago. There are four subgroups (depending on the evidence but that presented by Ross seems to capture both earlier methods) identified by Tryon and a Proto Malayo-Polynesian subgroup (Tryon, 2006) which splits into two as Western Malayo-Polynesian (including the Philippines, now Indonesian, Micronesian, and Malagasy) and Central-Eastern Malayo-Polynesian, the Eastern subgroup splitting into South Halmahera West New Guinea (SHWNG) (some found in the Bird's Head Peninsula and southern coast of New Guinea) and Oceanic subgroups. Unlike the Western group of languages that spread and differentiated, Proto Oceanic developed in a relatively restricted area of island New Guinea (PNG) before spreading. Our interest is in the Austronesian languages of Melanesia in particular in New Guinea and further east. The SHWNG subgroup includes the Halmahera and its satellites, northern coast of the Bird's Head and Yapen Island and satellites (based on Blust's work) (Addison & Matisoo-Smith, 2010).

Proto Oceanic is the ancestor of some 450 languages of Melanesia, Micronesia and Polynesia. Based on the sound changes of the Oceanic languages, Dempwolff's (e.g., 1938) later works distinguished the Oceanic group from that of the Central Malayo-Polynesian group. The Oceanic groups had words that were similar for aspects related to seafaring, the sea, horticultural terms such as related to the coconut, winds, food preparation, pottery, land animals such as pig, cuscus and marsupial rats, the bandicoot and cassowary, and relationships except for "father's sister" according to Pawley and Ross (2006). Recent work on animal biology, physical features and DNA, has suggested that the Lapita pottery group moved eastward and at some point established a homeland that was then influenced by a later migration of Austronesians via Micronesia to establish a newer version of Polynesia in Samoa about 1 500 years ago. Polynesian culture then spread both back towards the west and to more remote parts of Polynesia (Addison & Matisoo-Smith, 2010).

Thus languages and cultures developed by colonisation, intrusion, innovation and adaptation, integration and expansion. This is a more complex perspective on diffusionist theory in the Polynesian area than previously suggested. The fact that Austronesians in New Guinea may look different to Polynesians is not regarded as mainly due to marrying with the first wave of Papuan immigrants as surmised by Pawley and Ross (2006). They noted the southern Papuan, Wuvulu, and Aua genetic and visual similarities to Micronesians and Polynesians, but Addison and Matisoo-Smith (2010) maintained that it is more plausible to consider alternate migration roots as shown in Figure 1.6.

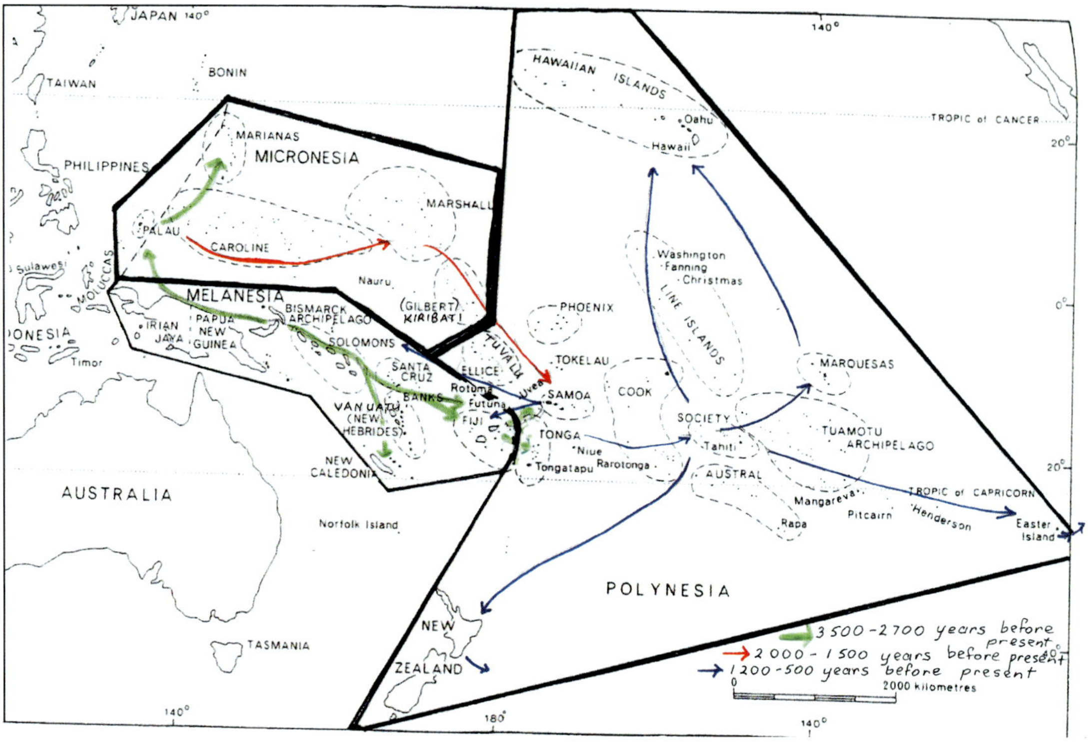

Figure 1.6. Probable expansion of Lapita culture.
(Based on Addison & Matisoo-Smith, 2010, p. 8).

Some reasons for differentiation of the languages is due to large ocean gaps but also mountain ranges or social ties or intervening Papuan speakers as found in Bougainville. However, the greatest innovation and hence diversity of lexicon and sound is found on the north coast of New Guinea (e.g. Morobe and Madang). According to Pawley and Ross (2006), this situation was likely to have arisen because of one or more of the following reasons:

> (i) imperfect acquisition of an Oceanic language by a non-Austronesian (Papuan) speech community, (ii) long-term bilingualism between Oceanic and Papuan neighbours, sustained by trade and intermarriage, (iii) very small size of speech communities, as the result of migration, political structures, etc., (iv) social pressures to develop a distinct language from one's neighbours, (v) taboos on using words coinciding with the names of chiefs or of the dead, (vi) changes of physical environment following migration, (vii) cultural changes generated internally, (viii) cultural changes generated by contact, and (ix) phonological change creating problematic (especially ambiguous) word forms. (p. 66)

As indicated earlier, the AN languages of West Papua comprise 30-odd non-Oceanic and 5 Oceanic languages. It is possible that the Oceanic languages represent the furthest western migration of the North New Guinea Cluster. It is unclear how long the non-Oceanic languages have been established in western West Papua.

It appears to be the case that the peopling of the remainder of Oceania began with an early migration out of the Proto Oceanic homeland and into Bougainville and the Solomon Islands. Further migrations from this region filtered down through Island Melanesia, that is Vanuatu and New Caledonia and eventually Fiji, by about 1 500 BCE "At about this time," Tryon wrote (1984), "a set of migrations apparently began in the northern/central Vanuatu region, one moving north, spreading the Austronesian languages throughout Micronesia (for which there is evidence of an east to west spread), another moving southeast to the Fiji group. From there, after a period of consolidation, the Polynesian languages evolved, moving out from the Tonga-Niue area sometime around 1 000 (BCE)" (p. 154). The great Polynesian migrations spread throughout Triangle Polynesia as well as back into Island Melanesia to form the Outliers. There was an eastward flow from Samoa to the Marquesas, the Society Islands, Mangareva and Easter Island; there was also a north-eastward movement to Hawaii. From the Tahiti area, there were migrations to the Cook Islands and eventually to New Zealand which was settled by 900 CE (Tryon, 1984, p. 154).

Another important piece of evidence for the Proto Oceanic and spread of Oceanic languages comes from the records of findings of Lapita pottery. This incised pottery is found across a wide range of the Oceanic area with datings and detail strongest around the New Britain north coast areas with lessening and almost plain pottery found in Tonga.

Prehistoric Trade and Technology

Allen, a well-regarded archeologist, puts the view that Papua New Guinea communities were small but generally supported themselves from their own territory which often included water, arable ground, and higher forested or rocky terrain but the country had a well established, complex set of trade routes. Hughes (1977), in his analysis of New Guinea stone-age trade, indicated that the existence of an extensive trade system which involved the exchange of various goods such as salts, pigments, mineral oils, pottery, stone tools, and sea shells. In addition to these interior trade routes there were also the well-known trading cycles of the coastal regions such as the Vitiaz Strait cycle (Harding, 1967), the Kula Ring (Malinowski, 1920, 1922) in the Trobriand Islands-Milne Bay area, and the Hiri exchange system between the Port Moresby area and the Gulf of Papua (Dutton, 1982). That there is considerable time depth to the existence of trade routes is supported by archaeological evidence. Bulmer (1975) indicated that the introduction of new technology, in the form of a particular type of stone axe/adze, occurred in the highlands of Papua New Guinea between 11 000 and 6 000 years ago. Evidence for widespread trade occurring during the period 6 000 to 3 000 years ago is clear.

In Island Melanesia, there is evidence of continued contact between fairly widespread Oceanic groups. For example, Kennedy (1983) indicated

> that more than haphazard and accidental contact between the Admiralties and other parts of Melanesia continued for most of the prehistoric period, marked by pottery styles widespread in Melanesia, first Lapita then incised-impressed relief, and by distributions of Lou Island obsidian. (p. 115)

Obsidian from the Talasea region of West New Britain has been found in various sites in Melanesia, the earliest recorded example, found in another part of West New Britain, is dated 11 400 BP, and a further sample, from New Ireland, some 600 km away, is dated 6 800 BP (Spriggs, 1984, p. 205). Various prehistorians (Allen, 1984; Spriggs, 1984) have associated Proto Oceanic and its early daughter languages with the Lapita cultural complex. There is evidence of a period of widespread ceramic production throughout Oceania which has some shared features and which suggests the spread of an identifiable cultural complex from, possibly, the Proto Oceanic homeland right through to Polynesia (Spriggs, 1984). For example, Davidson (1977) noted that

> the archaeological evidence suggests that most or all of the Western Polynesian islands as well as Fiji were settled during the second millennium B.C. by related people and thereafter were in sufficiently regular contact to share in a similar sequence of ceramic change. (p. 90)

Summary on Migration and Prehistory

The overall picture we obtain from linguistic reconstructions and from the archaeological record is a complex one. The main points that have a bearing on the subsequent argument are summarised here.

- There was at least one or perhaps several migrations by Australoids into the New Guinea-Australia continent beginning at least 60 000 years ago. Remnants of these Australoid groups may still have descendants in New Guinea and these may be language groups belonging to the Sepik-Ramu Phylum.
- There were at least two, perhaps more, migrations of NAN language groups into the New Guinea mainland, the first between 10 000 to 15 000 years ago. The main migration was that bringing speakers of the language(s) of the Trans New Guinea Phylum and this is thought to have occurred about 9 000 years ago. East Papuan Phylum speakers were, at some time, displaced from the mainland and moved into the islands to the east.
- Perhaps about 4 000 to 5 000 years ago, AN speakers settled in the New Britain region which became the homeland of Proto Oceanic. Various migrations originating from here brought AN speakers to the various coastal areas of Papua New Guinea, the remainder of Island Melanesia, Micronesia, and Triangle Polynesia.
- A significant amount of contact occurred between groups in Island Melanesia. Trade routes were established throughout various coastal regions and also between the coast and the interior highlands, and within the highlands region itself. Thus, in the prehistoric period, within the New Guinea mainland as well as in the islands of Oceania, groups were not entirely isolated from one another and there existed varying degrees of contact and flow of trade goods.

Organisation of the Book

Cycle Analysis of Counting Systems

In the chapters which follow, the counting systems of New Guinea and Oceania are discussed. The book follows Lean's contribution to this area in terms of the way of describing counting systems. The descriptive terms which were adopted for this study are those coined by Salzmann (1950) namely frame pattern, cyclic pattern, and operative pattern. They are briefly summarised here. Suppose we have a sequence of natural language numerals which we can associate, in a one-to-one correspondence, with the symbolic numerals of mathematical notation. This sequence will normally contain a unique set of number morphs together with a complementary set of complex number words which are partitioned into component number morphs. For each complex number word which can be broken down into its components, it is usually possible to infer a number sentence involving operations on two or more numbers. To take an example, let us suppose that we have analysed a sequence of numerals to have the form:

$$1,\ 2,\ 3,\ 4,\ 4+1,\ 4+2,\ 4+3,\ 2\times 4,\ (2\times 4)+1,\ (2\times 4)+2,\ (2\times 4)+3,\ 3\times 4,\ \ldots,\ (4\times 4)+3,\ 20,$$
$$20+1,\ 20+2,\ 20+3,\ 20+4,\ (20+4)+1,\ (20+4)+2,\ \ldots,\ 2\times 20,\ (2\times 20)+1,\ (2\times 20)+2,\ \ldots$$

That is, there are distinct number morphs for 1 to 4, and 20, and all other members of the sequence are composed of these. The numeral 5, for example, may be analysed as having the component number morphs 4 and 1. We infer that 5 is thus a complex number word which is represented by the number sentence $5=4+1$. Similarly 8 is analysable as having the component number morphs 2 and 4 and we infer that 8 is a complex number word representable by the number sentence $8=2\times 4$. Using Salzmann's terminology, the set (1, 2, 3, 4, 20), that is to say, the number morphs from which all other numerals in the sequence are generated, is called the frame pattern of the sequence. Once the sequence has been analysed into its symbolic form, its cyclic structure becomes apparent. For the example above, the sequence has a cycle of 4 and a superordinate cycle of 20. Salzmann calls this the cyclic pattern of the numeral sequence and denotes it by the set (4, 20). It is often the case that there is not always sufficient data available to determine the complete cyclic pattern of a numeral sequence. Finally, the operative pattern of a numeral sequence is essentially a summary of the various number sentences which indicate how the complex number words in the sequence are composed.

Lean grouped various types together according to their primary cyclic pattern rather than use any of the alternative typological systems outlined in Chapter 2 where we outline his project. However, he then discusses in some detail the variants which occur in a particular type. In Chapter 3, there is a brief summary of the existence and nature of 2-cycle systems as they occur in other parts of the world. Several variants of numeral systems which have a primary 2-cycle and their geographical distribution in the New Guinea area are discussed. This chapter includes the work of Mimica and Muke. One of the variations of the 2-cycle systems, for example, incorporates the (2, 5, 20) cycle systems, most of which are digit-tally systems. Other digit-tally systems are (5, 20) cycle systems covered in Chapter 5. Fingers are often used to represent numbers in various systems but the distinction is in having a 20 cycle. In Chapter 4, there is discussion of several varieties of body-part tally methods (as distinct from systems which use only the fingers and toes for tallying) which appeared to occur uniquely in Australia and New Guinea. However, these types of systems may now be recognised in other parts of the world.

In Chapter 5, the 5-cycle system and its variants are considered. The analysis there includes the common digit-tally system with a (5, 20) cyclic pattern. The occurrence of the various 5-cycle variants in both the NAN and AN languages is discussed and their geographical distribution in New Guinea and Island Melanesia is given. In Chapter 6, the 10-cycle system and its variants are discussed after providing a brief summary of their existence in other parts of the world. The distribution of these in New Guinea, Island Melanesia, Polynesia, and Micronesia is also discussed. In Chapter 7, the relatively unusual 4-cycle and 6-cycle systems are dealt with, and the chapter concludes with an overview of the various types of numeral system which occur in the region under consideration, together with a brief delineation of types which are not found.

While Chapters 3 to 7 are essentially the basis of the argument on the history of number from a Papua New Guinea and Oceanic perspective, they also prepare the ground for the remaining chapters of the book that discuss the spread of counting systems, their positioning within cultures, teaching in a cultural context and extending knowledge in the field of mathematics and mathematics education. Chapter 8 deals with counting and number in context and the range of circumstances in which number is used or is important in the cultures under examination; this is illustrated by a series of brief case studies of several foundational societies. In addition, several related matters such as numeral classification, and large numbers and the limits of counting are discussed.

Chapter 9 gives Lean's examination of aspects of the diffusionist view of the existence of various counting system types and their distribution throughout New Guinea and Oceania, especially a major work by Seidenberg (1960) who was a leading proponent of the diffusionist view. Although others may have discredited this theory or have other theories, it is still seen in histories of mathematics for school children but Lean provided a critique of diffusionist theory in the light of the data collected for his study.

In Chapter 10, an attempt is made to construct an outline of the prehistory of number for this region. Beginning with the NAN languages, the contemporary situation regarding the distribution of counting systems and tally methods in each of the major and minor phyla is surveyed. From this survey an attempt is made to infer the nature of the proto systems which may have been present in the early history of these phyla. Secondly, the fate of 10-cycle systems of the AN immigrants is traced as they spread out from the Proto Oceanic homeland into the coastal and island regions of PNG and into the remainder of Island Melanesia and the Pacific.

Building on Chapter 8, Chapters 11 to 13 consider the information found in the studies of Lean, Paraide, Saxe, and Owens in terms of the meaning of number and its relationships to societies. In particular, Paraide considers how different knowledges exist and develop using Tolai as an example.

She considers the reduction in knowledges as a result of power struggles and why fundamental Indigenous knowledges and ideas are important for education and how they can best be incorporated into education, particularly in PNG. The concluding chapter summarises the book's argument and recognises its relevance of this history of number for mathematics and mathematics education.

References

Adam, N. A. (2010). Mutual interrogation: A methodological process in ethnomathematical research. *Procedia, Social and Behavioral Sciences, 8*, 700–707.

Addison, D. J., & Matisoo-Smith, E. (2010). Rethinking Polynesians origins: A West-Polynesia Triple-I Model. *Archaeology in Oceania, 45*(1), 1–12.

Allen, J. (1984). In search of the Lapita homeland. *Journal of Pacific History, 19*(3–4), 186–201.

Allen, J. (1996). The pre-Austronesian settlement of Island Melanesia: Implications for Lapita archaeology. *Transactions of the American Philosophical Society, 86*(5), 11–27. doi:10.2307/1006618.

Ascher, M. (1994). *Ethnomathematics: A multicultural view of mathematical ideas*. New York, NY: Chapman & Hall.

Ascher, M. (2002). *Mathematics elsewhere: An exploration of ideas across cultures*. Princeton, NJ: Princeton University Press.

Berry, J., & Dasen, P. (1974). *Culture and cognition: Readings in cross-cultural psychology*. London, UK: Methuen.

Boyer, C., & Merzbach, U. (1989). *A history of mathematics* (2nd ed.). New York, NY: Wiley.

Bulmer, S. (1975). Settlement and economy in prehistoric Papua New Guinea: A review of the archaeological evidence. *Journal de la Société des Océanistes, 31*, 7–75.

Butler, M., & Dean, B. (Writers). (2013). *First footprints* (4 episodes). Australia: ABC Broadcasting.

Capell, A. (1969). *A survey of New Guinea languages*. Sydney, Australia: Sydney University Press.

Carrier, A. (1981). Counting and calculation on Ponam Island. *Journal of the Polynesian Society, 90*(4), 465–479.

Clark, G. (2003). Shards of meaning: Archaeology and the Melanesia-Polynesia divide. *The Journal of Pacific History, 38*(2), 197–215. doi:10.1080/0022334032000120530.

Closs, M. (Ed.). (1986). *Native American mathematics*. Austin, TX: University of Texas Press.

Cole, M., Gay, J., Glick, J., & Sharp, D. (1971). *The cultural context of learning and thinking*. London, UK: Methuen.

Cooke, R. L. (2011). *The history of mathematics: A brief course* (2nd ed.). New York, NY: Wiley.

D'Ambrosio, U. (2006). *Ethnomathematics: Link between traditions and modernity* (M. Borba, Trans.). Rotterdam, The Netherlands: Sense Publishers.

Dantzig, T. (1954). *Number: The language of science* (4th ed.). New York, NY: Macmillan.

Dasen, P., & de Ribaupierre, A. (1987). Neo-Piagetian theories: Cross-cultural and differential perspectives. *International Journal of Psychology, 22*(5/6), 793–832.

Davidson, J. M. (1977). Western Polynesia and Fiji: Prehistoric contact, diffusion and differentiation in adjacent archipelagos. *World Archaeology, 9*, 82–94.

Dempwolff, O. (1938). *Vergleichende des Austronesischen Wortschatzes. Austronesischen Wörterverzeichnis* (Vol. 3). Berlin, Germany: Dietrich Reimer.

Dutton, T. (Ed.). (1982). *The Hiri in history, Pacific Research Monograph No. 8*. Canberra, Australia: Australian National University Press.

Dwyer, P., & Minnegal, M. (2016). Counting systems of the Strickland-Bosavi languages, Papua New Guinea. *Language and Linguistics in Melanesia, 34*(1), 1–36.

Eglash, R. (1999). *African fractals*. New Brunswick, NJ: Rutgers University Press.

Eglash, R. (2010). *From ethnomathematics to ethnocomputing.* Paper presented at the Fourth International Conference on Ethnomathematics ICEm4, Towson.

Ernest, P. (1996). The nature of mathematics and teaching. *Philosophy of Mathematics Education, 9*, 40–45.

Ernest, P. (2012). What is our first philosophy in mathematics education? *For the Learning of Mathematics, 32*(3), 8–14.

Fisher, J. (~2010). *Enriching students' learning through ethnomathematics in Kuruti elementary schools in Papua New Guinea*. Lae, Papua New Guinea: ResearchGate.

Flegg, G. (1984). *Numbers: Their history and meaning*. New Orleans, LA: Pelican Books.

Flegg, G. (1989). *Numbers through the ages*. London, UK: Macmillan and the Open University.

Frobenius, L. (1900). *Die Mathematik der Oceanier, (Allgemein-verständliche Naturwissenschaftliche Abhandlungen, Heft 23)*. Berlin, Germany: F. Dümmler.

Gay, J., & Cole, M. (1967). *The new mathematics and an old culture*. New York, NY: Holt, Rinehart and Winston.

Gerdes, P. (1999). *Geometry from Africa: Mathematical and educational explorations*. Washington, DC: The Mathematical Association of America.

GLEC. (2008). *Glen Lean Ethnomathematics Centre*. Retrieved from http://www.uog.ac/glec/

Goetzfridt, N. (2008). *Pacific ethnomathematics: A bibliographic study*. Honolulu, HI: Japanese Cultural Centre and University of Hawaii Press.

Goetzfridt, N. (2012). Pacific ethnomathematics: The richness of environment and practice. *Journal of Mathematics and Culture, International Conference on Ethnomathematics, 4*(Focus Issue), 223–252.

Green, R. (2003). The Lapita horizon and traditons: Signature for one set of oceanic migrations. In C. Sand (Ed.), *Pacific archaeology: Assessments and prospects. Proceedings of the First International Conference for the 50th Anniversary of the First Lapita Conference* (Vol. 15). Noumea, New Caledonia: Les Cahiers de L'Archelogie en Nouvelle-Caledonie, New Caledonia Museum.

Harding, T. (1967). *Voyagers of the Vitiaz Strait: A study of a New Guinea trade system*. Seattle, WA: University of Washington Press.

Hughes, I. (1977). *New Guinea stone age trade: Terra Australis 3*. Canberra, Australia: Australian National University.

Hurford, J. (1987). *Language and number: The emergence of a cognitive system*. Oxford, UK: Basil Blackwell.

Ifrah, G. (1987). *From one to zero: A universal history of numbers*. Harmondsworth, UK: Penguin.

Izard, V., Pica, P., Spelke, E., & Dehaene, S. (2011). Flexible intuitions of Euclidean geometry in an Amazonian indigene group. *Proceedings of the National Academy of Sciences of the United States of America, 108*(24), 9782–9787. doi:10.1073/pnas.1016686108.

Joseph, G. (2000). *The crest of the peacock: The non-European roots of mathematics*. Princeton, NJ: Princeton University Press.

Keck, V. (Ed.). (1998). *Common worlds and single lives: Constituting knowledge in Pacific societies*. London, UK: Berg.

Kennedy, J. (1983). On the prehistory of western Melanesia. *Australian Archaeology, 16*, 115–122.

Kettenis, F. (1978). Traditional food classification and counting systems of Kilenge, West New Britain. *Papua New Guinea Journal of Education, 14*, 28–43.

Lancy, D. (1983). *Cross-cultural studies in cognition and mathematics*. New York, NY: Academic.

Lave, J. (1988). *Cognition in practice*. Cambridge, MA: Harvard University Press.

Lean, G. (1992). *Counting systems of Papua New Guinea and Oceania* (Unpublished PhD thesis). PNG University of Technology, Lae, Papua New Guinea.

Lynch, J., & Tryon, D. (1985). Central-Eastern Oceanic: A subgrouping hypothesis. *Austronesian linguistics at the 15th Pacific Science Congress (C-88)*. Canberra, Australia: Pacific Linguistics, ANU.

Malinowski, B. (1920). Kula: The circulating exchange of valuables in the archipelagoes of eastern New Guinea. *MAN, 20*, 97–105.

Malinowski, B. (1922). *Argonauts of the Western Pacific*. London, UK: Routledge and Kegan Paul.

Matang, R., & Owens, K. (2006). Rich transitions from Indigenous counting systems to English arithmetic strategies: Implications for mathematics education in Papua New Guinea. In F. Favilli (Ed.), *Ethnomathematics and mathematics education, Proceedings of the 10th International Congress on Mathematical Education Discussion Group 15 Ethnomathematics*. Pisa, Italy: Tipografia Editrice Pisana.

Matang, R., & Owens, K. (2014). The role of Indigenous traditional counting systems in children's development of numerical cognition: Results from a study in Papua New Guinea. *Mathematics Education Research Journal, 26*(3), 531–553. doi:10.1007/s13394-013-0115-2.

Mimica, J. (1988). *Intimations of infinity: The mythopoeia (cultural meanings) of the Iqwaye counting and number systems*. Oxford, UK: Berg Publishers.

Muke, C. (2000). *Ethnomathematics: Mid-Wahgi counting practices in Papua New Guinea* (Unpublished MEd thesis). University of Waikato, Waikato, New Zealand.

O'Connell, J., & Allen, J. (2004). Dating the colonization of Sahul (Pleistocene Australia-New Guinea): A review of recent research. *Journal of Archaeological Science, 31*(6), 835–853. doi: 10.1016/j.jas.2003.11.005.

Owens, K. (2001). The work of Glendon Lean on the counting systems of Papua New Guinea and Oceania. *Mathematics Education Research Journal, 13*(1), 47–71.

Owens, K. (2010, July). *Changing our view on space: Place mathematics as a human endeavour.* Paper presented at the annual conference of the Mathematics Education Research Group of Australasia, Perth.

Owens, K. (2014). The impact of a teacher education culture-based project on identity as a mathematics learner. *Asia-Pacific Journal of Teacher Education, 42*(2), 186–207. doi:10.1080/1359866X.2014.892568.

Owens, K. (2015). *Visuospatial reasoning: An ecocultural perspective for space, geometry and measurement education*. New York, NY: Springer.

Owens, K., & Kaleva, W. (2008). *Indigenous Papua New Guinea knowledges related to volume and mass*. Paper presented at the International Congress on Mathematics Education ICME 11, Discussion Group 11 on The Role of Ethnomathematics in Mathematics Education, Monterray, Mexico. http://dg.icme11.org/document/get/311

Paraide, P. (2010). *Integrating Indigenous and Western mathematical knowledge in PNG early schooling* (PhD thesis). Deakin University, Geelong, Victoria, Australia.

Pawley, A. (1981). Melanesian diversity and Polynesian homogeneity: A unified explanation for language. In K. Hollyman & A. Pawley (Eds.), *Studies in Pacific languages and cultures in honour of Bruce Biggs* (pp. 269–309). Auckland, NZ: Linguistic Society of New Zealand.

Pawley, A., & Ross, M. (2006). The prehistory of Oceanic languages: Current view. In P. Bellwood, J. Fox, & D. Tyron (Eds.), *The Austronesians: Historic and cultural perspectives*. Canberra, Australia: Pacific Linguistics, Australian National University.

Pica, P., Lemer, C., Izard, V., & Dehaene, S. (2004). Exact and approximate arithmetic in an Amazonian indigene group. *Science, 306*(5695), 499–503.

Pickles, A. (2013). *The pattern changes changes: Gambling value in highland Papua New Guinea*. PhD thesis, University of St Andrews, UK. Retrieved from https://research-repository.st-andrews.ac.uk/bitstream/10023/3389/6/AnthonyPicklesPhDThesis.pdf

Rosa, M., & Orey, D. (2012). Ethnomodeling: A pedagogical action for uncovering ethnomathematical practices. *International Congress on Mathematics Education.* Retrieved from http://www.icme12.org/upload/UpFile2/TSG/1799.pdf

Ross, M. (1988). *Proto Oceanic and the Austronesian languages of western Melanesia*. Canberra, Australia: Pacific Linguistics, C-98.

Ross, M. (1989). Early Oceanic linguistic prehistory. A reassessment. *Journal of Pacific History, 24*(2), 135–149.
Salzmann, Z. (1950). A method for analyzing numerical systems. *Word, 6*, 78–83.
Saxe, G. (1979). A comparative analysis of the acquisition of numeration: Studies from Papua New Guinea. *The Quarterly Newsletter of the Laboratory for Comparative Human Cognition, 1*(3), 37–43.
Saxe, G. (1981a). Body parts as numerals: A developmental analysis of numeration among the Oksapmin in Papua New Guinea. *Child Development, 52*, 306–316.
Saxe, G. (1981b) Numerical reasoning. Oksapmin: Development and change. *Educational Research Unit Occasional Paper 7*. Port Moresby, PNG: University of Papua New Guinea.
Saxe, G. (1981c). When fourth can precede second: A developmental analysis of an indigenous numeration system among Ponam Islanders in Papua New Guinea. *Journal of Cross-Cultural Psychology, 12*(1), 37–50.
Saxe, G. (1982). Developing forms of arithmetical thought among the Oksapmin of Papua New Guinea. *Developmental Psychology, 18*(4), 583–594.
Saxe, G. (1985). The effects of schooling on arithmetical understandings: Studies with Oksapmin children in Papua New Guinea. *Journal of Educational Psychology, 77*(5), 503–513.
Saxe, G. (1991). *Culture and cognitive development: Studies in mathematical understanding*. Hillsdale, NJ: L. Erlbaum Associates.
Saxe, G. (2012). *Cultural development of mathematical ideas: Papua New Guinea studies*. New York, NY: Cambridge University Press.
Saxe, G., & Esmonde, I. (2005). Studying cognition in flux: A historical treatment of *fu* in the shifting structure of Oksapmin mathematics. *Mind, Culture, and Activity, 12*(3), 171–225.
Seidenberg, A. (1960). The diffusion of counting practices. *University of California Publications in Mathematics, 3*, 215–300.
Sheppard, P., Walter, R., & Roga, K. (2010). Friends, relatives and enemies: The archaeology and history of interaction among Austronesian and non-Austronesian speakers in the western Solomons. In J. Bowden, N. Himmelmann, & K. Ross (Eds.), *A journey through Austronesian and Papuan linguistic and cultural space: Papers in honour of Andrew Pawley* (Vol. 615, pp. 95–112). Canberra, Australia: Pacific Linguistics, ANU.
SIL. (nd). *Ethnologue: Languages of the world*. Retrieved from http://www.ethnologue.com/
Smith, G. (1978). Counting and classification on Kiwai Island. *Papua New Guinea Journal of Education, Special Edition the Indigenous Mathematics Project, 14*, 53–68.
Smith, G. (1981). *Traditional mathematics in Morobe* (Working Paper no. 21). Lae, PNG: Papua New Guinea University of Technology.
Smith, G. (1984). *Morobe counting systems: An investigation into the numerals of the Morobe Province, Papua New Guinea* (MPhil). Papua New Guinea University of Technology, Lae, PNG.
Spriggs, M. (1984). The Lapita cultural complex. *Journal of Pacific History, 19*(3–4), 202–223.
Spriggs, M. (2006). The Lapita culture and Austronesian prehistory in Oceania. In P. Bellwood, J. Fox, & D. Tyron (Eds.), *The Austronesians: The historical and comparative perspectives* (pp. 119–142). Canberra, Australia: Australian National University.
Spriggs, M. (2010). "I was so much older then, I'm younger than that now": Why the dates keep changing for the spread of Austronesian languages. In J. Bowden, N. Himmelmann, & K. Ross (Eds.), *A journey through Austronesian and Papuan linguistic and cultural space: Papers in honour of Andrew Pawley* (Vol. 615, pp. 113–140). Canberra, Australia: Pacific Linguistics, Australian National University.
Spriggs, M. (2011). Archaeology and the Austronesian expansion: Where are we now? *Antiquity, 85*(328), 510–528.
Struik, D. (1987). *A concise history of mathematics*. New York, NY: Dover Publications.

Tryon, D. (1982). Austronesian languages. In R. May & H. Nelson (Eds.), *Melanesia: Beyond diversity* (pp. 241–248). Canberra, Australia: Research School of Pacific Studies, Australian National University.

Tryon, D. (1984). The peopling of the Pacific: A linguistic appraisal. *Journal of Pacific History, 19*(3–4), 147–159.

Tryon, D. (2006). Proto-Austronesian and the major subgroups of Austronesian. In P. Bellwood, J. Fox, & D. Tryon (Eds.), *The Austronesians: Historical and comparative perspectives* (pp. 19–42). Canberra, Australia: ANU Press.

Tryon, D., & Hackman, B. (1983). *Solomon Islands languages: An internal classification*. Canberra, Australia: Pacific Linguistics, C-72.

Van der Waerden, B., & Flegg, G. (1975a). *Counting 1: Primitive and more developed counting systems*. Milton Keynes, UK: Open University.

Van der Waerden, B., & Flegg, G. (1975b). *Counting 2: Decimal counting words, tallies and knots*. Milton Keynes, UK: Open University.

Voorhoeve, C. (1987). *Worming one's way through New Guinea: The chase of the peripatetic pronouns*. A world of language: Papers presented to Professor S.A. Wurm on his 65th birthday (C-100) (pp. 709–727). Canberra, Australia: Pacific Linguistics.

Wassmann, J. (1997). Finding the right path. The route knowledge of the Yupno of Papua New Guinea. In G. Senft (Ed.), *Referring to space: Studies in Austronesian and Papuan languages*. Oxford, UK: Oxford University Press.

Wassmann, J., & Dasen, P. (1994). Yupno number system and counting. *Journal of Cross-Cultural Psychology, 25*, 78–94.

White, J., O'Connell, J., & Koettig, M. (1982). *A prehistory of Australia, New Guinea and Sahul*. Sydney, Australia: Academic Press.

Wurm, S. (1982a). Papuan languages. In R. May & H. Nelson (Eds.), *Melanesia: Beyond diversity* (pp. 225–240). Canberra, Australia: Research School of Pacific Studies, Australian National University.

Wurm, S. (1982b). *Papuan languages of Oceania*. Tübingen, Germany: Gunter Narr Verlag.

Wurm, S., & Hattori, S. (Eds.). (1981). *Language atlas of the Pacific area. Part 1: New Guinea area, Oceania, Australia*. Canberra, Australia: Australian Academy of the Humanities.

Wurm, S., Laycock, D., Voorhoeve, C., & Dutton, T. (1975). Papuan linguistic prehistory, and past language migrations in the New Guinea area. In S. A. Wurm (Ed.), *New Guinea area languages and language study, Vol. 1: Papuan languages and the New Guinea linguistic scene* (Pacific Linguistics, C-38, pp. 935–960). Canberra, Australia: Australian National University.

Zaslavsky, C. (1973). *Africa counts: Number and pattern in African culture*. Boston, MA: Prindle, Weber & Schmidt.

Chapter 2
The Languages Studied by Lean and his Analysis of Counting Systems

Kay Owens and Glen Lean

Abstract The complexity and diversity of the languages of the Melanesian region are outlined in terms of their linguistic classification. The languages are divided into Austronesian (mainly Oceanic) and Non-Austronesian categories. Lean attained data from 883 of the 1 201 languages recognised in the 1980s, that is 74% of the languages. The trustworthiness of this book lies largely in the methodology meticulously undertaken by Lean in his study. This chapter provides details of the methodology and data sources that were used. In particular, the extensive nature of his data collection from first contact to field visits and students completing questionnaires are justified and presented as strengths of this book. Some limitations are discussed together with alternative ways of describing counting systems and the reason for selecting Salzmann's (1950) cycle system. In analysing systems, Lean has used the frame pattern (basic number words or morphs, recorded as numerals, from which other number words are formed), the cyclic pattern which is based on the frame words that are repeated in forming new number words, and the operative patterns which are the patterns for generating new number words. In addition, Lean recorded any body parts used for tallying in order to determine the overall nature of the counting system.

Keywords Alternative classifications of counting systems • Austronesian and Non-Austronesian languages • Counting System Questionnaire • Frame patterns in number systems • Operative patterns in number systems • Cyclic patterns in number systems • Digit tally systems • Body-part tally systems • Melanesian languages • Oceania • Papua New Guinea

Introduction

After arriving in Papua New Guinea (PNG) in 1968, Lean became fascinated that his students had multiple ways of counting and he began his journey to collect as many of the counting systems from the languages of PNG as he could and to categorise them in a useful way. Lean continued to collect data on PNG natural language numerals and counting systems from both primary and secondary sources until 1988. At the same time, he became aware that these data might contradict current theories of how counting systems developed and what in fact were key concepts in number. Having students from across the Pacific, he began to realise that Oceania languages could add to this story and he pursued that information when he left PNG in 1989.

K. Owens (✉)
Charles Sturt University, Dubbo, NSW, Australia
e-mail: kowens@csu.edu.au

G. Lean

K. Owens et al., *History of Number*, History of Mathematics Education,
DOI 10.1007/978-3-319-45483-2_2

The documentation of the accumulated data began in 1985 and a final revised version was completed at the beginning of 1991: this material is held as a counting system database at the Glen Lean Ethnomathematics Centre and was available on a website until recently. In all, data were acquired on the numerals, counting and tally systems of 532 languages, or just over 70% of the total languages of Papua New Guinea, then regarded as 750. Generally speaking, the languages for which it was difficult to obtain data were those with 500 speakers or less; in the Madang Province alone some 93 languages fall into that category. In the ensuing years, SIL (Summer Institute of Linguistics) and others have determined many more languages identified by the people as distinct languages. In Lean's work he had the difficult task of collating often incomplete and differently spelt data from informants on the Counting System Questionnaire (CSQ). Many came from different villages and subsequently many new languages have been identified since Wurm and Hattori's (1981) *Language Atlas of the Pacific Area* which Lean used. Owens also found that variations occurred even between villages and within the same village of the same language from her studies after 2000. Both Lean's and Owens' informants varied in their knowledge of their village language and the degree to which languages such as the lingua franca Tok Pisin and neighbouring languages influenced them. While some detail may be lost, Lean tried to select the best representation from his CSQ sources. It is not possible to compare new data with his original raw data as it is no longer available. For that reason, the language naming from Wurm and Hattori has been used in this book knowing that the Ethnologue provides additional languages. If we attempted to revise the language names, the integrity of Lean's work would be lost as he provided extensive comparative data between early sources, more recent sources and CSQ data. However, many of the languages that are now identified were dialects so a person familiar with the villages, recent changes in village migrations and marriages, and commonly used language names may be able to determine if the data should be given another language name. Examples of this analysis may be found in Appendices B and C of this book.

Details of Lean's Methods and Language Sources

Lean collected data from a large proportion of known languages. Thus out of a total of about 720 Non-Austronesian (NAN) languages (see Chapter 1 for more details), he acquired data for 430 or about 60%. The detailed summary of acquired data is provided in Table 2.1. The distribution of the NAN languages and their phyla are shown in Table 1.1 and Figure 1.2.

Table 2.1
The Number and Percentage of Acquired Language Data in Each NAN Phylum

NAN Language Phyla	Number of Acquired Language Data	Percentage of Languages
Trans New Guinea Phylum	298	>60%
Sepik-Ramu Phylum	44	45%
Torricelli Phylum	26	54%
East Papuan Phylum	23	100%
West Papuan Phylum	10	77%
Sko Phylum-level Stock	6	75%
Arai Phylum-level Family	6	100%
Geelvink Bay Phylum	3	50%
Kwomtari Phylum-level Stock	4	80%
East Vogelkop Phylum-level Stock	2	67%
Amto-Musian Phylum-level Family	2	100%
Isolates	6	75%

Note. Wurm (1982) gave 27 languages for the East Papuan Phylum but 4 were extinct by 1990 and for the West Papuan Phylum, Wurm gave 24 but with 11 of these spoken in North Halmahera, Maluku Islands, Indonesia, Lean only considered 13, of which he had data for 10.

The Trans New Guinea Phylum covers most of PNG and West Papua with small areas on the north for other phyla with the East Papuan in pockets of New Guinea and Papuan Islands and Island Melanesia.

Of the 38 AN languages shown in Table 2.2 for West Papua, 33 are non-Oceanic and the remaining 5 are Oceanic. All other languages are Central/Eastern Oceanic in the excluded areas. Of a total possible 479 AN languages (see Chapter 1 for more details), data have been acquired for 453, or about 95%. Appendix E provides Ross, Pawley, and Osmond's (2003) Proto Austronesian and Proto Oceanic tree for this region.

Table 2.2
The Number of Acquired AN Language Data Given by Cluster within Countries and Percent for Each Country

Country # Clusters	No. of AN Languages For Which Data Were Acquired	Percentage of Known Languages
A. Papua New Guinea	188	91%
1. Admiralties Cluster	24	
2. Western Oceanic		
2.1 North New Guinea Cluster	78	
2.2 Papuan Tip Cluster	41	
2.3 Meso-Melanesian Cluster	64	
3. St. Matthias Group	1	
B. West Papua	38	97%
Oceanic Austronesian	5	
Non-Oceanic Austronesian	33	
4. Central/Eastern Oceanic		
C. Solomon Islands	56	100%
4.1 Southeast Solomonic	22	
4.2 Eastern Outer Islands	6	
D. Vanuatu	105	100%
E. New Caledonia	28	100%
F. Fiji	2	100%
G. Rotuma	1	100%
H. Polynesia (see note)	23	96%
I. Micronesia	12	67%

Note. 1. Language clusters are not restricted to the borders of the countries in this region so it is not perfect to place the clusters under the countries or the countries under the clusters but the table provides some indication of the clusters and their locations.

2. The total number of languages regarded as Polynesian is 36; 22 languages are spoken in "Triangle Polynesia" and 2 others are situated in Micronesia. The remaining 12 languages are dotted throughout Island Melanesia and are included in the figures for Papua New Guinea (3), Solomon Islands (5), Vanuatu (3), and New Caledonia (1). These Polynesian languages found in Melanesia, plus the two in Micronesia, are known as the "Polynesian Outliers".

Combining the AN and NAN languages, there was a total of approximately 1 500 languages recognised as spoken in the region under consideration in 1980s. Lean acquired data for 1 157, 72%, which is a substantial proportion on which to develop his thesis.

Papua New Guinea

The sources of a particular set of numerals are often numerous and at other times very limited. Lean documented the data sources at the bottom of each data table. For some languages, he gave more than one table because the counting system data for some languages was quite varied. This detailed information upon which Lean's thesis was developed were given in the appendices of his thesis with a few hard copies of his original collation available in libraries in PNG and elsewhere. The sources are basically of two types: primary and secondary; each of these is discussed below. However, the data sources will be included in this book when referenced in the discussion in text or in Appendices B and C.

Primary sources. The primary sources are divided into four main groups. The first group comprises Lean's field notes which were taken in two different sets of circumstances. There are those field notes taken from informants living in villages which he visited. One-to-one interviews were held, wherever possible, with an older member of the village, usually, though not always, male. The language used was invariably Tok Pisin and he had no records of ever having to resort to an interpreter in the case, say, where an elderly informant did not speak Tok Pisin. Prior to 1980 notes were taken during the interview; subsequent to 1980 the interviews were recorded on a portable cassette recorder and the relevant material was later transcribed, normally using Roman orthography, from the tape within a day or two of the interview. From his first field trip in 1968 to his last in 1987, he had records of interviews with informants from 35 villages in eight provinces. The second set of circumstances in which he acquired field notes was when informants, particularly those living in relatively remote areas, visited Lae and he was able to arrange an interview with them. He had records of 29 interviews which fall into this category.

The second main group of primary sources, which in fact forms the major part of the complete database for PNG, comprises questionnaires given to three different populations (see Appendix A). First, there are those completed by incoming students at the PNG University of Technology during the period 1968 to 1983; these account for a total of 1 200. Second, students at four National High Schools (Sogeri, Aiyura, Keravat and Passam) completed questionnaires in 1982, 1984, and 1986; these account for a further 1 022 questionnaires. Third, in 1985, copies of a slightly revised questionnaire were sent to headmasters of 1 700 Community (primary) Schools in all provinces of PNG. In each case the headmaster was asked to assist in the completion of the questionnaire with an adult who spoke the local vernacular (in general, a single rural community school serves a single language group). These account for an additional 302, giving altogether a total of 2 524 usable, completed questionnaires.

The third group of Lean's primary sources comprises unpublished material gathered during the Indigenous Mathematics Project (IMP) during the period 1976 to 1979. The IMP was established in 1976 by the PNG Department of Education under the directorship of Dr David Lancy. The research phase of the project was primarily concerned with studying aspects of children's cognitive development and, in particular, possible cultural influences. In association with the latter, linguistic and anthropological data were collected at each village test site and these data usually consisted of information about the vernacular classification (or taxonomic) systems used in the society together with information about its counting system. Much of this information has subsequently been published (for example, Carrier, 1981; Kettenis, 1978; Lancy, 1983; Saxe, 1979, 1981a, 1981b, 1981c, 1982; Smith, 1978). The unpublished material deriving from the IMP comprises two sets of questionnaires: one by Lancy sent to Grade 10 students and a second by Deibler sent to Summer Institute of Linguistics (SIL) members working in the field each of whom was asked to provide details of the counting system of the language on which they were working. Altogether a total of 238 IMP questionnaires were made available for this study.

The fourth group of unpublished primary source material derives from survey word lists compiled by members of SIL. The PNG Branch of the Institute was established in 1956 and is concerned primarily with Bible translation and literacy. Its members, all of whom receive training in linguistics, have studied in considerable detail several hundred PNG languages; in addition, they have surveyed many more. The SIL library, at Ukarumpa in the Eastern Highlands Province, holds an estimated 500 completed and unpublished Standard Survey Word Lists of various languages from all provinces of PNG (there were 19 provinces until 2013 when two new provinces were made and more recently Bougainville became an autonomous region). The Survey Word List in use after 1965 contains 170 lexical items and 20 standard sentences. The completed Word Lists included data on the numerals 1 to 5, and 10 (and often further numeral data added by the surveying linguist). A complete search of these materials in the period 1984 to 1986 yielded a total of 362 Word Lists which contained some numeral data on various languages spoken in each of the PNG provinces.

In addition, Lean's questionnaire (Appendix A) was modified as an activity by Kaleva and other lecturers for the University of Goroka (UoG) students and by Owens during interviews with villagers. These data are interwoven as needed in this book. During Owens and Kaleva's measurement study from 2006, counting data were included in some reports by UoG students (studying Mathematics, Language and Culture from 1996 to 2010), in measurement questionnaires and interviews, and during village visits with data being collected by video-recording or written down by villagers. Owens and Kaleva's data as well as Lean's databases informs the discussions of this book but Lean's thesis forms the basis of Chapters 3-10. Additional data are now available from SIL and other linguistic sources and some are considered in this book where appropriate and it is possible to link to Lean's data. Chapters 11-12 by Paraide are examples of data that further inform our understanding building on Lean's initial study. However, there are likely to be discrepancies between recently recorded counting systems and the data sourced for this book as the languages of PNG since the introduction of English and Tok Pisin, money and businesses, and greater mobility and external contact have increased the changes in languages significantly. In addition, the degree of regular use of a language varies considerably along with the value placed on keeping the original form of the language.

Secondary sources. The secondary, published sources of data on PNG numerals, counting systems, or tally practices are numerous and listed in the tables collated into 17 volumes by Lean in 1992). Altogether, these sources contributed an estimated 600 or so items of data (one item of data being, for example, a set of numerals or a description of a counting or tally system for a particular language). Most but by no means all are in the reference list of this book.

In 1968, at the beginning of the data-gathering phase of this study, a search of the published literature revealed only a small number of articles which dealt specifically with aspects of numeration in PNG. The linguist S.H. Ray (1907) had included a chapter on Numeration and numerals in the Melanesian languages of British New Guinea in Volume 3 of the *Reports of the Cambridge Anthropological Expedition to Torres Straits*. The German scholar Fr. W. Schmidt (1929) included some PNG numeral data in an article on *Numeral systems* in the 1929 edition of the *Encyclopaedia Britannica*. Two SIL linguists working in the Southern Highlands Province of PNG, Karl and Joice Franklin, published an article on the *Kewa counting systems* in 1962 and subsequently in *Pacific Linguistics, C-53* (SIL website, Franklin & Franklin, 1962, 1978). Ted Wolfers (Wolfers, 1969, 1971), a journalist based in Port Moresby, wrote three articles on PNG counting systems, one of these appearing in the published *Encyclopaedia of Papua and New Guinea* (Wolfers, 1972). Contrary to Lancy (1978) who referred to the "scanty available literature" on PNG counting systems (and this, at the time, was probably the prevailing view of the small number of scholars with an interest in the field), Lean's exhaustive search of the ethnographic and linguistic literatures of PNG revealed a large amount of first- and early-contact data about natural language numerals and counting systems. Although the details of the sources of these data are provided in the appendices of Lean's thesis and the 1991 publications of the Provincial data, it is useful to provide a brief survey and summary here. The appendices of this book provide details of the sources for the specific languages chosen to give the background for the diverse languages and counting systems used in the argument of this book.

The earliest published examples of PNG numerals were collected by the Dutch explorers Le Maire and Schouten in 1616 during their voyage from the Netherlands to the East Indies. Brief vocabularies, including numerals, were taken at Moyse (i.e. Moses) Island and Nova Guinea; the locations of these were subsequently established, by German scholars (Schlaginhaufen, Sapper, & Freiederici, 1915–1916) to be respectively Tabar Island, situated off the east coast of New Ireland, and Muliama, situated on the south-east coast of New Ireland. More than 200 years were to elapse before further vocabularies were collected in the New Guinea region. In 1827, the French explorer Jules Dumont d'Urville sailed through the New Guinea islands region and collected a vocabulary of the Siar language at Port Praslin near the southern tip of New Ireland, published 1883. In 1849, Owen Stanley in H.M.S. Rattlesnake sailed through the islands of the Milne Bay area and along the southern coast of Papua; several vocabularies were taken during this voyage and were published by MacGillivray (1852).

With these few exceptions, the published sources of numeral data for the coastal PNG languages begin to appear in the 1870s after the settlement of Port Moresby and the Duke of York Islands and Rabaul region. Much of the data collection for the northern part of New Guinea took place during the period of the German Protectorate, i.e. from 1884/85 to 1914, after the administrative regions of Kaiser Wilhelmsland and the Bismarck Archipelago were established. These regions encompassed what are now the Provinces of East New Britain, New Ireland, Manus, Madang, Morobe, East Sepik and Sandaun (West Sepik); most of the first- and early-contact data on the languages of these provinces emanate from the work of German missionaries and scholars during this period, the publications appearing, by and large, in books and scholarly journals rather than the official colonial reports and gazettes. Strictly speaking, Kaiser Wilhelmsland also incorporated what are now the Provinces of the vast inland Highlands region; this, however, was not penetrated until the 1930s and remained a terra incognita during the German colonial period.

The southern half of the island was called British New Guinea during the period from 1884 to 1906, and Papua from 1906 onwards. Much of the first- and early-contact data on the languages of this region derive not so much from missionaries and explorers, as is the case for German New Guinea, but from the officers of the colonial administration: the published sources of data are not so much the scholarly Zeitschrift but rather the official Government reports, notably the Annual Reports on British New Guinea (1889-1906) and the Annual Reports for Papua (from 1907). That this comprehensive collection of data exists, mainly as vocabularies which appear in the appendices of the Reports, is due largely to the efforts of two men: Sir William MacGregor, Administrator of British New Guinea from 1888 to 1898, and Dr W. M. Strong, Chief Medical Officer for Papua, both of whom had an amateur interest in linguistics and who collected data themselves as well as organising members of the colonial service to do so. Between 1889 and 1917, 389 vocabularies were published in the Reports, some others, though not so many, appearing in the period 1918 to 1940; the large majority of these contained some data on numerals. Much of this material, allowing for some differences in orthography, shows good agreement with more recently collected data. It is quite clear that some care was exercised in the elicitation and recording of vocabularies. Dating from MacGregor's time, a standard vocabulary list existed (in fact what would now be called a Standard Survey Word List) which was compiled from a set of "questions on the customs, beliefs, and languages" originally drawn up by Sir James Frazer, author of The Golden Bough. A standardised orthography was recommended, based, it seems likely, on that used by the Royal Geographical Society.

It was during MacGregor's administration that the eminent Cambridge anthropologist A. C. Haddon organised two expeditions to the Torres Strait region, the first in 1889 and the second, the famous Cambridge Anthropological Expedition to Torres Straits in 1898, in which he was accompanied by the linguist Sidney H. Ray. Ray's published works on PNG languages (e.g., 1895, 1907, 1919a), and indeed many other languages of Melanesia and Polynesia, which appeared over the period 1891 to 1938, establish him as the most important linguistic scholar of the early colonial period, rivalled only by Fr. W. Schmidt (1926, 1929). The results of the expedition were published in seven volumes, one of which is Ray's contribution on linguistics.

Excluding the vocabularies published in the Annual Reports mentioned above, something like 200 publications on linguistics and ethnographic studies of PNG appeared in the period 1875 to 1940, many of which contained data on natural language numerals and counting systems. During the 1930s, a German scholar, Dr Theodor Kluge (1938, 1939, 1941), assembled data, mostly from published sources, on the numerals of a large number of languages spoken throughout the world. Several substantial typescripts were prepared during the period 1938 to 1941; two of these contain data on the numerals of languages of PNG and Melanesia generally. Due to the outbreak of the Second World War these materials were never put into final form and published; however the uncorrected typescripts survive and were consulted by Lean for the PNG part of this study.

Oceania

The sources of data for the six volumes on the Counting Systems of Oceania, which comprise Appendix D of Lean's study, are the published literatures of linguistics and ethnography of the geographical regions of Melanesia, Polynesia, and Micronesia. As was the case for the PNG data, there is only a small number of publications which deal specifically with numerals and counting systems. For West Papua, for example, there is a survey of counting systems by Galis (1960), together with an article by Briley (1977), an SIL member, which deals with several counting system types. For the Solomon Islands there is a study by Fox (1931) of "numerals and numeration" of the Arosi language. Macdonald (1893) had an article on the "Asiatic origin of the Oceanic numerals", although this deals in fact with a small number of Vanuatan languages. More recently, we have two further articles on numeration in the Vanuatan languages, one by Lynch (1977) on Tanna and another by Charpentier (1979) on the languages of south Malekula. For the whole of the Polynesian region, only a few studies deal specifically with numeration. Audran (1930) published a one-page summary of the numerals of several Polynesian languages. Best (1907) had some data on the counting system of New Zealand Maori and Fraser (1901, 1902) had some comparative data on the Polynesian numerals 1, 5, and 10. Heider (1926-1927) had material on the numerals and counting system of Samoan, Metraux (1936) had an article on Easter Island numerals, and Lemaître (1985) had a survey of the numerals of mainly the languages of eastern Polynesia.

As was also the case with the collection of the PNG data from published sources, particular care was taken to try to obtain as much first- and early-contact data as possible. Some of this material derives from vocabularies taken during the three expeditions under the leadership of Captain Cook in the years 1768 to 1780. Many previously unpublished vocabularies obtained during these voyages were published by Lanyon-Orgill (1979). While most of these deal with various languages of Polynesia, for example New Zealand Maori, Pa'umotu, Tahitian, Easter Island, Tongan, Marquesan, Hawai'ian, and Cook Islands Maori, there are several languages of the Melanesian region included as well which were recorded in New Caledonia and Vanuatu. During the nineteenth century much more information about the Pacific region became available as the islands were explored and settled by the Western colonial powers. Between the years 1838 to 1842, the United States Exploring Expedition under the command of Captain Charles Wilkes visited many of the islands in both Polynesia and Micronesia and the ethnographic and linguistic results of the expedition were written by Horatio Hale (1846). A memoir by Turner (1861), a missionary who had spent 19 years in parts of Polynesia, includes some comparative lexical data, including numerals, on six Polynesian languages. The German linguist Hans von der Gabelentz (1861, 1873) in his two-volume *Die melanesischen Sprachen* published a survey of a number of languages in the Solomon Islands, Vanuatu and New Caledonia. Codrington's (1885) *The Melanesian Languages* contains grammatical and lexical data on some 35 languages in the non-New Guinea Melanesian region, in particular it has a chapter devoted to numeration and numerals. A significant contributor to the study of the languages of the region was Codrington's protégé Sidney Ray. Ray published a series on the languages of the New Hebrides (Ray, 1893), on Polynesian linguistics (Ray, 1912, 1915, 1916, 1917, 1919c, 1920) and the Polynesian languages of Melanesia (Ray, 1919b). These all contain some numeral data as does his major work *A Comparative Study of the Melanesian Island Languages* (Ray, 1926).

With a few exceptions, much of the modern linguistic and ethnographic data derive from sources published after the Second World War. These will be referred to when specific languages are targeted but they are extensive and available from Lean's thesis (Lean, 1992). They will not be discussed further here but there are a few major sources of data which have been indispensable for this study and which should be noted (Christian, 1924). For example, Tryon and Hackman's (1983) *Solomon Islands Languages: An Internal Classification* has lexical data, including numerals, on all the languages spoken in the Solomon Islands. Similarly, Tryon's (1976) *New Hebrides Languages: An Internal Classification* has lexical data on all the languages of Vanuatu. Information on the languages of New Caledonia has been harder to access than the data on other languages of Melanesia, however a general survey by Leenhardt (1946), contains lexical, phonological, and grammatical data on all the languages and, for some languages, these are the only data available. For Polynesia, a series of anthropological studies published by the Bernice P. Bishop Museum in Hawaii, appeared during the period 1920 to 1950, several of these by the eminent Polynesian scholar Sir Peter Buck (Te Rangi Hiroa); some, though not all of these, contain useful data on numerals and counting (Beaglehole & Beaglehole, 1938; Buck, 1938; Christian, 1924; Handy, 1923). Finally, for Micronesia, a series of publications of the University of Hawaii Press and subsequent articles have been indispensable and have provided valuable information on Marshallese, Mokilese, Kosraean, Ponapean, and Woleaian (Bender, 1971; Harrison & Jackson, 1984; Lee, 1976; Rehg & Sohl, 1979).

It should be noted that since Lean completed his study, there have been numerous further publications of data. Goetzfridt (2008) provided brief commentary on hundreds of articles and a few specifically link to numeracy across the region. Te Māori mathematics project (Meaney, Trinick, & Fairhall, 2012; O'Sullivan, McKinley, Stewart, Richards, & Ball, 2003) in the 1990s ensured a consensus of words for counting and related concepts such as operations on numbers, groupings and so on. However, the main developments in terms of data have led at best to discussions especially in Chapters 8, 10, and 13.

The Description and Classification of Data

In order to present a coherent picture of the very large amount of number data that formed the appendices of his doctoral thesis, Lean adopted a number of descriptive terms which may be used to classify the various systems into a relatively small number of types. In some of the older linguistic literature concerned with the description of natural language numeral systems, it was common to use the descriptive term "base" when discussing the cyclic nature of the system. Thus we find counting systems variously termed "binary" (base 2), "ternary" (base 3), "quinary" (base 5), "decimal" (base 10), and "vigesimal" (base 20). Using a single number to characterise a counting system is reasonably adequate when we are dealing with, say, the English counting system which, with some irregularities, is essentially a base 10 one. The cyclic structure of many of the counting systems found in Melanesia is often more complex than the English system, in that a single system may have elements of base 2, base 5, and base 20; others have a structure in which we can discern elements of base 5, base 10, and base 20. This was recognised in the older literature in which we find reference to "mixed base" systems and such terms as "incomplete decimal" systems (that is, one which had elements of both base 5 and base 10).

As detailed in Chapter 1, pages 20–21, Lean adopted Salzmann's (1950) cycle system and the analysis of frame words and operative patterns to determine the primary and secondary cycle systems for each language.

No single typological system can do justice to the diversity and richness of the ways number words are combined to form other number words. Salzmann introduced his terms as a reaction to the oversimplistic labelling of numeral systems as found in the work of nineteenth and early twentieth century writers. Hymes (1955) criticised Salzmann's terminology as being inadequate for the description of the numeral systems of certain Native American Indian (First Nations) languages, however the criticisms do not appear to apply to the situation regarding the systems found in PNG and Oceania. As far as the typology of numerals found in the New Guinea languages is concerned, there have been several different variants applied. Ray (1907, pp. 463-478), in his discussion of numeration and numerals in the Melanesian languages of British New Guinea used the terms, as given above, for describing the "base" of the numeral system, that is "incomplete decimal", "decimal", "vigesimal", and so on. Galis (1960), in his study of the counting systems of West Papua, arrived at a sixfold classification: (1) body-part tally systems, (2) base 2 (or "binary") systems, (3) base 6 systems, (4) base 4 systems, (5) the "digit-tally" system (with a (5, 20) cyclic pattern), (6) the "Austronesian" type, i.e a base 10 system with 10 discrete numerals. Lancy (1978, pp. 6-8), introduced a four-type classification ranging from body-part tally systems (Type I); systems which have 3 to 4 discrete number words and a base of 2, 3, 4, or 5, and where objects may be used in carrying out tallies (Type II); the "quinary- vigesimal" system, that is a counting system having a (5, 20) cyclic pattern and which usually employs fingers and toes as an aid to tallying (Type III); and the "decimal" system, i.e a 10-cycle system which normally has no reference to body-parts and which has 6 to 10 discrete numerals (Type IV). Type II was based on the use of sticks or stones for tallying. However, this system often overlaps any of the other systems as mere representations. Smith (1984), in his study of the counting systems of the Morobe Province of PNG, distinguishes eight varieties of "counting procedure": (1) body-part tally systems, (2) systems with only two numerals, (3) systems which employ two numerals plus finger-and-toe "digit tally", (4) systems with three numerals only, (5) systems which employ three numerals plus "digit tally", (6) systems which employ four numerals plus "digit tally", (7) systems which have a numeral for 10, and (8) systems which have numerals for 10 and 20. In Chapter 3, 2-cycle systems includes Smith's categories 2, 3, 4, 5, and 6 (see Figure 3.1) but some of the digit tally systems are discussed in Chapter 5 if there was not a 2-cycle. In particular, it should be noted that the Salzmann system is more flexible than most of the above, particularly Lancy's system, and that discussing the systems in terms of base can be misleading as most of the systems do not follow a system of powers as in the base 10 Hindu-Arabic system. Furthermore, the use of Salzmann's terminologies assists in recognising some of the culturally related contexts, some of which were detailed by Lean but subsequent research has extended especially in Chapters 8 and 11.

Spatial Analysis of the Languages of Papua New Guinea

To illustrate the density of languages in PNG, small maps from the SIL Ethnologue have been used in Figure 2.1 to provide a spatial representation of density with names of languages or numbers (matching names in the tables on each map). It provides an appreciation of the density of the 850 PNG languages and thus the enormity of Lean's data collection and analysis.

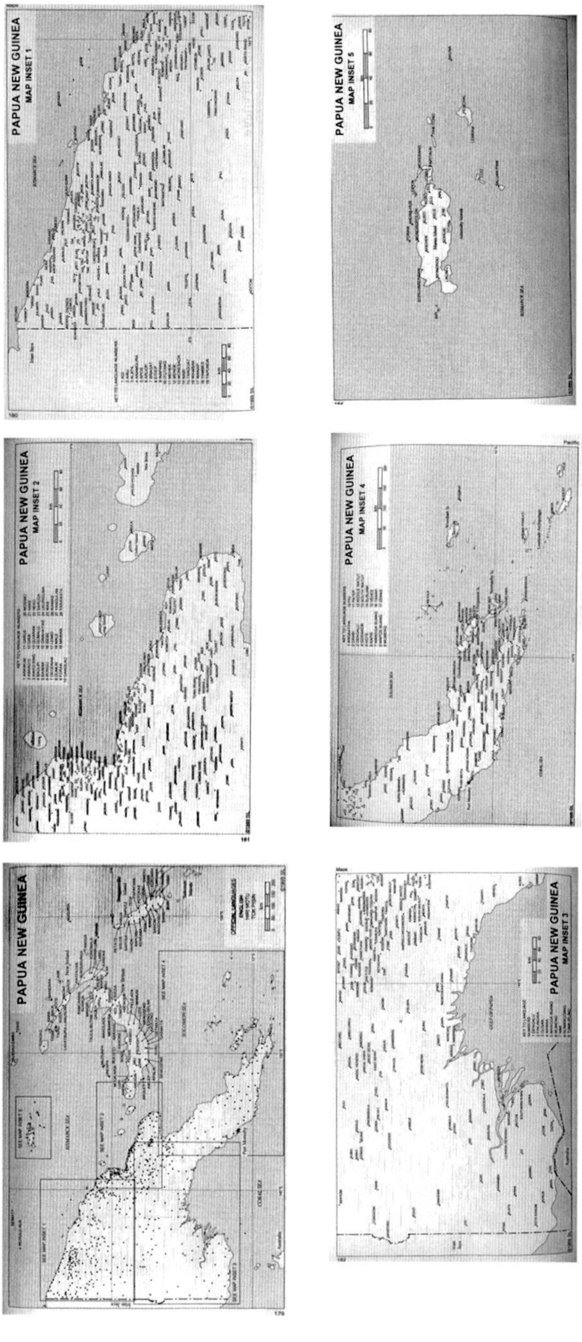

Figure 2.1. Spatial analysis of the languages of PNG.
Source. Maps from Ethnologue of languages in PNG. *Source.* (SIL, n.d.).

References

Audran, H. (1930). Numération polynésienne. *Bulletin de la Société des Études Océaniennes (Polynésie orientale), 35*(4), 25.

Beaglehole, E., & Beaglehole, P. (1938). Ethnology of Pukapuka. *Bernice P. Bishop Museum Bulletin, 150.*

Bender, B. (1971). Micronesian languages. In T. Sebeok (Ed.), *Current trends in linguistics, Current trends in linguistics, Vol. 8, Linguistics in Oceania* (pp. 426–465). The Hague, The Netherlands: Mouton.

Best, E. (1907). Maori numeration, the vigesimal system. *Journal of the Polynesian Society, 16*, 94–98.

Briley, J. (1977). Some counting systems of Irian Jaya. *Irian: Bulletin of Irian Jaya Development, 6*(3), 28–32.

Buck, P. (1938). Ethnology of Mangareva. *Bernice P. Bishop Museum Bulletin, 157.*

Carrier, A. (1981). Counting and calculation on Ponam Island. *Journal of the Polynesian Society, 90*(4), 465–479.

Charpentier, J.-M. (1979). *La langue de Port-Sandwich (Nouvelles-Hebrides).* Paris, France: SELAF.

Christian, F. (1924). Vocabulary of Mangaian language. *Bernice P. Bishop Museum Bulletin, 11.*

Codrington, R. (1885). *The Melanesian languages.* Oxford, UK: Clarendon Press.

Fox, C. (1931). Arosi numerals and numeration. *Journal of the Polynesian Society, 40*, 236–243.

Franklin, K., & Franklin, J. (1962). The Kewa counting systems. *Journal of the Polynesian Society, 71*, 188–191.

Franklin, K., & Franklin, J. (1978). *The Kewa dictionary, with supplementary grammatical and anthropological materials* (Pacific Linguistics, C-53). Canberra, Australia: Australian National University.

Fraser, J. (1902). The Polynesian numerals: One, five, ten. *Journal of the Polynesian Society, 11*, 1–8, 107–120, 171–177.

Fraser, J. (1901). The Polynesian numerals: One, five, ten. *Journal of the Polynesian Society, 10*, 171–177.

Galis, K. (1960). Telsystemen in Nederlands-Nieuw-Guinea. *Nieuw Guinea Studien, 4*(2), 131–150.

Goetzfridt, N. (2008). *Pacific ethnomathematics: A bibliographic study.* Honolulu, HI: Japanese Cultural Centre and University of Hawaii Press.

Hale, H. (1846). *United States exploring expedition* (Vol. 6). Philadelphia, PA: G. Sherman.

Handy, E. S. C. (1923). The native culture in the Marquesas. *Bemice P. Bishop Museum Bulletin, 9.*

Harrison, S., & Jackson, F. (1984). Higher numerals in several Micronesian languages. In B. Bender (Ed.), *Studies in Micronesian linguistics* (pp. 61–79). Canberra, Australia: Pacific Linguistics.

Heider, E. (1926–1927). Die Zahlwörter und Zahlansdrücke in der Samoanischen Sprache. *Zeitschrift für Ethnologie, 17.*

Hymes, V. (1955). Athapaskan numeral systems. *International Journal of American Linguistics, 21*, 26–45.

Kettenis, F. (1978). Traditonal food classification and counting systems of Kilenge, West New Britain. *Papua New Guinea Journal of Education, 14*, 28–43.

Kluge, T. (1938). *Die Zahlbegriffe der Australier, Papua und Bantuneger nebst einer Einleitung ueber die Zahl, ein Beitrag zur Geistesgeschichte des Menschen.* Berlin, Germany: Selbstverlag.

Kluge, T. (1939). *Zahlenbergriffe der Völker Americas, Nordeurasiens, der Munda, und der Palaioafricaner.* Unpublished manuscript. Berlin, Germany.

Kluge, T. (1941). *Die Zahlenbegriffe der Sprachen Central- und Südostasiens, Indonesiens, Micronesiens, Melanesiens und Polynesiens mit Nachträgen zu den Bänden 2–4.* Berlin, Germany. Ein fünfter Beitrag zur Geistesgeschichte des Menschen nebst einer principiellen Untersuchung über die Tonsprachen. Original unpublished manuscript held at Yale University, microfilm held at Australian National University, Canberra, Australia.

Lancy, D. (1978). Indigenous mathematics systems: An introduction. *Papua New Guinea Journal of Education, 14*(Special issue), 6–15.

Lancy, D. (1983). *Cross-cultural studies in cognition and mathematics*. New York, NY: Academic Press.

Lanyon-Orgill, P. (1979). *Captain Cook's south sea island vocabularies*. London, UK: Author.

Lean, G. (1992). *Counting systems of Papua New Guinea and Oceania* (Unpublished PhD thesis). PNG University of Technology, Lae, Papua New Guinea.

Lee, K.-D. (1976). *Kusaiean-English Dictionary. PALI Language Texts: Micronesia*. Honolulu, HI, USA: University Press of Hawaii.

Leenhardt, M. (1946). Langues et dialectes de l'Austro-Mélanésie. *Travaux et Mémoires de l'Institut d'Ethnographie, 9*.

Lemaître, Y. (1985). Les systèmes de numération en Polynésie orientale. *Journal de la Société des Océanistes, XLI*(80), 3–13.

Lynch, J. (1977). On the history of the Tanna numerals and number markers. *Te Reo: Journal of the Linguistic Society of New Zealand, 20*, 3–28.

Macdonald, D. (1893). Asiatic origin of the Oceanic numerals. *Journal of the Polynesian Society, 2*, 88–89.

MacGillivray, J. (1852). Comparative vocabulary of three of the languages of the south-east coast of New Guinea and the Louisiade Archipelago. In *Narrative of the voyage of H.M.S. Rattlesnake*. London, UK: Boone.

Meaney, T., Trinick, T., & Fairhall, U. (2012). *Collaborating to meet language challenges in Indigenous mathematics classrooms*. Dordrecht, The Netherlands: Springer.

Metraux, A. (1936). Numerals from Easter Island. *MAN, 36*, 190–191.

O'Sullivan, D., McKinley, E., Stewart, G., Richards, P., & Ball, T. (2003). *Maori knowledge, language and participation in mathematics and science education: Literature review*. Auckland, New Zealand: National Institute of Research Excellence for Maori Development and Advancement.

Ray, S. (1893). The languages of the New Hebrides. *Journal of the Royal Society of New South Wales, 27*, 101–167.

Ray, S. (1895). The languages of British New Guinea. *The Journal of the Anthropological Institute of Great Britian and Ireland, 24*, 15–39. doi:10.2307/2842475.

Ray, S. (1907). *Reports of the Cambridge anthropological expedition to Torres Straits* (Vol. 3). Cambridge, UK: Cambridge University Press.

Ray, S. (1912). Polynesian linguistics. *Journal of the Polynesian Society, 21*(65–76), 164–172.

Ray, S. (1915). Polynesian linguistics. *Journal of the Polynesian Society, 24*(62–64), 92–97.

Ray, S. (1916). Polynesian linguistics. *Journal of the Polynesian Society, 25*, 18–23, 44–52, 99–103.

Ray, S. (1917). Polynesian languages of the Solomon Islands. *Journal of the Polynesian Society, 26*, 36–43, 99–105, 170–179.

Ray, S. (1919a). The languages of northern Papua. *Journal of the Royal Anthropological Institute of Great Britain and Ireland, 49*, 317–341.

Ray, S. (1919b). The Polynesian languages of Melanesia. *Anthropos, 14*, 46–96.

Ray, S. (1919c). Polynesian linguistics. *Journal of the Polynesian Society, 28*, 168–177.

Ray, S. (1920). Polynesian linguistics. *Journal of the Polynesian Society, 29*(76–86), 207–214.

Ray, S. (1926). *A comparative study of the Melanesian island languages*. Cambridge, UK: Cambridge University Press.

Rehg, K., & Sohl, D. (1979). *Ponapean-English Dictionary. PALI Language Texts: Micronesia*. Honolulu, HI: University Hawaii Press.

Ross, M., Pawley, A., & Osmond, M. (2003). *The lexicon of Proto Oceanic: The culture and environment of ancestral Oceanic society. 2: The physical environment*. Canberra, Australia: Pacific Linguistics, Australian National University.

Salzmann, Z. (1950). A method for analyzing numerical systems. *Word, 6*, 78–83.

Saxe, G. (1979). A comparative analysis of the acquisition of numeration: Studies from Papua New Guinea. *The Quarterly Newsletter of the Laboratory for Comparative Human Cognition, 1*(3), 37–43.

Saxe, G. (1981a). Body parts as numerals: A developmental analysis of numeration among the Oksapmin in Papua New Guinea. *Child Development, 52*, 306–316.

Saxe, G. (1981b). Numerical reasoning. Oksapmin: Development and change. *Educational Research Unit Occasional Paper 7*. Port Moresby, PNG: University of Papua New Guinea.

Saxe, G. (1981b). When fourth can precede second: A developmental analysis of an indigenous numeration system among Ponam Islanders in Papua New Guinea. *Journal of Cross-Cultural Psychology, 12*(1), 37–50.

Saxe, G. (1982). Developing forms of arithmetical thought among the Oksapmin of Papua New Guinea. *Developmental Psychology, 18*(4), 583–594.

Schlaginhaufen, O., Sapper, K., & Freiederici, G. (1915–1916). Die lokalisation der claes Pieterz-Bucht in Neu-Irland. *Mitteilung der geographischen und ethnographischen Gesellschaft, 16*, 3–36; 1920–1921: 1921, 1935–1959.

Schmidt, W. (1926). *Die Sprachfamilien und Sprachenkreise der Erde*. Heidelberg, Germany: Carl Winter's Universitätsbuchhandlung.

Schmidt, W. (1929). Numeral systems. In *Encyclopedia Britannica* (14th ed., pp. 614–615). London, UK/New York, NY: Sears Roebuck.

SIL. (n.d.). *Ethnologue: Languages of the world*. Retrieved from http://www.ethnologue.com/

Smith, G. (1978). Counting and classification on Kiwai Island. *Papua New Guinea Journal of Education, Special Edition the Indigenous Mathematics Project, 14*, 53–68.

Smith, G. (1984). *Morobe counting systems: An investigation into the numerals of the Morobe Province, Papua New Guinea* (MPhil). Papua New Guinea University of Technology, Lae, Papua New Guinea.

Tryon, D. (1976). *New Hebrides languages: An internal classification* (Pacific Linguistics, C-50). Canberra, Australia: Australian National University.

Tryon, D., & Hackman, B. (1983). *Solomon Islands languages: An internal classification* (Pacific Linguistics, C-72). Canberra, Australia: Research School of Pacific Studies, Australian National University.

Turner, G. (1861). *Nineteen years in Polynesia: Missionary life, travels, and researches in the islands of the Pacific*. London, UK: John Snow.

von der Gabelentz, H. (1861, 1873). *Die melanesischen Sprachen* (Vols. 1–2). Leipzig, Germany: S. Hirzel.

Wolfers, E. (1969). Do New Guineans count? *Institute of Current World affairs Newsletter, EPW-18*, 1–20.

Wolfers, E. (1971). The original counting systems of Papua and New Guinea. *Arithmetic Teacher, 18*(1), 77–83.

Wolfers, E. (1972). *Counting and numbers* (Vol. 1). Melbourne, Australia: Melbourne University Press.

Wurm, S. (1982). Papuan languages. In R. May & H. Nelson (Eds.), *Melanesia: Beyond diversity* (pp. 225–240). Canberra, Australia: Research School of Pacific Studies, Australian National University.

Wurm, S., & Hattori, S. (Eds.). (1981). *Language atlas of the Pacific area. Part 1: New Guinea area, Oceania, Australia*. Canberra, Australia: Australian Academy of the Humanities.

Chapter 3
2-Cycle Systems Including Some Digit Tally Systems

Kay Owens, Glen Lean, and Charly Muke

Abstract During Lean's (1992) analysis of counting systems in Papua New Guinea (PNG) using Salzmann's (1950) notions of frame pattern, cycle, and operative pattern (see Chapter 1, pp. 20–21 for more details), he found that the most prominent cycle to be 2. Further analysis showed that there were a number of modified 2-cycle systems that he denoted by 2′ and 2″ as they had a non-compound word for 3 or 4 respectively. Furthermore, many of these systems had additional cycles especially 5- and 5- and 20- cycle systems. He also noted some that had 4- and 8- and 10- cycles. His map of the spatial occurrences of these types of systems is given in Figure 3.1. Further data (Mimica, 1988; C. Muke, 2000) and that obtained by Owens indicate that these systems were developed from cultural beliefs and practices to incorporate large numbers.

Keywords 1-2 counting • 2-cycle systems • Neo-2 cycle systems • (2, 5) cycle systems • (2, 5, 20) cycle systems • Digit-tally systems • Iqwaye counting systems • Large number counting in highlands of Papua New Guinea • Wahgi counting systems

The Existence and Nature of 2-Cycle Systems

Schmidt (1929), in an article on *Numeral Systems* in the 1929 edition of the *Encyclopaedia Britannica*, said:

> there is no language without some numerals; the notion of unity and plurality is expressed at least in the formation of "one" and "two", though "two" is often equal to "much", thus concluding a numeration that has only just started. (p. 614)

The type of numeration in which we have two, or perhaps three or four, numerals which are not compounded to form larger numbers and of which the largest is the limit of counting, Schmidt termed "systemless", and noted that it is doubtful that it really exists "as it is mostly reported of peoples that are vaguely known" (W. Schmidt, 1929, p. 614). Lean adopted the view, consistent with Schmidt's, that the basic criterion that must be met in order for a set of distinct numerals to qualify as a "counting system" is that there must exist some method for forming composite number words which extend the limit of counting beyond the numeral of greatest magnitude. The simplest such system is that for

K. Owens (✉)
Charles Sturt University, Dubbo, NSW, Australia
e-mail: kowens@csu.edu.au

G. Lean

C. Muke
St. Teresa's College, Abergowrie, QLD, Australia

K. Owens et al., *History of Number*, History of Mathematics Education,
DOI 10.1007/978-3-319-45483-2_3

which we have the numerals "one" and "two" and all larger numerals are compounds of these so that, for example, 3 is a compound of "two" and "one", 4 is a compound of "two" and "two", and 5 is a compound of "two", "two", and "one", and so on. This type of system is known variously in the literature as the "pure 2-system" (Seidenberg, 1960, p. 219), the "pair system", (W. Schmidt, 1929, p. 614) and the "pure 2-count"(Van der Waerden & Flegg, 1975, p. 15). In Salzmann's (1950) terminology, the system has a frame pattern (1, 2), a cyclic pattern (2), and an operative pattern which may be summarised as (3=2+1, 4=2+2, 5=2+2+1, ...). As we shall see subsequently, there are other systems which have a primary cycle of 2 but which have a secondary cycle as well. However in our search for the existence of "pure" 2-cycle systems we will require that the system has at least: 3=2+1, 4=2+2, and 5=2+2+1. This requirement does not of course preclude the possibility of a secondary cycle of magnitude greater than 5, e.g. 10, and we should bear this in mind if the data are insufficient to determine whether or not this is the case.

At this point it should be noted that Pica, Lemer, Izard, and Dehaene (2004) and Davis (2009), based on their studies in South America, noted that approximations were more appropriate than exact counting practices in some Indigenous cultures. So it is possible that some languages that seemed to have only 1 and 2 as numerals actually had a different approach to quantifying.

Seidenberg (1960, p. 281; 1976) listed four geographical locations where "pure 2-counting" is found: South America, Australia, South Africa and New Guinea (West Papua and Papua New Guinea). For South America, Schmidt noted that "it is found among the ethnologically oldest tribes - the Fuegian tribes" (W. Schmidt, 1929, p. 614), which are located near Tierra del Fuego at the southernmost tip of the American continent. Conant, in his book *The Number Concept* (Conant, 1896, p. 111), listed seven examples of 2-cycle systems ("binary scales") from South America, different from those given by Schmidt. Similarly, Seidenberg (1960, p. 281) provided nine examples from South America, all of which derive from Kluge (1939), and three of which are identical with those given by Conant. Between the various sources cited, a total of about 18 language groups are listed as possessing 2-cycle systems. By contrast, for the entire African continent, the same sources indicate that only the counting system of the Bushmen, located in southern Africa, possesses a 2-cycle system. Schmidl (1915, p. 195) gave an example of a system from one of the Bushmen dialects: this appears to have three numerals 1, 2, and 3, the last numeral being compounds of the first two so that 4=2+2, 5=2+(2+1), 6=2+2+2, and so on until 10 is reached.

Australian 2-Cycle Systems

In Australia, there is evidence of widespread use of the 2-cycle system. Schmidt said that it is found "among the tribes ethnologically the oldest - the Kurlin-Kurnai of the south-east, the Narrinyeri of the south; several of them count up to "ten" in this manner" (W. Schmidt, 1929, p. 614). Conant gave 35 examples of "Australian and Tasmanian number systems", each of which shows evidence of a 2-cycle (Conant, 1896, pp. 106-110). Only six of the examples, however, give data beyond the numeral 4 and thus it is impossible to establish whether the remaining 29 systems given have secondary 5-cycles or whether they are "pure" 2-cycle systems. Kluge (1938) had an extensive collection of numeral lists of Australian languages taken mainly from sources published in the nineteenth century. Of the several hundred lists only a few give data beyond the first four numerals. About 25 numeral lists clearly indicate examples of 2-cycle systems with 3=2+1, 4=2+2, and 5=2+2+1. Those lists which give only the first four numerals include about 150 which clearly indicate the existence of a 2-cycle in which we have either 3=2+1 or 4=2+2, or both. It is quite clear from the data presented in Kluge's manuscript that other types of counting systems exist in Australia; some of these will be discussed in the next chapter. It would nevertheless appear to be the case that variants of the 2-cycle system predominate in Australian language groups and that these take one of the three forms: (a) (1, 2, 2+1, 2+2), or (b) (1, 2, 3, 2+2), or (c) (1, 2, 2+1, 4). Only a small proportion of these types combine the numerals 2 and 1 to form numerals larger than 5. Both Dawson (1881, pp. xcvii-xcviii) and Kluge (1938, p. 52) gave examples of type (b) where a secondary 5-cycle operates (so that 6=5+1, 7=5+2, etc.), and Kluge also had many examples where type (a) also has a secondary 5-cycle. In summary, there are examples of language groups that possess a "pure" 2-cycle system but there is a larger number of groups that possess variants as given above, and these include systems with a (2, 5) cyclic pattern. These early references provide background for sections of the argument in Chapter 10 on the Prehistory of Number in Papua New Guinea and Oceania.

2-Cycle Systems in New Guinea

Various 2-cycle systems are found in mainland New Guinea (PNG and West Papua), the Torres Strait Islands, and the island of New Britain to the east of the mainland. Nowhere are they found in the rest of Island Melanesia, Polynesia or Micronesia. As was found in the case of the Australian languages, it is possible to distinguish several variants of the 2-cycle system. Similar variants are also possessed by New Guinea language groups and each of these will be discussed separately below. Altogether Lean distinguished five types:

(a) The "pure" 2-cycle system which has two basic numerals, 1 and 2, all other larger numerals being compounded from these two so that no secondary cycle exists,
(b) The (2, 5) or (2, 5, 20) type which is such that the first four numerals follow the 2-cycle pattern: 1, 2, 2+1, 2+2, but subsequently a secondary 5-cycle comes into play so that, for example, 6=5+1, 7=5+2, and 8=5+2+1, and so on. In some cases this sequence is continued until a tertiary 20-cycle comes into operation.

Lean characterised two variants, (c) and (d), of the 2-cycle system as "modified" or "quasi" 2-cycle systems:

(c) This type is such that we have at least three distinct numerals 1, 2, and 4; the numeral 3 is compounded as (2+1). This system usually has a secondary 5-cycle and occasionally a tertiary 20-cycle. To distinguish this variant the notation 2′-cycle is used and the cyclic patterns of the system are expressed as (2′, 5) or (2′, 5, 20).
(d) The other "quasi" 2-cycle system is such that we have at least three distinct numerals 1, 2, and 3; the numeral 4 is compounded as (2+2). This system, too, usually possesses a secondary 5-cycle and occasionally a tertiary 20-cycle as well. The type (d) variant is distinguished by the notation 2″ cycle and the cyclic patterns of the system are expressed as (2″, 5) or (2″, 5, 20).
(e) It was noted that the type (c) system, defined above, usually has a secondary 5-cycle. There is a single variant of the 2′-cycle system, however, which instead possesses a secondary 4-cycle so that the system may be represented as: 1, 2, 2+1, 4, 4+1, 4+2, 4+2+1, and so on. This variant, type (e), has a tertiary 8-cycle so that the sequence continues: 8, 8+1, 8+2, 8+2+1, 8+4, 8+4+1, and so on. The cyclic pattern of this type is expressed as (2′, 4) or (2′, 4, 8).

In establishing these categories, he is merging and crossing over Smith's (1988) classifications but it assists to present the arguments of Chapter 9 on the diffusion theory critique and Chapter 10 on the prehistory of number.

Type (a): The "Pure" 2-Cycle

It is recalled that the total number of Papuan (NAN) languages in Lean's counting systems database is 430; the total number of Austronesian (AN) languages in the database is 453, of which 226 are found in the West Papua and Papua New Guinea region. Of these totals, 42 (or less than ten percent)

Non-Austronesian (NAN) language groups possess "pure" 2-cycle systems and 2 Austronesian (AN) language groups also have them, that is less than one percent. In none of these cases is there evidence of a secondary 5-cycle; also, in most cases, it is not possible to determine whether there is a larger secondary cycle, 10 for example, as the data are usually not sufficient. Of the 42 NAN systems, 18 of these are associated with body-part tally systems which are discussed in later sections of this chapter and in Chapter 4. Several examples of this type of system, as it occurs on the New Guinea mainland, are given in Table 3.1.

Table 3.1
"Pure" 2-Cycle Systems in Three NAN Languages of New Guinea

	Southern Kiwai, Western Prov.	Giri, Madang Prov.	Gende, Madang Prov.
1	*na'u*	*ibabira*	*mapro*
2	*netowa*	*ppunini*	*oroi*
3	*netowa na'u bi*	*ppuni kagine*	*oro gu mago*
4	*netowa netowa*	*ppunini ppuninin*	*oroi oroi*
5	*netowa netowa na'u*	*ppunini ppunini kagine*	*oroi oroi mago*

Note. Southern Kiwai data are from Smith (1978), Giri from Stanhope (1972) and Gende from Kluge (1941, p. 280) and Counting System Questionnaire data.

It may be observed that none of the systems shown in Table 3.1 is precisely "regular" in that "one" in combination is not identical to the numeral "one". The pattern displayed here is quite common among those systems classified as Type (a). In some cases the numeral 3 may be translated as "two and another"; in other cases copula may be used to link the compounding numerals so that 4 may be translated as "two and two" rather than just "two two". Some systems have compound numerals which incorporate grammatical or syntactic elements. Nomad, in the Western Province of PNG (Lean's data based on SIL and IMP/SIL data in the 1970s) had:

2 *benau*
4 *benau-ili benau-ili*
6 *benau-ili benau-ili benau-ili*

where *benau* is the numeral "two" and *-ili* is the dual form of the third person plural, that is "they-2", which appears only in compound numerals.

However, 2-cycle systems have been developed innovatively into more complex systems. Sometimes speakers will refer to 6 in several ways first beginning with 2 + 2 + 1 for 5 + 1 or 2 + 1 + 2 + 1 possibly building on a three cycle. Donahue (2008) remarked that despite the restricted numerals,

> this does not mean that people are not capable of keeping careful track of precisely how much is owed to which parties in any transaction, with quantities reckoned routinely extending up to and beyond 50, indicating that the absence of verbal representation for numerals does not indicate their psychological absence (consistent with the discussion in Gelman & Gallistel, 2004). (p. 424)

It is useful to comment at this point on the fact that two AN language groups are found to possess "pure" 2-cycle systems. While it is not known whether specific counting system types were associated with the languages ancestral to the NAN languages now found in the New Guinea region, this is not the case for the Oceanic AN languages. There is good evidence that the reconstructed Proto Oceanic (POC) numerals form a 10-cycle system and indeed the putative form of these numerals derives from the many 10-cycle systems belonging to contemporary daughter languages of POC.

However the data for the AN languages Sissano, Sandaun Province, and Middle Watut, Morobe Province indicate that instead of there being at least ten distinct numerals we have in fact only two

(Table 3.2). It is apparent that, contrary to expectation, language groups which have a counting system with a given primary cycle may be influenced to change their system in the direction of a *reduction* in the magnitude of the primary cycle, in this case from 10 to 2. As we shall see, these are not two isolated cases and that throughout the New Guinea and Melanesian Islands region a large number of AN languages have counting systems which have a primary cycle of less than 10 (several examples are given in this chapter and Chapter 5).

Table 3.2
Examples of Two Austronesian Languages With 2-cycle Numeral Systems

	Sissano, Sandaun Prov.	Middle Watut, Morobe Prov.
1	*puntanen*	*morots*
2	*eltin*	*serok*
3	*eltin puntanen*	*serok a morots*
4	*eltin eltin*	*serok a serok*
5	*eltin eltin puntanen*	*serok a serok a morots*

Note. Sissano data come from early data given by M. Schmidt (1900, p. 360), Ray (1919, p. 330), Kluge (1938, p. 177) and included the village of Arop. Middle Watut data are based on Holzknecht (1989).

Type (b): With a (2, 5) or (2, 5, 20) Cyclic Pattern

This is the most common type of counting system found among the NAN language groups. Altogether 109 of these languages have systems which possess a primary 2-cycle and which have a secondary 5-cycle; for those languages for which sufficient data exist there is almost always a tertiary 20-cycle in operation as well. Included in this group are variants, similar to those mentioned in the previous section. One of these is such that the numeral 3, normally expressed as a simple combination of the numerals "two" and "one" (i.e., 2+1 and never, incidentally, 1+2), is expressed as "two and another" and in which the numeral "one" does not appear explicitly. Another variant which occurs in only one language group, the Fore people of the Eastern Highlands of PNG has 4=2+2 but 3 can be interpreted as having the construction "1+1+1". Examples of the Type (b) counting system from NAN languages – Sulka, East New Britain, Tairora and Fore from Eastern Highlands Province – are given in Table 3.3.

Table 3.3
Systems Which Have (2, 5, 20) Cyclic Patterns: Examples From Three NAN Languages

	Sulka	Tairora	Fore
1	*a tiang*	*vohaiqa*	*ka:*
2	*a lomin (lo)*	*taaraqanta*	*tara*
3	*kor lo tige*	*taaraqanta vohaiqa*	*kakaga*
4	*kor lo lo*	*taaraqanta taaraqanta*	*tara-wa tara-wa-ki*
5	*a ktiek*	*kauqu-ru*	*naya:ka-mu*
10	*a lo ktiek*	*kauqa-tanta*	*naya:tara-mu*
20	*a mhelum*	*vohaiqa vaiinta*	*ka:kina*

Note. Sulka sources are fairly similar e.g., Müller (1915/1916), pp. 82-3), Kluge (1941, p. 28, p. 196). Taitora and Fore data are from G. Scott's (1978) compilation for the IMP/SIL project.

Each of these systems has a secondary 5-cycle and a tertiary 20-cycle. This type of system is, in fact, a 2-cycle system augmented by a digit-tally. The number words for 5 and 10 each contain "hand" morphemes and those for 20 contain a "man" (or "being") morpheme. This type of system often contains grammatical or syntactic elements when compound numerals are formed; the compounds, too, may contain words which are tally prescriptions or directions. The Asaro dialect of Gahuku-Asaro illustrates both of these aspects (Lean's database on Gahuku-Asaro, from IMP/SIL questionnaire) as shown in Table 3.4.

Table 3.4
Translation of Tallying Terms of the Asaro Dialect of the Gahuku-Asaro Language

	Gahuku-Asaro	Translation
1	*hamo've*	one it is
2	*sita've*	two it is
3	*sito-hamo've*	two-one it is
4	*sita've sita've*	two it is two it is
5	*ade hela osu'live*	hand-our this at finished it is
6	*ade hela osu'livo hamo've*	hand-our this at finished being one it is
10	*ade hela hela osu'live*	hand-our this at this at finished it is
20	*evene'hamo'gizene ana osu'live*	person one foot-his hand-his finished it is

Source. Data from SIL/IMP questionnaire.

It is pertinent here to provide other Gahuku-Asaro or Alekano systems provided in 2004 and 2006. When visiting different villages, there were often different versions given for this counting system indicating people would "play" with the words when counting. The data in Table 3.5 indicate some interesting results in terms of school, language, Tok Pisin, and humour.

Table 3.5
Alternative Words for Gahuku-Asaro Counting from Gavehumuto (Asaro area) and Kaveve (Gahuku area) villages.

	Gavehumuto village (2004)
1	*hamo*
2	*losi*
3	*losive makole*
4	*losive losive*
5	*ligizani luga*
6	*ligizani luga hamo*
7	*ligizani luga losi*
8	*ligizani luga losive makole*
9	*ligizani luga losive losive*
10	*ligizani luga luga or asasi hamo*
11	*asasigi hamoki*
12	*asasigi losigi*
13	*asasigi losi hamo*
14	*asasigi losive losive*
15	*asasigi ligizani luga*
16	*asasigi ligizani luga hamo*
17	*asasigi ligizani luga losi*
18	*asasigi ligizani luga losive makole*
19	*asasigi ligizani luga losive losive*
20	*asasi losi*
21	*asasi losi hamo*
100	*asasi ligizani luga luga* (stick)
200	*go'hamo* (bilum)
1000	*mulisi* (hip=heap)

(continued)

Table 3.5 (continued)

	Kaveve village (2006)		
	From Stan Aize (Kaveve)	From Zuzai Hizole (Kaveve)	From Elder
1	*hamako*	*hamo*	*amoko*
2	*logosi*	*logosita*	*lohosive*
3	*luguha (logosigi moka)*	*logidigi hamoki*	*luguhave*
4	*logosigi logosi* or *logosi*^{2} *(logosi logosi* or 2 + 2*)*	*logosivi logosive*	*lohosi lohosi*
5	*logosigi luguhagi*	*logosigi logosi hamo* or *nigizani hamo asu igo* (one hand finished)	*lohosigi lohosigi makoki*
6	*luguha logosi*	*luguha luguha*	
7	*luguha logosigi makoki (segininaga)*	*luguha luguha hamoki*	
8	*logosi*^{4} *(*means *logosi logosi logosi logosi,* really *2 + 2 + 2 + 2)*	*nigizani hamo asu o'oko makotoka logsive hamo ol o malago*	
9	*luguhagi luguhagi luguhagi*	*nigizani hamo a su o'ko logosive logosive oli'o malago*	
10	*golaha*	*nigizani logosi asu igo (nagahuni hamo)*	
11	*golohaki hamakoki*	*nigizani logosi asu oko ligisaloka hamo oli'o malago (*2 hands finished and one on the leg*)*	
12	*golohaki logosigi*	*nigizani logosi asu oko ligisa loka logosi oli'omalago (*2hands finished and two on the leg*)* or *nagahuni makoki logosita*	
13	*golohaki luguhagi*	*nagahuni makoki logosigi makoki*	
14	*golohaki logosigi logosigi*	*nagahuni makoki logosi logosi*	
15	*golohaki luguhagi logosigi*	*nagahuni makoki logosi logosi hamo or nigizani logosi asu 'olo ligisa hamo asuigo (*2 hands finished-one leg finished*)*	
16	*golohaki luguhagi logosigi*	*nagahuni hamo luguha luguha*	
17	*golohaki segini nagaki*	*nagahuni hamo luguha luguha hamoko*	
18		*nagahuni hamo, luguha luguhagi logosi* or *nigizani logosi asu'oko ligisahamo asuiko mako toka luguha oli'o'mallago (*2 hands finished, lone leg finished and 3 on the other side*)*	
19		*nagahuni logosi, hamo hakene igo* (20-1) or *nigizani logosi asu okake nigisa mako asu oko makotoka logosigi logosi oli'o'malago*	
20		*nagahuni logosi* or *nigizani logosi aso'oko nigisa logosi asu igo (*2 hands finish, 2 legs finish*)*	

Note. See comment in text on the use of the power of 2. These data were obtained during field trips by Owens in 2004 and 2006.

The symbol for power of 2 is linked to reduplication which is common in many languages of PNG as well as the lingua franca Tok Pisin. Here it refers to addition rather than multiplication which is its use in school mathematics and perhaps indicates diminishing knowledges due to interference of schooling and the local languages. It also indicates the willingness and ability to "play with numbers" with reordering them and also a sense of ownership of what they are doing with mathematics. Gavehumuto data were from a teacher fluent in her home language unlike some of the other elementary school teachers. She did not hesitate to group count in threes even though the language had cycles of 2 and 5 suggesting that this is not a difficult task in any counting system. In Kaveve, both men were educated and hard working in different careers and positions. It is also likely that the neighbouring Bena Bena language influenced various people who tried to count in their home language in this village. In general, across Eastern Highlands the rapid increase of Tok Pisin among all age groups is influencing thought, knowledge of language, and counting systems so that a group of 30 teachers from this area in 2014 were not really sure of their language's counting system and the more prominent speaker, if not the most fluent in language, provided a simple 1, 2 system repeating the words without further meaning, e.g. 5 was *losive losive hamo* and 6 was *losive losive losive.*

There are 18 Austronesian (AN) language groups that possess this type of system and 13 of these belong to the Markham Family (Holzknecht, 1989) from the Markham Valley of the Morobe Province of PNG. The others are Sera (Sandaun), Roinji (Morobe), Dawawa (Milne Bay), Igora (Milne Bay) and Tomoip (East New Britain). (Tomoip details are also given in Appendix C). However, Roinji at least is based on pairing e.g. 2+2+2+1 for 7; 10=2x5. This provides further evidence of AN languages having had a change in the cyclic nature of their counting systems from, originally, a (putative) 10-cycle to systems with (2, 5) or (2, 5, 20) cyclic patterns. Three examples – Tomoip, East New Britain, Wampar and Duwet, Markham Valley, Morobe – are given in Table 3.6.

Table 3.6
Three Examples of AN Languages With Counting Systems Having (2, 5) Cyclic Patterns

	Tomoip	Wampar	Duwet
1	*denan*	*orots*	*taginei, taine, ta*
2	*ro huru*	*serok*	*seik*
3	*horum detu*	*serok orots*	*seik mba ta*
4	*horumu horum*	*serok a serok*	*seik mba seik*
5	*ko liem*	*bangi-d ongan*	*lima-ngg alinan*
10	*liem*	*bangi-d serok*	*lima-ngg a seik,*
20	*tamdil*	*ngaeng orots*	*lima-ngg a seik a mbei-ngg seik*

Note. The Tomoip data are from Parkinson (1907, p. 780), and Kluge (1941, p. 197) while Wampar and Duwet data are from Holzknecht (1989).

As was found in the case of the NAN systems, the 2-cycle system is augmented by a (5, 20) digit-tally system. The number words for 5 and 10 in each of the examples above contain a "hand" morpheme. For Tomoip and Wampar, the word for 20 contains a "man" morpheme while that for Duwet contains both a "hand" and a "foot" morpheme.

Type (c): With a (2′, 5) or (2′, 5, 20) Cyclic Pattern

This type of system, which has a compound numeral "three" such that 3=2+1 but has a distinct numeral "four", is relatively uncommon. No AN language group has this type of system while 27 NAN groups have it. Six of those also have body-part tally systems as well (Awin, South Kati, North Kati, Yongom, and Ninggirum, situated near the border in either Western Province, PNG or West Papua) while a further 14 have systems with (2′, 5, 20) cyclic patterns. Three examples of the Type (c) system – Kol-Sui in East New Britain, Biangai in Morobe, and Au in Sandaun – are given in Table 3.7. (See Lean's appendices for other data, also in Lean (1992)).

Table 3.7
Examples of the Type (c) 2′-Cycle Numeral System

	Kol-Sui	Biangai	Au
1	*'pusup*	*wame-nak*	*kiutip*
2	*te'tepe*	*na-yau*	*wiketeres*
3	*tetepe kosup*	*nayau keya nak*	*wikak*
4	*ke'a so*	*mango-bek-tau ono*	*tekyait*
5	*'a:meleng*	*mele-na-zik*	*his pinak*
10	*melem'be:ga*	*mele yau*	*his wien*

Note. Kol-Sui data are from SIL word list, 1981; Biangai is from IMP/SIL project (1978) and SIL word list (1975); Au from IMP/SIL project (1978).

Members of the (2′, 5, 20) group are largely located in the Oro Province and the southern part of the Morobe Province. It is possible that one language group, Biangai, may have a numeral 4 which has the meaning "one less than five" and of which we find several examples in the next type of system.

Type (d): With a (2″, 5, 20) Cyclic Pattern

This system, with a distinct numeral "three" but with the numeral "four" compounded such that 4=2+2, is not uncommon. Some 12 examples are found among the AN languages of PNG, five of these (Vehes, Mapos Buang, Manga Buang, Mumeng, and Piu) are members of the Buang Family, located in the Morobe Province (details are given in Appendix C), and a further 4, perhaps 5 (Anuki, Paiwa, Boianaki, Wedau, and, possibly, Taupota), are located in the Milne Bay Province. Among the NAN languages there are 40 examples of this system; three of these may not have a secondary 5-cycle and are associated with body-part tallying methods (Bine, Gidra in Western Province and Oksapmin in the south of Sandaun Province have 2″-cycle system). One AN and two NAN examples are given in Table 3.8.

Table 3.8
Examples of the Type (d) 2″-Cycle Numeral System

	Mumeng (AN)	Kwanga (NAN)	Usarufa (NAN)
1	*ti*	*findara*	*morama*
2	*yuu*	*frisi*	*kaayaqa*
3	*yon*	*lamor*	*kaomomo*
4	*yuu di yuu*	*frisi frisi*	*kaayate kaayate(qa)*
5	*vige vilu*	*tabanangki*	*mora tiyaapaqa*

Note. Mumeng (Morobe Province) data are from IMP/SIL project (1978); Kwanga (East Sepik Province) is from SIL word list (1960); Usarufa (Eastern Highlands Province) data are from SIL word list (1968). It should be noted that some data (IMP/SIL) from Mumeng suggest a 20-cycle but the older data available are used here.

Type (e): With a (2′, 4, 8) or (2′, 4, 8, 10) Cyclic Pattern

There is a rare example of this type of system and this occurs in the Melpa dialect of the Hagen (NAN) language. The structure of the system is: 1, 2, 2+1, 4, 4+1, 4+2, 4+2+1, 8, 8+1, (10), … There is an argument for including this system under the 4-cycle classification as the putative 2+1 construction for the numeral may not in fact be valid and there may be a distinct, uncompounded numeral 3. There are two representations for the numeral 10: one has the meaning "hands one", that is "the hands of one man"; the other representation is as the compound 8 + 2 as given in Table 3.9.

Table 3.9
Numerals of the Melpa Dialect of the Hagen Language

	Hagen (Melpa dialect)
1	*tenta*
2	*ralg*
3	*raltika*
4	*timbakaka*
5	*timbakaka pamb ti*
6	*timbakaka pamb ralg*
7	*timbakakagul raltika*
8	*engaka*
9	*engaka pamb ti*
10	*engaka pamb ralg pip*, **or**, *ki tenta*, **or**, *ki ti*

Note. Data are from IMP/SIL project in 1978 but are not unlike the Vicedom and Tischner (1948) data.

Lean provided other alternatives and gave numbers well beyond 10 indicating the 8 cycle system but this will be discussed in Chapter 7 and Appendix B.

Summary of 2-Cycle Data

If all the 2-cycle variants are combined we have a total of 218 NAN languages which have counting systems belonging to this category, just over half the total sample (430) of NAN languages. The 218 languages are distributed mainly among the major phyla, the Sepik-Ramu, the Torricelli, and the Trans New Guinea, as given in Table 3.10. Thus more than half (58%) of the total sample of the Trans New Guinea Phylum languages possess a 2-cycle system variant, as do 92% of the Torricelli Phylum sample and 36% of the Sepik-Ramu Phylum sample.

Table 3.10
The Distribution of 2-Cycle Variants Among Three of the Major Phyla of the NAN Languages

Type	Sepik-Ramu	Torricelli	Trans New Guinea
a	3	0	39
b	5	16	86
c	5	3	17
d	3	5	31
Totals	16	24	173
Total Sample	44	26	298

Altogether there are 32 AN languages which have a 2-cycle variant. All of these are Oceanic and are found in PNG. Twenty-four of these 32 languages belong to Ross's North New Guinea Cluster, 7 belong to the Papuan Tip Cluster, and 1 belongs to the Meso-Melanesian Cluster (see Appendix E, based on Ross, Pawley, and Osmond (2003)). We should note that only two of the AN systems appear to be "pure" 2-cycle systems while the remaining 30 all have a secondary 5-cycle, and in many cases a tertiary 20-cycle as well.

Figure 3.1 shows the distribution of the 2-cycle variants (a) to (d) in the New Guinea region for both the NAN and the AN languages. The distribution is widespread throughout the region and is found in both the coastal and inland highland areas. Type (e) is discussed in Chapter 7.

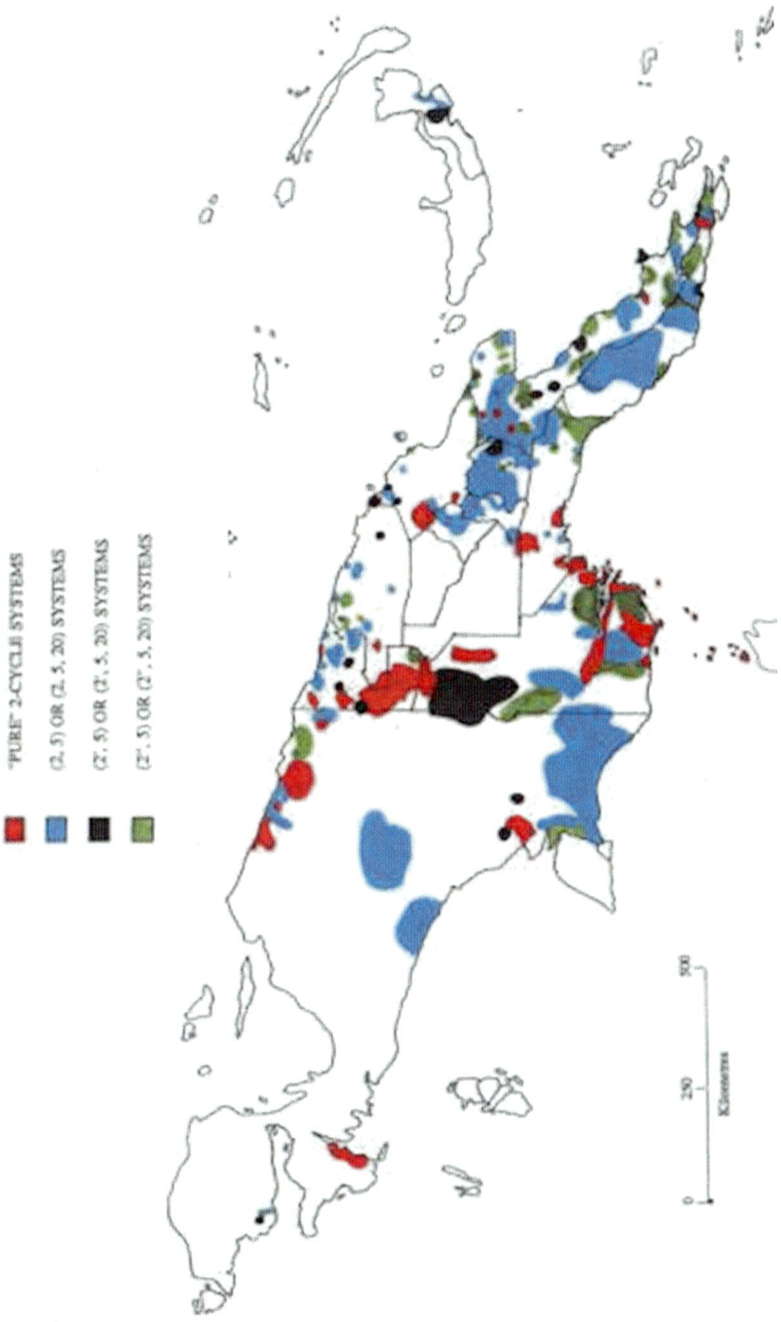

Figure 3.1. Distribution of 2-cycle variants in New Guinea.
Source. Lean's (1992) thesis.

Digit Tally Systems – Embedded in Culture and Extending Mathematical Thinking

However, there is more to say about digit tally systems as pointed out by Mimica (1988) for a Yagwoia language, Iqwaye. Yagwoia, a member of the Angan Stock-Level Family, is spoken in two separated mountainous areas in the extreme south-east of the Eastern Highlands Province and in the south-west of the Morobe Province in the Menyamya area. Wurm and Hattori (1981, Map 8) gave the total resident population of Yagwoia-speakers as a little over 6 000, based on statistics compiled in the early 1970s but Mimica gave the total of 9 000 in 1988 with the Iqwaye numbering about 2 000. Smith (1984, p. 264) had some Yagwoia number data in his study of the counting systems of the Morobe Province. Mimica (1988) had data on the Iqwaye (Ikwaye) counting system. The Yagwoia system is a digit-tally one with a (2, 5, 20) cyclic pattern. Mimica's description of the Iqwaye system (Lean, 1992, citing an earlier unpublished paper from Mimica) was that:

> Counting on digits is commonly done in such a way that a person starts with the thumb of either open hand, uttering the numerals simultaneously as he (she) does so and proceeds systematically to all the fingers of one hand (1-5), then through all the fingers of the other (6-10). Following this, the person then carries on by counting the toes of firstly one foot (11-15), then of the next (16-20). (Mimica, 1984, p. 9, cited in Lean, 1992)

The tally of 20 having been reached, the tally may continue to higher numbers by counting on the fingers and toes of another person. Mimica emphasised that:

> it does not follow that for a person, in order to be able to count or to express in numbers above 20, there should necessarily be the corresponding number of digits and persons … in principle no more than those of one person's, and for practical reasons, no more than those of two are required (Mimica, 1984, p. 10, cited in Lean, 1992).

Mimica indicated that the number word for 5 means "hand" (an arm morpheme but closely linked to gestures of counting involving fingers and the hand) and that for 10 means "two hands". The number word for 15 contains a "leg" morpheme *hyule* and has the gloss "half of the legs, all (digits)". The number word for 20 may be expressed either as "two hands two legs" or as "one person". Data from Lean's study are given in Table 3.11.

Table 3.11
(2, 5, 20) Cycle System of Iqwaye on the Eastern Highlands and Morobe Border

	Iqwaye Counting Words	Yagwoia
1	*ungwonangi*	*'kwananoi*
2	*huwlaqu*	*u'laako*
3	*huwlaqungwa* or *huwlaqanga*	*u'laangwa*
4	*hyaqu-hyaqu*	*'yaako 'yaako*
5	*hwolyempu*	*walyampu*
6	*hwolye indeumoni ungwonangi*	
7	*hwolye indeumoni huwlaqu*	
8	*hwolye indeumoni huwlaqungwa*	
9	*hwolye indeumoni hyaqu-hyaqu*	
10	*hwolye kaplaqu*	*'walya 'mplaako*
11	*hyule yengwonye ungwonangi*	
15	*hyule umance hyelaq*	*hyula walyampa*
20	*hwolye kaplaqu hyule kaplaqu* or *amnye ungwonangi*	*apni*
30	*anmye ungwonangi amnye ungwoli amnye ineumoni hwolye kaplaqu*	
40	*[amnye hyule hwolye hyepu], amnye huwlaqu*	
100	*amnye hwolyempu kokoleoule hwolye hyelaqapu*	

Note. Iqwaye data were provided by Mimica (1988, pp. 32-33), while the Yagwoia were supplied by Smith (1984, p. 264)

The deliberate use of digits when counting indicates that the fingers and toes are important representations of number that links counting to their whole bodily existence and the relationship with others when using the fingers of another. Even the numbers can incorporate this deictic of "this" and "that" or "mine" and "yours" or "another". Nevertheless, this close link with the physical body parts does not prevent the Iqwaye of thinking of numbers more abstractly. While the first two and a second twenty are useful for counting purposes, the Iqwaye are also able to use the set of fingers and toes to denote not only ones but each group of 20 as they are being counted and even to a third level, thus making it possible for the Iqwaye to think of large numbers (powers of 20) appropriately.

While the Iqwaye generally do not count large numbers except of cowrie shells, they are able to do it and explain their system. They also make use of cowrie sticks of 5 shells *ungye hilyce* so when counting these, for example, 3 or 7 *ungye hilyce,* they are referring to 15 or 35 individual shells. A rope of shells may also be used for comparison. This is important in bride price exchanges which are generally not accepted on the first offer. One-to-one matching of ropes and shells also occurs for other occasions such as provision of warriors, the rope representing a sufficient number of warriors for combat. This sense of wholeness of a rope and a sense of the body, namely the digits, as embodiments of the counting scheme also provide relationships between numbers that are significant for different purposes. Quantifying is not necessarily the end point but the decision required for which counting or its equivalent objects might assist. The objects and digit assist with sense of number size and comparison. Thus "the background of number use [is evident in] ... the Iqwaye perception of quantities in their appearances as deemed equivalent and different, substitutable, exclusive or commensurate" (Mimica, 1988, p. 18). Thus there is a flow of objects—money, food and other things—and the actual object (shell or crisp banknote) is valued. This qualitative structuring of substitutes and commensurability is evident too in the classifications of marriage arrangements (Mimica, 1988).

The body digits are important from birth and are involved in rhymes with each digit of the infant being associated with a line of verse and an action on an edible animal. The numerical series for which the body digits can be used provides for both abstraction of number and for large numbers. Thus the one digit may represent one, 20, 20 × 20, or 20 × 20 × 20 depending on the size of numbers under consideration. The Iqwaye thus express 400 without a numeral as

Aa' 'mnye, aa'mnye, toqwotni tepu hyelaqa kokoloule hyule hwolye hyelaqapu

Person person this-me this that-all their leg hand that-all

'[as] this many persons [as] me this [one] person [speaker] all their legs and hands'

... 1000 .... 'two persons [as] me this [one] person all their legs and hands and to another person's two hands (= ten persons) all their legs and hands' (Mimica, 1988, pp. 35-36)

The lingual form of numbers and the system is deictic. The lexemes *indeumoni* (to the next) is a horizontal shift on hands and feet; *ineumoni* (to the side of) is a shift to the next person; and *yengwonyemoni* (down-to) for a vertical shift from the hands to feet occur in the appropriate numbers. The word for two contains the morpheme *–aqu* (two) which occurs in other words such as the third person dual reference while *hyaqu-hyaqu* (four) also contains the demonstrative *hye* (that) (Mimica, 1988, p. 46). Furthermore, the pattern of numbers to 20 is divided into four segments of five related to the natural body segments but the numbers are in a specific series of one, two, two-one, two-two, hand and five is conceived not as the addition of 5 ones but as 2 twos and 1 giving 5 as an indivisible whole but also a marker of a pattern (Mimica, 1988, pp. 48-52). According to Mimica (1988), this system based on one and two, the links (e.g. in gestures of the hands) to joining together of two for procreation, the myth of the bow and arrow relationship, and the generating of 20 parts for a whole give the numeric system a mythopoeic perspective.

However, the body crouched with fingers and toes touching may be a macro-view of the world and culture, a part of which is the person with four sets of five digits that generates self-similarity of 20s. With the digits then being able to represent 20 persons, then 20 x 20 then 20 x 20 x 20 for an infinite appreciation of abstract number. It seems that out of the binary system of 1 and 2, the sense of self-generation over time resulted in the relationships that could be represented by powers of 20. It seems that a cosmic view linked to numbers can be associated with generating large numbers. It is not recorded for related Angan languages of the same area and is not necessarily present in other (2, 5, 20) digit tally systems as far as current data can provide (Mimica, 1988).

While Mimica recorded the general size equivalents often given for large numbers, he also noted the importance to the Iqwaye of precise money exchanges. The importance of this comparison is in placing attitudes to numbers and their values within a social context where numbers are valued for their relationships to culture rather than of themselves. This view of number in a cultural activity and worldview then provides a new perspective on number.

This perspective of generating numbers is also known in the Western Province where the act of doing many times or by many people creates the sense of "many" or large in a numeric sense. This is in contrast to "few" for small numbers of objects or times (personal communication with SIL linguist, 2013). The next section provides another example of representing large numbers in a (2, 5) cycle system.

A (2, 5, 20) Cycle System With Large Numbers

Muke's Study of Mid-Wahgi (Yu Wooi) Counting in a Cultural Context

Charly Muke (2000) used an ethnographic research design to explore how easily patterns of mathematical practices are recognised in a culture. He mainly used a two-page questionnaire gathering quantitative data to support the claims of recognition as well as a conversational interview over several nights with his father, an Elder and keeper of knowledge. He used data from 30 teachers from the Yu Wooi language and neighbouring groups (Kumai on the Simbu border, Yu Nimbang variant of mid-Wahgi and North Wahgi speakers). Seven different tribes were represented in the Yu Wooi group and four different tribes were included in the North Wahgi language group. There were 40 school students from eight Yu Wooi tribal groups and two from Yu Nimbang dialect. Two non-teaching members of the cultural group also completed the questionnaire. He observed people in their own territory and interacted with them in their own language and on their terms in order to study the knowledge of counting subsumed within a cultural context. Yu Wooi means real talk and constitutes accurate talk compared to parables, folded talk, jokes, exaggerated talk, swear talk, war talk, magical utterances, stories and legends, oratory talk, and lies as identified by J Muke (1993).

His father gave words of the ancestors: *endi, tak, takendeka, taksi taksi and angek yemdo* (1, 2, 3, 4, 5 respectively). The majority of teachers gave the same words (95%, 100%, 75%, 61%, 50% respectively) with *angek yemto* also given for 5 in a further 36% of cases. After 5, there was mainly 38% agreement except for 7 which dropped to 29% while 41% agreed 10 was *angek yem yem*. As the numbers increased, the number of variations tended to increase between 6 and 9. This variation reflects much of the work of Lean especially from his counting system questionnaire. Some variation is just in pronunciation or the use of additional or different phonemes or morphemes. The students all provided the same words for 1, 2 and 4 (*endi, tak*, and *taksi taksi)* with half of them giving 3 as *takendek* while nearly all the rest gave *takendika.*

Table 3.12 provides the most commonly used words. At least one variation used the names of fingers for adding onto the hand almost as a vestige of a body-part tally system. Furthermore, when it came to counting and having the opportunity to use hands, people often counted in twos. They would

Table 3.12
Yu Wooi (Mid-Wahgi) Number Words

Numeral	Yu Wooi	Comment
¼	*kekep*	quarter of an animal for food
½	*arhka*	half an animal for food
½	*yem*	half of other things
1	*endi*	
2	*tak*	
3	*takendeka*	2 and 1
4	*taksi taksi*	2 and 2
5	*angek yem*	hands half
6	*angek yemsi endi*	hands half one
7	*angek yemsi tak*	hands half two
8	*angek yemsi takendeka*	hands half three
9	*angek yemsi taksi taksi*	hands half two two (four)
10	*angek yem yem*	hands half half
11	*angek yem yem endi*	hands half half one
15	*angek yem yem simb yem*	hands half half leg half
20	*simb angek yem yem*	legs hands half half
27	*angek yem yemsi simb yem yemsi, angek yemsi, tak*	both hands + both legs + half hand + 2
40	*hi ende simb angek*	2nd man's legs and hands
80	*ala hi ende simb angek*	4th man's legs and hands
100	*ala hi ende simb angek*	5th man's legs and hands
250	*simb angek mongum pel angek yem yem kunum taksi angek yem yem*	20 x 10 + 2 x 20 + 10 hands and legs (20)

Note. The most commonly supplied words are used except for 250 for which the non-English was chosen. The data are from C. Muke (2000, pp. 199, 118-119)

fold down two fingers at a time saying *eraksi* meaning "take 2". This was accompanied by 2 then 2 on one hand followed by 2 and 2 on the other and bringing the four folded fingers of each hand together being *mam erak* followed by the 2 thumbs with the words *angek yem yem* being said (some also folded 2, 2, then thumb *eraksi eraksi el* and then 2 and 2 and thumb repeated before bringing together and saying *angek yem yem*). This use of 4 and 4 and then 2 to make 10 is reminiscent of the Hagen language as described in Chapter 7 in more detail and mentioned above. The counters in pairs would also continue with the toes.

Variations were analysed by Muke (2000) to fall into a number of categories:

(a) the use of the suffix *–si* for numbers from 6-9,
(b) use of the suffix *–to*,
(c) pronunciation especially those influenced by North-Wahgi dialect,
(d) shortening of the word for three especially for the number eight,
(e) some additional words used in phrases such as "and" and "you will do it" signaling that two numbers were used to form compound numbers,
(f) variations on 10 being "hand half half" because of suffixes, pronunciation, repeat use of the word "hand" before the second "half", or additional words like "both of them".
(g) However, in some cases, the word for "fingers" *mongum* was used in line with Lean's (1992) records.

Large Numbers

Like other counting systems as illustrated for the Iqwaye, Muke showed the idea of whole person tally. Some people chose to count items in hundreds by grouping them in tens and then tally each ten using their fingers and toes by twos. Thus by using fingers and toes they reached 200 and would then use another person's fingers and toes. One participant said "for six hundred pigs, they would say that they will kill pigs equal to the hands and legs of three man" (C. Muke, 2000, p. 134). Table 3.13 indicates how tallying occurs.

Table 3.13
Tallying of Quantities in Yu Wooi

Numeral	*Yu Wooi*	Explanation
2	*eraksi*	1st two (left hand first)
4	*eraksi*	2nd two
6	*eraksi*	3rd two
8	*eraksi*	4th two
10	*eraksi* or *angek yem yemsi elsi*	5th two or 1st ten (fingers are folded)
20	*elsi*	this one, 2nd ten
30	*elsi*	3rd ten
…	*elsi* for each ten	
100	*elsi* or *angek yem yemsi peng ngond*	10th ten or head (ordered from head to leg)
100	*peng*	head – 1st 100
200	*komuk*	ear – 2nd 100
300	*gnumb*	nose – 3rd 100
400	*gupe*	mouth – 4th 100
500	*angek woiro*	right hand – 5th 100
600	*angek daro*	left hand – 6th 100
700	*buk*	back – 7th 100
800	*kumbuk*	belly – 8th 100
900	*simb woiro*	right leg – 9th 100
1000	*simb daro* or *hi ende simb angek poro bekenj*	left leg – 10th 100 (whole body parts of one person)
2000	*hi tak*	two persons
3000	*hi takendeka*	three persons

Note. Data are from C. Muke (2000, p. 199).

For the larger thousands, people used the fingers and toes for groups of ten and when they had ten groups of ten, they gave each on a specific body part, starting from the head towards the legs. Muke's father said "the body parts were used as representation that helped us remember how many pigs we were asked to pay or spread the news of the event accurately" (C. Muke, 2000, p. 135). This is illustrated in Figure 3.2. When they had tallied 1000 items in this way they would say *hi end simb angek begenj* (I have tallied all body parts of a person) and so for two or three thousand they would use two or three persons and so on.

When deciding the number of pigs to be given by each person in a compensation claim, the leader asked people to take the number they would give from a bunch of small banana fruit. When everyone had offered as they wished, then the banana fruit were put together and tallied in groups of 10, each group matched with a digit tally parts starting with the fingers. In another ceremony a parcel of sweet potato would equal a quarter of a pig to be received in the big pig kill. In yet another ceremony, pigs were lined up one-to-one matching of pigs from each clan in two parallel rows.

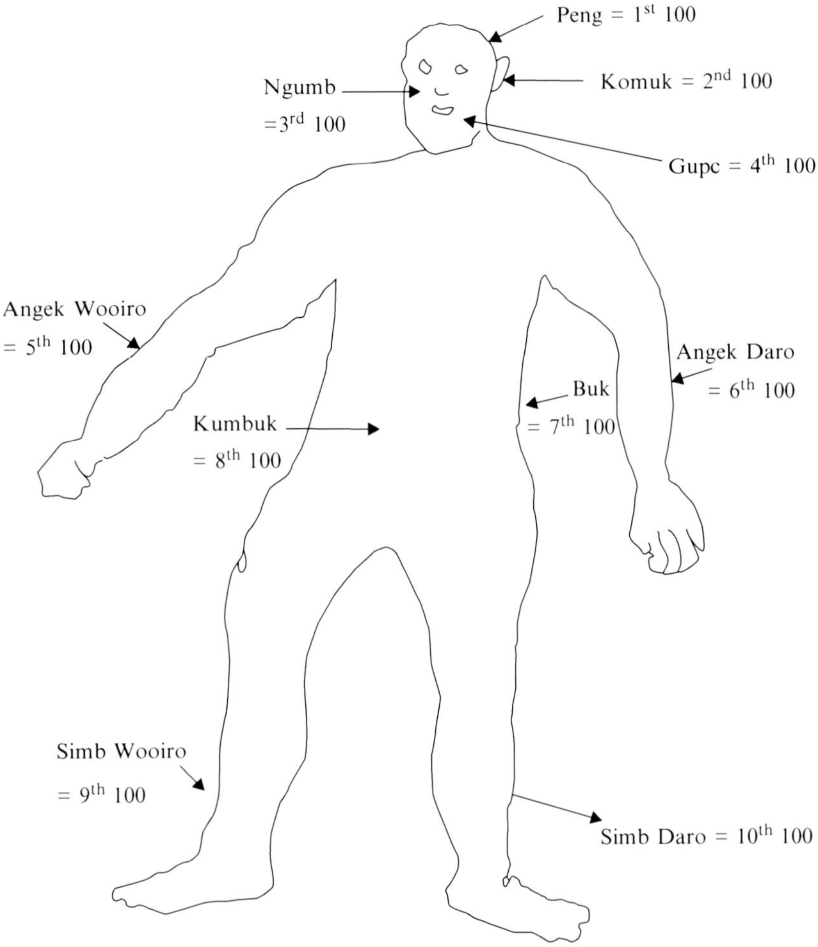

Figure 3.2. *Yu Wooi* (*Mid-Wahgi*) tallying of thousands using body parts. Data are from C. Muke (2000, p. 135).

Nevertheless, there was one idea introduced after colonisation when brideprices began to escalate. It was the use of tying knots. Knots might represent pigs then a knot was used to separate kinds of items such as feathers or shells. The rope could also be carried to the bride's parent to indicate the amounts being negotiated. The girl's family returned a rope indicating what they would give and these two ropes could be compared as a form of subtraction. Another newer method was that of having stakes for the pigs and poles for money, for example a pole might represent 100 or 1000 *kina*.

Cultural Beliefs Associated With Number

Yu Wooi speakers avoid using numbers, especially totals, when presenting goods to others. This is particularly the case when giving to maternal uncles. For this reason, they use alternative methods of tallying to indicate amounts and display these tallies. It was also important to show wealth and status. The desire to tally also encouraged large numbers as shown above. In each case, the tally point was a one-to-many matching as we found with the Iqwaye speakers.

Analysis of the language itself indicates that *si* means "take" and *eraksi* appears to mean "take these two" (*rak* sounding like *tak* and *e* meaning "these") but to use the morpheme *ka* would appear to be related to "good" or "base/root". It is however clear that like most digit-tally systems numbers from 6 to 9 and in this case 10 are the word for $5+n$ where $n=1, 2, 3, 4$, or 5. However, it is in the use of words greater than 20, that the ideas of tens and twenties as a composite unit became apparent and was conceptually understood allowing respondents to provide a meaningful expression for numbers such as 27, 250 and 1376 and that there were not too many variations suggesting a reasonable acceptance of these methods by the community. In de Abreu's (2002) terminology, this valued practice can be accepted as a valorisation of the community.

Another aspect of culture was the effect of the neighbouring languages. While the Ek Nii influenced the Yu Nimbang dialect with whom they shared a land border, the result was more about individual words rather than the system since the Ek Nii had a (5, 10) system while the Yu Wooi has a (2, 5) system. Within Yu Wooi, the dialect for one of the tribes borrowed the Ek Nii word for 3 giving the set of frame words as 1, 2, 3, 5, rather than 1, 2, 5. In Lean's categorisation, Ek Nii probably has a neo-2 cycle system with a special word for 3 but it also has a special word for 10 (containing the morpheme for one). In all other respects it was similar to Yu Wooi but the 10 cycle required a different classification in Lean's terms. The North-Wahgi language was also a (2, 5) system except that it used different words for 1, 2 and 5.

One other aspect of note is the use of different words for "half". This is associated with classification of objects and ideas, a relatively common occurrence in different parts of PNG in terms of counting numbers. In this case, the community use *kekep* and *arhka* for ¼ and ½ respectively for dividing up animals with four or two legs and these words come from the words "to cut". The animal name is used eg. *kong arhka* for half a pig. However, *yem* is used for half of other things, e.g., pair of hands as noted for five as *angek yem*. C. Muke (2000) considered whether classifiers may have been used for counting words but there is no evidence of this in this language group or neighbouring groups.

Finally, comparing was an important aspect of culture whether it was achieved through counting or displays of tallying.

In summary, the Yu Wooi language involved two forms of quantifying: they used number words for small group counting and they used a matching and tallying approach for larger numbers. Counting in pairs and counting first the fingers of both hands before adding the thumbs is reminiscent of Hagen counting (not far away along the Highlands Highway) in which the people also count in fours then eights but superimpose the decimal system marking the two thumbs being added. This is appropriate too in ceremonies for counting where a second person tallies the 10s.

Conclusion

This chapter presents evidence that 2-cycle systems may provide sufficient numeral representation to generate large numbers and ways of thinking systematically in terms of the number sense of quantity. This generation of systematic thinking about quantity is embedded in cultural ways of thinking. Thus it is important not to separate mathematical thinking from the context of counting if we are to appreciate the systematic ways of thinking. Number then is represented not only in words but also in gestures and begins a better appreciation of digit-tally systems that may not be limited to just (5, 20) cycle systems. The chapter certainly disputes suggestions that 2-cycle systems are "less sophisticated" systems than the base 10 system. In place of place value notation in non-written languages, the body provides an important recording tool visible for communication with others together with the associated language words. The chapter also indicates that innovation is highly likely in the Papua New Guinea and Oceania region rather than counting systems being diffused from far away.

References

Conant, L. (1896). *The number concept*. New York, NY: Macmillan.

Davis, W. (Writer). (2009). *The Wayfinders: Why ancient wisdom matters in the modern world*. Lecture 1, Season of the Brown Hyena, 2009 Massey Lectures. Toronto, ON, Canada: The House of Anansi Press.

Dawson, J. (1881). *Australian Aborigines: The languages and customes of several tribes of Aborigines in the Western District of Victoria, Australia*. Melbourne, Australia: George Robertson.

de Abreu, G. (2002). Towards a cultural psychology perspective on transitions between contexts of mathematical practices. In G. de Abreu, A. Bishop, & N. Presmeg (Eds.), *Transitions between contexts of mathematical practices* (pp. 170–189). Dordrecht, The Netherlands: Kluwer.

Donahue, M. (2008). Complexities with restricted numeral systems. *Linguistic Typology, 12*, 423–429. doi:10.1515/LITY.2008.043.

Gelman, R., & Gallistel, C. R. (2004). Language and the origin of numerical concepts. *Science, 306*, 441–443.

Holzknecht, S. (1989). *The Markham languages of Papua New Guinea* (Pacific Linguistics, C-115). Canberra, Australia: Australian National University.

Kluge, T. (1938). *Die Zahlbegriffe der Australier, Papua und Bantuneger nebst einer Einleitung ueber die Zahl, ein Beitrag zur Geistesgeschichte des Menschen*. Berlin, Germany: Selbstverlag.

Kluge, T. (1939). *Zahlenbergriffe der Völker Americas, Nordeurasiens, der Munda, und der Palaio-africaner*. Unpublished manuscript, Berlin, Germany.

Kluge, T. (1941). *Die Zahlenbegriffe der Sprachen Central- und Südostasiens, Indonesiens, Micronesiens, Melanesiens und Polynesiens mit Nachträgen zu den Bänden 2-4*. Berlin, Germany. Ein fünfter Beitrag zur Geistesgeschichte des Menschen nebst einer principiellen Untersuchung über die Tonsprachen. Original unpublished manuscript held at Yale University, microfilm held at Australian National University, Canberra, Australia.

Lean, G. (1992). *Counting systems of Papua New Guinea and Oceania* (Unpublished PhD thesis). PNG University of Technology, Lae, Papua New Guinea.

Mimica, J. (1988). *Intimations of infinity: The mythopoeia (cultural meanings) of the Iqwaye counting and number systems*. Oxford, UK: Berg Publishers.

Muke, C. (2000). *Ethnomathematics: Mid-Wahgi counting practices in Papua New Guinea* (Unpublished MEd thesis). University of Waikato, Waikato, New Zealand.

Muke, J. (1993). *The Wahgi opo kumbo: An account of warfare in the central highlands of New Guinea* (PhD). University of Cambridge, Cambridge, UK.

Müller, H. (1915/1916). Erster Versuch einer Grammatik der Sulka-Sprache, Neu-Pommern (Südsee). *Anthropos, 10*(75–97), 523–552.

Parkinson, R. (1907). *Dreissig jahre in der Südsee*. Stuttgart, Germany: Strecker und Schröder.

Pica, P., Lemer, C., Izard, V., & Dehaene, S. (2004). Exact and approximate arithmetic in an Amazonian indigene group. (Reports). *Science, 306*(5695), 499–503.

Ray, S. (1919). The languages of northern Papua. *Journal of the Royal Anthropological Institute of Great Britain and Ireland, 49*, 317–341.

Ross, M., Pawley, A., & Osmond, M. (2003). *The lexicon of Proto Oceanic: The culture and environment of ancestral Oceanic society. 2: The physical environment*. Canberra, Australia: Pacific Linguistics, Australian National University.

Salzmann, Z. (1950). A method for analyzing numerical systems. *Word, 6*, 78–83.

Schmidl, M. (1915). Zahl und Zählen in Afrika. *Mitteilungen der Anthropologischen Gesellschaft in Wien, 45*, 165–209.

Schmidt, M. (1900). Die sprachlichen Verhältnisse von Deutsch Neu-Guinea. *Zeitschrift für Afrikanische und Ozeanische Sprachen, 5*, 345–384.

Schmidt, W. (1929). Numeral systems. In *Encyclopedia Britannica* (14th ed., pp. 614–615). London, UK/New York, NY: Sears Roebuck.

Scott, G. (1978). The Fore language of Papua New Guinea. *Pacific Linguistics, B-47.*

Seidenberg, A. (1960). The diffusion of counting practices. *University of California Publications in Mathematics, 3.*

Seidenberg, A. (1976). *km,* a widespread root for 10. *Archives for History of Exact Sciences, 16*(1), 1–16.

Smith, G. (1978). Counting and classification on Kiwai Island. *Papua New Guinea Journal of Education, 14*, 53–68.

Smith, G. (1984). *Morobe counting systems: An investigation into the numerals of the Morobe Province, Papua New Guinea* (MPhil). Papua New Guinea University of Technology, Lae, Papua New Guinea.

Stanhope, J. (1972). The language of the Kire people, Bogia, Madang District, New Guinea. *Anthropos, 67*(1/2), 49–71.

Van der Waerden, B., & Flegg, G. (1975). *Counting 1: Primitive and more developed counting systems*. Milton Keynes, UK: Open University.

Vicedom, G., & Tischner, H. (1948). *Die mbowamb*. Hamburg.

Vicedom, G., & Tischner, H. (1943–1948). *Die Mbowamb: die Kultur der Hagenberg-Stämme im östlichen Zentral-Neuguinea* (Vol. 3 (1943 volume published by de Gruyter)). Hamburg, Germany: Museum für Völkerkunde und Vorgeschichte.

Wurm, S., & Hattori, S. (Eds.). (1981). *Language atlas of the Pacific area. Part 1: New Guinea area, Oceania, Australia*. Canberra, Australia: Australian Academy of the Humanities.

Chapter 4
Body-Part Tally Systems

Kay Owens and Glen Lean

Abstract Unique systems of counting found in Papua New Guinea (PNG), West Papua, Torres Strait Islands Australia and south-east Australia involving tallying against body parts. These differ from the digit-tally systems that only involve fingers and toes. In body-part tally systems, words for 1-4 are likely to be specific finger names, the word for 5 or thereabouts is thumb rather than hand and then other parts of the arm are used. Most of these systems reach a mid-point, usually on the head and then go down the other side in a symmetrical fashion. However, these systems vary considerably. Some are truncated, and the number in the cycles can vary considerably from 13 to 59. The last counting word for the cycle could be the last tally point or a word to indicate total. These words are believed to only involve the upper body parts with one system involving the outside of the legs. To count beyond the final tally point would require going to a second person, at least abstractly if not as a person. These systems often occur in 2-cycle system areas and there are some occasions where other systems exist. In terms of spread of counting systems, these appear to only occur in pockets in certain areas and not to be widespread.

Keywords Body-part tally systems • diversity of body-part tally across Papua New Guinea

The Existence of Body-Part Tally Systems

In his *Native Tribes of South-East Australia*, published in 1904, Howitt reported two unusual methods of "enumeration of the days", of the stages of a journey, which tallied an ordered sequence of parts of the body. One of these is given in Table 4.1.

Table 4.1
A South-East Australian Body-Part Tally System from Howitt (1904)

Number	Body Part Reference	Number	Body Part Reference
1	little finger	9	upper arm
2	ring finger	10	point of shoulder
3	middle finger	11	side of neck
4	fore finger	12	ear
5	thumb	13	point on the head above the ear
6	hollow between radius and wrist	14	muscle above the temple
7	forearm	15	crown of the head
8	inside of elbow joint		

K. Owens (✉)
Charles Sturt University, Dubbo, NSW, Australia
e-mail: kowens@csu.edu.au

G. Lean

K. Owens et al., *History of Number*, History of Mathematics Education,
DOI 10.1007/978-3-319-45483-2_4

"From this place the count goes down the other side by corresponding places" (Howitt, 1904, p. 698). If we take the crown of the head, being on the body's vertical axis of symmetry, as a unique mid-point, and with the 14 symmetrical counterparts of the ones given above, we have altogether 29 tally-points. Howitt also notes, in passing, that "this method seems to do away with the oft-repeated statement that the Australian aborigines are unable to count beyond four or at most five" (Howitt, 1904, p. 698). The second tally method had a further two symmetrical points at "the divergence of the radial muscles" giving a total of 31 tally-points altogether. Howitt indicates that this system was probably widespread in south-east Australia.

Some three years later, in 1907, Ray reported the existence of similar tally methods in the Torres Strait islands (Ray, 1907, p. 47, p. 86). In both the western and eastern islands (the former are Saibai, Murulag, Boigu, Mabuiag, Badu, and Moa, and the latter are the Murray Islands, Erub, and Ugar), Ray indicated that 2-cycle counting systems are found. He then described a tally system used in the western islands (Table 4.2). For the eastern islands, Ray gave examples of two other tallies, one with 29 points and another with 25.

Table 4.2
Western Islands of Torres Strait Body-Part Tally System

Number	Body-Part Reference
1	little finger
2	ring finger
3	middle finger
4	index finger
5	thumb
6	wrist
7	elbow joint
8	shoulder
9	nipple (left)
10	sternum
11	nipple (right)
and so on, continuing with the corresponding symmetrical counterparts in reverse order until the little finger is reached at 19th tally point.	

Note. Data are from Ray (1907).

Ray also provided data on a similar tally method used in the Papuan Gulf region of the New Guinea mainland by the Purari and which has a total of 23 tally-points (Ray, 1907, p. 331). The Purari also, according to Ray, have a 2-cycle numeral system. Another group in the Gulf region, Orokolo, are reported by Ray (1907, p. 323) as having a tally method which utilises 27 body parts. These reports, deriving from the turn of the century, of the coastal part of the mainland are supplemented by data collected subsequently in the interior highlands region. In 1940, the Government Anthropologist for Papua, F. E. Williams, first reported the existence of similar body-part systems in use in the Southern Highlands Province of PNG by the Foe people of Lake Kutubu. Williams indicated that their tally method utilises 37 tally-points and said that it "is used with rather remarkable accuracy, especially as providing a timetable for feasts, etc" (Williams, 1940–1941, p. 151). Their neighbours, the Fasu have a 35 point tally system in their language Namo Me as shown in Figure 4.1 and in Table 4.3.

It is worth noting that in the mid 1990s, May and Loeweke returned to the Fasu to revise the Bible translation because the language had changed so much in that time. Subsequently large mining corporations entered the area so it is likely that further changes have occurred in their counting practices and foundational knowledge may be at risk in everyday practice. This is a salient reminder of the imminent loss of foundational knowledge of many of these cultures and languages and another reason for the extraordinary work of Lean in attaining first contact data for the arguments of this book and an important addition to our understanding of the history of number. To this we will return in Chapters 11-13 in discussing Tolai of East New Britain Province and schooling.

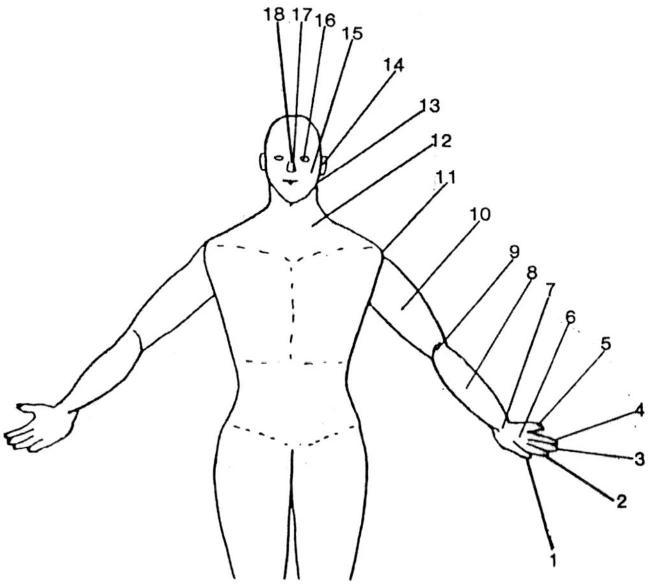

Figure 4.1. Fasu or Namo Me body-part tally system.
Source. Lean (1992). Data are from May and Loeweke (1981).

Table 4.3
Namo Me Tally System of the Fasu, Southern Highlands Province, PNG

Numeral	Fasu (Namo Me)	English Equivalent
1	*mēno*	little finger
2	*tetá*	ring finger
3	*isiá*	middle finger
4	*kitafá*	index finger
5	*kakórea*	thumb
6	*namá*	palm of hand
7	*yatipinu*	inside wrist
8	*kāri*	forearm
9	*tōkona*	inside wrist
10	*kaeyako*	upper arm
11	*kinu*	shoulder
12	*kenó*	collar bone
13	*fúfu*	neck (below ear)
14	*senaki*	ear
15	*parē*	cheek bone
16	*hī*	eye
17	*nō*	side of nostril
18	*terayia*	ridge of nostril (mid-point of cycle)
19	*tāku nō*	other side of nostril
20	*tāku hī*	other eye

Note. Data are from May and Loeweke (1981).

Further data on the body-part tallies of the Southern Highlands were provided by two other SIL members, the Franklins (Franklin & Franklin, 1962) with details of the *Kewa Counting Systems*. The eastern and western dialects of Kewa both have 47-cycle body-part tally methods and these are used, as with the Foe, for calendric purposes, and in particular for reckoning when a certain festival should occur within a cycle of festival dances lasting over many months. The Franklins also note that the Kewa have a 4-cycle counting system in addition to their tally methods and that it is this system which is used to express precise numerical quantities and not the tally method (Franklin & Franklin, 1962, p. 189). Further west in Hela Province, PNG, the Yuna (Duna) language group have truncated their tally system due to the tonal difference between the word for 14 and a word people do not wish to say. Interestingly, the counting system also provides a special link for the 7 which cycles so that 8 is equivalent to 1 relating to the kinship relationships. This asymmetric body-tally system has 12 *konane* for "ear" (SIL word list).

The available literature, then, provides evidence of the existence of this most unusual tally method and that it is found in the widely separated geographical locations of southern Australia, the Torres Strait islands, and the highlands and southern coastal areas of PNG. As far as we could ascertain, there is no record of this type of body-part tallying existing beyond the Australia-New Guinea region. We will now summarise the extent to which it is found on the New Guinea mainland and the location of those groups which use it. These systems are different to digit tally systems that make use of fingers and toes which seem to be widespread across Indigenous people around the globe (Dixon & Kroeber, 1907; Eells, 1913a, 1913b; Lipka & Adams, 2004).

Body-Part Tally Methods in New Guinea

This type of tally which uses the fingers and other tally-points which are situated on the upper part of the body are, on the New Guinea mainland, found only among certain NAN language groups and never among AN language groups. It is thus unknown in Island Melanesia, Polynesia and Micronesia. The data for New Guinea may be grouped into three categories:

(a) the case where the data are reasonably complete enough to determine the nature of the tally method,
(b) the case where the data are sufficient to determine that a language group possesses a body-part tally method but are insufficient to determine the total number of tally points used, and
(c) the data suggest that a language group has a body-part tally method but further data are required to confirm this.

In the last category we have 21 candidates which probably have a body-part tally method, 14 of which are language groups belonging to the Trans New Guinea Phylum, the remainder belonging to the Arai Stock, the Sepik-Ramu Phylum, and the Amto-Musian Family. The Trans New Guinea Phylum languages which probably have body-part tallies are: Kaluli, Beami, Konai, Podopa, Pawaian, Omati, Ipiko, Minanibai, Toaripi, Wabuda, Morigi, Lewada, and Kilmeri. The Arai Stock languages are: Rocky Peak, Bo, and Iteri (see Appendix C for details). The Amto-Musian Family comprises Amto and Musian (see Appendix C for details). In category (b) we have 12 language groups which definitely have body-part tally methods but for which we are unable to determine the magnitude of the complete cycle. Three of these language groups are members of the Sepik-Ramu Phylum and the remaining nine belong to the Trans New Guinea Phylum. The Trans New Guinea Phylum languages are Wiru, South Kati, Yongom, Bimin, Dem, Tao-Suamato, Gizra, Taikat, and Dera. The Sepik-Ramu languages are: Piame, Tuwari, and Waibuk.

In category (a) we have 40 language groups which definitely possess body-part tally methods and for which we have sufficient information to determine the cycle of each tally. Of these, four are members of the Sepik-Ramu Phylum, one (Yuri) is a Phylum-level isolate (see Appendix C), and the remaining 35 are members of the Trans New Guinea Phylum. There is no evidence that members of the East Papuan and Torricelli Phyla possessed this type of tally method.

The characteristics which define this type of body-part tally are as follows. The tally usually begins on the fingers of the left hand (this is not universal and occasionally the tally will be started on the right hand). The tally proceeds in order from the little finger through to the thumb, then to various points along the arm, usually beginning at the wrist (the 6th tally-point). Other points, after those on the arm, may be the neck, ear, eye, and nose; occasionally the breast and sternum may be used. In 31 out of 40 cases the tally has a unique midpoint which lies on the body's vertical axis of symmetry (and, unlike all other tally-points, does not have a symmetrical counterpart). Once the points on one side of the body are tallied, the sequence continues on the opposite side of the body, tallying the symmetrical counterparts of the original tally-points in reverse order until the little finger of the hand is reached. For those tallies which do have a unique midpoint, the complete tally cycle will, of course, be an odd number. This type of tally does not employ the toes or indeed parts of the anatomy below the navel.

The mathematical term for the tally cycle is, properly, modulus in that it is usual to start the tally from the beginning again once the full tally cycle has been completed. As was noted above, the language groups which possess these tally methods use them for calendrical reasons and, in particular, for establishing when a given festival should occur within a cycle of festivals. The magnitude of the tally cycles varies considerably between the different groups in Table 4.4.

Table 4.4
The Frequency of Occurrence of the Cycles of Body-Part Tallies

Cycle Size	Frequency	Cycle Size	Frequency
18	2	29	3
19	2	30	1
22	1	31	1
23	8	35	1
25	2	36	1
26	1	37	1
27	11	47	2
28	2	68	1

The modal cycle magnitudes are 27 and 23. No data are available which suggest why these two tally cycles should predominate. It is tempting to think that the 27-cycle may have some connection with the lunar month although this is, of course, 29.5 days and not 27.

One notable feature of these tally methods, which are used by widely separated language groups, is the similarity with which the first ten tally points occur. We have, altogether, 24 groups with tally methods which follow the sequence given in Table 4.5.

Table 4.5
Common Body Parts used in 24 of the Body-Part Tallies

Number	Body-Part Reference
1	little finger
2	ring finger
3	middle finger
4	index finger
5	thumb
6	wrist
7	forearm
8	elbow
9	upper arm
10	shoulder

It should be noted that the Aboriginal tally recorded in southeastern Australia, given earlier, also follows this sequence. Some six other language groups omit the forearm and upper arm and thus have:

6 wrist
7 elbow
8 shoulder

This latter type also occurs in the western Torres Strait islands. The language groups which deviate from the patterns given above are those which have adopted larger tally cycles, usually greater than 35, and these are all located in the Southern Highlands Province of PNG, the dialects of Kewa being examples (these may have other language names in the latest Ethnologue).

Several other methods of body-part tallying, not included in the data just discussed, exist. The Ama of the East Sepik Province (Lean, East Sepik data, source: SIL/IMP data, 1978) first tally the fingers of one hand, as for the other body-part methods except that a hand morpheme is used for five (which is unusual for body-part tally systems). However they then proceed:

6 navel
7 breast (1)
8 breast (2)
9 shoulder (1)
10 shoulder (2)
11 eye (1)
12 eye (2)

at which the tally terminates. The Sakam of the Morobe Province (PNG) (Lean's Morobe counting system data; Smith, 1984) have what is essentially a digit tally which is augmented by the two nostrils: tallying starts on the fingers of one hand and proceeds to the fingers of the other hand until a tally of 10 is reached. The left thumb is placed on the left nostril for the 11th tally point and then the right thumb is placed on the right nostril for the 12th tally point. After that the toes are tallied giving a complete cycle of 22. A further method is reported by an SIL member in surveying the Nagatman of the Sandaun Province (PNG) (Lean cited this system from a 1974 SIL Word List but only tally words for 1 to 4 are listed as probable – see Appendix C). The unpublished report indicates that "tallying begins on the left side of the body, proceeding up the left arm and then down the left side. The toes are tallied and upon reaching the big toe the tally stands at 36. The side of the left foot is then tallied and then the symmetrical points on the right side of the body." The complete cycle of this tally method is not given although it would have to be at least 74. One other unusual tally method is that of the Abau, also in the Sandaun Province and an endangered language (SIL, PNG). This is essentially a digit-tally augmented by the navel, the breast, and the eye. Tallying begins on the fingers of the left hand proceeding from the little finger to the thumb as shown in Table 4.6.

It is usually the case that a particular language group will possess a single tally method. Two groups, however, are exceptions. The Ketengban of West Papua have two tally methods. According to Briley (1977) "the main one is based on twenty-five while the second is based on twenty-eight" (p. 31) as indicated in Table 4.7. The 25-cycle method is given in full and this illustrates, in passing, the way in which left- and right-side symmetrical counterparts are distinguished verbally.

Table 4.6
Body Part Descriptions for Abau, Sandaun Province, PNG

Number	Body Part Description
6	hand one navel with
7	hand one breast two
8	hand one breast two navel with
9	one hand and the fingers of the other hand excluding the thumb
10	hands two
11	hands two navel with
12	ands two breast two
13	hands two breast two navel with
14	hands two breast two navel eye one
15	hands two foot one

The 28-cycle is obtained by augmenting the 25-cycle with the two eyes and the nose in the order: left eye (14th tally-point), nose (15th), right eye (16th), and then continuing with ear, neck, shoulder, etc., as given above. Another language group which has two different tally methods is reported by Bruce (1979, 1984). This is the Alamblak of the East Sepik Province (PNG) which has a 29-cycle tally method used exclusively by men and another (similar) method used exclusively by women using the breasts instead of points of the face. They also have a (2, 5, 20) digit tally system.

Table 4.7
Ketengban, West Papua, Body Tally Systems

	First Side		Second Side	Meaning
1	*tekin*	25	*telepcap*	(little finger)
2	*betene*	24	*danabeteniap*	(ring finger)
3	*weniri*	23	*danaweniriap*	(middle finger)
4	*dombare*	22	*danadome*	(index finger)
5	*pambare*	21	*danapame*	(thumb)
6	*napbare*	20	*dananap*	(wrist)
7	*tabare*	19	*danata*	(lower arm)
8	*pinbare*	18	*danapin*	(elbow)
9	*topnebare*	17	*danatopne*	(upper arm)
10	*taubare*	16	*danatau*	(shoulder)
11	*kumbare*	15	*danakum*	(neck)
12	*amulbare*	14	*danamul*	(ear)
13	*mekbare*			(crown of head)

Source. Data are from Briley (1977).

Those language groups which possess body-part tallies will, generally speaking, also possess a numeral system. Of the 52 NAN language groups known to have body-part tallies, the data for 16, or less than a quarter, are such that we are uncertain whether there is an accompanying numeral system as well. However, there are 33 groups which do have an accompanying numeral system: 28 of these have a "pure" 2-cycle system, or another group with only one side (non-symmetrical and finishing at a body mid-point), comprising Wiru and several Kewa dialects, are all located in the Southern Highlands Province (PNG) with Etolo given by Lean as a dialect of Beami, in his Western Province data but known to extend into Hela Province. (Lean had insufficient data to be sure that this was a body-part tally system.) Dwyer's 1979-1980 data (personal communication, 2015) showed a 17 part asymmetrical system finishing at the "tip of the nose". Dwyer and Minnegal (2016) showed that the language has words for 1 x 17, 2 x 17 etc continuing the counting at least to 17 x 17=289 (see Appendix B for more details of Dwyer's work on Etolo and Bedamuni). Interestingly Bedamuni was shown by Dwyer to have a symmetrical body-part counting system with some similar words. Onabasulu (Hela Province) had similar words beyond 18. (Lean had only recorded 1 and 2.) Dwyer's work indicated that people had a more abstract system for these two dialects (now taken as languages). For Bedamuni after 6, this was a suffix of the other counting system. Lean had made a similar judgement that Nomad and its dialects—Kubo, Samo, Honibu, and Bibo—were likely body-tally systems with his limited data. Dwyer and Minnegal's data (collected in 1986-1987) showed that Kubo also had a symmetrical body-tally system of 27 parts but their informants did not note higher numbers using groups like Etolo. Dwyer and Minnegal (2016) indicated that asymmetrical systems included Kovai.

These authors noted that giving first, second, third, etc. for birth order primarily relied on first born and last possible to be born with second last as a distinct word. Other terms depended on the number of children in total. There were also variations when counting days, often indicating days from expected end point. In most cases, when referencing the English decimal system, the language groups used their first 10 number names.

Dwyer and Minnegal make the point that the order of a counting system is clearly being used since sometimes the word is not the exact term of the part being pointed to (e.g. the word might be "bone" but the person might be pointing to the collar bone or forearm) and thus countering the argument that they were just one-to-one tallying methods without a sense of order of size of number. Dwyer and Minnegal are also collating the various systems building on Lean's initial work indicating the diversity of different types and possibly they will draw some conclusion regarding spontaneity or diffusions of the systems. For example, one of the Bosavi language group had a truncated one-sided system unlike the others. Similar examples occur in neighbouring Hela and Southern Highlands Provinces such as Yuna (Duna) and Kewa. Counting in Etolo and Bedamuni are different for counting objects and people or giving an abstract number especially for 1 to 5. For Etolo, Dwyer and Minnegal have also shown that higher numbers are possible based on the cycle of 17. Appendix D provides some data on Nomad (Kubo, now regarded as a separate language rather than a dialect), Ebolo (also named Etolo), and Bedamuni. Etolo people enjoyed counting and the younger men could quickly convert decimal numbers to the original and preferred foundational counting system.

It was found with Muke's (2000) data that higher numbers using tallying with body-parts was possible mostly as a memory tool (see Chapter 3) and in Oceanic and other Island Melanesia languages that powers could be associated with classifiers (see Chapter 6) and body parts given the use of fingers to accompany counting (Lincoln, 2010).

For the 28 groups discussed in Lean's (1992) study which have both 2-cycle system variants and body-part tallies, some 15 are such that the first four finger names of the tally (usually "little finger", "ring finger", middle finger", and "index finger") are displaced by the first four numerals, that is a hybrid of the tally method and the numeral system is formed. The remaining groups appear not to form a hybrid and the tally and the numeral system are kept separate and distinct.

The Distribution of Body-Part Tallies

Appendix D gives a Table D2 that summarises the majority of known body-part tally systems. The distribution of the body-part tallies throughout the New Guinea mainland is shown in Figure 4.2. Generally speaking, they are located in the central interior region and in the southern coastal areas. In PNG, the Provinces which have the majority of the tallies are the Southern Highlands, Hela, Western, Sandaun, and Gulf. There are traces as well in the Enga, Madang and Jiwaka Provinces. In the

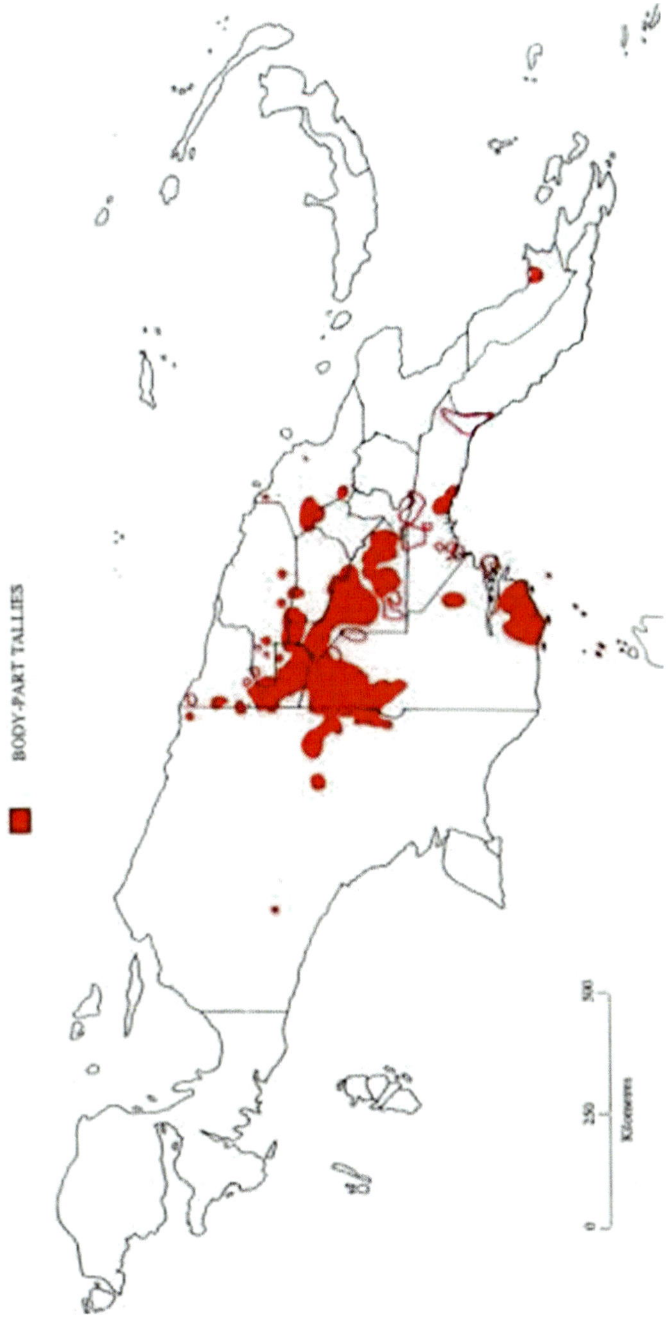

Figure 4.2. Distribution of body-part tallies in New Guinea (from Lean's, 1992 study).

northeast and eastern coast of PNG we have only a few examples: Giri, Murupi, and Baruga and a possible language Yupno (Isan) on the Morobe-Madang border as discussed below. In West Papua we have evidence that there is a predominance of tallies in the central border region; that there may be further tallies found to the west is likely as more data become available. Except for the Southern Highlands Province, the body-part tallies are usually associated with variants of 2-cycle numeral systems.

In 1994, Dasen and Wassman made a study in Gua village of the Yupno language (Wassmann & Dasen, 1994). They were interested in the variation of counting and arithmetic processes for different age groups. The 10 older and middle-aged men who had not attended school and apparently did not speak the lingua franca Tok Pisin gave body-part tally systems. The article did not give the actual words but Owens checked with two different people from the area to see if this village and Isan from which Smith (1984) and Lean (1992) obtained data spoke the same language and they did. The two older records had called the language Isan and they had words for a digit tally system with words for 1, 2, 3, 4 = 2 + 2, 5 having a hand morpheme, the rest of the numbers involved two hands, and then a foot morpheme but the system stopped at 20 as the last frame word. However, one of Smith's informants used big toe which is consistent with a body-part tally system. Wassman and Dasen noted that the older men used thumb for 5 consistent with a body-part tally system. However, it would seem that these were superimposed on the digit-tally system rather than the system to 20 coming from a body-part system. To continue beyond the hands and feet, body parts were grouped as 2, 2, 1 (a group of 5) for continuing to say 33 with pairs of eyes, breasts, nostrils and men's parts and the single part being nose, belly button, single man's part.

Owens checked the nature of the body-part tally system with others recorded by Lean and there were some inconsistencies. Lean found no evidence of languages possessing body-part tallies among the Finisterre-Huon Stock to which Yupna languages belong or other neighbouring stocks in the Trans New Guinea Phylum. Neighbouring languages do not have a body-part tally system and in fact Yupna is much further to the east than any known body-part tally system in the northern section of the island. The closest is Murupi, according to Lean, inland from Alexishafen (over 100 km north-west of Yupna with many languages in between). Wassman and Dasen's body-part tally system would be the only system which has grafted onto it the digit-tally system. Other variations from Lean's records included only one body-part tally system in which a 2-cycle system had been grafted on so 4 = 2 + 2 but variants of 2-cycle systems did occur in several places concurrently (e.g., Oksapmin). In some places, the body-part tally system was truncated. For example, Gende was truncated to 10 and Yuna to 13, Kewa dialects to 10, Etolo (Hela, Western Provinces) to 17 (Dwyer, personal communication, 2015) and possibly Samo to 16 stopping at the wrist and not counting with the fingers of the second hand. More recently Oksapmin truncated to 20 as the complete group although one informant called the complete group 30 (Saxe, 2012). Oksapmin also may have started with a thumb and when reaching the other hand counted from the thumb again and continued up the other side beyond 27 to may be 29 or 30.

Lean (1992) suggested that digit-tally systems may have existed before neighbouring body-tally systems were introduced. For example, Hewa in the Southern Highlands and Alamblak in Sandaun Province both have body-part tally systems like their neighbours. Both digit-tally and body-part tally systems are used in Alamblak. They are not grafted together, and the body-part tally does not switch sides. The closest account of a body-part tally and digit-tally system merging is provided by the limited data available for Kwomtari in the Sandaun Province, which appears to have been influenced by neighbouring languages. The informants suggested that there is a 10-cycle system with counting on two hands (but not a digit-tally system using a foot morpheme), as well as a body-part tally system.

The system recorded by Wassman and Dasen (1994) differs from other body-part tallying systems in that the people move from one side of the body to the other (which may follow from the side switching for digit tallying). There is one further unique feature of the Wassman and Dasen records and that is the use of lower body parts. Nagatman of the Sandaun Province appear to have a 74 body-part tally system requiring lower parts of the body to the toes but only the sides of the body. Tallying proceeds up the left arm and then down the left side. The toes are tallied and upon reaching the big toe the tally stands at 36. Then the side of the left foot is tallied and the symmetrical points on the right side of the body are tallied in reverse order (Lean, 1992). This was the only case recorded by Lean using lower body parts but only the side. Some informants have noted that lower parts of the abdomen are not used in tallying systems. Biersack (1982, p. 813, cited in Lean's appendix) observed that, with the Paiela, the lower part of the abdomen is associated with "abominations such as intercourse, menstruation, childbearing, defecation, and the like". "Tallying, thus, is an abomination-free process" (Lean, 1992's Appendix on Ipili). Hence it is rather surprising that the old men in switching from side to side included the testicles. However, the older men deferred to one man to give the last counting word as several were supplied in the course of the data gathering and he said it was "penis." There are other occasions on which different informants have given different words but none of Lean's data gave the last counting word as different.

Wassmann and Dasen reported that Yupno women do not count in public. Lean refers to only one report of a different counting system for women. Bruce (1984, pp. 321-322, cited in Lean, 1992) reported that Alamblak use three types of systems: One is a digit-tally system with a (2, 5, 20) cyclic pattern. Another, used exclusively by men, is a body-part tally system with a 29-cycle. A third, used exclusively by women, is similar to the men's system except that the breasts (not regarded as private parts) are used as tally-points instead of points on the face.

It is likely that data are incomplete for many language groups so it is important to consider these data from Wassman and Dasen as indicating that perhaps there were body-part tally systems for special purposes occurring more spontaneously or by certain people who had not been informants in Lean's and Smith's studies. It could also be that having Lean and Smith already asking about counting words and using a diagram of body-parts had encouraged conversations leading these old men to give Wassman and Dasen this information partly as humour. The fact that counting was a man's activity might suggest that many of the communities from which data have been analysed as digit-tally systems may also have rituals that develop body-tally systems. No data have been collected on this, although we do know that in some places traditional counting is used in ceremonies and not necessarily on other occasions.

The replacement of body-tally systems by a digit tally system in Yupno is in contrast to the argument by Dwyer and Minnegal (2016) that the body-tally system replaced (2) or (2, 5) cycle systems, perhaps because they were more efficient. In some cases, the (2) and (2, 5) systems disappeared but the numbers, especially 1 and 2 and in some cases variations on 3 and 4 (often 2 + 1 or related words) were incorporated into the body-tally system replacing the first few body part names and in some cases the last set of fingers with numeral words. It is also important to counter the arguments that body-part tallying was disparagingly referred to as "concrete" thinking because the parts were touched and visible. In many cases Dwyer and Minnegal noted that different people within the language group would touch alternative parts for numbers and that people were easily able to do calculations and to convert to the decimal system when it arrived with English using the first 10 number indicators as in the body-tally system. These authors also noted that there were some modifications to the symmetrical body-part system as people interacted with neighbours such as the Bosavi with the Huli as counting was part of the reflexive relationship building.

As has been noted above, body-part tallies which are virtually identical to those found in New Guinea are also found in the Torres Strait Islands and in south-eastern Australia. The close similarity of many of these quite complicated tallies which are located over a widespread geographical area suggests that the tally method has been diffused; in some cases local variations have occurred but there is nevertheless a marked uniformity between the various tally methods, particularly with regard to the first 10 tally points. The tallies are found in both NAN and Australian language groups. Given that this type of tally is not found at all among the AN language groups, we will assume that its entry into the New Guinea region was with one of the migrations which occurred prior to the pre-Proto-Oceanic movement into northeast New Guinea. Alternatively, it was a development within a group of languages that traded or otherwise interacted and hence body-tally systems spread to related groups. The question then arises whether the body-part tally entered the Australia-New Guinea region with the earliest Australoid migrations or whether it came subsequently with one of the NAN migrations. This issue will be taken up again in Chapter 9.

References

Briley, J. (1977). Some counting systems of Irian Jaya. *Irian: Bulletin of Irian Jaya Development, 6*(3), 28–32.

Biersack, A. (1982). The logic of misplaced concreteness: Paiela body-counting and the nature of primitive mind. *American Anthroplogist, 84*(4), 811–829.

Bruce, L. (1984). The Alamblak language of Papua New Guinea (East Sepik) (Pacific Linguistics, C-81). Canberra, Australia: Australian National University.

Bruce, L. (1979). *A grammar of Alamblak (Papua New Guinea)*. Canberra, Australia: Australian National University.

Dixon, R., & Kroeber, A. (1907). Numerical systems of the languages of California. *American Anthropologist, 9*, 663–690.

Dwyer, P., & Minnegal, M. (2016). Counting systems of the Strickland-Bosavi languages, Papua New Guinea. *Language and linguistics in Melanesia, 34*(1), 1–36.

Eells, W. (1913a). Number systems of the North American Indians. *American Mathematical Monthly, 20*(9), 263–272.

Eells, W. (1913b). Number systems of the North American Indians: Systems of numeration. *American Mathematical Monthly, 20*(10), 293–299.

Franklin, K., & Franklin, J. (1962). The Kewa counting systems. *Journal of the Polynesian Society, 71*, 188–191.

Howitt, A. (1904). *The native tribes of south-east Australia*. London, UK: Macmillan.

Lean, G. (1992). *Counting systems of Papua New Guinea and Oceania* (Unpublished PhD thesis). PNG University of Technology, Lae, Papua New Guinea.

Lincoln, P. (2010). Count not on substrata. In J. Bowden, N. Himmelmann, & M. Ross (Eds.), *A journey through Austronesian and Papuan linguistic and cultural space: Papers in honour of Andrew Pawley* (Vol. 615, pp. 225–244). Canberra, Australia: Pacific Linguistics, ANU.

Lipka, J., & Adams, B. L. (2004). *Culturally based math education as a way to improve Alaska native students' math performance* (Working Paper 20). Appalachian Collaborative Center for Learning, USA (ED484849).

May, J., & Loeweke, E. (1981). *Fasu namo me: Fasu lexicon and grammar*. Ukurumpa, Papua New Guinea: SIL.

Muke, C. (2000). *Ethnomathematics: Mid-Wahgi counting practices in Papua New Guinea* (Unpublished MEd thesis). University of Waikato, Waikato, New Zealand.

Ray, S. (1907). *Reports of the Cambridge anthropological expedition to Torres Straits* (Vol. 3). Cambridge, UK: Cambridge University Press.

Saxe, G. (2012). *Cultural development of mathematical ideas: Papua New Guinea studies*. New York, NY: Cambridge University Press.

Smith, G. (1984). *Morobe counting systems: An investigation into the numerals of the Morobe Province, Papua New Guinea* (MPhil). Papua New Guinea University of Technology, Lae, Papua New Guinea.

Wassmann, J., & Dasen, P. (1994). Yupno number system and counting. *Journal of Cross-Cultural Psychology, 25*, 78–94.

Williams, F. (1940–1941). Natives of Lake Kutubu. *Oceania, 11*, 121–157.

Chapter 5
5-Cycle Systems

Kay Owens and Glen Lean

Abstract In Chapter 3, (2, 5) and (2, 5, 20) systems and variants were explored, so in this chapter only those systems with a 5-cycle but no 2-cycle are discussed. The cycle approach to analysing systems clarifies what were often incomplete classifications of quinary, vigesimal or quinary-vigesimal systems. Most of these 5-cycle systems are Austronesian (AN) in Papua New Guinea (PNG) and Oceania. There are Non-Austronesian (NAN) languages that are near Austronesian languages that may have incorporated numerals for 1 to 4. These 5-cycle systems also occur among Indigenous groups such as in the Americas and Africa. Although Proto Oceanic was thought to have a 10-cycle system as a primary cycle, a careful analysis by Lean has shown that at least half of the AN (all but two are Oceanic) languages have a 5-cycle primary cycle. There are some AN languages with (5) and (5, 20) cycles but more NAN languages with these cycles. The AN languages are more likely to have a numeral for 10 and a 10 cycle, usually as a secondary cycle.

Keywords 5-cycle systems • (5, 10) cycle systems • (5, 10, 100) cycle systems • (5, 10, 20) cycle systems • digit-tally systems

The Existence and Nature of 5-Cycle Systems

Edward Tylor, Professor of Anthropology at the University of Oxford, wrote in 1871 that "the general framework of numeration stands throughout the world as an abiding monument to primaeval culture. This framework, the all but universal scheme of reckoning by fives, tens, and twenties shows that … the practice of counting on fingers and toes lies at the foundation of our arithmetical science" (Tylor, 1871, p. 271). As the data on counting systems of newly discovered Indigenous societies became available during the 19th century, Tylor, like many of his contemporaries, was struck by the ubiquitous nature of the quinary, or 5-cycle, systems. The dictum of the Brothers Grimm that "all numerals are derived from fingers" (Trumbull, 1894, p. 41, translated) appeared to hold true at least for the numeral 5 which, for a great many languages, was either identical to, or cognate with, the word for "hand". Tylor continued: "To count the fingers of one hand up to 5, and then go on with a second five, is a notation by fives, or, as it is called a quinary notation. To count by the use of both hands to 10, and thence to reckon by tens, is a decimal notation. To go by hands and feet to 20, and thence to reckon by twenties, is a vigesimal notation … In surveying the languages of the world at large … there prevails, with scarcely an exception, a method founded on hand counting, quinary, decimal, vigesimal, or

K. Owens (✉)
Charles Sturt University, Dubbo, NSW, Australia
e-mail: kowens@csu.edu.au

G. Lean

K. Owens et al., *History of Number*, History of Mathematics Education,
DOI 10.1007/978-3-319-45483-2_5

combined of these" (Tylor, 1871, pp. 260-261). It was commonly noted, too, that for the last type of system, the numeral 20 often had the meaning "one man", i.e. the fingers and toes of one man.

We thus find several types of 5-cycle numeral system reported in the literature. Schmidt (1929, p. 614) for example, discussed three variants: the "pure" quinary, the "quinary vigesimal", and the "quinary decimal". The first corresponds, in Salzmann's terms, to a 5-cycle system which does not possess a secondary cycle and which has the symbolic representation: 1, 2, 3, 4, 5, 5+1, 5+2, 5+3, 5+4, 2×5, (2×5)+1, … , 3×5, and so on. The quinary vigesimal system has a (5, 20) cyclic pattern which is such that it is pure quinary up to 20 after which a secondary 20-cycle comes into play and we have, for example, 40=2×20 and 100=5×20. A third type, the quinary decimal, also referred to as the "imperfect decimal" (Ray, 1907, p. 464), has a (5, 10) cyclic pattern so that the symbolic representation is: 1, 2, 3, 4, 5, 5+1, 5+2, 5+3, 5+4, 10, 10+1, … , 2×10, and so on. A further type, discussed by Eells in his study of North American Indian counting systems (Eells, 1913b, p. 295), is the quinary-decimal-vigesimal system, or in Salzmann's terms, one with a (5, 10, 20) cyclic pattern. Each of the four types mentioned here has a number of variants.

The 5-cycle systems are distributed throughout all the major continents. Schmidt's language atlas, associated with his *Die Sprachfamilien und Sprachenkreise der Erde* (Schmidt, 1926), contains a counting systems map which shows the distribution of two types of quinary system throughout the world. The (5, 20) type occurs "sporadically in almost all the Australian linguistic groups; in nearly all the NAN languages of the (north) east coast and exterior of New Guinea, in the oldest Melanesian languages of New Caledonia, the Loyalty Islands, etc" (Schmidt, 1929, p. 614). In the Australian languages, the literature indicates that the (5, 20) system is usually associated with a variant of the 2-cycle numeral system. Dawson (1881, pp. xcvii-xcviii), for example, provided data on two language groups found in western Victoria, each of which has a system which may be represented as follows: 1, 2, 3, 2+2, 5, 5+1, 5+2, … , 2×5, (2×5)+1, … , 20, and so on. Among Kluge's (1938) numeral lists there is ample evidence of similar systems in other Australian languages while Harris (1982) provided data on the existence of (5, 20) systems without the primary 2-cycle. The (5, 20) system is also found in the Americas. As Nykl (1926) noted

> on North American soil, as far as is known, the quinary-vigesimal system does not exist east of the Rocky Mountains with but one exception, while it can be found everywhere along the whole Pacific Coast from Alaska to Southern California, and from there it becomes almost universal as far as Brazil. (p. 169)

Dixon and Kroeber, in their 1907 article on the counting systems of California, also noted that for North America "as contrasted with the wide extension of thorough decimal systems, quinary-vigesimal systems occur but rarely. Outside of Mexico, they are to be found only among the Caddoan tribes [in mid-West United States] (Eells, 1913a), the Eskimo, and in parts of California" (Dixon & Kroeber, 1907, p. 672). Schmidl's (1915) article on African counting systems indicated the existence of (5, 20) systems which are found in the Ivory Coast and central African regions. The existence of quinary systems in Africa is also confirmed by data in Conant (1896) and in Pott (1847).

The (5, 10) system, according to Schmidt's (1926) counting systems map, is found throughout Africa, Southeast Asia and North America. Schmidt noted that

> in Australia and in the NAN languages it is found isolated and in rudimentary form; with the Melanesians it is fully displayed. In Asia it originally dominated the two great families of Austroasiatic and the Tibeto-Chinese languages … So too, in North America; but in neither Mexico, Central America, nor in South America has the system spread widely. (p. 614)

Schmidt's (1929) definition of the (5, 10) system was that

> the numbers of the second pentad are formed by composition with 'five' (6=5+1, 7=5+2, etc.) or by the pair system (especially 6=3×2, 8=4×2) or by subtraction (especially 9=10-1). (p. 614)

This is a broader definition than will be adopted here and indeed it is one which Seidenberg has criticised (1960, p. 241). Specifically, we will not regard as 5-cycle systems those which form the second pentad subtractively (as is found among the languages of the Manus Province in PNG or the Ainu language of Japan), or by doubling, as given by Schmidt (1929) for 6 and 8, and which is found in the Motu and related languages of the Central Province of PNG. By adopting this more stringent definition, the distribution of the (5, 10) system as given by Schmidt will no doubt be diminished, nevertheless the evidence from other sources indicates that the system is relatively widespread. Conant, for example, provides many instances of (5, 10) systems: some 35 from various African languages, 30-odd from North America, and several from Oceania (Conant, 1896, pp. 134-175).

The (5, 10, 20) system, rarely mentioned in the literature usually because it is not distinguished from either the (5, 10) or the (5, 20) systems, is however mentioned by Eells who notes that for North American Indian languages "several systems show a combination of three digital bases in their formation" (Eells, 1913b, p. 295). The normal structure for this type of system is: 1, 2, 3, 4, 5, 5+1, 5+2, 5+3, 5+4, 10, 10+1, 10+2, … , 10+5, 10+5+1, … , 20, 20+1, … , 20+10, … , 2×20, and so on. Eells mentions a variant in which the quinary nature of the system does not become apparent until the second decade. The Welsh counting system is, in fact, of this type and is such that it has ten distinct numerals and then 11 to 19 are constructed as follows: 1+10, 2+10, 3+10, 4+10, 5+10, 1+5+10, 2+5+10, 3+5+10, 4+5+10. There is a distinct numeral 20 and we have 30=20+10, 40=2×20, 50=10+(2×20), and 60=3×20.

The three types of system discussed above are each important variants of 5-cycle systems. A fourth type, the "pure quinary" system (without a secondary cycle), Schmidt said "is found only in Saraweka, a South American Arowak language. Everywhere else it is combined either with the decimal or the vigesimal system" (Schmidt, 1929, p. 614). There is also the possibility, in the absence of there being a sufficient amount of data, that some systems have been classified as pure quinary whereas they may in fact have a secondary or even tertiary cycle. However, we now consider the New Guinea and Oceania evidence.

5-Cycle Systems in New Guinea and Oceania

Commenting, in 1907, on numeration in British New Guinea, Sidney Ray contrasted the counting systems of the NAN languages and those of the "Melanesian", or Austronesian, languages. The former, he said,

> very rarely advances beyond five, and that as a rule only two, or at most three numerals are named … In the Melanesian languages without exception, numbers can be named at least as far as five, and counting can be performed beyond, by fives, tens, or twenties. (Ray, 1907, p. 463)

Ray also noted that, for the AN languages, "counting is performed with the fingers, and in some the toes also are counted" (Ray, 1907, p. 463). Similarly, Codrington, in 1885, commented on numeration in other parts of Island Melanesia and observed that

> three systems of numeration which are based on the counting on the fingers are found … In some of the islands of the New Hebrides group and in the Banks Islands the notation is quinary; in other islands of the New Hebrides, in Fiji and in the Solomon Islands, it is decimal; in the Loyalty Islands, New Caledonia and Anaiteum, the notation is, or was, vigesimal. (p. 220)

Thus the quinary nature of the counting systems of some AN language groups, and the widespread occurrence of such systems throughout New Guinea and Island Melanesia, has long been observed. As we shall see, Ray's comments about the non-quinary nature of the NAN languages do not generally hold true and that the various 5-cycle systems are found in AN and NAN languages alike. In the sections that follow, the main types of 5-cycle system mentioned above will be discussed. We will also deal with the variants in the way the second pentad is constructed.

Systems With a (5) or (5, 20) Cyclic Patterns

The pure 5-cycle system is a rare phenomenon in New Guinea and Oceania: for almost every case, where sufficient data are available, there is evidence of a secondary cycle. (Chapter 3 covers (2,5) cycle systems.) One exception appears to be Kaulong, spoken in the West New Britain Province (PNG) (see Table 5.1) and which is such that the numeral 20 has an implied 5×4 construction.

Table 5.1
A Possible Example of a Pure 5-Cycle System: Kaulong in West New Britain (PNG)

Numeral	*Kaulong*
1	*ten* or *tehen*
2	*ponwal*
3	*miuk*
4	*mnal*
5	*eip* or *ep*
6	*ten mesup* or *ta mesup*
7	*ponwal mesup*
8	*miuk mesup*
9	*mnal mesup*
10	*eip ne mesup* **or** *ep mesup*
20	*eipnal* **or** *epnal*

Note. Original data from Throop and Throop (1980, p. 231) and Wakefield and Whitacre (SIL Word List, 1980). The former also gave 21 as *eipnal ten mesup* and 100 as *eipunwal eipunwal.* Other sources only go to 10 as "5 fingers" from a Counting System Questionnaire (CSQ) or "hand finished hand finished" (Chinnery, 1928, p. 99).

The other type of 5-cycle system included here, the (5, 20) type, is however common throughout New Guinea and parts of Melanesia and is found in both the NAN and AN language groups. The (5, 20) system is also referred to as the "finger and toe" or the "digit tally" system. Typically, this is such that we have four distinct monomorphemic numerals 1 to 4. The word for "five" is usually identical to, or cognate with, the word for "hand" or "arm". Occasionally "five" is a tally direction which contains a "hand" ("arm") morpheme and may have the meaning "hand to one side", or something similar. The second pentad, particularly the numbers from 6 to 9 (excluding the numeral 10 for the moment), usually has the construction represented symbolically as: $6=5+1$, $7=5+2$, $8=5+3$, and $9=5+4$; the word for "hand" usually appears explicitly in these compounds. Several variants of how these compounds are formed will be discussed below.

The word for "ten" normally contains a "hand" ("arm") morpheme and has the meaning "hands two", or something similar. "Ten" may be tally directions and we may have, thus, "hand side side" or "hands finished". Less often, "ten" is formed as an additive compound continuing the pattern for the numerals 6 to 9; thus $10=5+5$, i.e "hand hand". In the true digit tally system, the numerals 11 to 19 contain a "foot" or "leg" morpheme indicating that tallying the toes is taking place. For example, 15 may be expressed as "hands two, foot one", or sometimes just "foot one" or "foot completed". On reaching 20 many systems have "one man" or "man completed", and so on, where it is understood that "man" means "the fingers and toes of one man". Indeed, some systems have this explicitly: "hands two, feet two" or "hands, feet completed", or occasionally "feet (legs) two" where "hands (arms) two" is implicit. The construction of higher numerals is either multiplicative, for example $40 = 2 \times 20$ and $100 = 5 \times 20$, or additive, for example $30 = 20 + 10$, i.e. "one man and two hands". Various examples of the digit tally system are discussed below together with several of their variants.

Two examples which are typical of the digit tally as described above are those of Sentani, located in West Papua, and Onjob, located in the Oro Province (PNG). Both languages are NAN and the data are given in Table 5.2. For each of the systems, the numerals 5 and 10 contain a "hand" morpheme – *me* and *yani* respectively and the numeral 20 contains a "man" morpheme, *u* in Sentani and *yote* in Onjob. The second pentads of each system have the construction "$5 + n$" where "n" takes the values 1 to 4. Onjob has a common variant where the numeral 6 does not explicitly contain the numeral 1 in its compound; in cases like this, "six" often can be translated as "one hand and another".

Table 5.2
Two Examples of Digit-Tally Systems Belonging to NAN Languages

	Sentani	Onjob
1	*mpai*	*kema*
2	*be*	*ameg*
3	*name*	*towa*
4	*qeli*	*rome*
5	*mehempai*	*yanisiwi*
6	*mehine mpai*	*yanisiforkema*
7	*mehine be*	*yanisiforkema ameg*
8	*mehine name*	*yanisiforkema towa*
9	*mehine qeli*	*yanisiforkema rome*
10	*me be*	*yani sisi*
15	*me be odo fe mpai*	*amandi i ta kema ameg*
20	*u mpai*	*yote kema*
40	*u be*	*yote ameg*

Note. Data for Sentani, which has several dialects, came from Cowan (Cowan, 1950–1951, 1951–1952), but earlier data by Moolenburgh (1906) collaborated these data. Data for Onjob came from Macdonnell (1917, p. 171), Ray (1938, p. 185), and Kluge (1941, p. 24).

In Table 5.3, three examples, each AN, illustrate some of the variant ways in which the second pentad is constructed. The languages are Seimat, located in the Manus Province (PNG), Gitua, located in the Morobe Province (PNG), and Faga-Uvea, a Polynesian Outlier located in New Caledonia (see Appendix C for more details). Each of the systems shown has four monomorphemic numerals 1 to 4, and 5 has a "hand" ("arm") morpheme *nim*, *lima*. The systems differ in the construction of the numerals 6 to 10 and we delineate each variant here. Seimat has, for 6 to 9, what we have termed a $5 + n$ construction. In this, 5 appears explicitly with each of the numerals 1 to 4 so that we have: $6 = 5 + 1$, $7 = 5 + 2$, $8 = 5 + 3$, and $9 = 5 + 4$; each of these compounds is formed without the use of copula or conjunctions. The numeral 10, in Seimat, has the construction 2×5 and does not continue the additive pattern of the numerals 6 to 9. In Gitua, 5 is *nimanda sirip* and has the meaning "hand half" or "hand side" and in the numerals 6 to 9 this appears explicitly. Each compound numeral has the form $5 + c + n$ where 5 is followed by a conjunction (in this case *volo*), and then the corresponding numeral 1 to 4, thus $6 = 5 + c + 1$, and so on. The numeral 10, in Gitua, does not continue this pattern and we have instead the construction 5×2. For Faga-Uvea, incidentally the only Polynesian language to have a 5-cycle counting system, we find that each member of the second pentad has the relatively unusual construction $n + c + x$, in which the numeral 5 does not appear explicitly, and where the n takes the values 1 to 4, c is a conjunction, and x is a second pentad indicator. Thus 6, for example, is the compound

1 + *ona* + *tupu* where *ona* is a conjunction and *-tupu* is the second pentad indicator. The numeral 10, for Faga-Uvea, continues the pattern established for 6 to 9 and has the construction $10 = 5 + c + x$.

Table 5.3
Three AN Systems With Variant Second Pentads

Numeral	Seimat	Gitua	Faga-Uvea
1	*tehu*	*eze*	*tahi*
2	*huohu*	*rua*	*lua*
3	*toluhu*	*tolu*	*tolu*
4	*hinalo*	*pange*	*fa*
5	*tepanim*	*nimanda sirip*	*lima*
6	*tepanim tehu*	*nimanda sirip volo eze*	*tahiatupu*
7	*tepanim huohu*	*nimanda sirip volo rua*	*luaonatupu*
8	*tepanim toluhu*	*nimanda sirip volo tolu*	*toluonatupu*
9	*tepanim hinalo*	*nimanda sirip volo pange*	*faonatupu*
10	*huopanim*	*nimanda rua*	*limaonatupu*

Note. Data for Seimat are early data from Dempwolff (1905; 1909, p. 197) and Kluge (1941, p. 199). Later data indicate different pronunciation and the fact that these data have at least 16 classes indicated by suffixes Z'graggen (collected by Smythe), (Z'graggen, 1975, p. 183-187). Gitua data are from a CSQ and are similar to Lincoln (1976) and Smith (1984) data *(volo = wolo)*, Smith provided data for decades showing a 20 cycle predominating. Faga-Uvea is a unique Polynesian case from New Caledonia (Leverd, 1922, p. 99; Ray, 1919b).

These three ways of constructing the second pentad do not exhaust the various possibilities; there is, nevertheless, across all the 5-cycle system types, a relatively small number of methods employed for compounding the numerals 6 to 9. One further and important method is illustrated in Table 5.4 taking one NAN and one AN example each. These are, respectively, Panim and Bilbil, both located in the Madang Province (PNG).

Table 5.4
Second Pentad Construction of the Form "x + n" in a NAN and an AN Numeral System

Numeral	Panim	Bilbil
1	*olufan*	*kete*
2	*elis*	*oru*
3	*ized*	*toli* or *tol*
4	*woalai*	*pali* or *pal*
5	*mamagai*	*nimanta*
6	*eben nahe*	*koku kete*
7	*eben elis*	*kokun oru*
8	*eben ized*	*koku toli*
9	*eben woalai*	*koku pali*
10	*ma magunum*	*nimanoru*

Note. Data are available after 10 and for 20. Panim, NAN language data were from Dempwolff (1905, p. 245), Ray (1919a, p. 329) and Kluge (1938, p. 172). The data for Bilbil, an AN language, were from Dempwolff (1909, p. 240) and are similar to other data. Owens field work indicated that these two groups exchanged pots from Bilbil for baskets of quality garden food.

In each case the second pentad numerals 6 to 9 have the construction $x + n$ where x is not identical to 5 and where n takes, respectively, the values 1 to 4. In the case of Bilbil, x is *koku(n)-*; whereas the "hand" morpheme *nim* appears in both 5 and 10, it does not appear explicitly in the intervening numerals. In the case of Panim, x is *eben* which is identical to the word for "hand"; this, however, does not appear in the word for 5 which is *mamagai*. The latter, in fact, contains a "thumb" morpheme, *mamag-*, which also appears in the word for 10.

Among the NAN languages of the New Guinea region, the (5, 20) digit tally system is the second most common type of counting system after the (2, 5, 20) type discussed in Chapter 3. Of the latter, there are 109 altogether; by comparison 79 NAN languages have (5, 20) systems, just under one-fifth of the total sample. These are distributed among the various phylic groups as shown in Table 5.5.

Table 5.5
Distribution of (5, 20) Systems Among the NAN Language Phyla

Language Phylum	Number
small phyla	7
West Papuan	0
East Papuan	1
Torricelli	2
Sepik-Ramu	17
Trans New Guinea	52
Total	79

The distribution of the NAN (5, 20) systems throughout the New Guinea region is shown in Figure 5.1. The majority are found in the coastal regions and in particular are located largely in the East Sepik, Madang, and Morobe Provinces of PNG.

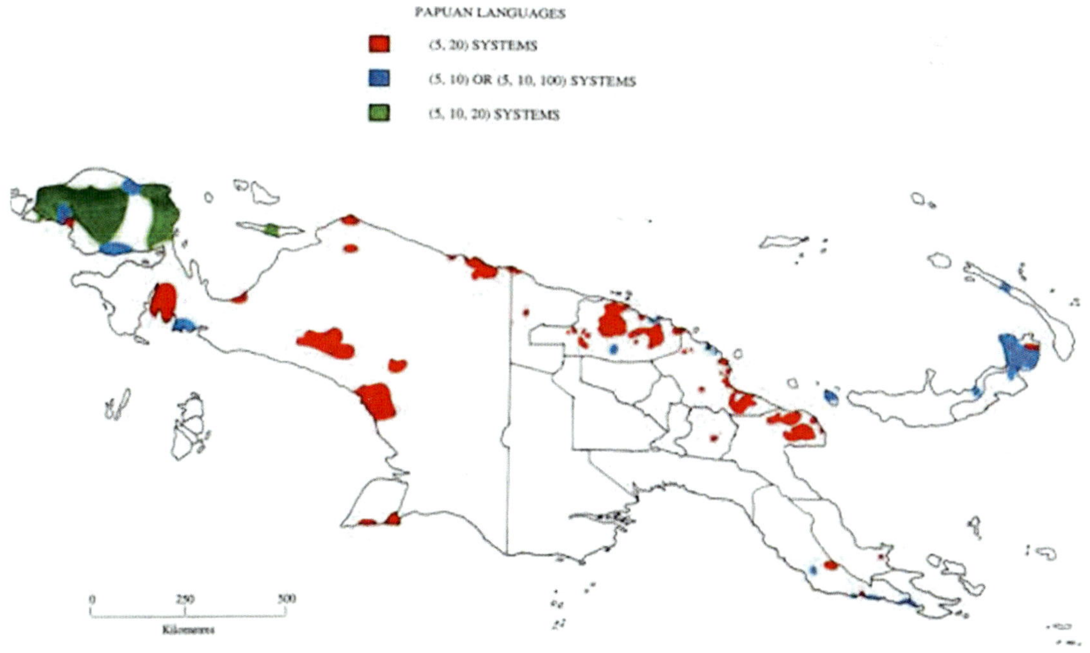

Figure 5.1. Distribution of NAN 5-cycle systems – New Guinea.
Source. Lean's (1992) thesis.

Among the AN languages of New Guinea we have 39 classified as having a (5, 20) system. These are distributed as given in Table 5.6. The majority of these are confined to the North New Guinea and Papuan Tip Clusters and are thus located, as can be seen from Figure 5.2, in the coastal and island regions of the Milne Bay, Morobe, Madang, and West New Britain Provinces. The non-Oceanic AN groups with (5, 20) systems are largely confined to the Maccluer Gulf region immediately to the south of the Vogelkop. In the remainder of Island Melanesia we find (5, 20) systems in both Vanuatu and New Caledonia.

Table 5.6
Distribution of (5, 20) Systems Among the AN Language Groups

Type	Cluster	Number
Oceanic	North New Guinea Cluster	19
	Papuan Tip Cluster	11
	Meso-Melanesian Cluster	1
	Admiralties Cluster	1
	West Papua	2
Non-Oceanic	West Papua	5
Total		39

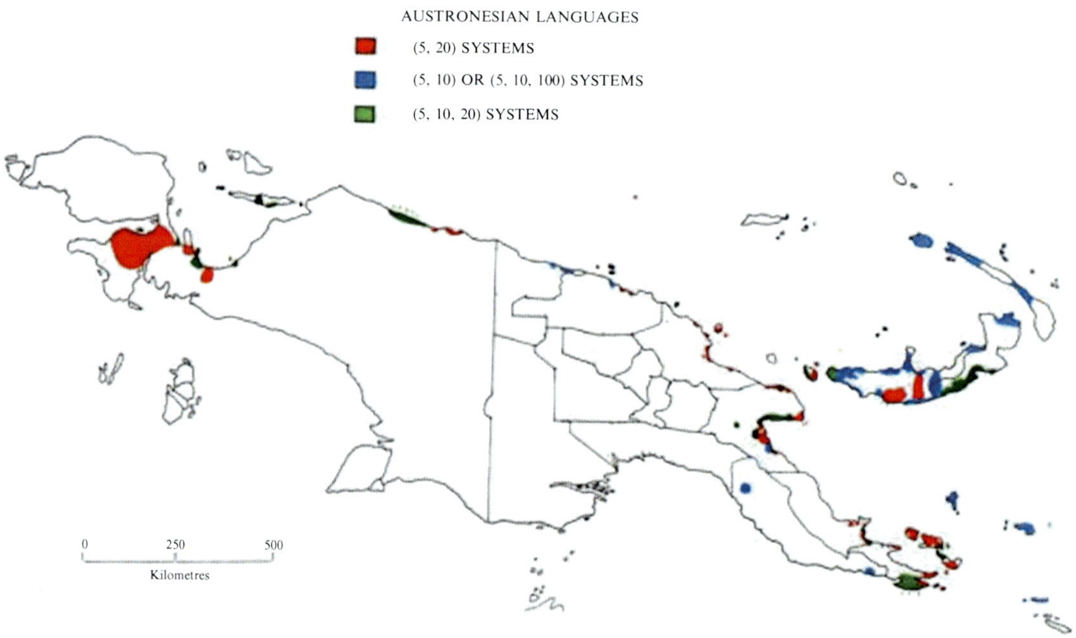

Figure 5.2. Distribution of AN 5-cycle systems – New Guinea.
Source. Lean's (1992) thesis.

In Vanuatu, the 15 language groups which possess this type of system are mainly found in the islands of Ambrym, Epi, Erromanga, Tanna, and Aneityum (see Figure 5.3).

In New Caledonia at least 9 languages have (5, 20) systems and these are mainly found in the south of Grande Terre and in the Loyalty Islands. One of these, Faga-Uvea, is Polynesian (see

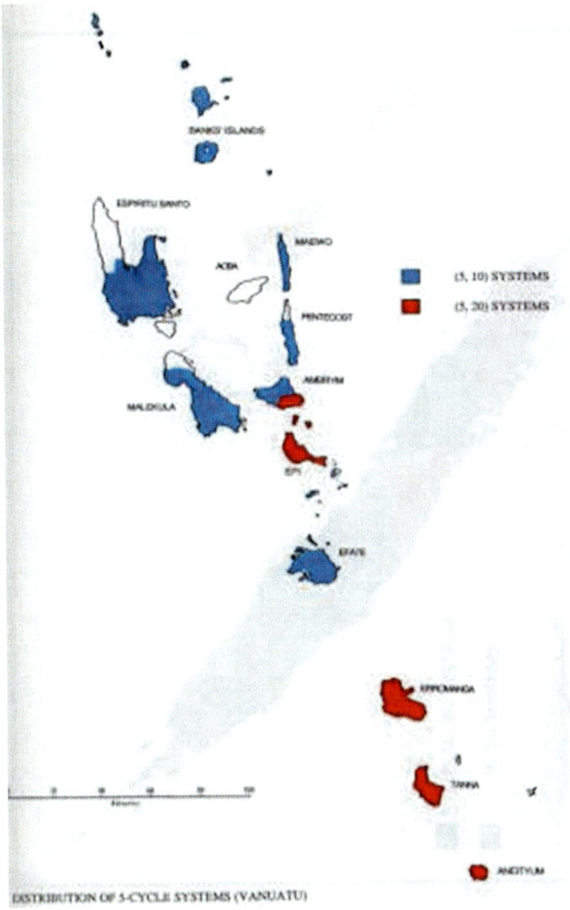

Figure 5.3. Distribution of 5-cycle systems – Vanuatu.
Source. Lean's (1992) thesis.

Figure 5.4 for these systems). Over the entire sample of 453 AN languages in this study we have, therefore, 63 which possess (5, 20) systems, or about 14%.

Among the 79 NAN systems and the 63 AN systems which possess this type of 5-cycle system there is, overall, a degree of uniformity in the etymological derivations of the words for 5 and 20. There are 51 NAN systems in which 5 contains a "hand" or "arm" morpheme; a further four systems contain a "thumb" morpheme and one has "palm". There are 26 NAN systems in which 20 contains a "man" morpheme; a further three have "hands two feet two". Among the AN systems at least 33 have a "hand" ("arm") morpheme in 5; there is a further seven for which 5 does not contain a "hand" morpheme but 10 does. There are 35 systems in which 20 contains a "man" morpheme and a further four which have "hands two feet two" or something similar.

The construction of the second pentad among all systems is restricted to a relatively small number of types and indeed the majority can be classified into just three as shown in Table 5.7.

Thus it can be seen most Non-Austronesian languages have a number other than 5 plus a number but fewer Austronesian languages have this construction. However, a surprising number have 5 plus a conjunction plus another with slightly fewer having 5 plus a number.

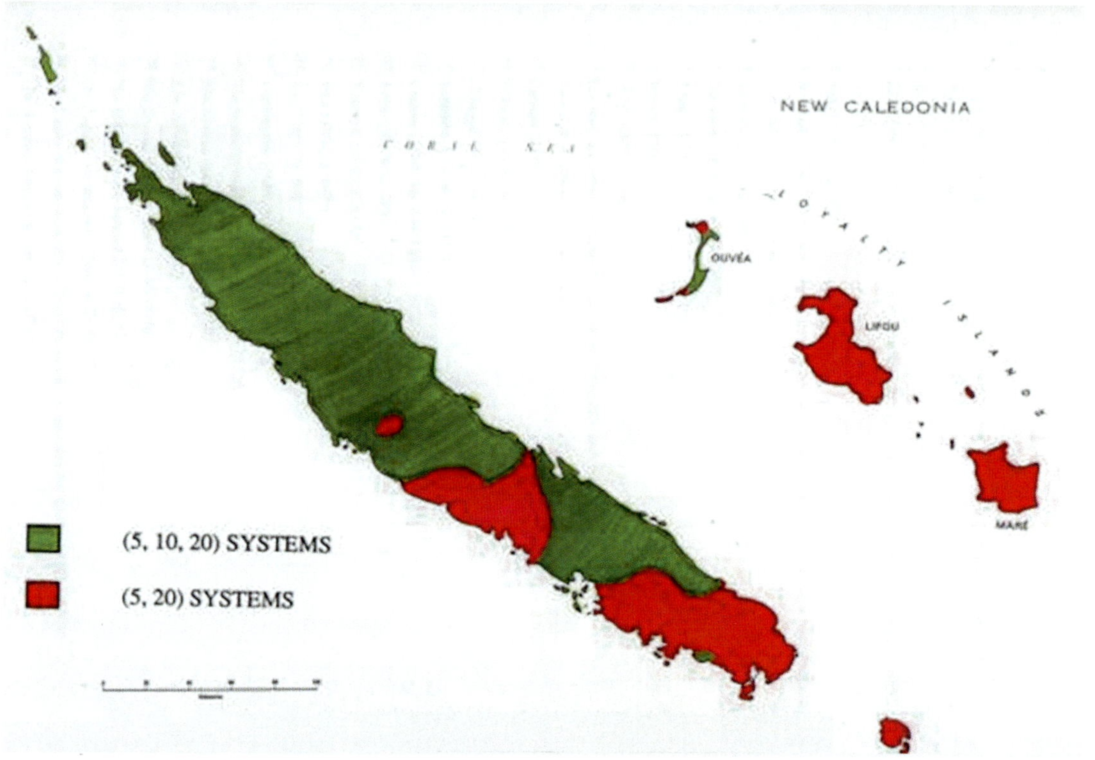

Figure 5.4. Distribution of 5-cycle systems – New Caledonia.
Source. Lean's (1992) thesis.

Table 5.7
Showing the Distribution of Second Pentad Constructions of (5, 20) Systems Among the NAN and AN Languages

Second Pentad Construction	$5+n$	$5+c+n$	$X+n$
Non-Austronesian (N=79)	14	17	26
Austronesian (N=63)	18	21	10

Note. n represents number morphemes for 1, 2, 3, 4 while c represents a conjunction (e.g. "and" or other morphemes). X refers to a frame word other than 5.

Systems With a (5, 10) or (5, 10, 100) Cyclic Patterns

The distinguishing features of the (5, 10) cycle system are that there are four monomorphemic numerals 1 to 4; 5 may be identical to, or cognate with, the word for "hand" or "arm"; the second pentad is formed additively although the word for 5 may or may not appear explicitly in the compounds for 6 to 9; there is a distinct numeral 10 and thereafter the system has a secondary 10-cycle and possibly a tertiary 100-cycle. Two examples, given in Table 5.8 below, one NAN and one AN, illustrate the nature of the (5, 10) system: these are Arandai, located in West Papua, and Kaliai, located in the West New Britain Province (PNG). In both systems 5 appears explicitly in the construction of the numerals of the second pentad. For Arandai, the construction takes the form $5+n$ where n takes the values 1 to 4 and there is no conjunction. Kaliai, however, has the construction $5+c+n$ for the second pentad where a conjunction is explicitly used. Both systems possess a distinct numeral 10 and the higher decades have an implied multiplicative construction, i.e. $20=2\times 10$ or 10×2 but not $10+10$.

Table 5.8
Examples of Numeral Systems from NAN and AN Languages With (5, 10) Cyclic Patterns

Numeral	*Arandai*	*Kaliai*
1	*onati*	*ethe*
2	*ogi* or *ouge*	*rua*
3	*aroi (ga)*	*tolu*
4	*Idati*	*pange*
5	*radi*	*lima*
6	*radi-onati* or *gendi*	*lima ga ethe*
7	*radi-ogi*	*lima ga rua*
8	*radi-aroi*	*lima ga tolu*
9	*radi-tob*	*lima ga pange*
10	*tobuti*	*sangaulu*
20	*ogi-tobuti*	*sangaulu rua*
30	*aroi-tobuti*	*sangaulu tolu*

Note. Arandai (NAN) data come from Galis (1960, p. 138) and the Kaliai (AN) data from CSQs (up to 10) are similar to Haywood and Haywood's (1980, p. 51).

Two further examples, both AN, illustrate the (5, 10) system and, in particular, the common $x + n$ construction for the numerals 6 to 9, where x is not identical to 5. The languages are Tinatatuna (Tolai or Kuanua), spoken in the East New Britain Province (PNG), and Mota, located in the Banks Islands in northern Vanuatu. These are given in Table 5.9.

Table 5.9
Two Examples of (5, 10) Systems With "x + n" Second Pentad Constructions

Numeral	Tinatatuna	Mota
1	*tikai*	*tuwale*
2	*aurua*	*ni rua*
3	*autul*	*ni tol*
4	*aivat*	*ni vat*
5	*ailima*	*tavelima*
6	*laptikai*	*lavea tea*
7	*lavurua*	*lavea rua*
8	*lavutul*	*lavea tol*
9	*lavuvat*	*lavea vat*
10	*avinun*	*sangavul*
20	*aura vinun*	*sangavul rua*
100	*a mar*	*melnol*
1000	*arip*	*tar*

Note. Tinatatuna (Tolai or Kuanua) data come from Glen Lean's field notes. He spoke the language. There are several dialects but generally similar to other data including Codrington's (1885, p. 225) early collection (Kabanga dialect recorded by Schnee (1901, p. 247-248) has several differences). See Chapter 11 for more details. The Mota data are from Codrington (1885, p. 235, 302-3).

In both Tinatatuna and Mota, the numerals are compounded with a second pentad identifier, respectively *lap-/lav-* and *lavea*, rather than the numeral 5 which does not appear explicitly. Both systems have distinct numerals 10, 100, and 1 000 and indeed the cyclic patterns of each of these systems is (5, 10, 100, 1 000). The Tinatatuna (Tolai or Kuanua) numeral 5, *ailima* contains a "hand" morpheme *lima*, however Mota has *tave lima* for 5 which does not contain a "hand" morpheme *pane-*. It is a distinctive feature of this type of system that 10 is normally a monomorphemic numeral unrelated to the words for "hand" or "arm". Further details on Tinatatuna are given in Chapter 11 and for Mota in Appendix B.

The (5, 10) system is relatively uncommon among the NAN languages. The distribution between the various phylic groups is as shown in Table 5.10. We have, thus, a total of 21 such systems or less than 5% of our sample. Generally, the (5, 10) systems of the NAN languages of PNG are located in the coastal and island regions; five are located on Bougainville Island in the North Solomons Province and two are located in New Britain. In West Papua, several are located in the Vogelkop region (see Figure 5.1).

Table 5.10
Distribution of (5, 10) Systems Among the NAN Phylic Groups

Phylum	Number
small phyla	0
West Papuan	2
East Papuan	12
Torricelli	0
Sepik-Ramu	3
Trans New Guinea	4
Total	21

Among the AN languages the (5, 10) system is reasonably common. In PNG at least 45 language groups have it and it is distributed between the various clusters as shown in Table 5.11. These are located largely in the New Britain, New Ireland and Milne Bay Provinces (see Figure 5.2). There is no evidence of this type of system occurring in the AN (either Oceanic or non-Oceanic) languages of West Papua. Nor is there evidence that this type of system occurs in New Caledonia. In Vanuatu, however, 68 of the 105 AN languages spoken there possess (5, 10) systems and these are located mainly in the Banks' Islands, Maewo, Pentecost, Efate, Malekula, and in most of Espiritu Santo (see Figure 5.3).

Table 5.11
Distribution of (5, 10) Systems Among the PNG Oceanic AN Clusters

Type	Number
North New Guinea Cluster	20
Papuan Tip Cluster	7
Meso-Melanesian Cluster	17
Admiralties Cluster	1
Total	45

As was the case with the (5, 20) systems, the (5, 10) systems are such that the construction of the second pentad numerals 6 to 9 fall largely into three main types as shown in Table 5.12.

The large number (74) of AN languages which have an $x+n$ construction for their second pentad is due mainly to the contribution of the Vanuatu systems, 64 of which have this type of construction.

Table 5.12
Distribution of Second Pentad Constructions for the (5, 10) Systems of NAN and AN Languages

Second Pentad Construction	$5+n$	$5+c+n$	$X+n$
Papuan (N=21)	2	5	6
Austronesian (N=113)	10	19	74

Among the (5, 10) systems we find that a proportion are such that the word for 5 contains a "hand" (or "arm") morpheme. Of the 21 NAN languages five have this property, although one, Magi, has borrowed the numeral 5 from an AN source (see Appendix B for more details). Of the 113 AN languages, 40 are such that 5 contains a "hand" morpheme, that is about 35%. In some cases it is clear that for some languages the words for "five" and "hand" were, at some time during the past, identical but that over time a phonological shift has occurred. Several Milne Bay Province (PNG) languages are examples of this: Budibud has respectively *-lima* for five and *nima-* for "hand"; Kilivila has, respectively, *-lima* and *yama-*; Muyuw has respectively *-nim* and *nama-*. In other languages there does not appear to be any relation between the two words.

Systems With a (5, 10, 20) Cyclic Pattern

The (5, 10, 20) system has features in common with both the (5, 20) and (5, 10) systems. There are four monomorphemic numerals 1 to 4 and the word for 5 normally contains a "hand" (or "arm") morpheme. The second pentad numerals 6 to 9 are additive compounds of the form $5+n$ where n represents the numerals 1 to 4 respectively; there are variants of this as well. There is a distinct numeral 10 which does not usually contain a "hand" or "arm" morpheme. There is a distinct word for 20, usually containing a "man" morpheme, or something similar. Three AN systems are given below to illustrate this type of system: these are Manam, located on Manam Island in the Madang Province (PNG), Mengen, located in the East New Britain Province (PNG), and Pwapwâ, located on Grande Terre (New Caledonia). The data are given in Table 5.13.

Table 5.13
Examples of (5, 10, 20) Numeral Systems for Three AN Languages

Numeral	Manam	Mengen	Pwapwâ
1	*teke*	*kena*	*cac*
2	*rua*	*lua*	*celuk*
3	*toli*	*mologi*	*poecen*
4	*watti*	*tugulu*	*poeovec*
5	*lima*	*lima*	*nim*
6	*lima teke*	*lima va kena*	*ni cac*
7	*lima rua*	*lima va lua*	*ni celuk*
8	*lima toli*	*lima va tugulu*	*ni cen*
9	*lima wati*	*lima va tugulu*	*ni ovec*
10	*kulemoa*	*tangalelu*	*paidu*
20	*tamoata*	*giaukaena*	*cac i kac*

Note. The Manam data were from 7 CSQs but are similar to earlier data and SIL word lists as given by Lean in the appendices of his thesis. Mengen are for the Mengen No. 1 dialect from Panoff (1969, p. 359) and Pwapwâ from Leenhardt (1946, pp. 496-501).

Each of the systems above has a distinct numeral 10 and a distinct word for 20 meaning "man" or "being". The second pentad constructions differ however and represent the three main types of construction found: Manam has $5+n$ for the compounds 6 to 9 while Mengen has $5+c+n$ and Pwapwâ has $x+n$.

Among the NAN languages this type of system is relatively rare with only 13 examples in our sample. These are distributed between the phylic groups as shown in Table 5.14.

Table 5.14
Distribution of (5, 10, 20) Systems Among the NAN Phylic Groups

Phyla	Number
Small phyla	3
West Papuan	6
East Papuan	0
Torricelli	0
Sepik-Ramu	0
Trans New Guinea (2 in PNG, 2 in West Papua)	4

There is no evidence that this type of system occurs in the East Papuan, Torricelli or Sepik-Ramu Phyla. The majority of the systems, 11 all together, are located in West Papua in the Vogelkop region, while two examples are found in the Madang Province (PNG) (see Figure 5.1).

Among the AN languages we have a total of 46 which possess this type of system. Of these, 18 are located in PNG and are distributed among the clusters as shown in Table 5.15. These languages are largely confined to the New Britain, Morobe and Milne Bay Provinces (see Figure 5.2). There is no evidence that this type of system occurs in either the Admiralties or Meso-Melanesian Clusters. In West Papua, one Oceanic language has a (5, 10, 20) system while 8 non-Oceanic AN language groups do. In the remainder of Melanesia only in New Caledonia do we find this type of system with some 19 languages possessing it, the majority of these being located in the northern half of Grande Terre (see Figure 5.4).

Table 5.15
Distribution of (5, 10, 20) Systems Among the PNG Oceanic AN Clusters

Cluster	Number
North New Guinea Cluster	13
Papuan Tip Cluster	5
Admiralties Cluster	0
Meso-Melanesian Cluster	0

As indicated above, there is a limited number of variants in the way the second pentad of this type of system is constructed. As with all 5-cycle systems there is a small number in which the numerals from 6 to 9 are compounded using the construction $n+x$ or $n+5$ where n appears first in the compound and takes, respectively, the values 1 to 4 (thus $6=1+5$, $7=2+5$, and so on). However, generally speaking, the majority of the (5, 10, 20) systems have second pentad constructions which fall into just three main types: the $5+n$, $5+c+n$, and $x+n$, as previously defined. Among the 59 languages with this type of system, the distribution of second pentad constructions is as given in Table 5.16. Out of all 59 systems there are 30 in which the word for 5 contains a "hand" (or "arm") morpheme.

Table 5.16
Distribution of Second Pentad Constructions of (5, 10, 20) Systems for NAN and AN Languages

Language Type	5+*n*	5+*c*+*n*	X+*n*
Papuan (N=13)	4	2	6
Austronesian (N=46)	12	22	7

Note. Only languages that could be definitely determined from the data are given.

Summary

Combining the various types of 5-cycle system we have a total of 113 NAN languages which possess one of the variants, that is just over a quarter of the total sample of 430. Of the AN languages, there are 222 which possess a 5-cycle variant, almost half the total sample of 453. As shown in Table 5.17, the most common variant among the NAN group is the (5) or (5, 20) system, while among the AN group the most common variant is the (5, 10) system. It may be recalled from Chapter 2, in discussing earlier typologies, that it is thought that the Proto Oceanic numerals formed a 10-cycle system and yet we find that almost half of the languages in the AN sample no longer have a 10-cycle system but have a 5-cycle system instead. Almost 36% of the sample do, however, retain a distinct numeral 10 and a 10-cycle although this is a secondary rather than a primary cycle.

Table 5.17
Summary Data for the 5-Cycle Systems: Their Distribution Among Both NAN and AN Language Groups

Phylum or Cluster	(5)/(5, 20)	(5, 10)	(5, 10, 20)
Non-Austronesian			
Small Phyla	7	0	3
West Papuan	0	2	6
East Papuan	1	12	0
Torricelli	2	0	0
Sepik-Ramu	17	3	0
Trans New Guinea	52	4	4
Totals	79	21	13
Austronesian			
North New Guinea	19	20	13
Papuan Tip	11	7	5
Meso-Melanesian	1	17	0
Admiralties	1	1	0
Solomons	0	0	0
Vanuatu	15	68	0
New Caledonia	8	0	19
Polynesia	1	0	0
West Papua	7	0	9
Totals	63	113	46

While there are more (5) and (5, 20) among Non-Austronesian languages than the Austronesian languages in New Guinea and Oceania, these are still prevalent in Austronesian languages. Furthermore, among the Austronesian languages, there are more languages having a numeral for 10 and a cycle of 10, not necessarily as the main cycle for larger numbers but this is far from being universal for these kinds of languages.

References

Chinnery, E. (1928). Certain natives in south New Britain and Dampier Straits, Territory of New Guinea. *Anthropological Report, 3.*

Codrington, R. (1885). *The Melanesian languages*. Oxford, UK: Clarendon Press.

Conant, L. (1896). *The number concept*. New York, NY: Macmillan.

Cowan, H. (1950–1951). Notes on Sentani grammar. *Oceania, 21*, 227–228.

Cowan, H. (1951–1952). Notes on Sentani grammar. *Oceania, 22*, 315–316.

Dawson, J. (1881). *Australian Aborigines: The languages and customes of several tribes of Aborigines in the Western District of Victoria, Australia*. Melbourne, Australia: George Robertson.

Dempwolff, O. (1905). BeitrŠge zur Kenntnis der Sprachen von Deutsch-Neuguinea, 8, 182–254. *Mitteilungen des Seminars fŸr orientalische Sprachen zu Berlin, 8*, 182–254.

Dempwolff, O. (1909). BeitrŠge zur Kenntnis der Sprache von Bilibili. *Mitteilungen des Seminars fŸr orientalische Sprachen zu Berlin, 12*(1), 221–261.

Dixon, R., & Kroeber, A. (1907). Numerical systems of the languages of California. *American Anthropologist, 9*, 663–690.

Eells, W. (1913a). Number systems of the North American Indians. *American Mathematical Monthly, 20*(9), 263–272.

Eells, W. (1913b). Number systems of the North American Indians: Systems of numeration. *American Mathematical Monthly, 20*(10), 293–299.

Galis, K. (1960). Telsystemen in Nederlands-Nieuw-Guinea. *Nieuw Guinea Studien, 4*(2), 131–150.

Harris, J. (1982). *Facts and fallacies of Aboriginal number systems*. Work papers of SIL-AAB, Series B.

Haywood, G., & Haywood, I. (1980). The languages and communities of north-western New Britain. In R. Johnston (Ed.), *Language, communication and development in New Britain* (pp. 41–70). Ukurumpa, Papua New Guinea: Summer Institute of Linguistics.

Kluge, T. (1938). *Die Zahlbegriffe der Australier, Papua und Bantuneger nebst einer Einleitung ueber die Zahl, ein Beitrag zur Geistesgeschichte des Menschen*. Berlin-Steglitz, Germany: Selbstverlag.

Kluge, T. (1941). *Die Zahlenbegriffe der Sprachen Central- und Südostasiens, Indonesiens, Micronesiens, Melanesiens und Polynesiens mit Nachträgen zu den Bänden 2-4*. Berlin, Germany. Ein fünfter Beitrag zur Geistesgeschichte des Menschen nebst einer principiellen Untersuchung über die Tonsprachen. Original unpublished manuscript held at Yale University, microfilm held at Australian National University, Canberra, Australia.

Lean, G. (1992). *Counting systems of Papua New Guinea and Oceania* (Unpublished PhD Thesis). PNG University of Technology, Lae, Papua New Guinea.

Leenhardt, M. (1946). Langues et dialectes de l'Austro-Mélanésie. *Travaux et Mémoires de l'Institut d'Ethnographie, 9*.

Leverd, A. (1922). Polynesian language of Uvea, Loyalty Islands. *Journal of the Polynesian Society, 31*, 94–103.

Macdonnell, F. (1917). Vocabularies—Cape Nelson Station, North-Eastern Division. *Papua, Annual Report for 1914-15, Appendix 4,* (pp. 171–174).

Moolenburgh, P. (1906). Woordenlijst van het Sentanisch. *Bijdragen tot deTaal Land-en Volkenkunde (van Nederlandsch Indie), 59*, 658–661.

Nykl, A. (1926). The quinary-vigerimal system of counting in Europe, Asia and America. *Language, 2*, 165–173.

Panoff, M. (1969). The notion of time among the Maenge people of New Britain. *Ethnology, 8*, 153–166.

Pott, A. (1847). *Die quinare und vigerimale Zählmethode bei Völken aller Welttheile*. Weisbaden, Germany: Dr Martin Sändigohg.

Ray, S. (1907). *Reports of the Cambridge anthropological expedition to Torres Straits* (Vol. 3). Cambridge, UK: Cambridge University Press.

Ray, S. (1919a). The languages of northern Papua. *Journal of the Royal Anthropological Institute of Great Britain and Ireland, 49*, 317–341.

Ray, S. (1919b). The Polynesian languages of Melanesia. *Anthropos, 14*, 46–96.

Ray, S. (1938). The languages of the Eastern and South-Eastern Divisions of Papua. *Journal of the Royal Anthropological Institute of Great Britain and Ireland, 68*, 153–208.

Schmidl, M. (1915). Zahl und Zählen in Afrika. *Mitteilungen der Anthropologischen Gesellschaft in Wien, 45*, 165–209.

Schmidt, W. (1926). *Die Sprachfamilien und Sprachenkreise der Erde*. Heidelberg, Germany: Carl Winter's Universitätsbuchhandlung.

Schmidt, W. (1929). Numeral systems. In *Encyclopedia Britannica* (14th ed., pp. 614–615). London, UK/New York, NY: Sears Roebuck.

Smith, G. (1984). *Morobe counting systems: An investigation into the numerals of the Morobe Province, Papua New Guinea.* MPhil thesis, Papua New Guinea University of Technology, Lae, Papua New Guinea.

Schnee, H. (1901). Beitrag zur Kenntnis der Sprachen im Bismarck-Archipel, 4(3), 229–279. *Mitteilungen des Seminars für orientalische Sprachen zu Berlin, 4*(3), 229–279.

Seidenberg, A. (1960). The diffusion of counting practices. *University of California Publications in Mathematics, 3*.

Throop, C., & Throop, L. (1980). The languages and communities of the Kandrian inland region. In R. Johnston (Ed.), *Language, communication and development in New Britain* (pp. 225–240). Ukurumpa, PNG: Summer Institute of Linguistics.

Trumbull, J. (1894). On numerals in the American Indian languages, and the Indian mode of counting. *Transactions of the American Philological Association, 5*, 41–76.

Tylor, E. (1871). *Primitive culture: Researches into the development of mythology, philosophy, religion, language, art and custom*. London, UK: John Murray.

Wakefield, D., & Whitacre, S. (1980). *Word lists from the Arove survey*. Ukurumpa, PNG: Summer Institute of Linguistics Library.

Z'graggen, J. (1975). Comparative wordlists of the Admiralty Islands languages collected by W. E. Smythe. *Workpapers in Papua New Guinea Languages, 14*, 117–216.

Chapter 6
10-Cycle Systems

Kay Owens and Glen Lean

Abstract There are among the Austronesian (AN) and Non-Austronesian (NAN) languages of Papua New Guinea and Oceania, examples of 10-cycle systems that do not have a 5-cycle or 20-cycle system. Of these, there are a number that have (10, 20) or (10, 20, 60) cycle and given the ancient origins of these cultures, it is evident that the Babylonians were not the only people to have 60-cycle systems nor that they were diffused from Babylon but rather were unusual, unique local developments. In addition, some systems have a replacement for 10 in higher numbers and others regard the specific group of 10 as more important than the number 10 itself leading to a range of words for 10 as in Motu. Lean (1992) also indicated that pairs may dominate and there are some systems in which 10 is likely to refer to 10 pairs. In addition, a number of systems refer to numbers such as 8 as 4×2, $9 = 4 \times 2 + 1$ while others in the second pentad refer to the number required to reach the complete group of 10 such as 8 = 10-2. Furthermore, a 10-cycle system is by no means the norm for AN languages. Interestingly, classifiers for number groups are also found in a range of languages from those in Manus (AN), to NAN languages in Bougainville to Polynesian languages of Micronesia as well as in the Solomons (Chapter 8 has more details.). Comparative data indicate the probability of Proto Polynesian language having this kind of classifier rather than it occurring discretely.

Keywords Non-Austronesian 10-cycle systems • Oceanic Austronesian 10-cycle systems • Micronesian 10-cycle systems • Polynesian 10-cycle systems

The Existence and Nature of 10-Cycle Systems

Several types of 10-cycle systems have been documented in the literature on natural language numerals. The most common type is such that there are distinct words for the numbers 1 to 10; the second decade numbers 11 to 19 are additive compounds which usually take the form $10 + n$ (or $n + 10$) where n takes the values 1 to 9 respectively so that $11 = 10 + 1$, $12 = 10 + 2$, and so on; the decades from 20 to 90 are normally multiplicative compounds of the form $n \times 10$, where n takes the values 2 to 9 respectively so that $20 = 2 \times 10$, $30 = 3 \times 10$, and so on. There is often a distinct word for hundred and the system may therefore possess a secondary 100-cycle. Such "pure" 10-cycle systems are found in Asia, the Middle East, Africa, Europe, and North America. Various Chinese languages and Japanese possess this type of system as do the Semitic languages. Most of the languages belonging to the Indo-European

K. Owens (✉)
Charles Sturt University, Dubbo, NSW, Australia
e-mail: kowens@csu.edu.au

G. Lean

K. Owens et al., *History of Number*, History of Mathematics Education,
DOI 10.1007/978-3-319-45483-2_6

family possess 10-cycle system variants. Menninger (1969, pp. 92-99), for example, listed the 10-cycle numerals of 30 daughter languages of Proto Indo-European which itself has a 10-cycle system. Many of these languages have systems with irregularities of various kinds and which deviate to some extent from the standard defined above. The Germanic languages, for example, have suppletive forms of 11 and 12. The Romance languages, in forming the compounds 11 to 19, have the peculiarity in which, up to 15 or 16, the units are placed before the 10 morpheme, and thereafter the units are placed after it. More significant, however, is the appearance, in the counting systems of various languages, the remnants of a 20-cycle: French, for example, has $80=4\times20$ and $90=(4\times20)+10$; Breton has $40=2\times20$, $60=3\times20$, and $70=10+(3\times20)$. The non-Indo-European language, Basque, has a counting system with a fully formed (10, 20) cyclic pattern which Schmidt (1929, p. 614) said is also found "in the Munda languages of India and in the Tibeto-Chinese groups of the Himalayas, in Nicobarese, in the north (and south) Caucasian languages".

In the North American Indian languages, Eells (1913, pp. 293-295) reported that the majority of the counting systems in his survey were "decimal systems", that is having a 10-cycle, and, more rarely, "decimal-vigesimal", that is having a (10, 20) cyclic pattern. Hymes (1955), in her analysis of the Athapaskan Indian numeral systems, distinguished several variants of the 10-cycle system: (a) the "decimal", with distinct terms for 1 to 10; (b) the "decimal-pairing", with distinct terms for 1 to 7, and 10, and with 8 and 9 having the "pairing" structure $8=2\times4$ and $9=(2\times4)+1$; (c) the "decimal-subtractive", in which there are distinct terms for 1 to 7, and 8 and 9 contain 2 and 1 respectively, plus a morpheme or group of morphemes signifying roughly "it lacks". Hymes (1955) noted that "No pure subtractive system (6 containing 4, 7 containing 3, and 8 and 9 containing 2 and 1) has been found among the Athapaskan languages, but it exists elsewhere" (p. 27). Similarly, Dixon and Kroeber (1907, p. 666), in their discussion of Californian (Native American) numeral systems, indicated the existence of several variant 10-cycle systems: "A few Californian languages show a decimal system throughout, even to being based on hundreds from one hundred up … Sometimes a decimal system changes above twenty to a vigesimal one, for which an analogy is not far distant in French" (Dixon & Kroeber, 1907, p. 666). Dixon and Kroeber also noted the use of the "duplicative" method of forming compound numerals (cf. Hymes "pairing") in which 6 has an implied 2×3 structure and 8 is 2×4.

One notable system, which has several of the features mentioned above, is that of the Ainu people who at one time were located in Hokkaido, Sakhalin Island, and the Kurile Islands. The data, which derive from Laufer (1917, pp. 192-195) and from Peng and Brainerd (1970, pp. 391-395), indicate that the Ainu numerals are such that there are distinct terms for 1 to 5, and 10, and that the numerals from 6 to 9 are formed by subtraction from 10 so that, for example, 6 has an implied construction 10-4, $7=10\text{-}3$, $8=10\text{-}2$, and $9=10\text{-}1$. Laufer (1917) noted that "the numbers from eleven to nineteen are formed on the scheme $1+10$, $2+10$, … The unit of all higher counting is represented by the figure 20" (p. 194). Thus $40=2\times20$, $60=3\times20$, $80=4\times20$ and $100=5\times20$. The odd decades are formed subtractively so that, for example, 30 is "ten less than two twenties." The system thus possesses elements of a 10-cycle system and of a 20-cycle system together with with the subtractive formation, or what Menninger (1969, p. 74) calls "back-counting", for the numerals 6 to 9, and the odd decades. The use of the subtractive construction for certain compound numerals is not uncommon. Armstrong (1962) said of the Nigerian Yoruba that

> one of the most characteristic features of Yoruba numerals is the use of subtractive numerals in the higher decades … The compound word *eedogun* says literally "five reduces twenty". Mathematically, it means 20-5. Similarly, 16 is *erin din logun*, which means literally, "four of a reduction from twenty", or 20-4; 17, 18, and 19 are formed similarly. This pattern is preserved almost up to two hundred. The early members of a decade are formed by addition; the last five members are formed by subtraction. (p. 6)

Conant (1896, p. 44) also gave examples of the subtractive construction used in counting systems in various parts of the world, notably that of the spoken forms in Classical Latin of the numerals 18 and 19 which are, respectively, *duodeveginti* and *undeveginti*, that is 2 from 20 and 1 from 20.

One final example of what is essentially a 10-cycle system and which is relevant to our subsequent discussion is the numeral system of the Babylonian Sumerians (Menninger, 1969, p. 74). For the first few decades this system possesses an irregular (10, 20) cyclic pattern so that $30=3\times 10$, $40=2\times 20$, and $50=(2\times 20)+10$. However, at 60, a new distinct word is introduced ("the great unit") and thereafter we have an additional 60-cycle so that $120=60\times 2$, $180=60\times 3$, and $600=60\times 10$. Thureau-Dangin (1939) noted, however, that "the whole system of Sumerian numeration which is, properly speaking, not a sexagesimal system, but a system built up on two alternative bases of 10 and 6" (p.104).

In summary then, the literature reveals a number of important variants of the 10-cycle numeral system. In some of these, there are unusual methods of compounding the numerals 6 to 9 (see Chapter 5 for systems that have 5 as primary cycles and their variants for the numerals 6 to 9). One method of compounding is the subtractive one in which we have at least 8 = 10-2 and 9 = 10-1. In some, systems 7 and even 6 are also expressed subtractively. It should be noted that in these compounds the "ten" morpheme may not appear explicitly and 8, for example, may have the translation "it lacks two", or something similar. A further way of compounding the numerals 6 to 9 is the "pairing" method in which we may have, for example, $6=2\times 3$ and $8=2\times 4$. In such systems there may be a distinct word for 7 (and also for 9), however it is relatively common to find $7=(2\times 3)+1$ and $9=(2\times 4)+1$. In the construction of the numerals of the second decade, that is from 11 to 19, we usually find additive compounds of the form $10+n$ or $n+10$. Such a pattern may not persist for the whole decade: we find some systems which may have 11 to 15 expressed as, say, $n+10$ and then 16 to 19 expressed in a different fashion, either $10+n$ or 20-n. The construction of the decades 20 to 90 is usually multiplicative, that is $2=2\times 10$ and $50=5\times 10$, and so on. We do find, however, systems possessing secondary (and tertiary) cycles greater than 10 and less than 100, for example those systems with (10, 20) and (10, 20, 60) cyclic patterns. One other irregularity not uncommon among 10-cycle systems is what Hurford (1987, pp. 56-57) calls "base-suppletion" in which we have the numeral 10 differing from the 10 morpheme used in compounds. As we shall see below, these 10-cycle variants are also present in the numeral systems of New Guinea and Oceania.

10-Cycle Systems in the Papuan (NAN) Languages

The occurrence of any of the 10-cycle variants among the NAN languages is very rare and indeed there are only 16 instances in our sample of 430, that is just under 4% of the total. Three types of 10-cycle system are distributed among the various phylic groups as given in Table 6.1.

Table 6.1
Distribution of 10-cycle Systems Among NAN Phylic Groups

Phylic Groups	(10, 100)	(10, 20)	(10, 20, 60)
Small Phyla	-	-	-
West Papuan	1	2	-
East Papuan	8	-	-
Torricelli	-	-	-
Sepik-Ramu	1	-	-
Trans New Guinea	2	1	1
Totals	12	3	1

The three West Papuan languages with 10-cycle systems are Karon-Pantai, Moi, and Madik. Their numerals are given in Table 6.2. Each of these is located on the Vogelkop (Bird's Head) in West Papua and it seems likely that their possession of 10-cycle systems has occurred as a result of the influence of neighbouring AN languages. Indeed Moi has clearly borrowed several AN numerals while Karon-Pantai and Madik both have several numerals (4, 7, and 9) which appear to have been influenced by an AN source. The inference that may be drawn is that the three NAN language groups have augmented their numeral lexis by borrowing from AN language sources and it may be, that by doing this, each language now has a 10-cycle system it did not originally possess.

Table 6.2
Numerals of Three NAN Languages of the West Papuan Phylum Which Have 10-Cycle Systems

	Karon-Pantai	Moi	Madik
1	*dik*	*mele, mere*	*dik*
2	*we, uwe*	*ali*	*uwi, ue*
3	*kri, gri*	*tolu(k)*	*dili*
4	*at*	*fat*	*at*
5	*mek, wek*	*mafu(t)*	*mek*
6	*mat*	*(t)anim*	*mohemat*
7	*fit, mik*	*fotu*	*momit*
8	*mengwo, mko*	*walu*	*mosembu*
9	*mesi, misi*	*(walu)si*	*mosi*
10	*mesju, musiu*	*sitit, utmere*	*motu*
20	*musiu-we*	*ut-ali, na-mere-igi*	*moro*
30	*musiu-kri*	*ut-toluk, fe-toluk*	*moru-mutu-tongal*
40	*musiu-at*	*ut-fat, najali-igi*	*moru-uwi*

The East Papuan Phylum languages show the greatest incidence of 10-cycle systems with a total of eight languages possessing them. Three of these, Anem, Buin and Yele, are located in islands east of the New Guinea mainland. Their numerals are given in Table 6.3.

Table 6.3
Numerals of Three NAN Languages of the East Papuan Phylum (PNG) Which Have 10-Cycle Systems

	Yele	Anem	Buin
1	*ngme*	*udeta*	*nonumoi*
2	*mio*	*ngiak*	*keitako*
3	*pyiile*	*bik*	*paigami*
4	*paadi*	*tanol*	*korigami*
5	*limi*	*esi*	*upugami*
6	*weni*	*kamli*	*tugigami*
7	*pyidu*	*kisa*	*paigami tuo*
8	*waali*	*damil*	*keitako tuo*
9	*tyu*	*damli*	*kampuro*
10	*y:a*	*le*	*kiburo*
20	*myo y:a*	*le ngiak*	*kikoko*

Note. Anem is in the West New Britain Province (PNG) and data are from SIL Word List in 1967 and Haywood and Haywood (1980, p. 54). Buin is in Bougainville Province (PNG) and the data are from 25 CSQs from 11 different villages; these data agree with early 1907 and 1910 records except for the English letters used for sounds (see Lean's appendices). Lean (1992) also provides considerable data on tens and hundreds. Yele is spoken in the Milne Bay Province (PNG) and the data from Henderson (1975) are supported by CSQs.

The Yele numerals clearly show the effect of AN influence; Compare, for example, the numerals 5 to 9 with the corresponding Proto Oceanic numerals shown in Table 6.7. This is not readily apparent for either the Buin or Anem numerals, nor indeed is there any obvious lexical influence to be detected in any of the remaining five numeral systems East Papuan Phylum, all of which belong to languages spoken in the Solomon Islands.[1] Three of these are given in Table 6.3. Each of the Solomon Islands NAN languages, together with Buin and Anem possess AN neighbours and it seems likely that the AN influence has been such as to effect a change in the cyclic patterns of the NAN systems without direct lexical borrowing occurring. There is little uniformity between the various systems shown in Tables 6.3 and 6.4. While most of these have distinct numerals for 1 to 10 it can be seen that both Buin and Nanggu have subtractive constructions for at least the numerals 7 and 8 (which contain, respectively, the numerals 3 and 2). Furthermore, the complete group at 9 was indicated for Uisai, a Buin Proper (Kaleva, personal communication, 2003). Lean acquired data for 23 languages of the East Papuan Phylum. The majority (13) of these have 5-cycle numeral systems; the eight discussed here have 10-cycle systems, and two possess systems which have 2-cycle variants. It is, of course, impossible to know what the original numeral system of the East Papuan Phylum languages may have been. What seems likely, however, is that due to their removal from the influence of other NAN languages and to their relative isolation in areas which are predominantly AN, they have had their original systems changed in some way; for the nine languages discussed here the change has been such that they now possess 10-cycle systems (see Table 6.3 and 6.4).

Table 6.4
Numerals of Three NAN Languages of the East Papuan Phylum (Solomon Islands) Which Have 10-Cycle Systems

	Mbaniata	Savosavo	Nanggu
1	*tufi*	*ela*	*töti*
2	*eri*	*endo*	*tüli*
3	*hie*	*igiva*	*tütü*
4	*avo*	*agava*	*tupwa*
5	*sondu*	*ara*	*mööpwm*
6	*tumbi*	*pogoa*	*temuu*
7	*ohio*	*pogora*	*tutüü*
8	*mbihio*	*kui*	*tumulii*
9	*mbovoho*	*kuava*	*tumatee*
10	*to*	*atale*	*napnu*
20	*eri to*	-	*napnu li*

Three other NAN languages appear to possess 10-cycle systems. These are Wapi, of the Sepik-Ramu Phylum, several dialects of Enga, and Lembena, the latter two belonging to the Trans New Guinea Phylum (see Table 6.5 for the numerals of these languages). Each of these languages is situated in the Enga Province (PNG). Both the Wapi and Lembena numeral sytems appear to have been strongly influenced by the dominant Enga language, the Wapi numerals, indeed, having been displaced by the Enga numerals. The original Enga counting system, to be discussed subsequently under 4-cycle systems, is a complex one with a (4, 60) cyclic pattern. This system has been modified by the process of truncation of the original numeral sequence at 10: from 1 to 9 the numerals are identical to the (4, 60) system; a new term *akalita* has been introduced for 10: this means "man" and thus 10 has the meaning "the fingers of one man". It is likely that the change in the Enga system affected both the Wapi and Lembena systems and that the adoption of a 10-cycle system in each of these languages is a relatively recent innovation.

[1] The five NAN languages spoken in the Solomon Islands and which have 10-cycle systems are: Nanggu, Mbilua, Mbaniata, Lavukaleve, Savosavo. Kazukuru , a further NAN language is now extinct – in Lean's (1992) appendices.

Table 6.5
Numerals of Three NAN Systems: Enga (Traditional), the Layapo Dialect of Enga, and Wapi

Numerals	*Enga* (Traditional)	*Enga (Layapo d.)*	*Wapi*	Lembena
1	*me(n)dai*	*mendai*	*mendai*	*wamena*
2	*lapo*	*lapo*	*lapo*	*lamana*
3	*tepo*	*tepo*	*tepo*	*kipakita*
4	*kitome(n)de*	*kitumende*	*kitumende*	*kitumende*
5	*yungi*	*kondape*	*kingi yangi*	*kikomete*
6	*tokage*	*yangi mange*	-	-
7	*kalage*	*sakaita*	-	-
8	*tukulapo*	*tukulapo*	-	-
9	*tukuteponya me(n)dai*	*mange mendai wakitao*	-	-
10	*tukuteponya lapo*	*akalita*	*akalita*	*kalisa*
11	*tukuteponya tepo*	*akalita kisa mendai*	-	-
12	*tukuteponya gato*	*akalita kisa lapo*	-	-

Note. Enga (Traditional) comes from Lean's own field notes from Mai dialect informants. Layapo dialect data are from 9 CSQs from the Wapenamanda area. Wapi is from SIL word list (1979) from Yangis village.

One final NAN 10-cycle system is considered here. This is the unusual and notable Ekagi (or Kapauku) system as shown in Table 6.6.

Table 6.6
Numerals of the Ekagi Counting System

Numeral	Ekagi	Numeral	Ekagi
1	*ena*	11	*enama (gati)*
2	*wia, wisa*	12	*wiama (gati)*
3	*wido*	20	*gatima (gati)* **or** *mepina*
4	*wi, wie*	30	*jokagati* **or** *amonato*
5	*idebi*	40	*mepija, mepiia*
6	*benomi*	50	*gati beu*
7	*pituo*	60	*bado* **or** *muto*
8	*waruwo*	90	*bado wado jokagati*
9	*ije, ise*	120	*muto wia*
10	*gati*		

Note. Data are from de Solla Price and Pospisil (1966).

The Ekagi, located in the Wissel Lakes region of West Papua, possess a counting system which has essentially a (10, 20, 60) cyclic pattern. We have, for example, $40 = 2 \times 20$, 50 has the subtractive construction "ten without" or 60 - 10. There is a distinct term for 60 and $120 = 60 \times 2$. There are two terms for 30: "child of ten" (*jokagati*) or "a half" (*amonato*). The cyclic structure of this system is clearly similar to that of the Babylonian Sumerians discussed earlier. Indeed this similarity prompted de Solla Price and Pospisil (1966) to suggest that perhaps the Ekagi system was "a survival of Babylonian arithmetic in New Guinea" (p. 30), a suggestion which subsequently caused the authors to be taken to task by Bowers and Lepi (1975) who deplored "the racist implications of Price and Pospisil's fanciful effort to derive Kapauku numeration from Babylonia" (p. 322). In response to a defence of their position by Pospisil and de Solla Price (1976), Bowers (1977) asserted that

> too much store need not be placed on the 60-base of Kapauku numeration; body-count totals, sometimes functioning as bases, have been recorded for 19, 23, 25, 27, 33, 37, 44, 47, 68, etc., in NAN languages. The sexagesimal system is really not all that unique (p. 112)

Bowers' (1977) point about the existence of unusual bases in New Guinea is a valid one. Such bases, as she points out, are usually associated with body-part tally systems of the sort described in Chapter 6 but there is no evidence that the Ekagi system is or was a body-part one; indeed the data appear to indicate elements of a digit-tally system with both 10- and 20-cycles. The sexagesimal system is certainly not unique to the Babylonian Sumerians: at least two instances exist in New Guinea: the Ekagi itself and the Mae dialect of Enga, the latter possessing a (4, 60) cyclic pattern. While 60-cycle systems are therefore not unique to Babylonia, nevertheless the point needs to be made that this type of system is very unusual. (There may be a south coast language too with a 6-cycle that favours 60 rather than $6 \times 6 \times 6$). Yet there is no need to posit the diffusionist view of a distant Babylonian source for the Ekagi system if it can be shown, as seems likely, that it was a local development. If it were diffused, then there would be most of the counting systems like this. Bowers quotes Father Drabbe's view that the Ekagi system "was essentially senary", that is (6-cycle), "with the decimal aspects arising from Indonesian influence" (Bowers, 1977, p. 113). Bowers noted that "there is no basis for positing an original 6-base in highland West Papua counting systems" (Bowers, 1977, p. 113). However 6-cycle systems do exist in languages situated on the southern coast of West Papua and it may be the case that the Ekagi were influenced by the 6-cycle counters, as indeed they were influenced by a 10-cycle AN source. As Thureau-Dangin (1939) noted, in discussing the Babylonian system, the 60-cycle was a result of the hybrid of two bases: 10 and 6, and it may well be the case that the Ekagi system is, similarly, as Drabbe (1926) suggested, a hybrid of a 6-cycle (NAN) system and a 10-cycle (AN) system. It would seem possible, then, that in this particular example, an independent inventionist view, rather than a diffusionist one, may account for the facts. Throughout this book, we have noted the diversity of inventions in different language groups so this is not an isolated case.

10 Cycle Systems in the AN languages of New Guinea

The reconstruction of the languages which are the ancestors to the AN languages spoken in New Guinea today has indicated that the numeral systems of both Proto Austronesian (PAN) and Proto Oceanic *(*POC*)* were 10-cycle systems. Ross (1988, pp. 459-464) for example, has reconstructed the POC numerals which are given in Table 6.7.

Table 6.7
The Numerals 1 to 10 of Proto Oceanic (POC) As Reconstructed By Ross

Numeral	POC Morphemes
1	**kai, *sa, *tai*
2	**rua*
3	**tolu*
4	**pat, *pati*
5	**lima*
6	**onom*
7	**pitu*
8	**walu*
9	**siwa*
10	**sangapulu*

Note. Ross (1988) argued for a Proto-Oceanic language.

As we have seen in the previous chapters, many AN languages do not possess 10-cycle systems: 32 AN languages spoken in New Guinea now have variants of 2-cycle systems and 222 AN languages spoken in New Guinea and Island Melanesia have variants of 5-cycle systems. Indeed, of the 453 AN languages in our sample, there are only 187 which have only 10-cycle systems, or 41% of the total. Some 92 of these are located in the New Guinea region, 16 in West Papua (all of which are non-Oceanic AN) and 76 in PNG and the north-west Solomon Islands (all of which are Oceanic AN). Table 6.8 shows the distribution of the 10-cycle systems among the various Oceanic AN clusters.

Table 6.8
Distribution of 10-Cycle Systems Among the Oceanic AN Clusters

Cluster	(10, 100)	(10, 20)
North New Guinea Cluster	0	0
Papuan Tip Cluster	9	0
Admiralties Cluster/ St Matthias	23	0
Meso-Melanesian Cluster	41	3
Totals	73	3

Of the 78 languages altogether in the North New Guinea Cluster, none has a 10-cycle system: 52 possess a 5-cycle variant, 2 possess 4-cycle systems, and 24 possess a 2-cycle variant. In the Papuan Tip Cluster, of the 41 languages for which data were obtained, 7 possess a 2-cycle variant, 23 (and possibly a further 2) possess a 5-cycle variant, and only 9 have a 10-cycle variant. Of the 24 languages spoken in the Manus Province (and which belong to the Admiralties Cluster), 2 possess a 5-cycle variant and the remaining 22 have 10-cycle systems. The Meso-Melanesian Cluster comprises 20 languages of the North-West Solomonic Group, all of which have 10-cycle systems, and a further 44 languages within PNG. Of these, 24 possess 10-cycle systems while 18 possess a 5-cycle variant, and one language, Tomoip, has a 2-cycle variant (see Appendix C). Thus, of the four clusters, the only ones which have retained something of the original POC 10-cycle nature for their numeral systems are the Admiralties and Meso-Melanesian Clusters.

The POC system, as shown in Table 6.7, possesses ten distinct terms for the numerals 1 to 10. This is not true for many of its daughter languages which often have compounds rather than distinct terms for the numerals in the second pentad. The main type of 10-cycle system, of which the POC system is one example, has ten distinct terms for 1 to 10, a further term for 100, and often a further term for 1 000; such a system has a cyclic pattern of (10, 100, 1 000). There are two major variants of this system. The first variant Lean termed the "Manus" type, largely because this is found almost exclusively in the Manus Province (PNG). This type has distinct terms for the numerals 1 to 6, and 10 (and normally 100 as well). The numerals 7, 8, and 9 are, however, compounds which contain, respectively, the numerals 3, 2, and 1, that is the construction of the compounds is subtractive. In two cases (Likum and the Levei dialect of Levei-Tulu), the subtractive construction extends also to the numeral 6. Some examples of the Manus type of 10-cycle system are shown in Table 6.9, together with Ross' reconstructed numerals of Proto Eastern Admiralties (PEAd).

Of the 22 languages of the Admiralties Cluster which possess 10-cycle systems, 21 are of the Manus type, the exception being Wuvulu-Aua, the westernmost language of the Cluster and which is discussed further below. The only other AN language to possess the Manus type of system, both in New Guinea and Oceania, is the Mioko dialect of Duke of York in the East New Britain Province (PNG).

In addition, Buin (Uisai) in Bougainville is an NAN language that shows the system of subtraction or rather number to complete the group of 9 regarded as the complete group (Kaleva, personal communication, 2003). The construction of 7 and 8 are, respectively, *paigami tuo* and *kiitako tuo*, that is "three less than ten" and "two less than ten", so that a subtractive process operates for these numerals as is the case for the Manus-type counting systems. The numeral 9, *kampuro*, does not continue this pattern but, as Laycock (1975, p. 226) and Kaleva pointed out, is a separate word meaning something like

Table 6.9
Numeral Systems of the "Manus" Type Including Those of Proto Eastern Admiralties

Numeral	*Titan*	*Papitalai*	*Likum*	PEAd
1	*esi*	*ti*	*esi*	**si-*
2	*eluo*	*moruah*	*rueh*	**ru-*
3	*etalo*	*motalah*	*taloh*	**tolu-*
4	*ea(h)*	*mohahu*	*hahu*	**fa-*
5	*elima*	*molimah*	*limeh*	**lima-*
6	*ewono*	*mowonoh*	*chohahu*	**ono-*
7	*andratalo*	*moadotalah*	*chotaloh*	**(a)nto-tolu*
8	*andraluo*	*moadoruah*	*chorueh*	**(a)nto-ru*
9	*andrasi*	*moadoti*	*choesi*	**(a)nto-si*
10	*eakou*	*moasengul*	*senoh*	

Note. The Titan data came from Schnee (1901) and with similar spelling from other older references, 26 CSQs and Lean's field notes. It also has classifiers. Papitalai data come from 3 CSQs and differ only in having the *mo* and *h* on each word compared to older data. Likum data come from 1 CSQ but SIL word list (1975) from the village Likum collaborates 1 to 5, 10 except for "h". Proto Eastern Admiralties (PEAd) numerals are given by Ross (1988, p. 344). Further details are given in Lean's (1992) appendices.

"completed". Laycock continued: "Ten is then the first number in a number-set for counting tens, as hundred is the first in a set for counting hundreds" (p. 226). The word *kampuro* completes the decades number-set, in this case meaning 90, and also completes the hundreds number-set, in which case it means 900. This may be seen in Table 6.10. More commentary on Buin (Uisai) can be found in Appendix B.

Table 6.10
Decade and Hundred Sets as Numeral Classifiers for Uisai (Buin)

"Decades" Set		"Hundreds" Set	
10	*kipuro*	100	*pore*
20	*kikoko*	200	*kiporigo*
30	*paimaku*	300	*paiporegi*
40	*korimaku*	400	*koriporegi*
50	*upumaku*	500	*upuporegi*
60	*tugimaku*	600	*tugiporegi*
70	*paimaku tuo*	700	*paiporegi tuo*
80	*kikoko tuo*	800	*kiporigo tuo*
90	*kampuro (lopore)*	900	*kampuro*

Note. Data are from Laycock (1975). The language was also studied early by German researchers and later under the British.

The construction of the numerals 11 to 19 follow a consistent pattern. The numeral 11 is *noikei lugo* where *lugo* means "on top", i.e 11 is "one on top of", the "ten" being understood. The numeral 15 is *upugami lugo*, i.e "five on top of", and so on. A similar pattern is seen for the numerals 21 to 29 (and 31 to 39, etc). The numeral 21 is *kikoko girai noikei lugo*, i.e "twenty and one on top of"; the numeral 25 is *kikoko girai upugami lugo*, i.e "twenty and five on top of", and so on. There are distinct words for the numerals 1000 and 10 000, the former being *kukurei* and the latter *tarina*; *kukurei*, has the meaning "domestic fowl" as it does as well in Siwai and Nasioi but borrowed for another power of 10 in a Solomon island cluster. The classifier for numerals is also discussed below for Micronesian numerals.

One other important feature of the languages of the Admiralties Cluster, which is discussed at greater length in Chapter 8, is the use of numeral classification. This is found in all the languages of Manus Island and the islands peripheral to it. Essentially, numeral classification consists in this: the universe of countable objects is divided into (largely) discrete classes; in counting the objects of a particular class or in stating the cardinal number of objects of a particular class (in, for example, a noun phrase), use is made of the construction "numeral root + classifier" (rather than just a numeral). Thus one does not use the phrase, for example, "three spears" but rather "three-(classifier for long thin objects) spears". Some languages have in excess of 40 classes. There are usually numeral roots for 1 to 9; in certain languages, however, numeral classification is only used with the numeral roots 1 to 4, after which a general set of numerals is used. Data on the Gele' dialect of the language Ere-Lele-Gele'-Kuruti were collected by a medical officer and amateur linguist Dr W. E. Smythe (available from SIL) in the 1940s who reported 43 classes and gave the "numeral root + classifier" constructions for each class for 1 to 9; this unpublished report constitutes one of the few comprehensive records of a rapidly disappearing linguistic phenomenon in the Manus languages. It is corroborated by 23 CSQs, contained in Z'graggen's (1975) report, and Fisher's (~2010) paper on numbering, for example, different parts of the banana plant. Numeral classification is not a phenomenon found exclusively in the languages of the Admiralties Cluster: it is found in other AN language groups, both in PNG and Oceania, notably those of the Massim Group in the Papuan Tip Cluster (Kilivila in the Trobriands, Muyuw, Budibud, Misima, Nimowa and Sud-Est – all detailed in Lean's (1992) appendices) as well as in the languages of Micronesia (see later in this chapter). It is also found in various NAN languages. In particular, we note Nasioi, Siwai, Buin (Uisai) and Nagovisi which are four NAN languages spoken in Bougainville (PNG) and given in Lean's (1992) appendices. See Appendix B.

A second major variant, which Lean termed the "Motu" type, is largely found in the AN languages of the coastal region to the east and west of Port Moresby in the Central Province (and National Capital District) of PNG. These languages (Nara, Doura, Roro, Gabadi, Motu, Sinagoro, and Keapara) are members of the Central Papuan Branch of the Papuan Tip Cluster (in Ross's (1988) terms). Taking the example of the numerals of Motu itself (Table 6.11), we have distinct terms for the numerals 1 to 5, 7, and 10. The numerals 6, 8, and 9 are each compounds formed by what has been termed above as the "pairing" or "duplicative" method, that is $6=2\times 3$, $8=2\times 4$, and $9=(2\times 4)+1$. The Motu-type may have, as do the Nara, Gabadi, Roro, Keapara (generally), and Sinagoro (Balawaia dialect) systems, a compound for the numeral 7 such that $7=(2\times 3)+1$. There is also some evidence that one of the Keapara dialects has subtractive compounds for both 7 and 9 such that 7=8-1 and 9=10-1 while still retaining $6=2\times 3$ and $8=2\times 4$. Table 6.11 shows these Motu-type variants.

Table 6.11
Variants of the "Motu" Type of 10-Cycle System

Numeral	Motu	Nara	Keapara
1	*ta*	*kaonamo*	*ka*
2	*rua*	*lua*	*lualua*
3	*toi*	*koi*	*koikoi*
4	*hani*	*vani*	*vaivai*
5	*ima*	*ima*	*imaima*
6	*tauratoi*	*kalakoi*	*kaulakoi(koi)*
7	*hitu*	*kalakoi ka*	*mapere aulavaivai*
8	*taurahani*	*kalavani*	*kaulavaivai*
9	*taurahani ta*	*kalavani ka*	*mapere ka gahalana*
10	*gwauta*	*ouka*	*gahalana*

Note. Motu data were provided by 75 CSQs from Hanuabada, Porebada, etc. and agree with Lawes' (1885) data and Ray's (1895) data except they give *tamona* for 1 and spell the word for 10 different. However, the accompanying text about 10 is important to keep in mind. Nara was given in a British New Guinea report (1892) and other early reports. Keapara data are from Guise (1892), Ray (1895, pp. 35-38) and Kluge (1941, p. 218). Later data vary from this set but still show the prominence of doubles.

While Lean did not include it as a defining feature of the Motu-type, several systems of that type possess, using Hurford's (1987) term, "base-suppletion" in which the word for the numeral 10 is not the same as the "ten" morpheme used in multiplicative compounds. Two examples, those of Nara and Motu, are given in Table 6.12.

Table 6.12
Examples of Base-Suppletion in the "Motu" Type Numeral Systems

Numeral	Motu	Nara
10	*gwauta*	*ouka*
11	*gwauta ta*	*ouka ka*
20	*ruahui*	*lua na vui*
30	*toi ahui*	*koi na vui*
40	*hani ahui*	*vani na vui*

The Motu type of numeral system is not found outside the Central Province (PNG). Two systems which have elements of the Motu type, however, are Wuvulu-Aua, of the Admiralties Cluster, and Marshallese, an Isolate Oceanic AN language found in Micronesia. These are given in Table 6.13.

Table 6.13
Two Numeral Systems With Elements of the "Motu" Type

	Wuvulu-Aua	Marshallese
1	*ai, aiai*	*juon*
2	*guai*	*ruo*
3	*oduai*	*jilu*
4	*guineroa*	*emen*
5	*aipan*	*lalim*
6	*oderoa* (3 × 2)	*jiljino* (3 + 3)
7	*oderomiai* (3 × 2 + 1)	*jiljilimjuon* (3 + 3 + 1)
8	*vaineroa* (4 × 2)	*rualitök* ("give two", i.e. to make 10)
9	*vaineromiai* (4 × 2 + 1)	*ruatimjuon* (8 + 1?)
10	*vapa ani*	*jongoul*

Note. The Wuvulu-Aua data for Wuwulo were recorded by Dempwolff and Kluge and it differs from the Awa data recorded by Z'Graggen after Smythe. The Marshallese data were given in Lean's Appendix D on Oceania.

The Manus and Motu types of 10-cycle variant discussed above occur in the Admiralties and Papuan Tip Clusters. The 44 10-cycle systems which are found in the languages of the Meso-Melanesian Cluster (with the exception of the Mioko dialect of Duke of York) are all normal 10-cycle types without compounds for the numerals of the second pentad.

These systems, most of which have numerals which are clearly reflexes of the POC numerals, are found mainly in the West New Britain, New Ireland, and Bougainville Provinces (PNG), as well as the north-west part of the Solomon Islands. It is not uncommon, particularly for the Solomon Islands languages, for the numerals 1 to 9 to have a prefix (in some cases this may be a fossilised classifier). Three examples of systems belonging to languages of the Meso-Melanesian Cluster are given in Table 6.14. Nakanai is from West New Britain Province with five dialects and interestingly the CSQs from one dialect had a (5, 10) cycle (see also Johnston, 1980) but other dialects had abandoned this for the one given in Table 6.14. Kandas is on the west coast of southern New Ireland Province. Mbareke is from the New Georgia Group of the Solomon Islands.

None of the five Oceanic AN languages situated in West Papua possesses a 10-cycle system. Of the 33 non-Oceanic AN languages, however, 16 possess 10-cycle systems: 8 of these have (10, 100) cyclic patterns and 8 have systems with (10, 20) cyclic patterns and indeed these constitute about half

Table 6.14
Numerals of Standard 10-Cycle Systems of Three Languages of the Meso-Melanesian Cluster

Numeral	Nakanai	Kandas	Mbareke
1	*isasa*	*takai*	*meka*
2	*ilua*	*aru*	*karua*
3	*itolu*	*utul*	*hike*
4	*ivaa*	*uvat*	*kambuto*
5	*ilima*	*tilim*	*kalima*
6	*iwolo*	*vonom*	*kaonomo*
7	*ivitu*	*mavit*	*kanjuapa*
8	*iwalu*	*tival*	*kavesu*
9	*walasiu*	*lisu*	*kasia*
10	*savulu sa*	*singino*	*nangguru puta*

Note. Nakanai data were from 15 CSQs, IMP data and are similar to Johnston (1980) except he wrote *i-sasa*, etc. Kandas data were from 2 CSQs with some differences to earlier data in spelling and sounds from Stephan and Gräbner (1907) but both are 10 cycle with ten basic numerals.

of all the (10, 20) systems found in the 453 AN languages in our sample.[2] At least 6 occur in the Polynesian languages and a further 3 occur in the North-West Solomonic group of the Meso-Melanesian Cluster. The geographical distributions of both types of 10-cycle system for the NAN and AN languages in the New Guinea area are shown in Figure 6.1.

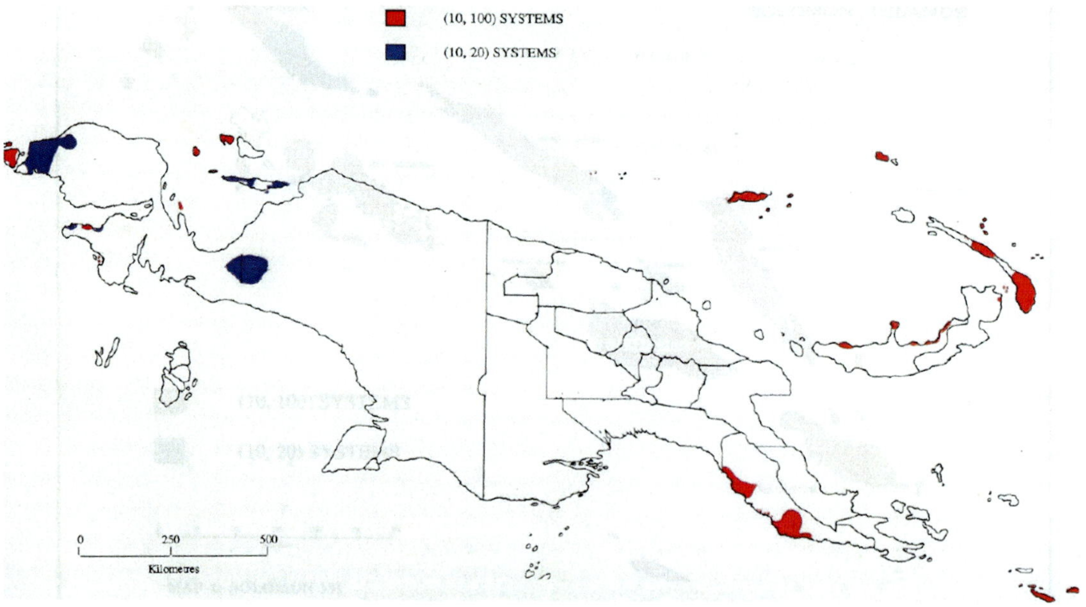

Figure 6.1. Distribution of 10 cycle systems in New Guinea among NAN and AN languages.
Source: Lean's (1992) thesis on *The Counting Systems of Papua New Guinea and Oceania.*

[2] The 8 non-Oceanic AN languages of West Papua which possess (10, 100) systems are: Biak, Maya, Laganyan, Maden, Ron, Sekar, Arguni and Uruangnirin. The 8 languages which possess (10, 20) systems are: Onin, Bedoanas, Woi, Pom, Marau, Munggui, Ambai and Wabo. The data for all of these may be found in Lean's Appendices.

10-Cycle Systems in Island Melanesia

The distribution of 10-cycle systems in the languages of Island Melanesia (excluding the Polynesian Outliers which are discussed subsequently) is given in Table 6.15. The number of AN languages in the Solomon Islands which possess 10-cycle systems is 51, however 23 of these belong to the North-West Solomonic group and have been included in our discussion above of the Melanesian Cluster. (The distribution of 10-cycle systems in the Solomons is shown in Figure 6.2).

Of the remaining 28 languages in the southern part of the Solomon Islands, all of which possess systems with (10, 100) cyclic patterns, 22 belong to the South-East Solomonic group and 6 belong to the Eastern Outer Islands. In Vanuatu, 19 of a total of 102 (non-Polynesian) Oceanic AN languages

Table 6.15
Distribution of 10-Cycle Systems in Island Melanesia

Country	(10, 100)
Solomon Islands	28 (51%)
Vanuatu	19
New Caledonia	0
Fiji/Rotuma	3
Total	50 (73%)

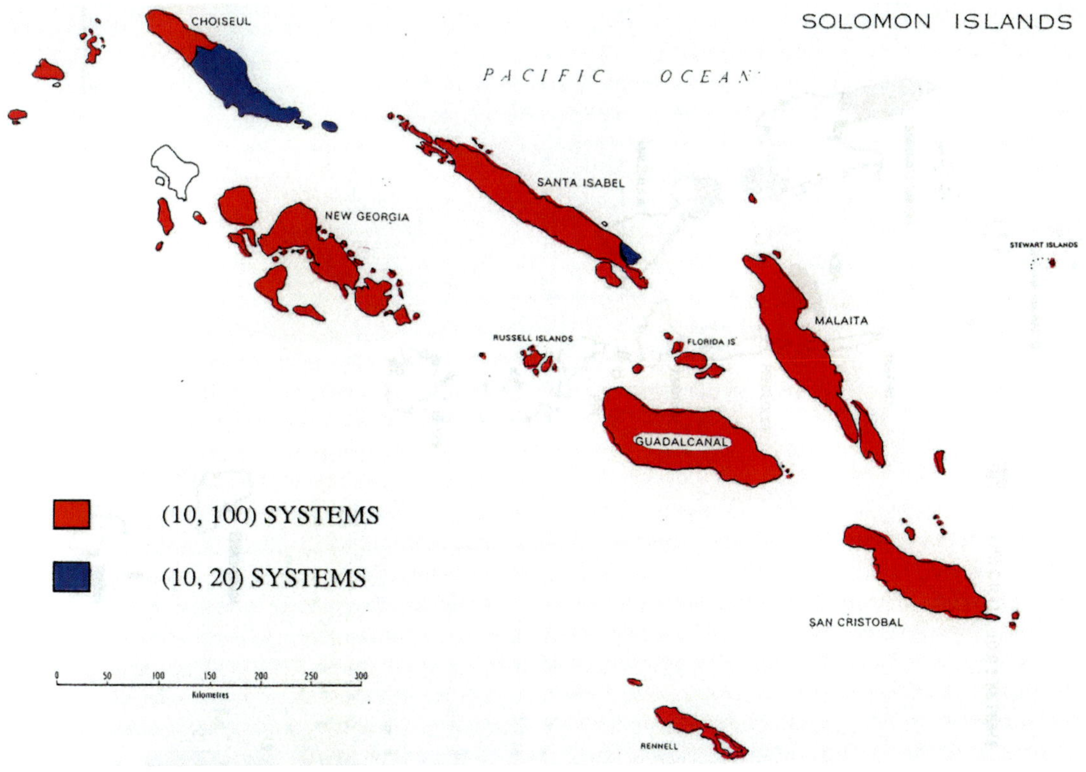

Figure 6.2. Distribution of (10, 100) and (10, 20) systems in the Solomon Islands.
Source: Lean's (1992) thesis on *Counting Systems of Papua New Guinea and Oceania.*

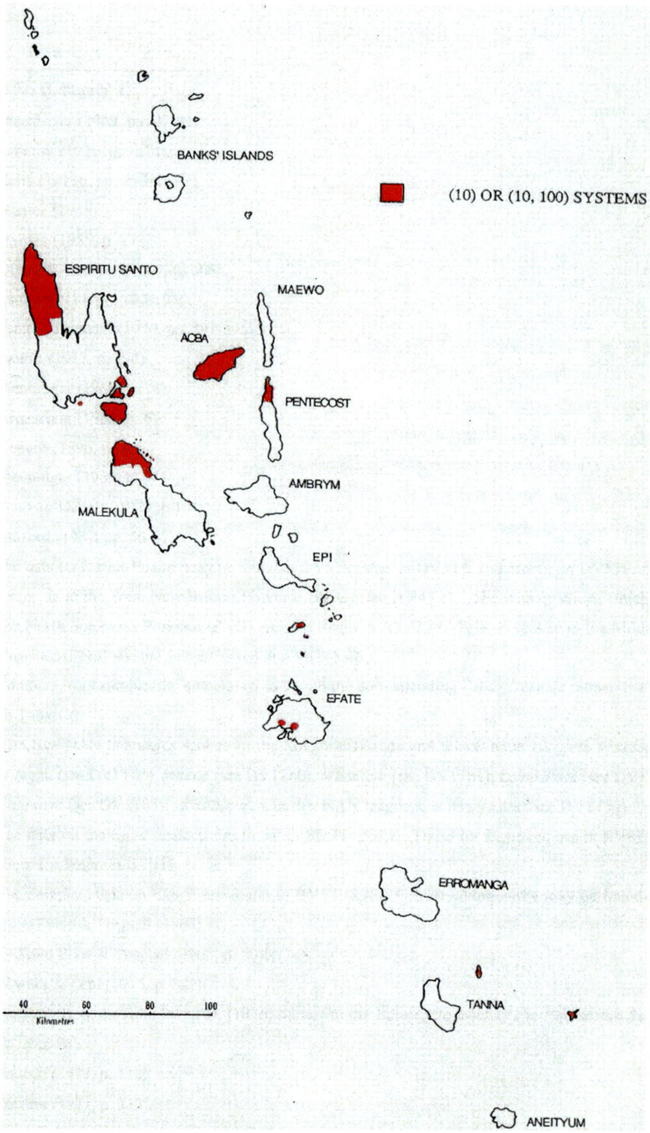

Figure 6.3. Distribution of 10-cycle systems in Vanuatu.
Source. Lean's (1992) thesis.

possess 10-cycle systems (see Figure 6.3), the remaining 83 possessing systems with 5-cycle variants. In New Caledonia, none of the 27 Oceanic AN languages possesses a 10-cycle system; each of these has instead a 5-cycle variant. In Fiji, the numeral systems of the Eastern and Western dialect chains of Fijian each has a (10, 100) cyclic pattern as does that of Rotuma. None of the 10-cycle systems of this region are of the Manus or Motu types; they conform rather to the standard 10-cycle type as shown in Table 6.16 where three examples are given: Nggela (South-East Solomonic), Valpei (North-Central Vanuatu), and the Mbau dialect of Fijian.

Table 6.16
The Numerals 1 to 10 for Three 10-Cycle Systems from Island Melanesia

Numeral	Nggela	Valpei	Fijian
1	*sakai*	*tewa*	*ndua*
2	*rua*	*rua*	*rua*
3	*tolu*	*tolu*	*tolu*
4	*vati*	*vati*	*va*
5	*lima*	*lima*	*lima*
6	*ono*	*ono*	*ono*
7	*vitu*	*pitu*	*vitu*
8	*alu*	*alu*	*walu*
9	*hiua*	*tchiwa*	*thive*
10	*hangavulu*	*sanawulu*	*tini*

Note. Nggelu in the Solomon Islands was first recorded by Codrington in 1883 and those data did not differ except for a couple of sounds from the data given here from Tryon and Hackman (1983, pp. 124,128, 132) who also gave words for 100 and 1000. Valpei data were recorded by Tryon (1976, pp. 406-416).

10-Cycle systems in Polynesia

Data were acquired in this study for 35 of a possible 36 Polynesian (PN) languages. These comprise: (a) the 14 PN Outliers which, with the exception of two located in Micronesia, are scattered throughout Island Melanesia, and (b) 22 languages located in "Triangle" Polynesia. Of the 35 PN languages, only one, Faga-Uvea, spoken in the Loyalty Islands of New Caledonia, does not have a 10-cycle numeral system. Faga-Uvea, one of the Outliers, possesses a 5-cycle variant. The data on the PN languages are such that it is not possible in every case to distinguish whether a numeral system has a (10, 100) or a (10, 20) cyclic pattern. These systems, however, are remarkably homogeneous in character and do not exhibit the diversity so apparent in the AN numeral systems of New Guinea and Island Melanesia. In Table 6.17, the reconstructed numerals of Proto Polynesian (PPN) are given, together with those of Takuu (an Outlier located in PNG), Samoan and New Zealand Maori (both located in Triangle Polynesia).

Table 6.17
Numerals 1 to 10 for Proto Polynesian and Three Polynesian Languages

Numeral	PPN	Takuu	Samoan	N. Z. Maori
1	**ha, *taha*	*tasi*	*tasi*	*tahi*
2	**rua*	*lua*	*lua*	*rua*
3	**tolu*	*tolu*	*tolu*	*toru*
4	**faa*	*fa "a*	*faa*	*whaa*
5	**lima*	*rima*	*lima*	*rima*
6	**ono*	*ono*	*ono*	*ono*
7	**fitu*	*fitu*	*fitu*	*whitu*
8	**walu*	*varu*	*valu*	*waru*
9	**hiwa*	*sivo*	*iva*	*iwa*
10	**hangafulu*	*sinahuru*	*sefulu*	*ngahuru*

Note. PPN (Proto Polynesian) data are from Pawley (1972, pp. 52-54); Takuu data from 6 questionnaires at Nukutoa village, Mortlock Island (Bougainville Prov, PNG), Samoan from earliest records to recent (Hale, 1846; Krupa, 1982; Macdonald, 1893).

Base-suppletion is not uncommon in the PN numeral systems. It is often the case that the numeral 10 is used to construct the compounds 11 to 19 but, for the decades 20 (or 30) to 90, a different "ten" morpheme is used. We have, for example, Takuu, Luangiua and Hawai'i use the constructions as given in Table 6.18.

Table 6.18
Examples of Base-Suppletion in Three Polynesian Languages

Numerals	Takuu	Luangiua	Hawai'i
10	*sinahuru*	*sengahulu*	*"umi*
11	*sinahuru ma tasi*	*sengahulu na kahi*	*"umi kuma kahi*
20	*mata rua*	*kipu lua*	*(iwakalua)*
30	*mata toru*	*kipu kolu*	*kana kolu*
40	*mata fa*	-	*kana ha*

Note. See comment on Takuu under Table 6.17; Luangiua data were from Woodford (1906) cited in Ray (1917), and Hawai'i from Elbert and Pukui (1979, pp. 158-159), cited in Lean (1992) Appendix on Oceania containing Polynesian numerals.

One possible explanation for this phenomenon is that the decades of the systems for the languages shown, as well as for some other PN languages, were constructed using numeral classification, i.e the irregular 10 morphemes are in fact "decade" or "group of ten" classifiers to which a numeral root is suffixed. For example, in Kapingamarangi, one of the PN Outliers located in Micronesia, the numeral 10 is *ehoru*, but there exists a "group of ten" classifier *mada-* to which the numeral roots for 2 to 9 are suffixed in order to form the decades 20 to 90. Thus, 20 is *mada-lua*, (30 is irregular: *mo-tolu*), 40 is *mada-ha*, 50 is *mada-lima*, and so on. Similarly, there exist "group of ten" classifiers in Nukuoru (*mada-*), Nuguria and Nukumanu (both *tipu-*), Takuu, Tikopian and Sikaiana (all *mata-*), Luangiua (*kipu-*), and Hawai'i (*kana*). While this method of constructing decades is not uncommon among the Outliers, it is relatively uncommon among the Triangle languages which do not generally exhibit base-suppletion and in which the numeral 10 does appear in the decades 20 to 90 which are multiplicative compounds. For example, in Tongan we have 10 *hongo-fulu*, 20 *uo-fulu*, 30 *tolungo-fulu*, and 40 *fango-fulu*, and so on, where *-fulu* is the 10 morpheme.

A number of (mainly) Triangle PN languages have numeral systems which possess a distinct term for 20 and which have a secondary 20-cycle. This term, *tekau* and its reflexes, often has the sense "ten pairs", the phenomenon of counting objects in twos and fours being common throughout most of Polynesia. In New Zealand Teo Māori, *tekau* is recorded in some vocabularies as 10 and in others as 20. In Tongan, *tekau* is used for 10 but only in counting pairs; in Tongareva, *tekau* is a numeral classifier used when counting coconuts. At least 8 and possibly 9 Triangle languages have *tekau* (*takau, ta'au*) for 20 (or 10 pairs) and at least 6 of these have systems with (10, 20) cyclic patterns. These are Minihiki-Rakahanga, Tahitian, New Zealand Maori, North Marquesan, Pa'umotu, and Mangareva.

10-Cycle systems in Micronesia

Each of 13 Micronesian (MC) languages for which data were obtained by Lean (1992) possesses a 10-cycle numeral system. Of these languages, 11 exhibit the phenomenon of numeral classification while two, Marshallese and Kosraean, have largely lost it. There are several varieties of classifiers (Song, 1997). In the enumeration of objects of a particular class each language makes use of the construction "numeral root+classifier", the same construction, in fact, as that found in the languages of the Admiralties Cluster and exactly the opposite of that found in the PN languages. The classifiers are suffixed to the numeral roots for 1 to 9. Each of the 11 languages has quantitative classifiers, or countable bases, for enumerating tens, hundreds, and thousands. There is, typically, a tens classifier which is suffixed to the numeral roots for 1 to 9 in order to form the decades 10 to 90. The systems are thus all 10-cycle ones (including those of Marshallese and Kosraean) and in fact have (10, 100, 1000) cyclic patterns. Examples of the classifier method of enumerating decades, hundreds, and thousands are given for three MC languages in Table 6.19.

Table 6.19
Numeral Classifier Constructions for Tens, Hundreds, and Thousands for Three Micronesian Languages

Numerals	Kiribatese	Woleaian	Mokilese
10	*te-bwi*	*se-g*	*ei-jek*
20	*ua-bwi*	*reu-g*	*riei-jek*
30	*ten-bwi*	*seli-g*	*jilih-jek*
100	*te-bubua*	*sa-pagui*	*e-pwki*
200	*ua-bubua*	*ru-pagui*	*rie-pwki*
300	*ten-bubua*	*selu-pagui*	*jili-pwki*
1000	*te-nga*	*so-ngeras*	*kid*
2000	*ua-nga*	*ru-ngeras*	*ria-kid*
3000	*ten-nga*	*seli-ngeras*	*jil-kid*

Note. Kiribatese (Groves, Groves, & Jacobs, 1985), Woleaian (Sohn, 1975) and Mokilese (Harrison, 1976; Harrison & Jackson, 1984) are provided to illustrate numeral classification.

Numeral classification is normally displayed in noun phrases; when serial counting of a set of objects takes place, each language possesses a special set of numerals which are not suffixed with a classifier. Several examples of serial counting numerals are given in Table 6.20. Serial counting is also carried out in other (non-MC) languages which possess numeral classification. In these cases we have numeral roots suffixed by a classifier which indicates that the objects being counted belong to a general class and these are used for serial counting rather than having a special set of unsuffixed numerals. Each of the MC languages also possesses a "general" class but has, in addition, the numerals used just for serial counting. Song (1997) indicated these processes in both Micronesian and non-Micronesian languages indicating a link between the languages.

Table 6.20
Numerals Used for Serial Counting in Three Micronesian Languages

Numerals	Ulithian	Woleaian	Trukese
1	*yood*	*yet, yut*	*eet*
2	*ruy*	*riew, ru*	*uruuw*
3	*yeel*	*yel*	*een*
4	*faag*	*fang*	*faan*
5	*liim*	*lim*	*niim*
6	*wòòm*	*wol, wul*	*woon*
7	*fiis*	*fis*	*fuus*
8	*waal*	*wal*	*waan*
9	*diiw*	*tiw, tiu*	*ttiiw*
10	*se-yexe*	*seg*	*engoon*

Note. Ulithian (Sohn & Bender, 1973), Woleaian (Sohn, 1975) and Trukese (Dyen, 1965; Goodenough & Sugita, 1980) data are provided.

It is not certain whether serial counting is an innovation of the MC languages. Harrison and Jackson (1984) note that Codrington (1885) reports the "existence of serial counting in several languages of the Solomons. Such systems are also found in Roviana and Rotuman" (p. 72).

Song (1997) argued that a discontinuity of the spread of languages was unlikely given the use of numeral classifiers in MC languages and other Oceanic languages.

> It is much simpler, and indeed makes more sense, to take the position that the Micronesian languages simply retained the POC feature: the exploitation of the possessive classifier system for benefactive marking (Song, 1997, p. 29).

Summary of 10-Cycle System Data

10-cycle numeral systems exist in New Guinea and Oceania. By far the majority (182) of these systems have a (10, 100) cyclic pattern while 20 have (10, 20) cyclic patterns. The distribution of these types among the AN and NAN language groups is given in Table 6.21.

Table 6.21
Distribution of 10-Cycle Systems Among the AN and NAN Languages of New Guinea and Oceania

	(10, 100)	(10, 20)	(10, 20, 60)
NAN (N=430)	12	3	1
AN (N=453)	170	17	0
Totals	182	20	1

We have suggested that in the case of the 16 NAN languages it seems likely that the possession of a 10-cycle system is an innovation. That is, these languages, or languages from which these are descended, did not originally possess 10-cycle systems and that as a result of the influence of languages which do possess 10-cycle systems (AN or tok pisin or English) the NAN groups augmented their original systems, in some cases by lexical borrowing, to form systems with a 10-cycle. Eight of the systems belong to the NAN languages of the East Papuan Phylum which are situated in islands east of the New Guinea mainland and which are occupied by predominantly AN-speaking groups. Three NAN languages, spoken in the Enga Province (PNG) and which have 10-cycle systems, have probably acquired these as a result of AN influence (see Swadling's (2010) comment on loanwords) but reinforced as a result of the recent influence of the 10-cycle systems of Tok Pisin or English.

The situation with the AN languages is complex. On the one hand we have the results of comparative linguistics in which the reconstructed languages of Proto Austronesian and its daughter Proto Oceanic both possess 10-cycle numeral systems. We might thus reasonably expect all 453 of the AN languages considered here (which includes 33 non-Oceanic AN and 420 Oceanic AN languages) to possess 10-cycle systems as well. On the other hand, the evidence indicates that this is not the case and that only 187 of these AN daughter languages now possess 10-cycle systems. The remainder have either a 2-cycle variant (32), a 5-cycle variant (222), or a 4-cycle system (4), with some 8 languages not having a definite classification. None of the AN languages of the North New Guinea Cluster or those of New Caledonia has a 10-cycle system; the large majority of the languages of the Papuan Tip Cluster and of Vanuatu similarly do not possess 10-cycle systems. The majority of the 187 languages which do have 10-cycle systems belong to the Admiralties and Meso-Melanesian Clusters, the South-East Solomonic group, and the Polynesian and Micronesian groups, none of which is located on the New Guinea mainland. Those AN languages situated on or near the mainland and which have 10-cycle systems comprise 7 Papuan Tip languages (all of the Motu type) and 16 non-Oceanic languages which are situated in coastal or island West Papua.

Several variants of the (10, 100) type of system have been delineated. The first of these Lean (1992) termed the "Manus" type which is characterised by having a subtractive construction for the numerals 7, 8, and 9 (and sometimes 6 as well). This type of system is largely found in the AN languages of the Admiralties Cluster which are located on Manus Island and the small islands proximate to it. The exceptions are the Mioko dialect of the AN Duke of York language and also Buin and Nanggu, both belonging to the NAN East Papuan Phylum. The second important variant is the Motu type characterised by the pairing or duplicative construction such that $6=2\times3$, $8=2\times4$, $9=(2\times4)+1$ (and sometimes $7=(2\times3)+1$). This type is largely found in the AN languages located east and west of Port Moresby which belong to the Papuan Tip Cluster. These two variants, the Manus and Motu types, with their unusual second pentad constructions, both deviate from the standard POC numeral system which has distinct numerals for 6 to 9. Both types of system would thus appear to be localised innovations.

Generally speaking, the 10-cycle systems discussed here have a $10+n$ construction for the numerals 11 to 19, where n takes, respectively, the values 1 to 9. One exception is the language Kuot, the only NAN language spoken in the predominantly AN-speaking New Ireland Province (PNG). This has, unusually, a $10+5+n$ construction for the numerals 16 to 19 (where n takes the values 1 to 4 respectively) even though the numerals 6 to 9 do not have a 5-cycle construction. The construction of the decades for many of the 10-cycle systems is of the form $10\times n$ or $n\times10$. There are some important exceptions to this which exhibit what Hurford terms "base-suppletion" in which the *te* morpheme used in constructing the decades 20 to 90 is not the same as the morpheme for 10. This is true particularly of those languages which possess, or show remnants of, numeral classification and which have "decade" or "group of ten" classifiers which are affixed with a numeral root.

References

Armstrong, R. (1962). Yoruba numerals. *Nigerian School and Economics 1*. Oxford, UK: Oxford University Press.

Bowers, N. (1977). Kapaukuu numeration: Reckoning, racism, scholarship and Melanesian counting systems. *Journal of the Polynesian Society, 86*, 105–116.

Bowers, N., & Lepi, P. (1975). Kaugel Valley systems of reckoning. *Journal of the Polynesian Society, 84*(3), 309–324.

British New Guinea. (1892). Aboriginal vocabulary of Nada. *Annual report for 1890-91* (pp. 139–142, Appendix GG135).

Codrington, R. (1885). *The Melanesian languages*. Oxford, UK: Clarendon Press.

Conant, L. (1896). *The number concept*. New York, NY: Macmillan.

de Solla Price, D., & Pospisil, L. (1966). A survival of Babylonian arithmetic in New Guinea? *Indian Journal of History of Science, 1*(1), 30–33.

Dixon, R., & Kroeber, A. (1907). Numerical systems of the languages of California. *American Anthropologist, 9*, 663–690.

Drabbe, P. (1926). Spraakkunst der lamdeensche taal. *Verhandelingen van het Koninklijk Bataviaasch Genootschap van Kunsten en Wetenschappen, 67*(2), 1–93.

Dyen, I. (1965). *A sketch of Trukese grammar*. New Haven, CO: American Oriental Society.

Eells, W. (1913). Number systems of the North American Indians: Systems of numeration. *American Mathematical Monthly, 20*(10), 293–299.

Fisher, J. (~2010). *Enriching students' learning through ethnomathematics in Kuruti elementary schools in Papua New Guinea*. Lae, Papua New Guinea: ResearchGate.

GLEC. (2008). *Glen Lean Ethnomathematics Centre*. Retrieved from http://www.uog.ac/glec/

Goodenough, W., & Sugita, H. (1980). *Trukese: English dictionary*. Philadelphia, PA: American Philosophical Society.

Groves, T., Groves, G., & Jacobs, R. (1985). *Kiribatese: An outline grammar and vocabulary* (Pacific Linguistics, D-64). Canberra, Australia: Pacific Linguistics.
Guise, R. (1892). Vocabulary of Bula'a dialect, spoken in the Central district of British New Guinea, by some coast tribes. *Annual report on British New Guinea from 1890 to 1891* (pp. 108–114, Appendix GG101).
Hale, H. (1846). *United States exploring expedition* (Vol. 6). Philadelphia, PA: G. Sherman.
Harrison, S. (1976). *Mokilese reference grammar*. Honolulu, HI: University Press of Hawaii.
Harrison, S., & Jackson, F. (1984). Higher numerals in several Micronesian languages. In B. Bender (Ed.), *Studies in Micronesian linguistics* (pp. 61–79). Canberra, Australia: Pacific Linguistics.
Haywood, G., & Haywood, I. (1980). The languages and communities of north-western New Britain. In R. Johnston (Ed.), *Language, communication and development in New Britain* (pp. 41–70). Ukurumpa, Papua New Guinea: Summer Institute of Linguistics.
Henderson, J. (1975). Yeletnye, the language of Rossel Island. In T. Dutton (Ed.), *Studies in languages of central and south-east Papua, Pacific Linguistics* (Vol. C-29, pp. 817–834).
Hurford, J. (1987). *Language and number: The emergence of a cognitive system*. Oxford, UK: Basil Blackwell.
Hymes, V. (1955). Athapaskan numeral systems. *International Journal of American Linguistics, 21*, 26–45.
Johnston, R. (1980). Nakanai of New Britain: The grammar of an Oceanic language. *Pacific Linguistics, B-70*.
Kluge, T. (1941). *Die Zahlenbegriffe der Sprachen Central- und Südostasiens, Indonesiens, Micronesiens, Melanesiens und Polynesiens mit Nachträgen zu den Bänden 2-4*. Berlin, Germany. Ein fünfter Beitrag zur Geistesgeschichte des Menschen nebst einer principiellen Untersuchung über die Tonsprachen. Original unpublished manuscript held at Yale University, microfilm held at Australian National University, Canberra, Australia.
Krupa, V. (1982). *The Polynesian languages: A guide, Languages of Africa and Asia* (Vol. 4). London, UK: Routledge & Kegan Paul.
Laufer, B. (1917). The vigesimal and decimal systems in the Ainu numerals. *Journal of the American Oriental Society, 37*, 192–208.
Lawes, W. (1885). *Grammar and vocabulary of the language spoken by the Motu tribe*. Sydney, Australia: C. Potter, Government Printer.
Laycock, D. (1975). Observations on number systems and semantics. In S. Wurm (Ed.), *New Guinea area languages and language study. Vol. 1: Papuan languages and the New Guinea linguistic scene* (pp. 219–233). Canberra, Australia: Pacific Linguistics.
Lean, G. (1992). *Counting systems of Papua New Guinea and Oceania* (Unpublished PhD thesis). PNG University of Technology, Lae, Papua New Guinea. Retrieved from http://www.uog.ac.pg/glec/
Macdonald, D. (1893). Asiatic origin of the Oceanic numerals. *Journal of the Polynesian Society, 2*, 88–89.
Menninger, K. (1969). *Number words and number symbols: A cultural history of numbers*. Cambridge, MA: M.I.T. Press.
Pawley, A. (1972). On the internal relationships of Eastern Oceanic languages (from 1969 MS). In R. Green & M. Kelly (Eds.), *Studies in Oceanic culture history 3: Pacific Anthropological Records 13* (pp. 1–142). Honolulu, HI: Bishop Museum.
Peng, F., & Brainerd, B. (1970). A grammar of Ainu number names. *Lingua, 25*, 381–397.
Pospisil, L., & de Solla Price, D. (1976). Letter to the Editor. *Journal of the Polynesian Society, 85*(3), 382–383.
Ray, S. (1895). The languages of British New Guinea. *The Journal of the Anthropological Institute of Great Britian and Ireland, 24*, 15–39. doi:10.2307/2842475.
Ray, S. (1917). Polynesian languages of the Solomon Islands. *Journal of the Polynesian Society, 26*, 36–43, 99–105, 170–179.

Ross, M. (1988). *Proto Oceanic and the Austronesian languages of western Melanesia* (Pacific Linguistics, C-98). Canberra, Australia: Australian National University.

Schmidt, W. (1929). Numeral systems. In *Encyclopedia Britannica* (14th ed., pp. 614–615). London, UK/New York, NY: Sears Roebuck.

Schnee, H. (1901). Beitrag zur Kenntnis der Sprachen im Bismarck-Archipel. *Mitteilungen des Seminars für orientalische Sprachen zu Berlin, 4*(3), 229–279.

Sohn, H. (1975). *Woleaian reference grammar*. Honolulu, HI: University Press of Hawaii.

Sohn, H., & Bender, B. (1973). A Ulithian grammar. *Pacific Linguistics, C-27*.

Song, J. J. (1997). The history of Micronesian possessive classifiers and benefactive marking in Oceanic languages. *Oceanic Linguistics, 36*(1), 29–64.

Stephan, E., & Gräbner, F. (1907). *Neu Mecklenburg (Bismarck Archipel)*. Die küst van umuddu bis kap St. Georg: Forschungsergebnisse bei den vermessungssfahrten von S.M.S "Möwe" im Jahre 1904.

Swadling, P. (2010). The impact of a dynamic environmental past on trade routes and language distributions in the lower-middle Sepik. In J. Bowden, N. Himmelmann, & M. Ross (Eds.), *A journey through Austronesian and Papuan linguistic and cultural space: Papers in honour of Andrew Pawley* (Vol. 615, pp. 141–159). Canberra, Australia: Pacific Linguistics, ANU.

Thureau-Dangin, F. (1939). *Sketch of a history of the sexagesimal system*. Ex officina Sanctae Catharinae.

Tryon, D. (1976). New Hebrides languages: An internal classification. *Pacific Linguistics, C-50*.

Tryon, D., & Hackman, B. (1983). Solomon Islands languages: An internal classification. *Pacific Linguistics, C-72*.

Z'graggen, J. (1975). Comparative wordlists of the Admiralty Islands languages collected by W. E. Smythe. *Workpapers in Papua New Guinea Languages, 14*, 117–216.

Chapter 7
4- and 6-Cycle Systems

Kay Owens and Glen Lean

Abstract Although there is evidence of 4 and 6 cycle systems in America, the 6-cycle systems in Papua New Guinea and Oceania only occur in areas of the southern side of New Guinea, namely a large island of West Papua (Indonesia) and crossing the border region into Western Province of Papua New Guinea. However, these systems have powers of 6. The 4-cycle systems are found with Non-Austronesian languages of some highlands provinces, often associated with 8-cycle and among pockets of languages on the northern coast of Papua New Guinea together with some Austronesian languages of West Papua's northern coast. The chapter concludes with a brief commentary of the survey covered in Chapters 3 to 7.

Keywords 4-cycle systems • 6-cycle systems • counting practices

The Existence and Nature of 4-Cycle Systems

Among the world's languages the occurrence of 4-cycle numeral systems is comparatively rare. Schmidt (1929) said "the quaternary system forms the numerals above four by composition: $5=4+1$, $7=4+3$, $8=4+4$ (or 2×4), $9=(2\times4)+1$, $16=4\times4$. In this consequent type, however, it is but seldom met, for example in California with the Salina and traces of it with the Chumash" (p. 614). Eells (1913), in discussing the Amerindian numeral systems indicates that "fairly well defined quaternary systems reaching to eight may be found among the Montaignais of the far north, the Foxes of Wisconsin, the Iowas and Missouris of the Plains - but they find their best and fullest development into true systems in various California tribes"(pp. 295-296). Dixon and Kroeber (1907), in their article on Californian numeral systems, confirm this: "Counting by fours is a striking feature of Californian languages" (p. 667). Beeler (1964) also documented the existence of 4-cycle systems among the Ventureño of California. Generally speaking, the 4-cycle system has not been documented in any other region except Africa where only one or two instances of numerals being compounded such that $8=2\times4$, $9=(2\times4)+1$, and $16=(2\times4)\times2$ have been recorded. Zaslavsky (1973) cited "the Haku language of Central Africa" (p. 46) as having a numeral system with these features although the system is not purely 4-cycle having, for instance, $7=6+1$ and a distinct word for 10. Beeler (1964) remarked that there is a vestigial trace of a 4-cycle in Proto Indo-European in that the numeral 8 is the dual form of 4; in the Indo-European languages as a whole, however, there is no sign now of 4-cycle features (p. 1).

K. Owens (✉)
Charles Sturt University, Dubbo, NSW, Australia
e-mail: kowens@csu.edu.au

G. Lean

K. Owens et al., *History of Number*, History of Mathematics Education,
DOI 10.1007/978-3-319-45483-2_7

Eells (1913), in discussing how 4-cycle systems might arise, said that "perhaps a few tribes, for reasons best known to themselves, did not use their thumbs in counting" (p. 296). He also speculated that the number four held special ritual significance for many Indian tribes, as does Schmidt (1929) "in California the four quarters of the sky play an important part in religion, mythology and custom" (p. 614). Dixon and Kroeber (1907), however, dismissed this latter view: "it is probably not connected to any extent with ritualism … the Californian Indians are distinctly unritualistic" (pp. 667-668). They also indicated that the Californian 4-cycle systems often have terms which refer to fingers, implying that they are a type of digit tally and they added that "it does not follow that because people count by their fingers they count by fives" (Dixon & Kroeber, 1907, p. 668). This view is supported by the data reported below.

In Papua New Guinea, there were only 14 languages of those collected with 4-cycle systems. The 10 NAN languages are: Vanimo, Rawo, the Melpa and Kaugel dialects of Hagen, various dialects of Kewa, Wiru, Nafri, the Island dialect of Boiken, Enga, the Angal Heneng dialect of Mendi, and Mountain Arapesh. The four AN languages are: Bam, Wogeo, Ormu, and Yotafa. Some of these are discussed below. See Figure 7.1 for their distribution.

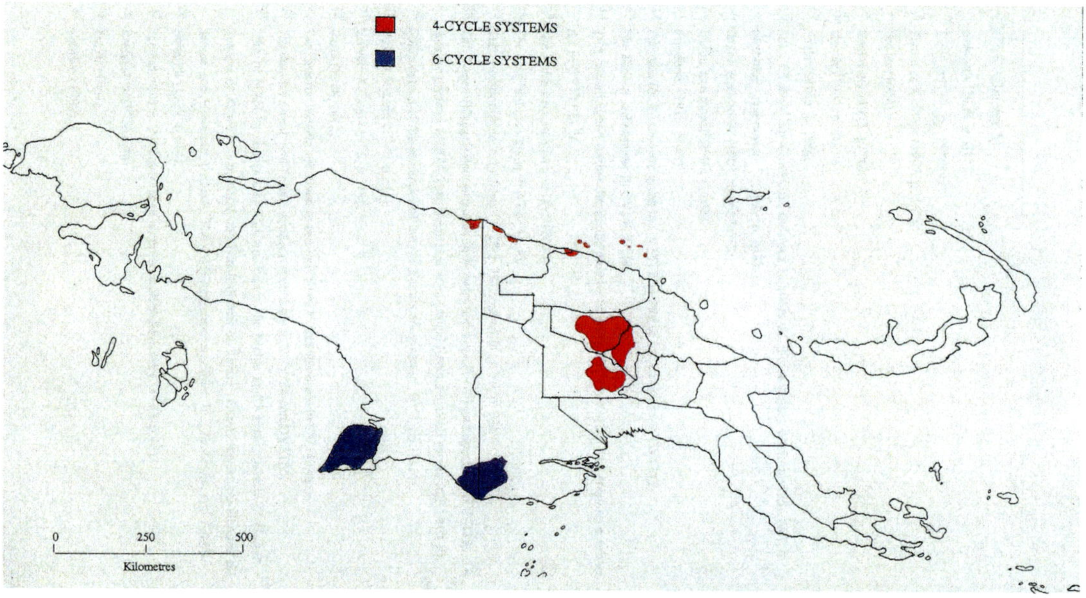

Figure 7.1. Distribution of 4-cycle and 6-cycle counting systems in New Guinea.

4-Cycle Systems Among the NAN Languages

Among the NAN languages of the New Guinea region, 4-cycle systems are rare: only about 10 or 11 languages possess numeral systems which exhibit, in whole or in part, features which are clearly 4-cycle in nature. These languages are located mainly in two geographical areas: the northern coast of New Guinea and the central highlands of PNG (see Figure 7.1). Two languages, Vanimo and Rawo, both of which belong to the Sko Phylum, and which are spoken in the West Sepik Province (PNG), had their numeral systems recorded by Friederici (1913, p. 42). According to this report, both systems were 4-cycle. The data are given below in Table 7.1. In both of these systems we have three monomorphemic numerals 1 to 3 while 4 is a "hand" morpheme *no-* or *nu-* (which normally has possessive suffixes). The numbers are tallied on the fingers of one hand excluding the thumb.

The Vanimo data indicate an alternative for 10, *moti*, a distinct term, and data recently collected by the author indicate that the Vanimo system no longer uses the $(4 \times 2) + 2$ construction for 10 and has instead a secondary 10-cycle.

Table 7.1
Numerals of Vanimo, Rawo, and Boiken Island Dialect

Numeral	*Vanimo*	*Rawo*	*Boiken*
1	*opa*	*opa*	*nese*
2	*yumono*	*yemeno, yumunu*	*siti, yiti*
3	*enu*	*eno, ino, inomumu*	*sumungkulung*
4	*no*	*no, nu-o*	*nawata*
5	*no meneau*	*no meu*	*(nawata) hifuase*
6	*no meneyu*	*no me yu* $4+1$	*(nawata) hifuayiti*
7	*no menehendu*	*no me 'no* $4+2$	*(nawata) hifuamungkukung*
8	*nuyu*	*no yo* 4×2	
9	*nuyu meneau*	*no yo meu* $4 \times 2 + 1$	
10	*nuyu meneyu, moti*	*no you meyu* $(4 \times 2) + 2$	

Note. 1. Vanimo and Rawo data were taken from Friederici (Friederici, 1913). Boiken dialect from Tarawai Island and Walis Island were sourced from 6 Counting System Questionnaires (Lean's Appendix on Sandaun). 2, *u* in Rawo is a suppletive form of 1.

Further to the east, in the Sandaun Province (PNG), the Island dialect of Boiken, spoken in the Tarawai and Walis Islands, also has a 4-cycle system (Table 7.1). The remaining six dialects of Boiken, however, all of which are located on the mainland, have 5-cycle systems. Adjacent to the Boiken region and located on the mainland coastal region are the Mountain Arapesh who have an unusual numeral system first described by Fortune in 1942. This has two monomorphemic numerals 1 and 2; 3 is $2+1$ and the system thus possesses a modified 2-cycle. There is a distinct term for 4, *nybat,* which, interestingly, means "dog". Fortune (1942) stated that

> counting from four to twenty-four proceeds on two roots for numerals … as follows: four one, four two, four three, *biogu nybat* (two fours or eight), eight one, eight two, eight three, *biogu atut* (three fours or twelve [the *nybat* being understood]), twelve one, twelve two, twelve three, *bigi biogu* (two eights or sixteen), and so on to *biogu atuga biogu* or twenty and from there similarly to *anauwip* or twenty-four. (p. 59)

To the west of the Vanimo-Rawo region, in West Papua, Nafri is spoken (Galis, 1960, p. 144). This is a member of the Sentani Family and its numeral system shows both 4- and 5-cycle features. It has four monomorphemic numerals 1 to 4 while 5 contains a "hand" morpheme *me*; 6 has a $5+1$ construction. The compounds for 8, 9, and 10 each show a 4-cycle construction and are, respectively, 4x2, (4x2)+1, and (4x2)+2. None of the other members of the Sentani Family exhibit traces of a 4-cycle in their numeral systems. There are, however, two other languages adjacent to Nafri which possess numeral systems showing 4-cycle features; these are Ormu and Yotafa, both of which are Oceanic AN and which are discussed in the next section.

The only other coastal or island NAN language which shows some trace of a 4-cycle in its numeral system is Yele, spoken on Rossel Island in the Milne Bay Province (PNG). While the numerals 1 to 10, and the decades and hundreds, show no trace of a 4-cycle, we find, however, that the thousands, used for counting shell money, do (Armstrong, 1928). These are given below in Table 7.2.

In the sequence given in Table 7.2 there is an $x+n$ construction for the numerals 5 000 to 8 000 where x is *mwa-* and n takes, respectively, the values 1 000 to 4 000. Normally, for a 4-cycle system, 8 000 would have a multiplicative construction (4 000x2) rather than an additive one. The 4-cycle construction does not continue after 8 000, and 10 000 has the surprising construction (9 000) (8 000). Armstrong, who studied the Rossel Islanders in the 1920s, said that this is explained in a legend about a mythological being, Wonajo, who invented counting and "having counted up to 9 000 grew weary, and, unable to think of a fresh word for 10 000 adopted the novel, if unmathematical, device of using in juxtaposition the words for the last two thousands" (Armstrong, 1928, p. 78).

Table 7.2
The Yele Numerals 1 000 to 10 000

Numeral	*Yele*
1000	*yili*
2000	*dwong*
3000	*teme*
4000	*dab*
5000	*mwa yili*
6000	*mwa dwong*
7000	*mwa teme*
8000	*mwa dab*
9000	*mwa di*
10000	*mwa di mwa dab*

The remaining 4-cycle systems are all located in the PNG highlands and several of these are remarkable and unique. Two standard 4-cycle systems, however, are exhibited by Wiru and the various dialects of Kewa, both languages spoken in the Southern Highlands Province (PNG). The numerals are given in Table 7.3.

Table 7.3
4-Cycle Systems of Wiru and Kewa

Numeral	Wiru	Kewa
1	*ondene*	*pameda*
2	*takura*	*laapo*
3	*tebolo*	*repo*
4	*tuyono, lu-u*	*ki*
5	*lu ke ondene*	*kode, kina kode*
6	*lu ke takura*	*kode laapo*
7	*lu ke tebolo*	*kode repo*
8	*lu-u takura*	*ki laapo*

Both systems have the numeral 4 containing a "hand" morpheme, *yono* in Wiru and *ki* in Kewa. The use of the "hand" morpheme, for both systems, indicates that the fingers but not the thumb are tallied. The nature of the Kewa system becomes apparent when we consider the semantics of the numbers 5 to 12, the data deriving from an article by the SIL linguists Karl and Joice Franklin and given in Table 7.4. The Wiru and Kewa 4-cycle systems are not the only means of enumeration for these language groups: both possess body-part tally systems, that for Kewa having a 47-cycle.

Table 7.4
The Semantics of the 4-Cycle System of (East) Kewa

Numeral	Kewa	Body-Part Translation
5	*kode*	the thumb
6	*kode laapo*	two thumbs, i.e. one hand and two thumbs
7	*kode repo*	three thumbs, i.e. one hand and three thumbs
8	*ki laapo*	two hands
9	*ki laapo na kode*	two hands, one thumb
10	*ki laapo kode laapo*	two hands, two thumbs
11	*ki laapo na kode repo*	two hands, three thumbs
12	*ki repo*	three hands

Three unusual 4-cycle systems which employ a similar principle of enumeration are exhibited in the languages Enga (Mai dialect), Hagen (Kaugel dialect), and Mendi (Angal Heneng dialect) which are spoken, respectively, in the Enga, Western Highlands, and Southern Highlands Provinces (PNG). Typically, these systems are such that counting proceeds initially from 1 to 8. Beginning at 9, there comes into play a means of proceeding by fours which employs what Lean termed "cycle units". In counting a particular sequence of four, for example 13 to 16, the cycle unit for that sequence is a word or phrase, perhaps a mnemonic, which is employed to form the compounds for 13 to 16 in the following way:

13 cycle unit + 1
14 cycle unit + 2
15 cycle unit + 3
16 cycle unit completed

In enumerating the next sequence of four, a different cycle unit is used but the same constructions are retained. To illustrate this, Table 7.5 shows enumeration of the sequences 13 to 16 and 17 to 20 for each of the three languages.

Table 7.5
Examples of Cycle Units in the 4-Cycle Systems of Enga, Angal Heneng, and Kaugel

Numeral	Enga	Angal Heneng	Kaugel
1	*mendai*	*pombor*	*telu*
2	*lapo*	*kap*	*talu*
3	*tepo*	*tep*	*yepoko*
4	*kitome(n)de*	*makl*	*kise*
8	*tukulapo*	*tukap*	*enggaki*
12	*tukutepon(ya) gato*	*tutep*	*rurepo*
13	*mapunya mendai*	*moklaopun pombor*	*malapungga telu*
14	*mapunya lapo*	*moklaopun kap*	*malapungga talu*
15	*mapunya tepo*	*moklaopun tep*	*malapungga yepoko*
16	*mapunya gato*	*tu moklaopu*	*malapu*
17	*yupunya mendai*	*supun pombor*	*supungga telu*
18	*yupunya lapo*	*supun kap*	*supungga talu*
19	*yupunya tepo*	*supun tep*	*supungga yepoko*
20	*yupunya gato*	*tu supu*	*supu*

Note. Kaugel data were from Bowers and Lepi (1975) and similar to 25 Counting System Questionnaires (CSQ); Angal Heneng data from 17 CSQs; Enga data from Lean's field visit to Mai dialect informants and given by Lang (1973).

For the Enga data shown, the cycle unit for the sequence 13 to 16 is *mapun(ya)* which means "sweet potato", while that for the sequence 17 to 20 is *yupun(ya)* which has the gloss "ground, earth". Bowers and Lepi (1975) indicated, for the Kaugel system, and the same is true for the other two, that

> unlike English or Pidgin numeration, when a set of four has been completed, the next three items are considered parts of the following named set rather than addition to the just-completed set, for example *supungga talu* "2 of 20" i.e., 18 (p. 313).

The enumeration in fours terminates at a different value for each system: Enga at 60, Kaugel at 32, and Angal Heneng at 48. In the case of Kaugel, 32 does not constitute a secondary cycle: the term for 24, *tokapu*, is used to form compounds for larger numbers so that 48 = 24 x 2, that is *tokapu talu*, and 72 = 24 x 3, that is *tokapu yepoko*.

This type of 4-cycle system is not documented in the literature of numeral systems for any other region in the world. Its uniqueness implies that the principle of the cycle unit as a means of enumeration is an independent and localised invention. It seems likely, however, that the 4-cycle nature of the systems has the same basis as other 4-cycle systems which exist in this region, that is treating the "hand" as four fingers. Bowers and Lepi (1975), indeed, suggested that, for the Kaugel system, "the everyday system was probably derived from finger counting" (p. 314). Both numerals 4 and 8, for Kaugel, contain a "hand" morpheme *ki*. Similarly, the number 4 in Enga, *kitomende*, also contains a "hand" morpheme *ki*, a feature shared by Wiru and Kewa, discussed earlier, as well as by the Melpa dialect of Hagen in which 8 contains a "hand" morpheme.

4-Cycle Systems Among the AN Languages

Only four AN languages possess numeral systems which exhibit 4-cycle features. Two of these, Wogeo and Bam, both of which belong to the Manam/Kairiru chain of the North New Guinea Cluster (Ross, 1988, p. 122), are spoken in the Schouten Islands in the East Sepik Province (PNG). The data for these systems are given in Table 7.6.

Table 7.6
Two 4-Cycle Systems of Oceanic AN Languages

	Wogeo	Bam
1	*ta*	*tini*
2	*ru*	*ru*
3	*tol*	*tuol*
4	*kwik*	*kiki*
5	*kwik bo koba*	*kikik be kubua*
6	*kwik ba rago*	*kiki be areg di ru*
7	*kwik be tol*	*kiki be areg di tuol*
8	*kiki ru*	*kiki ru*
9	*kiki tu bokoba*	*kiki ru be kubua*
10	*kiki ru barago*	*kiki ru be areg di ru*

Note. The Wogeo data were sourced from Friederici (1913, p. 42), Ray (1919, p. 328) and Kluge (1938, p. 177); Bam data from 2 CSQs and Laycock (1976, p. 416).

The Bam system is regular and shows clearly the pattern: 6 = 4 + 2, 7 = 4 + 3, and 8 = 4 × 2. While the numerals 1 to 3 are recognisable as reflexes of the Proto-Oceanic (POC) numerals but the numeral for 4 is not. It is not known whether, as is the case for the NAN 4-cycle systems, counting is done on the fingers but not the thumb of the hand. The word for "hand" in Bam is *lima*, identical to the POC numeral, which does not appear in the number 4; if it did this might lend some semantic weight to the idea that counting is done on the fingers only.

Two other Oceanic AN languages have numeral systems which possess 4-cycle properties. These are Ormu and Yotafa, both located on the north coast of West Papua adjacent to Nafri, a NAN language with a 4-cycle system discussed in the previous section. The data for these are given in Table 7.7.

Ormu displays a 4-cycle system with both 5 and 9 having "one", 6 having "two" and 8, as 2 presumably × 4. For Yotafa; 6 has "two", 9 is 8 + 1, *meniam* implies 4 and "one" which is similar to Vanimo's 4 + 1 (in Friederici's (1913) data 5 is *no menau* but other Vanimo data implies 4 and just

Table 7.7
4-Cycle Systems of Two Oceanic AN Languages in West Papua

	Ormu	Yotafa
1	*nitji*	*tei*
2	*rohi*	*ros, roti*
3	*toru*	*tor(u), ossor*
4	*awa*	*au(a)*
5	*ore-nitje-ma*	*meniam*
6	*man-rohi-ma*	*mandosim, ma-roti-ma*
7	*samecho-don-rohi-ma*	*tamecho-none-roti-ma*
8	*don-rohi-ma*	*none-roti*
9	*nen-rohi-fraja-nitje-ma*	*none-roti-fraja-tei*

Note. Ormu data were sourced from Galis (1960, p. 145) and Yotafa from Galis (1960, p. 144) and Batten (1894, p. 29).

gives 1 like Yotafa data). Interestingly, 7 contains the morphemes for 8 in both languages. This might suggest one more needed for the complete group of 8. As both of these languages do not possess systems with a 10-cycle which we might expect because both are daughter languages of POC, it is not known whether the 4-cycle properties apparent in their numeral systems is a result of influence by neighbouring 4-cycle NAN systems, that is those of Nafri and of Vanimo and Rawo to the east. Some sources for Vanimo have a distinct word for 10 suggesting a secondary 10-cycle. For Rawo, Friederici's (1913) data also give a 4-cycle but later CSQs give a distinct word for 5 and subsequently give $6=5+1$, $7=5+2$, $8=5+3$, $9=5+4$, $10=2\times4$.

Summary of 4-Cycle System Data

There is a total of 10 NAN languages (11 if we include Yele) and 4 AN languages which have numeral systems displaying 4-cycle features. Figure 7.1 indicates the distribution of these which is restricted to two main regions: 1) the northern New Guinea coast and islands, and 2) the central highlands of PNG. No 4-cycle numeral systems are found in Island Melanesia, Polynesia, or Micronesia, although counting objects in groups of four is quite common in these regions. Only several of the systems display "pure" 4-cycle features as described by Schmidt, that is $5=4+1$, $7=4+3$, $8=4+4$ or 2×4, and $9=(2\times4)+1$. Ormu and Yotafa are AN languages have an unusual construction for 7, and the neighbouring NAN 4-cycle systems also have some data suggesting both 4 and 5 cycles. Mountain Arapesh and the Melpa dialect of Hagen both have a primary 2′-cycle as well as a secondary 4-cycle. Yele exhibits a 4-cycle structure for the thousands only while the Enga, Kaugel, and Angal Heneng systems have their unique 4-cycle sequences beginning at 9.

The evidence is clear that some 4-cycle systems arise due to the practice of treating the four fingers, but not the thumb, as one "hand". This is certainly the case for the NAN languages Vanimo and Rawo and those of the central highlands. There is, however, no lexical evidence suggesting that this is also true for the four AN systems. While it would appear that counting on the fingers in this way may well be a localised innovation for the central highlands languages, it is not at all clear whether there is a similar common thread connecting the nine 4-cycle systems of the northern coast and islands, that is whether these are largely small groups of 4-cycle counters which have developed their systems independently or, as seems more likely, whether the 4-cycle system was developed in one location and was then diffused, possibly by trading, along the northern seaboard.

The Existence and Nature of 6-Cycle Systems

6-cycle, or senary, systems are comparatively rare and the literature indicates their existence in only two major regions: Africa and North America, as indeed was found in the case of the 4-cycle systems. Schmidt (1929) noted that "it has a rather limited dispersion in north-west Africa, e.g., in the Huka, the Bulanda, the Apko" (p. 614). In her 1915 article on African numeration, Schmidl (1915) indicated several languages which use the principle of "composition with 6" and her map of the distribution of various systems shows some seven locations where this occurs. Zaslavsky (1973, p. 46), quoting Schmidl's material and map, gave several instances of systems with 6-cycle properties in which, for example, $12=6\times2$ and $24=6\times4$. The Balante (Schmidt's "Bulanda") have numerals 7 to 12 compounded as $7=6+1$, $8=6+2$, and so on.

In North America, 6-cycle systems appear to be limited to a few examples in the languages of California. Beeler (1961a) published an account of the Wintun, Nomlaki, Patwin and Maidu systems, each of these being members of California Penutian. None of these systems is pure 6-cycle in nature, that is in a pure system we would expect 7 to 11 expressed as $6+n$ where n takes the values 1 to 5 respectively, and $12=6\times2$. They do, however, exhibit some 6-cycle features in that $7=6+1$, $8=6+2$, and $9=6+3$. Gamble (1980), similarly, in his analysis of Chunut, also located in California, indicated vestigial traces of a 6-cycle numeral system.

6-Cycle Systems in the NAN Languages

Lean had no evidence of the existence of 6-cycle numeral systems among the AN languages of New Guinea and Oceania. There is, however, evidence that such systems exist in a small group of NAN languages which are located in two regions on the south coast of New Guinea (see Map 15). Three languages, Kimaghama, Riantana, and Ndom, are all spoken on Kolopom Island adjacent to the south coast of West Papua, immediately to the west of the West Papua/PNG border. The numeral systems of each of these languages possesses a primary 6-cycle; Kimaghama has a secondary 20-cycle while Ndom has a (6, 18, 36) cyclic pattern. The data for these systems are given in Table 7.8.

Table 7.8
Three Examples of 6-Cycle Numeral Systems of the NAN Languages

	Kimaghama	Ndom	Riantana
1	*növere, nubella*	*sas*	*mebö*
2	*kave*	*thef*	*enava*
3	*pendji*	*ithin*	*pendö*
4	*jando*	*thonith*	*wendö*
5	*mado*	*meregh*	*mata*
6	*turo, ibolo-nubella*	*mer*	*törwa*
7	*iburo-növere*	*mer abo sas*	*mebö-me*
8	*iburo-kave*	*mer abo thef*	*enava-me*
9	*iburo-pendji*	*mer abo ithin*	*pendö-me*
10	*iburo-jando*	*mer abo thonith*	*wendö-me*
11	*iburo-mado*	*mer abo meregh*	*mata-me*
12	-	*mer an thef*	*törwa-me*

Note. These data were sourced from Galis (1960, p. 148) and Drabbe (1926, pp. 6-7) with agreement from Boelaars (1950, p. 34) for Kimaghama.

Although these three languages are related, each belonging to the Kolopom Sub-Phylum-Level Family, the numerals of each system differ markedly from those of the others. Nevertheless, each exhibits a 6-cycle structure, that of Ndom being the most regular. In addition to the data shown, Ndom has distinct terms for 18 and 36; 72 and 108 are compounds of the latter, that is $72=36\times2$ and $108=36\times3$.

To the east of Kolopom Island, two other languages are located in the Western Province PNG, both of which possess 6-cycle systems. These are Kanum around the border of West Papua and PNG and Tonda further east. Both seem to belong to the same Sub-Family of the Trans-Fly Stock. Although the data for Tonda are incomplete, so that it is uncertain how the numerals 7 to 12 are constructed, nevertheless we do have $12=2\times6$ and $18=3\times6$. The data for Kanum are more complete: there are distinct terms for 1 to 6 and the numerals 7 to 12 have an $x+n$ construction where n takes, respectively, the values from 1 to 6. The alternative construction for 12 is 2×6, 18 is 3×6, and 24 is 4×6. There is a distinct term for 36, as we found with Ndom, and this forms a secondary cycle with $72=2\times36$. These various systems discussed here form the most fully displayed 6-cycle systems to be found anywhere in the literature. The data for Kanum and Tonda are given in Table 7.9.

Table 7.9
The 6-Cycle Systems of Kanum and Tonda

Numeral	Kanum	Tonda
1	*namper*	*nabi*
2	*jempoka*	*yalmbe*
3	*juau*	*yala*
4	*eser*	*hasar*
5	*tampui*	*tambui*
6	*tarawo*	*trawa*
7	*pesmeri-emper*	-
8	*pesmeri-jalmpö*	-
9	*pesmeri-jela*	-
10	*pesmeri-eser*	-
11	*pesmeri-tampui*	-
12	*pesmeri-tarawo*	*yalmbe trawa*
18	*juau-tarawo*	*yala trawa*
36	*nimpe*	-
72	*jempoka-nimpe*	-

Note. Kanum data were sourced from Galis (1960, p. 149) and Boelaars (1950, p. 38), Tonda from 1 CSQ.

Later data from Hammarström (2009) suggest there are several languages in the Kanum and Tondo groups with the latter seeming at this stage to include Tondo, Nambu, and Yei spread over a considerable area around the Morehead river. He noted that they all have words for 1 and 2, and then

> above 2, the vocabularies wholly diverge. Some vocabularies show a restricted system with $3=2+1$ and $4=2+2$, whereas others have a monomorphemic 3, $4=2+2$, and evidence of base-5 above that. In other words, proto-Nambu must have had a restricted system whereas some modern Nambu varieties show base-5, or incipient base-5 systems.

This is interesting as Lean (1992) considered a similar issue of a 4 and 5 cycle systems for some north coast of West Papua border languages (Ormu and Yofata). Both Lean and Hammarström note that Williams (1936) had recorded that the Keruka, a Nambu speaking group, used a 6-cycle system borrowed from the Tondo for counting a small variety of yam *taitu* but otherwise used a 5 cycle system. Their frame words were 1, 2, 36, 216, and 1 296 showing powers of 6 were used.

Williams (1936) noted that such systems should develop for yam counting (two groups of 3 yams) and Hammarström noted the use of the thumb parts suggesting a hand link as was found for the

5-, 10-, and 4-cycle systems although the semantic evidence is not so clear. There is, however, a special counting system employed by the Mountain Arapesh which offers a clue to the origin of 6-cycle systems. The Mountain Arapesh possess a 4-cycle system which was discussed in the previous section; they also possess another unusual system which has a (3, 6) cyclic pattern. The procedure used in this type of counting has been described by Fortune; after counting to 3, the count proceeds: "three one, three two, *anauwip* or six. Then repeat to a second *anauwip,* or twelve. This is supposed to be the count on the hands, five fingers and a thumb joint as well, to make six for each hand" (Fortune, 1942, p. 59). Thus, for this particular system, the 6-cycle does arise from finger tallying. We have found, therefore, that tallying on the hand gives rise not only to 5-cycle variants, but also to 4-cycle systems in which the thumb is excluded from the count, and to 6-cycle systems in which the thumb joint augments the four fingers and the thumb.

Donahue (2008) explained one of the more interesting developments in counting systems for Kanum. He noted there were in fact three systems, simple, moderate and complex (Table 7.10). In the

Table 7.10
Three Levels of Kanum Counting Systems

	Simple	Moderate	Complex
1	*naempr*	*aempy*	*aempy*
2	*yempoka*	*ynaoaempy*	*ynaoaempy*
3	*ywaw*	*ylla*	*ylla*
4	*eser*	*eser*	*eser*
5	*swabra*	*tampwy*	*tamp*
6	*'swy*	*traowao*	*ptae*
7		*psymery aempy* 6 + 1	*aempy ptae* 1 + 6
8		*psymery ynaoaempy* 6 + 2	*ynaoaempy ptae* 2 + 6
9		*psymery ylla* 6 + 3	*ylla ptae* 3 + 6
10		*psymery eser* 6 + 4	*eser ptae* 4 + 6
11		*psymery tampwy* 6 + 5	*tamp ptae* 5 + 6
12		*psymery traowao* 6 + 6 or *yempoka traowao* 2 × 6	*tarwmpao* 12
13			*aempy tarwmpao* 1 + 12
14			*ynaoaempy tarwmpao* 2 + 12
15			*ylla tarwmpao* 3 + 12
16			*eser tarwmpao* 4 + 12
17			*tamp tarwmpao* 5 + 12
18			*ntamnao* 18
19			*aempy ntamnao* 1 + 18
20			*ynaoaemy ntamnao* 2 + 18
24			*wramaekr* 24
25			*aempy wramaekr* 1 + 24
30			*ptae wramaekr* 6 + 24
31			*aempy ptae wramaekr* 1 + 6 + 24
36	$= 6^2$		*(ntaop) ptae* (big) 6
37			*aempy (ntaop) ptae* 1 + (big) 6
50			*ynaoaempy tarwmpao (ntaop) ptae* 2 + 12 + 36
100			*eser wramaekr ptae ynaoaempy* 4 + 24 + 36 × 2
216	$= 6^3$		*tarwmpao* 216
1296	$= 6^4$		*(ntaop) ntamnao* (big) 18
7776	$= 6^5$		*(ntaop) wramaekr* (big) 24

Source. Donahue (2008, p. 426)

complex system there is an economy of words in that the words for 12 and are reused in larger numbers. The multiplication is placed after the "units" in the higher numbers.

> Large numbers, such as '500' or '1976', can easily be formed: *ynaoaempy tarwmpay ynaoaempy ptae wramaekr ntaop ptae*, '(2 × 216) + (2 + 6 + 24 + 36)', and *ntamnao tarwmpay ylla ynaoaempy ptae wramaekr*, '1296 + (216 × 3) + 2 + (6 + 24)', respectively, though it should be noted that some younger speakers are reinterpreting *ntamnao* '1296' as '1000' when counting, almost certainly under the influence of dealings with Indonesian currency (for which the 1000 is the lowest banknote of value, resulting in a system in which almost all products are priced in multiples of 1000). This means that *ntamnao tamp* is effectively ambiguous between '5000' (1000 × 5; new reading) or '6480' (1296 × 5; old reading), although only the latter is prescriptively correct. (Donahue, 2008, p. 427)

Further data are available from Evans (2009) for another 6-cycle system Nen which belongs to the Morehead-Maro rather than Kanum branch of the Morehead-Wasur languages. It also had powers of 6 to 6^5 and possibly 6^6. He commented that the Kanum, Tondo and other Morehead-Wasur languages probably had a 6-cycle system with powers in Proto-Morehead-Wasur a long time ago. He quoted Williams (1936) who described the counting for another of the languages Keraakie. He described the custom of two men taking three yams each to a display with a counter noting the 6 and each group of 6 marking 36 with a yam and then the counters started to count the next $36 = 6^2$. This counting would continue as each power of 6 was reached. The practice seemed to indicate the necessity of having 6^4 stored. He provided an image of the 6 small yam radiating in a circle with the narrow ends touching in the centre (like a daisy). He also noted that the Yei who relied on sago rather than yams did not have a 6-cycle system. Thus a cultural context appears to have encouraged the counting with 6 being represented by the hand, the part of the hand for the number 6 varied with pointing to wrist protuberance (Nen) and encircling base of the hand (Kanum). $4 = 2 + 2$ and 5 is either the 5th finger or the back of the hand in the various languages. His informants also suggested abstract counting and that this counting was not just for counting yam although they also used two-hands for a 10 system in talking about the number of children. According to Evans (2009), further linguistic and anthropological data were needed to decide if the exponential system was diffused or belonged in the Proto Morehead-Wasur system.

Summary of 6-Cycle System Data and General Commentary

The situation regarding 6-cycle systems in New Guinea is similar to that pertaining in Africa and North America, that is that such systems are rare and are only found in a few locations. There are five NAN languages which possess systems with a primary 6-cycle; three of these are located on Kolopom Island in southern West Papua and two are located near the south end of the West Papua/PNG border. Apart from these two neighbouring regions, 6-cycle systems do not exist anywhere else in New Guinea or Oceania, and in particular are not found in any of the AN languages. As indicated above, one other NAN language, Mountain Arapesh, has a special counting system which has a primary 3-cycle and which terminates on a count of 6 (and thus does not actually have a secondary 6-cycle). The notable feature of the 6-cycle systems located in southern New Guinea is that they are the most fully formed 6-cycle systems found anywhere in the literature.

In concluding the survey given in the last five chapters of the various types of numeral systems and tally methods which exist in New Guinea and Oceania, it is clear that these possess a considerable, but not unlimited, diversity. Evidence has been presented to indicate the existence of numeral systems which have primary cycles of 2, 4, 5, 6, and 10. The special Mountain Arapesh system has a primary 3-cycle and there are two other systems, those of Bumbita Arapesh (Sandaun Province) (see Appendix B) and Wasembo (upper Markham River, Morobe Province), which appear to have features of a secondary 3-cycle. The secondary cycles of the numeral systems range through 3, 4, 5, 10, 18, 20, 36, 60, to 100. The system of the Melpa dialect of Hagen, which has a "hand" morpheme for 4, also has a distinct word for 8 and a consequent 8-cycle. Despite this diversity of cyclic structures, it is important to note that there is no evidence that any language in the region under consideration possesses a numeral system with either a primary or secondary cycle of 7 or 9 in the first decade although both numbers feature in Papuan languages where 10 completes a cycle (e.g. one of the Engan systems and Uisai in Bougeanville), or of 11, 12, 16, and 19 in the second decade. Many of the other numbers in the second decade occurred with an asymmetrical or truncated system.

Tallying on the fingers of the hand (and, by extension, the toes) gives rise to the common "digit-tally" system with a primary 5-cycle and a secondary 20-cycle. However, we have also found that digit-tallying can give rise to both 4- and 6-cycle systems depending, respectively, on whether the thumb is omitted from the tally or whether the thumb joint or wrist joint augments it. There is no evidence, however, that pure 2-cycle systems have their genesis in digit-tallying. Tallying on body parts other than digits as a primary way of developing a counting system appears to be rare around the world except in New Guinea and Australia. Of the 40 body-part tallies documented for New Guinea, the least complete cycle is 18, the greatest complete cycle is 68, and the modal cycle is 27. We therefore have a considerable range of unusual cyclic structures, each of which has its genesis in parcelling out the body for tallying. There are numerous systems that are truncated or asymmetrical in using the body-parts. There is evidence that such tallies are, or were, used for calendrical purposes, in particular for establishing when a certain feast should occur during a cycle of ceremonial feasts but other groups used these systems for many purposes. Language groups which possess body-part tallies almost invariably possess, in addition, a relatively simple numeral system, that is either a 2-cycle variant or, as we find in the Southern Highlands Province (PNG), a 4-cycle system. The body-part tallies functioned as ordinal devices, using the name of a body part to indicate the position of a point in an ordered sequence of points; the names of the body parts were not generally used (in noun phrases, say) for stating the number of objects in a set and for this the numeral system was used and which thus had an independent purpose and existence.

Another important point to note are the body-part tallies for multiples of powers of 10 (e.g., Yu Wooi, Mid-Waghi, Jiwaka Province) and 20 (e.g., Iqawaye, Eastern Highlands Province). Meanwhile, other representations and cultural contexts encouraged powers of numbers such as 6 (southern New Guinea) or 10 times powers of 2 (Oceania) or powers of 10 times powers of 12 (e.g. Tinatatuna or Tolai, East New Britain). This diversity indicates that large numbers were expressed and a sense of infinity was often associated with the systems of powers. This is in contrast to people who noted in recent field trips by Owens, the numbers "finished" at a particular number, usually 20 or systems with a 2-cycle system (NAN languages around Bogia, Madang Province), or with size (few, many) determined by the number of people or occurrences by which an action was taken (Western Province languages). Thus the cultural context often determined ways of thinking of number that were systematic but not limited to western arithmetic systems and perhaps not fully appreciated by western-thinking people. Nevertheless, recent data have shown a greater richness in the counting systems, adding to Lean's (1992) and Smith's (Smith, 1984, 1988) original extraordinary surveys and analyses.

References

Armstrong, W. (1928). *Rossel Island: An ethnological study*. Cambridge, UK: Cambridge University Press.

Batten, G. (1894). *Glimpses of the Eastern Archipelago*. Singapore: Singapore and Straits Printing Office.

Beeler, M. (1961). Senary counting in California Penutian. *Anthropological Linguistics, 3*, 1–8.

Beeler, M. (1964). Ventureño numerals. In W. Bright (Ed.), *Studies in Californian linguistics* (pp. 13–18). Berkeley and Los Angeles, CA: University of California Press.

Boelaars, J. (1950). *The linguistic position of south-western New Guinea*. Leiden, The Netherlands: Brill.

Bowers, N., & Lepi, P. (1975). Kaugel Valley systems of reckoning. *Journal of the Polynesian Society, 84*(3), 309–324.

Dixon, R., & Kroeber, A. (1907). Numerical systems of the languages of California. *American Anthropologist, 9*, 663–690.

Donahue, M. (2008). Complexities with restricted numeral systems. *Linguistic Typology, 12*, 423–429. doi:10.1515/LITY.2008.043.

Drabbe, P. (1926). Spraakkunst der lamdeensche taal. *Verhandelingen van het Koninklijk Bataviaasch Genootschap van Kunsten en Wetenschappen, 1*(2).

Eells, W. (1913). Number systems of the North American Indians: Systems of numeration. *American Mathematical Monthly, 20*(10), 293–299.

Evans, N. (2009). Two *pus* one makes thirteen: Senary numerals in the Morehead-Maro region. *Linguistic Typology, 13*, 321–335.

Fortune, R. (1942). Arapesh. *American Ethnological Society, 19*, 1–60.

Friederici, G. (1913). Wissenschaftliche ergebnisse einer amtlichen forschungsreise nach dem Bismarck-Archipel im Jahre 1908. 3: Untersuchungen über eine melanesische wanderstrasse. *Mitteilung aus den Deutschen Schutzgebieten, Ergänzungsheft 7*.

Galis, K. (1960). Telsystemen in Nederlands-Nieuw-Guinea. *Nieuw Guinea Studien, 4*(2), 131–150.

Gamble, G. (1980). The "old time" Chunut count. In K. Klar, M. Langdon, & S. Silver (Eds.), *American Indian and Indoeuropean studies: Papers in honour of Madison S. Beeler*. The Hague, The Netherlands: Mouton.

Hammarström, H. (2009). Whence the Kanum base-6 numeral system? *Linguistic Typology, 13*(2), 305–319. doi:10.1515/LITY.2009.014.

Kluge, T. (1938). *Die Zahlbegriffe der Australier, Papua und Bantuneger nebst einer Einleitung ueber die Zahl, ein Beitrag zur Geistesgeschichte des Menschen*. Berlin, Germany: Selbstverlag.

Lang, A. (1973). *Enga dictionary* (Pacific Linguistics, C-20). Canberra, Australia: Australian National University.

Laycock, D. (1976). Austronesian languages: Sepik Provinces. In S. Wurm (Ed.), *New Guinea area languages and language study, vol 2: Austronesian languages, Pacific Linguistics C-39* (pp. 399–418). Canberra, Australia: ANU.

Lean, G. (1992). *Counting systems of Papua New Guinea and Oceania* (Unpublished PhD thesis). PNG University of Technology, Lae, Papua New Guinea.

Ray, S. (1919). The languages of northern Papua. *Journal of the Royal Anthropological Institute of Great Britain and Ireland, 49*, 317–341.

Ross, M. (1988). *Proto Oceanic and the Austronesian languages of western Melanesia* (Pacific Linguistics, C-98). Canberra, Australia: Australian National University.

Schmidl, M. (1915). Zahl und Zählen in Afrika. *Mitteilungen der Anthropologischen Gesellschaft in Wien, 45*, 165–209.

Schmidt, W. (1929). *Numeral systems. In Encyclopedia Britannica* (14th ed., pp. 614–615). London, UK/New York, NY: Sears Roebuck.

Smith, G. (1984). *Morobe counting systems: An investigation into the numerals of the Morobe Province, Papua New Guinea* (MPhil). Papua New Guinea University of Technology, Lae, Papua New Guinea.
Smith, G. (1988). Morobe counting systems. *Papers in New Guinea Linguistics* (Vol. A-76, pp. 1–132). Canberra, Australia: Pacific Linguistics, ANU.
Williams, F. (1936). *Papuans of the Trans-Fly*. Oxford, UK: Clarendon Press.
Zaslavsky, C. (1973). *Africa counts*. Westport, CT: Lawrence Hill and Co.

Chapter 8
Number and Counting in Context, Classifications and Large Numbers

Kay Owens and Glen Lean

Abstract This chapter begins the development of the importance of number in Indigenous societies. It takes a number of case studies to illustrate the role of counting in these societies and the influences of neighbouring tribes with different, e.g. Papuan language contexts on an Austronesian language. Counting is part of mathematical activities and is used for decision-making, comparison, and relationships. One aspect that the collation and analysis of counting systems in this region has highlighted is the notion of qualitative decisions about number. Another is the link with classifiers used in several areas and having considerable variation across systems. Other issues such as large numbers, alternate representations, and changes in language and practice over the years are addressed.

Keywords Cultural contexts • classifiers • quantity as part of qualitative decision-making • importance of counting • large numbers • numerical abstraction

Introduction

The foregoing survey of the counting system situation in New Guinea and Oceania has concentrated on delineating the various types of systems, their structural features, and their geographical distribution throughout the region in question. This focus on the nature of the systems themselves, abstracted from the cultures in which they are embedded, has been deliberate and is preliminary to addressing the chief questions with which this work is concerned, that is how and when the counting system situation, which is apparent now, came about. Any student of foundational numeral systems, however, is acutely aware that the description of their structural features does not in itself do justice to the richness of the notions of number held in a given culture: the human face of number. The highly abstract view of number, which is now held by mathematicians and logicians in the modern technological societies, is a relatively recent development, deriving from the work of Frege (1950, 1967), Russell, Whitehead, Gödel, and others (van Heijenoort, 1967). This exceedingly refined view of number was regarded by Wilder (Wilder, 1974, 1981) as the end result of a long process of cultural evolution and indeed it is far removed from the views of number which we may infer existed in foundational societies. In order to obtain some idea of what these views are, in this chapter we shall attempt to look at various aspects of what Crump (1990) called "the anthropology of numbers".

K. Owens (✉)
Charles Sturt University, Dubbo, NSW, Australia
e-mail: kowens@csu.edu.au

G. Lean

K. Owens et al., *History of Number*, History of Mathematics Education,
DOI 10.1007/978-3-319-45483-2_8

In the following sections, there are arguments from specific cases and some subsidiary questions on the following main questions:

1. Are there foundational societies in which number has a privileged position and, conversely, do societies exist in which there are alternative organising systems and principles which relegate number to a relatively unimportant position?
2. What is, and what is not, counted in various foundational societies and in what circumstances does counting occur? Are there particular types of foundational exchange systems or economies which emphasise counting more than others?
3. To what extent is the phenomenon of numeral classification found in the languages of New Guinea and Oceania? Are the ways in which the foundational societies of this region categorise their universe of countable objects similar? Is there any evidence that numeral classification may have been present in Proto Oceanic?
4. What, in the foundational societies of this region, are the terms for large numbers? Do the languages of New Guinea and Oceania differ with respect to the resources available for expressing large numbers? For the Oceanic AN languages, do we have evidence to suggest what the historical development of terms for large numbers might have been and whether such terms were likely to have been present in Proto Oceanic?

The Importance of Number in Foundational Societies

In modern technological societies, number plays a central and important role and pervades virtually all aspects of everyday life: there is a pronounced emphasis on enumeration as well as the use of the arithmetical processes and their physical applications. Foundational societies, however, differ from these, and indeed from each other, in the degree to which number holds a privileged position in everyday life. Between modern and traditional societies we have the obvious distinction, respectively, of numbers being used primarily in their written forms as opposed to their spoken forms. The former has clear advantages in terms of the efficient recording of totals, the carrying out of calculations using place-value notation, and the minimisation of the effect of the constraints of memory capacity. We would expect those societies which use numbers in their spoken, unwritten, forms, even if the results of enumeration are recorded by some form of tally, to differ in significant ways from those which use, primarily, the written forms. However, it is also apparent that traditional societies differ from each other in the degree to which number plays a role as an organising construct within each society. These differences will be illustrated by a series of brief case studies drawn from the linguistic and anthropological literatures.

The Adzera People of the Morobe Province (PNG)

The languages of the Markham Valley in the Morobe Province (PNG), which includes that of the Adzera people, have been studied by Holzknecht (1989). The Markham Family are all AN languages descended from POC for which a 10-cycle numeral system has been reconstructed. These language groups have moved away from the POC homeland (to the east in the coastal region of the Willaumez Peninsula of the West New Britain Province) and are now located inland in areas which are predominantly NAN-speaking. Holzknecht notes that "it appears from the lack of terms for maritime objects and activities that the Markham languages moved away from the sea early in their history" (Holzknecht, 1989, p. 207). One important accompaniment of this shift away from a maritime culture to an inland cultivation culture with a strong NAN influence is the change in the numeral systems of the languages of the Markham Family. Holzknecht said that "under the influence of their Papuan-speaking neighbours the system became eroded to what we find today, a system with two numerals, 'one' and 'two', combined with body-tallying (sic)" (Holzknecht, 1989, p. 127). Appendix B provides Lean's data on this system. We thus have evidence that various factors, such as a change of environment, of economy, and of the influence of the non-maritime NAN-speaking language groups, has brought about a change in the means by which the Adzera, and other members of the Markham Family, enumerate their world (Figure 8.1 and Appendix B).

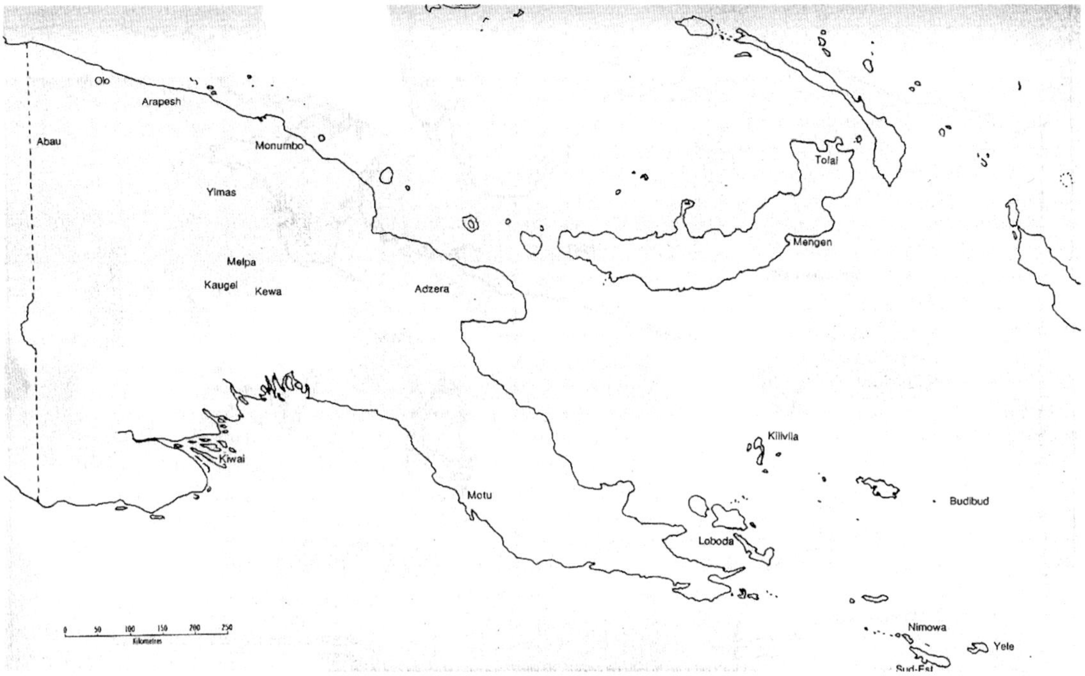

Figure 8.1. Location of societies cited in text (PNG).

In Smith's (1981) study of the foundational methods of enumeration in the Morobe Province, he noted that

> in many Morobe societies, as in other parts of Melanesia, status was achieved by distribution of wealth. Often this distribution implied a reciprocal obligation on the part of the recipients to pay back on some future occasion as much or more than the amount received. Hence it was important to know how much was received. (p. 7)

Smith indicated that for the Adzera there were certain ceremonial occasions during which huge quantities of banana bunches were given away. The counting of the exact number of bunches, however, was less important than the way in which the bunches were displayed: "a broad ladder is sometimes built to the top of a coconut tree, and this must be completely covered with bunches of bananas, or the sponsors of the distribution will be scorned" (Smith, 1981, p. 7). Even in the case of relatively small-scale gift-giving there is reciprocal obligation and, for example, on receipt of a joint of pork an Adzera man would measure it with a piece of bark string and this tally of its size would be kept for future reference. The main point, however, is that while the Adzera are able to count, precise enumeration is less important than satisfying the ceremonial obligation of providing an impressive display. Also, in the absence of written numerals, physical tallies such as a piece of knotted bark string are adequate substitutes for maintaining a record of gift presentation.

According to Holzknecht, Adzera trade was usually conducted between individuals in a trading relationship and such partners were either true kin or quasi-kin. However, trade transactions often transcended language boundaries: "For example, many Musom people have kin ties with

Nabak (Papuan) neighbours, because trade ties were so advantageous between those two groups. In exchange for mountain food and game from the Nabak, the Musoms gave salt, grindstones for sharpening axes and bamboo for bowstrings" (Holzknecht, 1989, p. 46). The Adzera and their neighbours did not possess monetarised economies and goods were exchanged rather than purchased.

The Grand Valley Dani of West Papua

The Grand Valley Dani live in the central mountain ranges of New Guinea to the west of the PNG border in West Papua. They speak a NAN language which belongs to the Trans New Guinea Phylum (see Appendix C on systems from Chapter 8). The Dani were studied by the American anthropologist Karl Heider in the early 1960s. Heider (1991) noted that some of the features of this horticultural society were "the complexities of a moiety-based kinship system; a political organization of confederations and alliances of great size but little power; and in the psychological realm, such traits as a five-year postpartum sexual abstinence without stress, a counting system which goes up to three, and tremendous conservatism in the face of great pressures to change their way of life" (p. 1). Data on the Dani numeral system, other than Heider's, indicate that it is a 2-cycle variant with the basic numerals 1, 2, and 3 and 4=2+2, after which digit-tallying may occur (Figure 8.2).

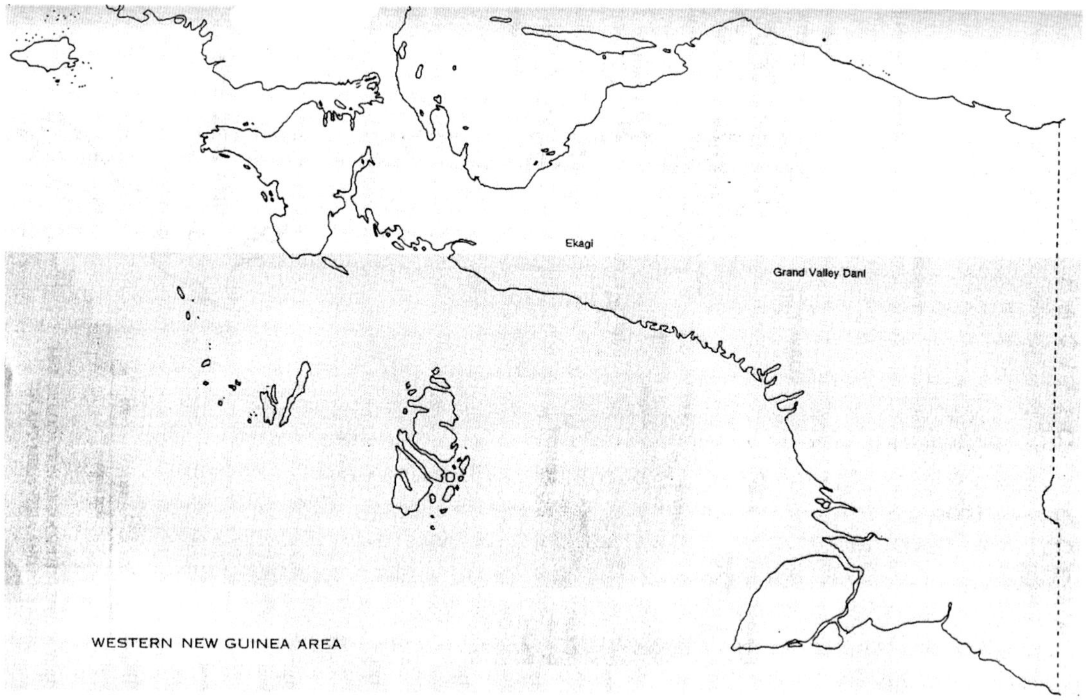

Figure 8.2. Location of societies cited in text (West Papua).

Heider (1991) indicated that the enumeration of objects, even in ceremonial gift-exchange, is not regarded as important in Dani society:

> the point is that the Dani get along very well without quantifying their environment. In other cultures, numbers are useful for tallying masses of identical things like monetary units. But the Dani do not use money. Or numbers can be useful for dealing with masses of similar things if one is willing first to consider them as identical and then tally them. If a man has a

> herd of pigs, it is rarely more than a dozen or two, and he can keep in mind the age, sex, and markings of his individual pigs. At a funeral, the people who were concerned could keep track of what sort of pig was brought by whom and add it to their knowledge of the current state of gifts and debts within the group. I would go to a funeral and sit counting the number of pigs, and in the end I would say "this was a 24-pig funeral". A Dani would know other, richer things about the funeral, but he would not be interested in such a tally. (p. 33)

The circulation of goods and the occasions for exchange occur in three circumstances: family consumption, ceremonies (both within-society exchanges), and the between-society exchange of outright trade. With the first two, the commodities exchanged are those things which everyone possesses and thus the exchange of these is an affirmation of the social rather than the economic basis of exchange. External trade, on the other hand, is carried out to obtain imported goods which are not present in the society; these are "adze stones, exchange stones, fine furs, feathers, and wood in exchange for goods like salt, pigs, and nets which are produced in the Grand Valley" (Heider, 1991, p. 33). Such transactions are normally carried out by barter in the absence of traditional money.

The Loboda of the Milne Bay Province (PNG)

Loboda village is situated at the northern end of Normanby Island in the Milne Bay Province (PNG). The people of Loboda speak Dobu, an AN language belonging to the Papuan Tip Cluster and, even though this is a descendant of POC, it possesses a numeral system with a (5, 20) cyclic pattern rather than the POC 10-cycle pattern (see Appendix B). In the mid-1970s, the Loboda people were studied by the anthropologist Carl Thune as part of the Indigenous Mathematics Project. Thune (1978) said of the traditional counting system:

> the highest number I ever heard mentioned in the Loboda numerical terminology was *tau nima*, "five twenties" which is one hundred. However it should be clear that it is in theory possible for the numerical system to go somewhat higher than this … Nevertheless, today at least, use of such higher numbers is only a theoretical possibility. … Outside of situations in which people were teaching me the traditional counting system, I never heard it used for numbers above five. (p. 71)

Thune characterised the Loboda as an example of a "non-numerically oriented culture" and adds that "although enumeration and counting are possible for the Loboda people, in fact they occur only rarely and are, for the most part, unnecessary" (Thune, 1978, p. 61). In the foundational spheres of life, people might report, in discussing a feast, the number of pigs killed or the number of yams distributed; also, the number of days in a journey or the number of months to a particular festival might be stated, "nevertheless, today higher numbers are rarely if ever used, and I suspect this was always the case" (Thune, 1978, p. 78). Thune (1978) attributed this lack of interest in the enumeration of totals to

> the fact they do not think of their world in terms of units which are easily organised or represented with numbers. There are, for example, no traditional units for measuring distance or height which could be counted in the way the English (sic) units of meters or kilometers can be counted. (p. 72)

A length of shell money is described, not against some absolute standard, but by relative and uncountable measures, for example comparison against a particular person's arm. In the description of people's ages, broad time-periods are used rather than a numerical specification; thus there are terms for infant, child, adolescent boy or girl, and so on. In ceremonial gift-giving, a large amount of yams may be given to a recipient group who will, at some later date, give back an equivalent amount. Thune (1978) pointed out,

> In fact, no one ever counts the yams of the first gift to ensure that a numerically equal repayment can later be made. Rather, all the yams making up the gift will be heaped together in one pile and it is this collective gift … which must later be repaid. (p. 75)

The gift which is repaid must be the same visually judged size as that given.

In any feast exchange, there are usually several different categories of gifts. Thus, in addition to yams, there may be betel nuts, tobacco, a pig, and so on. These are non-interchangeable and distinct categories which cannot be subsumed into a single total. Each category of gift must later be repaid by the same category and an equivalent amount. The amounts to be repaid are recalled by naming the individuals among whom the original gift was divided.

> By thinking on the one hand of the gift as composed of a number of categories of items to be repaid in kind, and on the other as divided among a series of specific individuals, counting is rendered not only unnecessary but irrelevant. (Thune, 1978, p. 75)

One final point made by Thune is that the nature of Dobu, the language of Loboda, to some extent precludes the extensive use of numerals as adjectives to qualify nouns. The essential feature is that many nouns are grammatically treated as mass nouns which are neither counted nor pluralised. Thune points out, for example, that the word for yam, *bebai*, is treated as a mass noun and is, for grammatical purposes, unable to be modified by a number or an indefinite article. Indeed, this grammatical treatment of the yam noun mirrors the way in which yams are treated in real life, that is as an indivisible mass rather than as a collection of countable, discrete yams, a fact that has mildly Whorfian overtones.

The Ekagi of the Wissel Lakes (West Papua)

The counting system of the Ekagi was described previously in Chapter 6, particularly because of its possessing an unusual (10, 20, 60) cyclic pattern. Ekagi, or Kapauku, is a NAN language belonging to the Trans New Guinea Phylum although, as discussed earlier, its numeral system shows indications of AN influence (see Table 6.6). The Ekagi inhabit the Wissel Lakes region of the west-central part of the highlands of West Papua. In the mid-1950s they were studied by the American anthropologist Leopold Pospisil. Ekagi culture, Pospisil (1963) indicates, is wealth oriented:

> This means that the highest prestige in this society and the highest status of political and legal leadership are achieved not through heritage, bravery in warfare, or knowledge and achievements in religious ceremonialism, but through the accumulation and redistribution of capital. The major and often the only source of capital, generally in the form of shell money, is successful pig breeding. (p. 5)

The Ekagi subsist by the cultivation of sweet potatoes which are also used as food for pigs and, thus, agriculture in an indirect way "not only creates wealth but also provides a basis for acquiring political and legal powers" (Pospisil, 1963, p. 5). The cowrie shell money is a scarce resource which is not produced by the people themselves but has to be acquired through trade with the lowland, coastal people.

Unlike the societies discussed in the previous case histories, the Ekagi, according to Pospisil, have a quantitative rather than a qualitative world view.

> Their highly developed decimal counting system enables them to count even into thousands … Not only is counting used in their economic transactions, but the people show a peculiar

> obsession for numbers and a craving for counting. They count their wives, children, days, visitors at feasts, and, of course, their shell and glass bead money. (Pospisil, 1963, p. 94)

Value is placed on large numbers, whether of people, pigs, or objects. Pospisil (1963) said that a

> pig feast is an impressive affair. On such occasions there may be as many as 2 000 visitors, and the slaughtered pigs may be counted in the hundreds. The trade turnover, in terms of shell money, may be quite considerable. (p. 77)

On another occasion, an Ekagi youth, who had counted the glass beads acquired by Pospisil (1963), reported that

> You have 6 722 beads in your boxes. That means that you have spent 623 beads since Gubeeni counted your money three days ago. I would suggest that you order more beads in about thirty days so that you do not run out of funds. (p. 94)

The Melpa of the Western Highlands Province (PNG)

The people speaking the Melpa dialect of the Hagen language live south-east and north of the township of Mount Hagen in the Western Highlands Province (PNG). Hagen is a NAN language belonging to the Trans New Guinea Phylum. The Melpa people were studied by the anthropologist Strathern in the 1960s and, in particular, he has published an account of the complex ceremonial exchange system called the *moka*, "an institution linking groups together in alliances and … a means whereby men try to maximise their social status" (Strathern, 1971, p. xii). The Melpa, like the Ekagi, subsist "by sweet potato horticulture, supplemented by cultivation of bananas, sugarcane, taro, yams, maize and cassava" (Strathern, 1971, p. 9). The wealth of a Melpa man is largely determined by his possession of two commodities, pigs and pearl shells, both of which are used in the *moka*, although other forms of wealth may also supplement these: live cassowaries, long bamboo tubes of oil for decorating the body, packs of salt, cassowary eggs, pandanus fruits, and, nowadays, live steers and cash (p. 94).

In the *moka* itself, and in the preliminaries leading up to it, the counting of both pigs and shells is important. However, during the *moka*, particular attention is paid to the display of wealth items, the shells, for example, being arranged in specially decorated display houses. Prior to the *moka*, the donor and recipient groups meet to negotiate the number of pigs and shells to be given. Wooden stakes are placed in the ground, each one representing a pig which, during the *moka*, will be tethered to the stake. The stakes are counted "in twos, making these into sets of eight or ten, and a supporter keeps a record on his fingers of how many sets are taken: for each set he bends one finger down" (Strathern, 1971, p. 116). The Melpa counting system is a somewhat complex one showing features which arise from treating the four fingers as a "hand", the thumb being tallied separately as 5. There is a distinct term for 8, *engaka* "man", that is the eight fingers of a man. The word for 10 is "two thumbs", that is the 8 fingers augmented by 2 thumbs (see Lean's Appendix data which are incorporated into Appendix B of this book). In the *moka,* the chief donor aims at giving a "grand set" of 8x8, 8x10, or 10x10 pigs. Shells are counted in a similar way in "sub-sets of two or four making up sets of eight or ten" (Strathern, 1971, p. 120). The *moka*, then, as a central feature in the social, economic, and political life of the Melpa, requires that the ongoing process of exchange be accounted for numerically: the number of pigs and shells involved must be counted and recorded by tally by both the donor and recipient groups for future reference. However, while the counting of gifts is important, so are other aspects of the ceremonial exchange such as the visual display of wealth, a feature shared with the Adzera in their gift presentation of bananas.

The Mountain Arapesh of the East Sepik Province (PNG)

The Mountain Arapesh are located in the East Sepik Province (PNG) in villages situated along the coast and in the region stretching southwards to the mountains. Mountain Arapesh is a NAN language belonging to the Torricelli Phylum and its counting systems were discussed in the previous chapter. The main system has a (2′, 4, 24) cyclic pattern but there is also a special system possessing a 3-cycle (see Appendix C, languages related to Chapter 8). The Mountain Arapesh people were studied in the 1930s by the anthropologist Fortune (1942).[1] Prior to this, Fortune had also studied the Dobu-speaking people in the Milne Bay Province of which the Loboda people, discussed earlier, are a part.

Fortune noted that in using their 2-cycle variant numeral system, with its secondary 4-cycle and tertiary 24-cycle, tallying of each 24-cycle is recorded with a stick or peg before proceeding to the next. Only certain types of objects are counted using this system: "coconuts, small yams, taro, arm-rings, dogs' teeth, house poles, house posts, breadfruit, sago frond sheaths, pots plates, spears, arm-bands, bamboo lengths, sugarcane lengths, eggs, birds, lizards, grubs, fish, [etc.]". However, a 3-cycle system is used for counting "betel nuts, thatch shingles, coconut, sago, betel nut palms, big yams, packets of sago, wild game, sheets of sago bark, packages of vegetable greens, braids of tobacco, packets of tobacco [etc.]" (p. 60). It is not uncommon, particularly in the nineteenth century literature on traditional numeral systems, to find somewhat disparaging remarks about low-cycle systems and of the ability of the people possessing such systems to count. Fortune (1942), however, indicated that with the Mountain Arapesh

> counting is done with great facility and ease with this conventionalisation of very few special roots. To suppose that the paucity of the Papuan languages in root words for numerals makes counting difficult to the Papuan is quite incorrect. The Arapesh people count rather more quickly than the Melanesian Dobuans, who use a decimal system with many more root terms. An unambigious conventional method, with or without many root terms, is all that is necessary. (p. 59)

The Arapesh engaged in three types of exchange. Oliver (1989) said:

> first were the random and informal transactions - barter, loans, "gifts", and so forth - whereby goods passed between kinsmen and acquaintances … Second were the trade expeditions … undertaken by small parties of men to exchange local goods for those not obtainable at home … These exchanges were by barter, accompanied by haggling …The third type of external exchange took place through formally constituted partnerships; and far from bartering, each such transfer was viewed … as a spontaneous "gift". (pp. 532-533)

None of these transactions, however, involved the use of traditional money although rates of exchange between various commodities were established.

The Woleai People of Micronesia

The previous case studies have derived from societies located on or near the New Guinea mainland. With the exception of the Loboda, none of the societies dealt with has a primarily maritime culture. We deal here, then, with a maritime culture par excellence, that of the people of Woleai atoll in Micronesia who speak an AN language (Table 6.19) and whose way of life probably approximates some aspects of its ancestral POC culture rather more closely than those of AN groups, like the Adzera, which have largely abandoned their maritime culture and economy.

The people of Woleai, part of the Caroline Islands, make use of numeral classification in common with other groups in Micronesia according to American anthropologist William Alkire (1970). The

[1] Gerstner published a grammar of it in the Micro-Bibliotheca Anthropos (microfilm) series in 1963.

Woleai categorise their universe of countable objects into at least 22 classes; the defining characteristics of the classes are, for the most part, related to certain physical attributes of the objects counted or to particular ceremonial uses to which the objects are put. A delineation of various members of the classes gives some indication of the wide range of countable objects: fish of many kinds, turtles, domestic animals, people, canoes, roof thatch, chickens, eggs, stalks of bananas, garlands, shell belts, paddles, shells, leaves, stones, palm fronds, and so forth (Alkire, 1970, pp. 9-10). While the enumeration of these objects is a common part of everyday life, Alkire indicated that counting, and indeed counting in rather special ways, acquires a particular importance in ceremonial circumstances, as we have found with several of the highlands societies on mainland New Guinea.

Alkire wrote that "On Woleai the ceremonial accumulation and redistribution of mature coconuts (*cho*) is part of all funeral rites". People from several islands are usually involved in the funeral rite of exchanging coconuts. "The exchange, however, is purely symbolic for the nuts are never moved from one island to another … There is no need to transport the 'real' nuts from the respective islands across the lagoon since the totals are the same and one would simply end up with the same number of nuts" (Alkire, 1970, p. 11). The Woleai possess a 10-cycle counting system. In counting coconuts in normal, everyday circumstances, the nuts are grouped in tens and a "group of ten" numeral classifier is used in their enumeration. However, in the ceremonial funeral exchange, this procedure changes: "In this context a Group of Ten in reality contains only eight nuts … The nuts are segregated into lots of eight and each of these is called a Group of Ten" (p. 12). Alkire gave the example of 1 529 Groups of Ten being counted and which, under normal circumstances, would mean that 15 290 nuts had been accumulated, but in these special circumstances the actual total is $1\,529 \times 8$ or 12 232. This unusual treatment of the number 8 in Woleaian ceremonials is one example of the special significance attributed to that number in many Pacific cultures. Indeed, Biggs (1990), a noted Polynesian scholar, described the large number of instances in which the number 8, and its association with sacred matters, supernatural power, high rank, and large size or totality, is a common feature of Pacific societies. Alkire (1970) also observed that "on many ceremonial occasions the number four (its multiples and divisions) is of great importance" (p. 16). Interestingly, 8 is also important for relationship reasons in Duna in Hela Province, PNG, a NAN language with an asymmetrical body-tally systems with words up to 13 (avoiding 14); 8 is often named 1 as a new cycle of relationships begin (Sakopa, personal communication, 2015).

Commentary

These case studies provide a representative survey of the various ways in which the foundational societies, existing in the region under consideration, place importance on number and enumeration. It is clear from the examples given that there is considerable diversity in the degree to which number holds a privileged position in these societies. In the NAN-speaking, horticultural society of the Ekagi, for example, number and counting play a prominent part indeed, whether in normal, everyday life or in ceremonial occasions. It will be recalled that their numeral system with its (10, 20, 60) cyclic pattern is probably not their original system in that it shows evidence of borrowing from an AN source and it may well be that this augmentation of their numeral lexis has been prompted by the prominent position that the Ekagi accord to number.

The other NAN-speaking societies in our sample do not show the same degree of preoccupation with number as the Ekagi do. The Melpa's complex exchange system, the *moka*, provides the context for the counting and display of wealth to enhance the status and prestige of highlands Big Men. Even though the Melpa numeral system is a 2-cycle variant augmented by digit tallying, this is sufficient for the task of counting the *moka* gifts, even in the case where the "grand sets" of 64, 80, and 100 are achieved. The Grand Valley Dani, who possess a 2-cycle variant system, do not, according to Heider, place any special emphasis on enumeration, even in ceremonial contexts. On the other hand, the Mountain Arapesh also have a 2-cycle variant numeral system which, as Fortune pointed out, is sufficient for counting a wide variety of objects, if necessary up to multiples of 24. Thus, while the Ekagi may well have adopted a 10-cycle system (and thereby abandoning a lower-cycle one) in order to satisfy their preoccupation with number, the possession of a 2-cycle system, or a variant, is not necessarily an indication that the societies having such systems are unconcerned with number, or that its members are unable to count with facility.

The AN-speaking societies discussed above, the Adzera, Loboda, and Woleai, each derive from a common ancestor, the society of POC-speakers. The maritime Woleaian culture probably resembles the culture of its distant ancestor more closely than those of the other two do. Number and counting play an important part in both the everyday and ceremonial aspects of Woleai life and large numbers of objects are counted with facility using their 10-cycle system. The Loboda, however, while still retaining some aspects of a maritime culture, have not retained a POC-derived 10-cycle system and now have a (5, 20) one. Even though this system would be quite sufficient for the enumeration required of a maritime economy, in fact the Loboda place little importance on counting and instead organise their transactions in largely non-numerical ways. In ceremonial circumstances, the Loboda place emphasis on the visual display of wealth; while the repayment of gifts has to satisfy quantitative requirements, quantities are judged by visual inspection rather than by counting. The tendency to regard quantities as indivisible masses is reflected in the grammar of their language in which many nouns are treated as group nouns which cannot be modified by numerals. By contrast with the other two societies, the Adzera have long ago left their original maritime environment and have acquired an economy which is heavily influenced by their inland, NAN-speaking neighbours. The Adzera numeral system has also undergone a change; the POC-derived 10-cycle system is no longer evident and a 2-cycle system (with digit-tallying) is used. In ceremonial circumstances, the counting of gifts is less important than providing an impressive visual display of wealth.

It is apparent, then, that there is no simple way of characterising the foundational societies dealt with here with regard to the way in which they accord importance to counting and number. Also, the degree of sophistication of the numeral system of a given society is not necessarily an index to the extent to which enumeration is important or can be carried out with facility. The possession of a large primary cycle system may reflect this, however the converse is not true: the possession of a low primary cycle system does not necessarily indicate a lack of interest in number, nor in the ability to count. Finally, it is apparent that, in each of the societies discussed here, ceremonial institutions involving displays of wealth play an important social, economic, and political role. The ability to amass large quantities of wealth items accords status and prestige to clans and individuals. The judgment of quantity, however, varies in each society from the impressionistic to the precise. With the Melpa and Woleai, certain numbers such as 8 or 10 are particularly significant in ceremonial situations. However, to invest numbers in this way with sacred and power and metaphysical connotations is not necessarily a feature of all societies in New Guinea and Oceania. In this respect, the evidence does not universally support Crump's (1990) assertion that

> as to the differences between traditional and modern thought, it is not so much that the former is concerned with application, and the latter with internal consistency, but rather that there is, as between traditional and modern societies, a pronounced preference for metaphysical as opposed to physical applications. (p. 287)

What Is, And What Is Not, Counted In Foundational Societies?

The primary focus of the previous section was the degree of importance with which number and counting are held in a sample of foundational societies. In this section, we focus instead on the following questions:

(1) What are the countable objects in a particular society? and
(2) In what circumstances are they counted?

In addition, various other questions related to these will be considered, inter alia, where data are available; these are:

(3) Do societies differ in the type of economies that they have and the nature of exchange in which they engage, and do these affect the degree to which counting is carried out?
(4) Are different types of objects counted in different ways?
(5) Are there ways of recording totals?
(6) Are there certain things which are not counted?

In order to address these questions, brief case studies illustrate the ways in which a sample of societies deal with enumeration. A series of NAN-speaking societies, located in both the highlands and in the coastal/islands region of New Guinea, will be considered first, followed by a number of AN-speaking societies located in both New Guinea and Oceania.

The Kaugel Valley People of the Western Highlands Province (PNG)

The people of the Kaugel Valley in the PNG highlands speak the Gawigl or Kakoli dialect of the NAN Hagen language. Their "systems of reckoning" were the subject of an article by Bowers and Lepi (1975). The Kaugel people have an agriculture-based economy not unlike that of the Melpa, discussed earlier, another sub-group of Hagen-speakers. Also like the Melpa, the Kaugel people have a complex exchange system in which the counting and tallying of wealth items is of central importance. Bowers and Lepi indicated that "Kaugel kin groups … formally distribute large numbers of goldlip pearlshells, pigs, game animals, and other valuables. There are two forms of pig distributions; in both, thousands of animals may change hands in a single day" (p. 312). During the ceremony, pigs are tied to stakes in long rows, sometimes half a mile long: "the entire prestation is recorded by one or more core donors: important men who publicly run along the rows in a stylised manner, counting objects in the collective gift" (p. 312). A particular set of terms is used in carrying out this formal, ceremonial count.

The terms used in the formal count are also used to count large numbers of objects in normal, everyday circumstances: the posts needed to build a house or the number of people coming to a feast. These formal counting terms are not, however, the only ones used. There is, for example, a related but somewhat different set of terms used for counting pandanus fruits. Also, Bowers and Lepi noted that "different, mainly sequential, numbering systems are used in the Kaugel for ordering or reckoning certain qualities, quantities or social features" (p. 319) (see Table 7.5). There are terms for designating the birth-order sequence of siblings, particularly males. Months are, similarly, counted as a sequence of six pairs of "first born" and "last born". The number of days before or after the present can be reckoned even though the days themselves are not assigned numbers. Finally, Bowers and Lepi made the important point, valid of many of the foundational societies discussed here, that "counting does not exist in isolation. It quantifies and qualifies relations between people, objects and other entities" (p. 322).

The Kewa of the Southern Highlands Province (PNG)

The Kewa people of the Southern Highlands Province (PNG) speak Kewa, a NAN language with three major dialects: East, West, and South. Their language has been studied extensively by two SIL linguists, the Franklins (Franklin & Franklin, 1962, 1978). Each of the Kewa-speaking dialect groups possesses a 4-cycle numeral system (an example of which was given in Chapter 7 and Table 7.3) together with a body-part tally method (Table 7.4) of the type discussed in Chapters 3 and 7; the East and West groups have tallies which utilise 47 body-parts and the South dialect group has a tally utilising 35 body-parts. While the 4-cycle numeral systems are used in both everyday and ceremonial circumstances for counting pigs, shells, horticultural produce and other items, the body-part tallies are not used for specifying exact numbers: their main function is calendrical. The Kewa have festival and dance-cycles, the preparations for which take place over many months in a strictly ordered sequence: the passing of each month, that is full moon, is tallied on the body-parts and thus these function as ordinal systems rather than counting systems. The Franklins gave examples of such festival cycles for the East and West Kewa groups which extend over periods in excess of 40 months.

The People of Kiwai Island, Western Province (PNG)

Kiwai Island, about 100 kilometres long, is situated in the mouth of the Fly River in the Western Province, on the south coast of PNG. The people of the Island speak a dialect of Southern Kiwai, a NAN language (see Chapter 3 and Table 3.1), and possess a "pure" 2-cycle numeral system. The numerals of this system are used in the enumeration of relatively small numbers of objects. Smith (1978) said that

> the most common method of keeping count of large numbers in traditional Kiwai society was the use of tallies. Tally sticks each representing an object could be kept in a bundle or tied to a string. Tallies were used to represent the number of heads captured by a man in battle, or the pigs killed in the bush, or any other significant number. (p. 56)

The Kiwai had ceremonial feasts in which prestations of food and various kinds of wealth items were made. In these,

> presents were often displayed on the wooden structure called a gaera and one old informant told me how he once kept a tally of the number of bunches displayed at a feast so that his group could repay slightly more when the return feast was made. (p. 57)

In tallying the number of days before a return feast was to be held, the donor and recipient groups would part, each having a bundle with the same number of sticks. Each day, both groups would discard a stick until the bundle was exhausted and the pre-arranged day for the feast had thus been reached.

The People of Rossel Island, Milne Bay Province (PNG)

Rossel Island lies south-east of the New Guinea mainland in the Milne Bay Province (PNG). The people of the Island speak Yele, a NAN language belonging to the East Papuan Phylum, and possess a 10-cycle numeral system (discussed previously in Chapter 6, Table 6.3) which shows some AN influence. The Rossel Islanders were studied by the Assistant Government Anthropologist for Papua, W.E. Armstrong (1924). Armstrong (1928) indicated that there exist two types of shell money: "The one, known as *ndap*, consists of single pieces of Spondylus shell, ground down and polished … The other kind of money, known as *nkö*, consists of ten disks of shell, perforated and strung together" (p. 59). These two types are divided into a number of "denominations": the *ndap* has 22 main values and the *nkö* has 16 values. Armstrong (1928) said,

> Payments of money are, perhaps, the most important constituents of marriage rites, mortuary rites, and many other ceremonial activities. I use the term "money" advisedly, for the objects [described] are systematically related as regards value; and any commodity or service may be more or less directly priced in terms of them. (p. 59)

There is a fixed and limited amount of both types of shell money on the island and there is a good deal of lending and borrowing, sometimes carried out by an intermediary broker. The rules for repayment of loans are complex and involve, say, a given denomination of *ndap* being lent and then subsequently repaid by a higher denomination according to the time period of the loan. Armstrong (1928) noted that, in keeping account of such transactions,

> it became obvious that an amount of counting was done … The frequent ceremonial of stringing *nkö* together, so as to make long ropes of money, is the reason for the Rossel Islanders' proficiency in counting, for counting up to the thousands would hardly be required in any other department of life … An investigation of this latter point showed that the native had no difficulty in counting up to 10 000, or even further. (pp. 76-78)

The Ponam Islanders of the Manus Province (PNG)

Ponam Island is located immediately to the north of Manus Island in the Manus Province (PNG). The islanders speak Ponam, an AN language belonging to the Admiralties Cluster. They possess a 10-cycle numeral system of the "Manus" type, discussed in Chapter 4, and also possess numeral classification in which numeral roots are suffixed by a classifier (see Appendix B). The counting and calculating approaches of the Ponam Islanders were studied by Achsah Carrier (1981) as a contribution to the Indigenous Mathematics Project. Carrier notes that "In Ponam … there is only one basic counting system … The Ponam language does contain a system of numeral classifiers, but … these are not an alternative counting system" (p. 466). Carrier also indicated that the Ponam have considerable skill in arithmetic: "Almost everyone on the island can count from one to ten thousand in both Ponam and Pidgin" (p. 468). Generally speaking, counting, by adults, is not accompanied by digit-tallying: "Young children are sometimes taught to count on their fingers, though more usually the process is illustrated by counting objects" (p. 468). In many NAN languages it is often the case that certain number words are derived from the names of body-parts. Even in the POC 10-cycle system the word for 5, **lima*, is identical to the word for hand. Carrier (1981) noted, however, that

> Ponams do not associate numbers with the body, even minimally … Although in Ponam the number five (*limef*) is almost identical to the word for hand (*lime-*), the connection is not perceived as significant. No one would ever translate *limef* as "hand" (p. 468)

Fisher (~2010) noted that for Kuruti, another language on Manus Island, there are classifier suffixes for different parts of the banana tree. He also noted the habit of rounding down to speed addition and representing with fingers, for example, one finger for 100 000.

Carrier (1981) reiterated the point made by Bowers and Lepi about the Kaugel, that counting does not exist in isolation, divorced from social relations. Carrier indicated that "all important social events on Ponam are the occasion for an exchange" (p. 472). In the exchange of gifts, the Ponam, in common with many other foundational societies, place importance on the visual display of wealth items. Counting of individual contributions made within the donor group is carried out; the total amount of the gift given to the recipient group is, however, not usually announced and often no one knows it. Carrier (1981) said

> the size of a gift reflects the donating group's sense of its relation to the focal donor and also reflects the group's individual members' sense of relation with their various kinsmen … the numbers announced in a formal count thus expresses the quality of relationships between groups and individuals rather than the quantitative strength of particular groups or individuals. (pp. 474-475)

While the Ponam do count a wide variety of objects, in both everyday and ceremonial circumstances, they do not, however, count people nor do they show any interest in doing so: "apparently these quantifications tell them nothing interesting about social relations" (Carrier, 1981, p. 471).

Frances Kari whose doctoral study was on attitudes to ethnomathematics also developed a program in EXCEL for showing the repetitive pattern of counting to large numbers in his Manus language of Baluan (personal communications, 2003). No doubt the idea of abstract numbers was absorbed into their mathematical thinking from a cultural background which were transferable to his modern programming skills.

The Mengen People of the East New Britain Province (PNG)

The Mengen, or Maenge, people live in the Pomio region of the south coast of East New Britain. They speak an AN language, Mengen, which belongs to the North New Guinea Cluster and, in common with all the members of this Cluster, no longer have a POC-derived 10-cycle numeral system, but now possess a system with a (5, 10, 20) cyclic pattern (see Appendix B). A French anthropologist Michel Panoff (1969, 1970) made the point that

> in the old days, informants insisted, all things were counted - the booms of outrigger canoes as well as the pegs inserted in the float below, the sticks used in gardening magic as well as the leaves collected for divination purposes, and of course pigs and the shell rings recognised as ceremonial money. (Panoff, 1970, p. 364)

In addition to these items, Panoff mentions the counting of coconuts in groups of four, taro tubers in groups of twenty, a large catch of fish distributed in fives, tens, and twenties, and cured tobacco leaves.

Panoff (1970) reported that the Mengen, unlike the Ponam, engage in digit-tallying:

> Their word for "counting" is *sising* … which primarily denotes the process of tallying off a number with the help of fingers and toes … Mental counting was practised and called *lau (long, lona …) -sisi*, that is "inside counting", but it was much less frequent than the use of the finger and toe method.… (In carrying out enumeration) the most common operations were, of course, addition and subtraction, but multiplication was by no means exceptional. (These operations were carried out mentally) at least when the result arrived at was not higher than 20 … In contrast with these three operations, division, which is called t*avoang*a ("portioning"), can only be done by counting with the aid of markers. (pp. 362-364)

The Arosi of the Solomon Islands

The Arosi live on the island of San Cristobal in the Solomon Islands and speak the AN language Arosi, a member of the South-East Solomonic Group (see Appendix B). The numeral system, like all those of the AN languages of the Solomons, is a 10-cycle one and there are distinct words for 100 and 1000. Fox (1931) said that, in addition to the basic numeral system,

> almost every children's game has its counting song with quite different numerals, for example they may all of them be names of trees; and also many special objects are counted in a different way … It would appear that objects were once divided into a number of categories; with some things counting only proceeded as far as ten, with some to a hundred, and in the case of coconuts to twenty million. (p. 238)

In one category there are yams, taro, bananas, stones, and mangoes, all of which are usually counted in groups of five: there are distinct words for groups of 5, 10, 25, 50, 100, 1000, 10 000, 100 000 and 1 000 000. The members of other categories which are counted somewhat differently include coconuts (counted in pairs), banana shoots, sago palm fronds for thatching, pigs and dogs, opossums, fish, eels, breadfruit, dogs' teeth, bats' teeth, porpoise teeth (the various teeth are counted in fours), and shell money. The last is strung and a length of four fathoms is regarded as a unit; there are special terms for groups of 10, 25, 50, 100 and 1 000 units of shell money.

> (Fox talks of) exogamous clans with matrilineal descent … and each clan has a totem, which is generally a bird (Fox 1919:101). Clan names are shared across some language borders: Fox said "the interesting point is that here we have clans quite new to San Cristoval" (Fox 1919:153). The Russells, too, have a new tribe (clan)—the process of adoption of new clans and tribes seems to be widespread, and due to movement patterns like those described by Bathgate (1985). (Terrell, 2011, pp. 316-317)

With many aspects of material culture also shared across islands and language groups, it is understandable that there is significant exchange of counting systems across NAN and AN groups as well as within AN groups. One aspect of trade includes the groupings of commercial crops so that words refer to a recognised size of certain objects such as coconuts. Groups of ten are often matched to pulling off one leaf of a fern and put in heaps of 100. The meat of 300 fill a bag (personal communication, Banga Island teachers, 2015). Roviana has a distinct word for 20 while most other decades are multiples of ten (Table 8.1) suggesting some influence from NAN digit tally counting but more

Table 8.1
Roviana Decades

	Roviana
10	*manege puta*
11	*manege eke*
20	*hiokona*
30	*tolo ngavulu*
40	*mande ngavulu*

likely related to pairs counting which is found in some Polynesian systems as pairs are classified in counting as mentioned below (Kokeqolo teachers, personal communication, 2015; Ray (1926, p. 560)).

Some, but not all, Polynesian languages have cyclic patterns at 2×10^2 and 2×10^3 (not a base system at 20^2 and 20^3). This development has developed through cultural practices of counting in pairs, often to large numbers (Bender & Beller, 2006; Best, 1907). When counting in pairs using the words used for counting single items, it was understood that the counting of those objects was in pairs. It was a common practice, also found in some AN languages to count collected birds or other objects in pairs. In the next chapter, we will also look at other collection counting by the Tolai of East New Britain, PNG. A Polynesian example, Maori, is given in Table 8.2 and Pukapuka in Appendix B. Further Micronesian examples are given in Table 8.3.

Rennell and Bellona are two islands in the southern part of the Solomon Islands. The inhabitants are Polynesian and speak Rennellese, a Polynesian Outlier (see Appendix B). The American linguist Samuel Elbert has produced both a dictionary and a grammar of Rennellese, the latter having some cultural notes on counting. Elbert (1988) noted that, traditionally,

> (much of a chief's life) consisted of fishing and raising fine gardens, and presenting the fruit of the land and the sea, carefully counted, first to the gods with impressive rituals and then to relatives and allies. A chief's prestige was gauged by the size of the offerings he was able to amass. (p. 186)

Prior to the Second World War, the Rennellese often engaged in competitive gift giving which, in one recorded instance, resulted in the distribution of 10 000 coconuts and 7 600 banana bunches. Another use of counting was in establishing the dimensions of houses, canoes, gardens, and mats. Not everything, however, was counted. Elbert (1988) said "No one knew or was at all interested in his own age … Years and generations were not counted at all. Time was told by looking at the sky" (p. 286).

The Rennellese, as is common with all Polynesians except those speaking Faga-Uvea, possess a 10-cycle numeral system. There are 15 categories of countable objects and the terms for 10, 100,

Table 8.2
Single Mode and Dual Mode of Counting in the Foundational Māori Number System

Number	Single Mode *tatau takitahi*	Dual Mode *tatau tōpū*	Composition
1	*tahi*		
2	*rua*	*tahi pū*	1 ·2
3	*toru*		
4	*whā*	*rua pū*	2 ·2
5	*rima*		
6	*ono*	*tor pū*	3 ·2
7	*whitu*		
8	*waru*	*whā pū*	4 ·2
9	*iwa*		
10	***ngahuru***	*rima pū*	5 ·2
20	*tekau*	***ngahuru** pū*	10 ·2
100	*tahi **rau** takitahi*	*hokorima[pū]*	50 ·2
200	*rua rau (takitahi)*	*tahi **rau** [pū]*	100 ·2
1000	*tahi **mano***	*rima rau [pū]*	500 ·2
2000	***[tini]***	*tahi **mano** [pū]* ***[tini]***	1000 ·2

Notes. Data are from Bender and Beller (2006, p. 385) and (Best, 1907). Prefixes are omitted for easier comparison; power terms are highlighted bold faced.

Table 8.3
Syntactic and Numerical Characteristics of Numeral Classifiers in Samoan

	Number terms (with 2 [*lua*] and 20 as example) Classifier Category N-C	C-N	Factor
TYPE I			
sautua- lines, rows, thicknesses, layers (-fold)	-	*sautualua: sautua-2*=2(fold)	-
tau- things in bunches or clusters like coconuts	-	*taulua: tau-2*=2	-
tua- rows, lines, layers, thicknesses (-fold)	-	*tualua: tua-2*=2(fold)	-
TYPE 2			
aft- packages of small fish wrapped in leaves	-	*afflua: afi-2*=20	-
toi- balls or lumps of food	-	*potoilua: potoi-2*=20	-
tu'e- crabs, lobsters	-	*tu'elua: tu'e-2*=20	-
-*'au-* bananas, yams, etc. (in bunches)	*lua'au: 2-'au*=2	*'aulua: 'au-2*=20	-
-*fua-* breadfruit, coconuts, fowls, some shellfish	*luafua: 2-fua*=2	*fualua: fua-2*=20	-
-mata- taro	*luamata: 2-mata*=2	*matalua: mata-2*=20	-
-*lau-* large fish (i.e., of a size that makes them suitable for cooking in leaf wrappings)	*lualau: 2-lau*=2	*laulua: lau-2*=20	-
'ofu- items of food (except fish) wrapped in leaves	*lua'ofu: 2-'ofu*=2	*'ofulua: 'ofu-2*=20	-
-*tino-* skipjack	*luatino: 2-tino*=2	*tinolua: tino-2*=20	-
TYPE 3			
-*aea* coconuts	*luaea: 2-aea*=40	-	-20
-*'aui* skipjack	*lua'aui: 2-'aui* =20	-	-10
-*oa* coconuts, young pigs	*luaoa: 2-oa*=4	-	-2

Note. Data are from Bender and Beller (2006,. p. 394) who adapted from Milner (1966) and Mosel and Hovdhaugen (1992, pp. 246-250). The classifier is suffixed to the numeral in "N-C" compounds, and prefixed in "C-N".

1 000, 10 000, and 100 000 are not all identical for each category. A listing of some of the members of the categories provides an indication of the wide range of countable objects: fish, birds, crustaceans, coconuts, taro, sweet potatoes, bunches of bananas (counted in fours), trees, fathoms of shell money, canoes, and spears. A variety of measurements, each relative to a person's body, exist and are countable: the fathom, for example, is the distance between the fingertips of a person's outstretched arms. Elbert noted that the large numbers, that is those greater than a thousand, "used in food distributions have never been taken too literally but symbolise unfathomably large quantities, which are so admired in counting food" (Elbert, 1988, p. 187).

The Pukapuka of Central Polynesia

The final example in this series of brief counting ethnographies concerns the people of Pukapuka which is located in the central part of Triangle Polynesia, north-east of Samoa. In the 1930s, two American anthropologists, the Beagleholes, studied the culture of this representative Polynesian group and published a monograph in 1938 which included some data on counting and measuring (Beaglehole & Beaglehole, 1938). They reported that the Pukapuka numeral system is a 10-cycle one with distinct terms for 100, 1 000, and 10 000 (see Appendix B). The Beagleholes indicated that there are terms for numbers larger than these and one word, *ye*, that corresponds to infinity; they added, however, that such terms "indicate not so much a definite number as a definite progression of increasing greatness that is more sensed or felt than definitely apprehended" (p. 354). As with the Rennellese, the Pukapukans subdivide their universe of countable objects into at least 14 categories. Some of the members of these categories are: people, coconut shells, fishhooks, oven stones, fishline sinkers, pandanus leaves, mats, fish, taro, crayfish, crabs and shellfish. Coconuts are counted in pairs and the terms used for this are in some cases identical to the ordinary numerals so that, for example, *lua*, that is 2, means 4 coconuts. A special term for 5 pairs of coconuts is used, *yepulupulu,* which is not identical to the numeral 5, *lima*. As is also common with the Rennellese, the Pukapukans have a series of countable length measurements which are made relative to a person's body and which are commonly used to measure fishlines: a fathom, for example, is *ngawa* and 10 fathoms is *kumi*.

Commentary

In discussing the place of counting and number among the Iqwaye people of the Eastern Highlands Province (PNG), Mimica (1988) argued that practical reasons and social systems are not the only reason for people developing natural numbers. To him, there are more primordial roots for numerical expression and its meanings. It seems possible that, on the New Guinea mainland, some foundational societies have existed in relative isolation over several millennia, more or less untouched by outside influence, and whose languages have been subject only to the inevitable and gradual effects of normal linguistic change. Yet it is clear that, for some of the societies discussed here, "the pragmatic exigencies of social existence" have had an effect on the structure of their counting systems and the degree to which these generate a series of natural numbers. Various AN groups, for example, no longer possess numeral systems with the same cyclic pattern as that of their POC ancestor: this is true of the Adzera, Mengen, and Loboda people. Similarly, there are NAN groups such as the Yele and the Ekagi people who have augmented their numerical lexis and have 10-cycle systems generally untypical of NAN languages. Changes to foundational economies, to the manner in which goods are exchanged, and the concomitant changes in the attitudes as to whether it is important to count marketable commodities, appear to be important factors which may influence the structure of a particular society's counting system, inducing innovations of various kinds.

The counting of important items of wealth serves a social function in providing a numerical indication of prestige, power, and obligation. Many of the foundational societies discussed here engage in within-group ceremonial exchange of valued commodities: the prestige of the donor individual or group depends on the quantity of goods amassed; similarly, the obligation of the recipient group must be determined. Societies vary, however, in the way in which quantities of wealth items are determined. The visual display of wealth is an important element in the exchange ceremonials of virtually all the foundational societies of Melanesia, Micronesia, and Polynesia. In certain of them, the Adzera, Loboda, and Kiwai for example, no particular importance is attached to the precise numerical determination of quantity which is, instead, judged by visually apprehended, impressionistic means. Other societies, however, such as those of the Kaugel, Melpa, Woleai, and Rennellese people, place particular importance on the precise enumeration of wealth items, whether these are pigs, shells, coconuts, or yams. Indeed this is usually achieved by carrying out a formal, ceremonial count of each category of wealth item. It is these totals on which the prestige of a group may depend and on which reciprocal obligation is determined. With the Ponam people, such totals reflect the important kinship interrelationships within the donor group.

Within-society exchange of commodities in ceremonial circumstances is but one example of conditions which may occasion counting in a particular society. The non- ceremonial, but still within-society, marketing of goods is another example. However, in this case, we need to distinguish those societies which rely primarily on the bartering of goods, as opposed to those which have economies that are, to some extent, monetarised. Normally, market bartering does not occasion the counting of commodities but involves instead the establishment of a one-to-one correspondence between different collections of commodities. Salisbury (1970) described the way in which bartering occurs among the Tolai women of Rabaul:

> the manner of trade among women was then [traditionally] and is now [1961] a silent one. Each woman sits demurely near her produce, which is divided into separate units - heaps, bundles, packages, or strings. When a trade is made the other party puts down a standard unit of similar value and picks up the exposed goods … The units are related to one another numerically, most of them on a one-to-one ratio. (p. 177)

The Tolai, like many other AN-speaking groups in Melanesia, have a monetarised economy and, in marketing, goods could be purchased with shell money (*tabu*) as well as obtained by barter. Salisbury (1970) indicated that "traded items had standard values in terms of shell money - for example, a *kure* of taro equaled a string of shell money about 6 inches long" (p. 180). The rates of exchange, however, between various commodities and shell money varied according to the supply of, and demand for, the commodities; Tolai men, though not women, engaged in trading (*a nivura*) in order to make a profit. In such transactions, shell money could be counted as individual shells or as multiples or subdivisions of the "standard" length, the fathom (*a pokono*).[2] We discuss Tolai counting systems and culture further in Chapter 11.

All of the foundational societies of New Guinea and Oceania, for which ethnographic data are available, engage in within-society marketing and the most common form of exchange involves bartering in which, on the whole, counting plays little, if any, part. There are, however, societies in which foundational varieties of money, usually shells of some kind, play an important role in market exchange as well as in ceremonial rites connected with birth, initiation, marriage, and death. In market transactions and ceremonial distribution, the counting of individual shells or lengths of strung shells, is an essential feature. Broadly speaking, those societies in which shell money constitutes a principal medium of exchange are AN-speaking groups located away from the New Guinea mainland. Those

[2] Lean spoke Tolai, and spent many vacations over 20 years staying in various villages in and around Rabaul. Various observations on Tolai culture are included in his *An introduction to Tolai* published in 1983 at the PNG University of Technology, and in an enlarged edition published in 1986.

societies which rely largely on bartering in their within-society marketing and which have economies that are not monetarised to any great degree, are, by and large, the NAN-speaking groups located on the New Guinea mainland. This, then, suggests a dichotomy between the AN- and NAN-speaking groups so far as the degree to which counting occurs during non-ceremonial, market exchange.

Such a broad generalisation must necessarily have important exceptions. For example, two societies discussed above, the Ekagi and the Yele, are both NAN-speaking but are such that shell money plays an important and central role in their economies. Oliver (1989), in discussing the Ekagi (Kapauku), said:

> Like all Oceanians, the Kapauku engaged in several types of exchange including services for services, objects for objects, and objects for services, but the noteworthy thing about Kapauku exchange was the large degree to which it was monetarised - money having been used to purchase not only objects, but labour, use-rights in land, and so forth ... Moreover, while most of the goods purchased had fairly fixed monetary values, immediate factors of supply and demand were on occasion influential enough to encouraging haggling. (p. 528)

Similarly, the Yele economy was also monetarised to a high degree and was such that it was common for money to be lent and repaid with interest under certain fixed rules (Armstrong, 1928). In both of these societies, the counting of shell money is of central importance and both, as has been remarked earlier, have well-developed 10-cycle numeral systems generally untypical of NAN-speaking groups.

In foundational societies, it is not uncommon to find that there exists more than one method or set of numerals for counting various objects. The Kaugel people have a primary counting system which is used for most purposes but use somewhat different terms for counting pandanus leaves. The Kewa people have a 4-cycle numeral system which is used to count various wealth items but, in determining the number of months which must elapse before a festival occurs, they use instead a body-part tally method. The AN-speaking Arosi and Rennellese each possess a 10-cycle numeral system similar to that of their POC ancestor. However, in counting certain categories of objects the terms for large numbers, for example 100, 1000, and so on, vary according to the category being counted. The Arosi also have a set of "numerals" used in children's counting games consisting of the names of trees. Such counting-game sets are common in AN-speaking societies and have been recorded by Lean for the Tolai and Duke of York (PNG) languages and by Codrington for Mota in Vanuatu (Codrington, 1885, p. 305). It is also common among AN-speaking groups for there to be a basic numeral system for counting objects singly but that, in counting collections of objects, for example pairs of coconuts or fish strung in fours, somewhat different terms may be used.

While, in each of the societies discussed above, counting of horticultural produce, wealth items, and so on, occurs to a greater or lesser degree, it is also the case that societies vary according to what is not counted. The Rennellese, for example, do not enumerate time periods. The Ponam, on the other hand, do not count people: whether this is the result of a *tabu* against counting people is uncertain. The *tabu* against counting various objects has been noted for foundational societies in other parts of the world. Frazer (1923, pp. 308-309) and Zaslavsky (1973), for example, gave many instances from African societies in which there are *tabu*s against counting men or animals: to count either of these would bring misfortune to them. Seidenberg (1962, pp. 14-16) also listed various societies in Africa, North America, and Europe in which counting prohibitions of certain kinds exist. Indeed, Seidenberg suggested that non-verbal digit-tallying may have originated as a means of circumventing *tabu*s on the verbal counting of people or objects. The evidence for the existence of counting *tabu*s in the societies of New Guinea and Oceania is far from conclusive. It is certain, however, that there are many societies, both AN and NAN, in which there is no prohibition on counting people; whether or not this is generally the case is unknown and further data are required.

In various types of ceremonial exchange it is necessary that both donor and recipient groups are able to recall, at some future time when the exchange is to be reciprocated, the quantities of commodities involved in the original presentation. How, in the absence of written records, is the memory of such transactions kept alive? While we have recorded two basic ways of tallying using either the fingers and toes or other body-parts, such tallies are the means of enumerating quantities or of determining the number of days or months elapsed from some initial starting point. In particular, Muke recorded systems in Mid-Wahgi (Chapter 3). It would seem that some other forms of recording tallies which are to act as a means of recording totals are clearly necessary and indeed there is some evidence for the existence of these in several societies. Wolfers (1972), for example, said that

> methods of tallying and recording quantities include cutting notches in a stick, piling sticks or stones together, or tying knots in lengths of twine. Each notch, stick, stone or knot then represents a certain number of units of whatever is being counted … The Parevavo, inland from the Gulf of Papua, tie a knot in a length of twine for each man killed in battle and unravel the knots as each man is avenged. Some Chimbu groups reputedly kept scoreboards in red pigment on the walls of rockshelters, one mark for each important man killed in battle. (p. 217)

Gwilliam (1982) reported an unusual method of recording the number of voyages made by a Motuan man in the *hiri* trading system which involved the tattooing of his wife: "the tattoos covered the whole of her body including her eyelids" (p, 42). It needs to be observed, however, that the keeping of such physical tallies was not necessary for the subsequent recall of totals. Strathern (1977) told us, with regard to the Melpa, that

> As a record of the shell-*moka* gift sets which they have amassed and given away men wear tallies made out of slats of bamboo or cane. Each slat represents a set of eight or ten shells given. The tallies are a sign or demonstration to the public in general of the extent to which their wearers have made *moka*. They are not an *aide-memoire*. (p. 16)

The *moka* exchange transactions are remembered not by these physical tallies but by regular recall and discussion: "Such gifts are the subject of frequent conversations at intervals of time in between bouts of public activity, so the memories are kept alive" (p. 16).

Numeral Classification

The linguistic phenomenon of numeral classification is found in many languages on all the major continents of the world. In this section, we will discuss some of the salient features of numeral classification and the ways in which countable objects are categorised. The distribution of this phenomenon in various parts of the world is summarised prior to an investigation as to its occurrence in New Guinea and Oceania, first among the NAN languages, and second, among the AN languages of Melanesia, Micronesia and Polynesia.

What is Numeral Classification?

How quantity is expressed varies in important ways in different languages, in particular in the way nominal phrases are constructed. In languages like English, the expression of quantity usually has the construction "numeral + noun" as in "three men" or "five houses". There are other languages, however, in which such quantifying expressions contain not only a numeral or quantifier and a noun but also an obligatory classifier which indicates the specific class or category to which the noun belongs. In such languages, the universe of countable nouns is categorised into a number of classes, each noun being assigned to a class according to some criterion, although some nouns may be members of more than one class. In quantifying a given noun both the numeral or quantifier and the classifier, which indicates the class to which the noun belongs, must be stated. Thus suppose in a particular language the noun "man" belongs to the class of "animate beings" then an expression quantifying the noun "man" might take the form "quantifier (Q) + classifier (C) + noun (N)", for example "three + [classifier for animate beings] + man", the English gloss of which is, of course, "three men". The word order in this construction, i.e QCN, which occurs in various Amerindian languages, is not the only possibility: Japanese, for example, uses the order NQC. Allan (1977), in his survey of more than fifty classifier languages, indicated that these two constructions, together with an additional two, CQN and NCQ, appear to be the only permissible ones and that the noun never interrupts the nexus of quantifier and classifier so that, for example, the orders CNQ and QNC do not occur (p. 288). Not all the permissible constructions have the same frequency of occurrence, indeed for all the classifier languages of east and south-east Asia for which we have been able to acquire data, the classifier invariably follows the quantifier, a construction which we shall find exists in certain of the Oceanic AN classifier languages as well.

After identifying the classifier languages of New Guinea and Oceania, we will consider the following questions:

1. To what extent is numeral classification used in the act of counting the members of a particular class?
2. What is the nature of the quantifiers which occur in quantifying expressions and, in particular, are these the same as the counting numerals?
3. Is numeral classification employed in all quantifying expressions or only some, for example for a limited range of numerals or quantifiers?
4. For various languages, what is the word order used in quantifying expressions? Is there a typical word order, say, in the AN classifier languages?
5. Does the use of numeral classification imply that there is a different counting system for each class of countable objects?
6. Do different classifier languages categorise their universe of countable objects in similar ways?

The Geographical Distribution of Classifier Languages

There is a widespread distribution of languages which employ numeral classification throughout the major continents of the world. Allan (1977), for example, reported their existence in Africa and cites the languages Bantu, Swahili, Loka, Luyana, Luganda, Fula, Tiv, and a further ten or so located in different parts of the continent. Various authors have reported on the classifier languages of east and south-east Asia; some examples are Thai, Burmese, Vietnamese, Chinese, Japanese, and the languages of the Mon-Khmer, Nicobarese, and Aslian sub-families.[3] Also, various non-Oceanic AN languages such as Malay, Iban, and those of Indonesia display numeral classification (Omar, 1972). Among the classifier languages of North America, Allan cited as examples Nootka, Ojibway, and Navajo, among others. In Central and South America there are a number of examples, notably the languages belonging to the Mayan Family such as Tzeltal (Berlin, 1968), Chontal (Keller, 1955), and Yucatec. Allan (1977) also indicated the existence of classifier languages in northern Australia (p. 285).

[3] The sources for these are: Thai (Haas, 1942); Burmese (Becker, 1975; Burling, 1965; Haas, 1951); Chinese (Schafer, 1948), Japanese (Brainerd & Peng, 1968); Vietnamese (Hoa, 1957); Khmer (Adams, 1989; Jacob, 1965).

The Nature of Categorisation

Lakoff (1987), in his study on systems of human categorisation and what they reveal about human cognition, argued that diverse language groups categorise their worlds in different ways and by different linguistic means. Allan (1977) concluded his survey of classifier languages by remarking on the similarities which do appear to exist between the classificatory systems of quite disparate cultural groups. The number of classes employed differs considerably between languages. For example, the Australian Aboriginal language Dyirbal, studied by Dixon and cited in Lakoff, has four: (1) human males and animals; (2) human females, water, fire, fighting; (3) non-flesh food; and (4) everything else not in the other classes. On the other hand, Keller's (1955) study of Chontal lists 78 classes and Berlin (1968) indicated that Tzeltal may have several hundred. Despite this wide variation regarding the number of classes employed, Allan's (1977) analysis of some 50 languages yields that seven major categories of classifiers can be identified, most languages having some or all of these. He describes the categories as being primarily based on: (1) material, (2) shape, (3) consistency, (4) size, (5) location, (6) arrangement, and (7) quanta (including groups, collections, measures) (p. 306). While it is possible to categorise a large number of classifiers into a relatively small set of semantic domains, it is not always possible to determine precisely what the criterion is for class membership on the basis of semantics alone. For example, in the Dyirbal case, the members of the class of women, fire, and dangerous things hardly appear to have a common semantic basis and indeed Lakoff (1987) indicated that it is necessary to understand Dyirbal mythology in order to see the connections between these apparently disparate things and why they are members of a common class (pp. 92-96).

The Occurrence of Numeral Classification Among the NAN Languages

There is evidence for the existence of a number of NAN classifier languages. At least four languages belonging to the East Papuan Phylum and which are located in central and southern Bougainville (PNG) possess numeral classification; these are Nasioi, Nagovisi, Siwai, and Buin. Foley (1986) said of these that

> the most extensive system of nominal classification in Papuan languages is found in Nasioi and perhaps other Papuan languages of southern Bougainville. The system of Nasioi … parallels in certain respects the numeral classifier systems of south-east Asia … The classifiers number over 100, and are very specific semantically. (p. 83)

In Nasioi the classifiers are suffixed to numeral roots (the quantifiers): 1 is *na-*, 2 is *ke-*, 3 is *bee-*, and 4 is *kare-*; 5 is invariant in any counting sequence and is unsuffixed: *panoko*. Hurd's (1977) analysis of the Nasioi semantic categories together with the number of classifiers in each category given in parentheses are: 1. social groupings of people (22); 2. body-parts (14); 3. animals and food (8); 4. trees, wood, leaves, feathers (13); 5. ropes and vines (3); 6. bamboo (4); 7. bananas (7); 8. taro (3); 9. coconuts (4); 10. fruit (2); 11. houses, furniture, building materials (14); 12. containers (7); 13. clothes, bags, nets (3); 14. money (2); 15. implements, weapons (7); 16. physical or geographical features (16); 17. locations, areas, paths (8); 18. temporal periods (7); 19. fractions, parts, sections, groups (16) (pp. 115-.123). Recent categories include one for money (D. Ope, personal communication, 2003). Further details are given in Appendix B.

Other languages which exhibit classification are located mainly in the East and West Sepik Provinces, several belonging to the Torricelli Phylum (Mountain Arapesh, Southern Arapesh, Monumbo, Olo), the Upper Sepik Stock (Abau) and the Lower Sepik Family (Yimas). Foley (1986) said of these that "the noun classification systems of the Torricelli and Lower Sepik languages … are among the most complex in the world and represent an extreme development in New Guinea" (p. 88).

Numeral Classification Among the AN Languages of PNG and Oceania

Among the AN languages of PNG, numeral classification occurs most noticeably in the languages of the Admiralties Cluster and the (Peripheral) Papuan Tip Cluster, located respectively in the Manus and Milne Bay Provinces. With regard to the Admiralties Cluster it seems likely that all 24 languages belonging to this group and for which data were acquired exhibit numeral classification (these are given in Lean's (1992) Appendix on Manus Province, examples of Ponam and Gele' can be found in Appendix B on languages of Chapter 8. Fisher (~2010) provided another example on Kuruti). The number of classes which are employed varies between languages: Ponam, for example, has at least 28 (Carrier, 1981) while the Gele' (or Kele) dialect of Ere-Lele-Gele'-Kuruti has at least 43 (Smythe, n.d.). In fact, Smythe discussed the various classificatory and other features of Gele' in terms of their similarities to other Austronesian Oceanic languages (Smythe, 1970). Fisher, as noted above, discussed the counting for different parts of the banana in Kuruti—single banana, hand of bananas, stalk of bananas, leaf, and tree—indicated by a suffix. He provided one table giving the words for 1 to 10 for 50 different classifications as given in Appendix B. In addition, his list of large numbers indicates the use of prefixes, e.g. *po-* for 100 000 as numeric classifiers (see Appendix B). Two features which appear to be common among all the languages of the Admiralties Cluster for which data exist are that, first, the word order in all classifier constructions is "quantifier + classifier" (that is QC), where the quantifier is a numeral root, and, second, that such quantifying expressions are used to state quantities but are not used in serial counting. In determining a quantity by serial counting, a basic set of unsuffixed numerals is used; once the number of objects is known, this may be stated using a classifier construction. The languages differ in regard to the number of quantifiers that must take the obligatory classifier: in Gele´, for example, classifiers are suffixed to the numeral roots for 1 to 9 while, in Ponam, classifiers are suffixed only to the numeral roots for 1 to 4, after which the counting, unsuffixed, numerals are used (Carrier, 1981, p. 471).

The classifier languages of the Peripheral Papuan Tip are Kilivila, Muyuw, Budibud, Sud-Est, and Nimowa, all of which are spoken on islands eastwards from the south-east tip of the mainland. (Details of Kilivila, Muyuw and Budibud are given in Appendix B.) The languages vary according to the number of classes employed: Kilivila has at least 42 while Sud-Est has about 22. The word order used in quantifying expressions is different from that found in the Admiralties Cluster and takes the form CQ, i.e the classifier is prefixed rather than suffixed to the numeral root. However, Kilivila, like the Admiralties languages, has a set of numerals used for serial counting; these do not have a classifier prefix and are used for the counting of yams. Kilivila, as well as the other languages of this group, have a "general" classifier which is frequently used instead of specific classifiers and numeral roots prefixed by this are usually the means by which serial counting of any class of objects (excluding yams) is carried out.

Outside of PNG, numeral classification is exhibited most noticeably in the languages of Micronesia, although Marshallese and, to some extent Kosraean, appear to have lost it (Lean's (1992) appendix on Oceania, gives some details of most languages, pp. 50-56). The way in which the classificatory systems operate in the Micronesian (MC) languages more closely resembles that of the Admiralties languages than that of the Papuan Tip group. While the number of classes employed varies between the Micronesian languages, from 62 in Trukese to only several in Kosraean, we find that all quantifying nominal expressions have the word order QCN, that is the numeral roots are suffixed by classifiers (and are followed by the noun). Each of the Micronesian classifier languages also has an unsuffixed set of numerals which are used for serial counting; once the number of objects in a set has been determined, this may be stated using a classifier expression.

While these three widely separated groups of AN classifier languages vary considerably in the number of classes into which their universe of countable nouns is subdivided and while some of these classes are clearly culture-specific, this does not necessarily mean that the way in which things are categorised differs fundamentally between languages or groups of languages. The categories which occur in all three groups of the classifier languages are listed here: 1. animate entities (Kilivila distinguishes two animate categories: males versus females and other animals; 2. various body-parts; 3. blades and cutting instruments; 4. long wooden objects, trees, canoes; 5. flat, two-dimensional objects; 6. round, three-dimensional objects; 7. ropes, strings, belts; 8. containers and hollow objects; 9. pools, streams, water courses; 10. paths, tracks; 11. collections of food, bundles; 12. fractions, parts, pieces; 13. measurements, particularly of length; and 14. countable bases, that is tens, hundreds, thousands, etc. It is interesting to observe that all of these categories are mentioned by Allan (1977) in his summary of the categories which occur in classifier languages in other parts of the world and of which he said "the recurrence of similar noun classes in many widely dispersed languages from separate families, spoken by disparate cultural groups, demonstrates the essential similarity of man's response to his environment" (p. 307).

The fact that three separate groups of Oceanic AN languages exhibit numeral classification and that their systems have a number of salient features in common, raises the question as to whether they have independently developed their systems as quite separate innovations which have occurred subsequent to the breakup of their common POC ancestor and that the common features which are apparent are indeed due, as Allan suggests, to "the essential similarity of man's response to his environment". Alternatively, it might be suggested that the similarities between these groups of daughter languages of POC are due to numeral classification being originally present in POC and that, while the majority of the descendants of POC have largely lost this feature, these three groups have retained it. The question as to why they should be conservative in this respect may perhaps be partially explained by the fact that they are all island communities which have undergone a degree of separation and isolation from outside influence. If numeral classification was an original feature of POC it would seem reasonable to expect to find vestiges of it remaining in some daughter languages other than those of the three groups in which it is fully displayed. And indeed there is evidence of such vestigial remains among a number of the Oceanic AN languages. Several of the Polynesian languages including Nuguria and Takuu (Bougainville Prov., PNG), Nukuoro, Kapingamarangi, Pukapuka, and Rennellese (Solomon Islands) show evidence of possessing "group of ten" classifiers which are prefixed to numerals or numeral roots and which are used to construct the decades from 20 onwards.[4] Similarly a number of the AN languages of the North Solomons Province (PNG), for example Solos, also appear to exhibit numeral classification to a limited extent and are such that the classifier is prefixed to the quantifier. In Chapter 6 that the phenomenon of base-suppletion, which occurs in some AN 10-cycle numeral systems, appears to be due to the use of "group of ten" classifiers which are affixed to numerals. This is apparent in a number of Polynesian languages, generically as a group of ten in the languages of the Solomon Islands such as Roviana, and in several languages of the Papuan Tip Cluster which have "Motu" type numeral systems. Indeed, according to Lawes (1885, p. 9), Motu itself has a number of classifiers, including one each for males and females and another for long wooden objects such as spears; these are prefixed to numerals (see Appendix C). The validity of the conjecture that numeral classification existed in POC and the question as to whether such classifiers are reconstructable, can only be settled by the methods of comparative linguistics and is beyond the scope of this work.

Classifiers in Micronesian Languages Compared to Polynesian Languages

In summarising a number of authors, Bender and Beller (2006) have suggested four categories:

1. Repeaters (set labels for individual classes of objects, they repeat the phonological form of the sole noun with which they co-occur),
2. Quality classifiers/qualitative or sortal bases (referring to salient features of the objects),

[4] Nukuoro, Kapingamarangi, Nuguria, Takuu, Rennellese (Elbert, 1988, p. 190), Pukapuka (Beaglehole & Beaglehole, 1938, pp. 354-355).

3. Quantifiers/quantitative or mensural bases (indicating a quantitative measurement), and
4. Digital classifers/numerative or ten-power bases (referring to the powers of the mathematical base 10). (Bender & Beller, 2006, p. 387)

In fact, the number of classifiers can range from two in Kosraean to 100 in Kiribati or Chuukese. Polynesian languages also have power and quality classifiers. Most categories fit into the areas of shape, nature and generality. For example, in Woleian, there are different words for *–yáf/yef* bundle of 10 ripe coconuts; *-ccoc*10 small pieces of breadfruit pudding; strings of 5 fish, bundles of 20 or more (about 30) breadfruit, bunches of 20 coconut; *-sópw* bundles of from 10 to 19 breadfruit; and *-ttit* string of 10 breadfruit. These are remarkably similar to Polynesian.

Among the digital or numeral classifiers are those that have a multiplying function such as by 2. This is the case for all Micronesian languages except for modern Marshallese suggesting that this is possible in Proto Micronesian (PM). Thus it is expected that PM includes the qualifiers for general objects, animate objects, thin flat objects and two power classifiers of units of tens and units of hundreds. Local variation may be to 80 rather than 100 for divination (Alkire, 1970). A similar system for multiples occurs also for Kiribati and Chuukese. Polynesian languages also have power and quality classifiers. In Proto Polynesian the term for 20 (40 in Hawaiian) seems to mean group and is not present in Micronesian as a classifier.

Associated with the multiplying function by two, some Micronesian languages have paired counting. This feature may be connected to Polynesia but there are insufficient data (Bender & Beller, 2006). Similarities tend to occur for important items such as fish, coconuts and breadfruit. In Proto Polynesian, there are six identified, with Proto Micronesian having ten. However, it is worth remembering that two Non-Austronesian languages in the highlands of Papua New Guinea, pairing to count was prevalent as shown by Strathern for the Melpa (Lancy & Strathern, 1981). One Melpa counting practice was using two fingers on the left hand, two more, followed by two and two on right hand, and then the "two thumbs down" as they counted 2, 4, 6, 8, 10 as an alternative to 1, 2, 3, 4, 4+1, 4+2, 4+3, 8, 8+1, 8+2. As mentioned in Chapter 3, Muke (2000) noted the counting of pairs in Wahgi allowing for groups of 40. However, neither use classifiers.

One difference between the Micronesian and Polynesian groups of languages is that classifiers are not essentially used in Micronesia (and not at all in modern Marshallese). Another is that in Micronesia the order of morphemes are always numeral-classifier in Micronesia but it varies in Polynesia as six languages have classifier-numeral order. Furthermore certain morphemes such as *fau* vary. In the former it refers to fruit but in the latter to 100 paired counting for three different types of objects when order is changed. Numeral-classifier order for "fathoms" denotes change in counting. Table 8.3 illustrates how Samoan, as an example, uses the order of the numeral and classifier for different number sizes involving multiples.

There are similarities of this kind between Tongan and Samoan so this is found basically in western Polynesian Outliers. However, Bender and Beller also suggested that:

> The Samoan expression refers to just 2 coconuts whereas the corresponding article in Tongan (2) multiplies this amount by 10-score (200), thus yielding 400 coconuts. It is only when numeral and classifier change their position (as *infua-lua*) in Samoan that a numerical change occurs (from2 to 20), but as Clark (1999) has argued, this would be a ten-deletion rather than a multiplication. Nevertheless, a process similar to the Tongan multiplication can also be observed in Samoan, and remarkably for the same object (coconuts) again. (Bender & Beller, 2006, p. 396)

Rennelles, a Polynesian Outlier, in the Solomon Islands also shows multiplication for large numbers so the practice was widespread.

Following their study of classifiers in both Micronesian and Polynesian languages and analysis in terms of the order of numeral and classifier as well as specific words, Bender and Beller (2006) suggested that the existence of classifiers was likely to exist in Proto Oceanic. They seem to be stronger in societies with chiefs and when the need for objects of exchange of large quantities was needed;

classifiers provided a short cut representation of large numbers. Furthermore, Bender and Beller showed that:

> The high numerals in many Micronesian and Polynesian languages have often raised doubts among scholars about the genuineness of their numerical value (Clark, 1999; Elbert, 1988; Elbert & Pukui, 1979). However, as high numerals are a recurrent pattern across two language groups, attesting a widespread interest in high numbers and possibly even mathematics, we should take the numerical interpretation seriously ... We have no doubt that the indigenous interest in high numbers in Polynesia and Micronesia inspired people to systematically incorporate numerical classifiers into an originally regular decimal system. (Bender & Beller, 2006, pp. 399-401)

The practical purposes such as tributes to chiefs, keeping track of gifts for reciprocal and status purposes, and even divination required calculations and not just counting. Furthermore, the classifiers assisted in the development of smart ways of communication and reducing cognitive difficulties when calculating. This is particularly significant given the emphasis on oral rather than recorded information that might not remain over time. The use of large numbers will be considered further in the next section of this chapter.

In terms of the links between the languages of Oceania, it is worth noting Song's (1997) argument that the possessive classifiers indicated a continuity between Proto Oceanic and Micronesian languages rather than a discontinuity and influence from elsewhere. He noted that Gilbertese, an anomaly, but its type is found in Oceanic languages and it could have been influenced by Polynesian rather than there being a discontinuity. Four classifiers are common across the Micronesian and Polynesia groups as well as the numerals themselves. Widespread among the Polynesian languages was counting by various means such as pairs, fours, tens, and scores using the same decimal numerals while Micronesians tended to vary the numeral classifiers for different kinds of objects (Bender & Beller, 2006).

Summary

Terrell (1986) noted that nouns are more likely borrowed than other lexical aspects of language. Numeral classification, a sub-category of the more general phenomenon of noun classification, occurs among several groups of NAN and AN languages in New Guinea and Oceania. Among the NAN languages it is found most noticeably among the members of the East Papuan and Torricelli Phyla, the ancestral languages of which were probably established in New Guinea prior to the migration of the speakers of the Trans New Guinea Phylum. This tends to suggest that noun classification found in the NAN languages is likely to be an archaic feature rather than an innovation, although the degree to which it has been elaborated may be innovative. Similarly, we have suggested that numeral classification may have been present in POC and this is, thus, a general archaic feature rather than an innovation which has occurred among a few groups of languages. While the majority of the Oceanic AN languages have subsequently lost numeral classification as a linguistic feature, a few groups of largely island communities have retained it either fully or in part.

It is apparent that at least among the AN classifier languages, numeral classifier expressions are not normally used in the counting of relatively small numbers of objects and for this purpose a set of serial counting numerals is used. These numerals are not affixed with a classifier. However, classifier constructions are employed when a nominal phrase is used to state the number of objects of a particular class once these have been counted. In counting a larger number of objects, in fact numbers greater than or equal to 10, it is necessary to use classifier constructions in that decades, hundreds, and usually thousands, are expressed by means of classifiers. In the Micronesian and Admiralties languages, for example, 40 is expressed as "numeral root for 4 + decades classifier" while 500 is expressed as "numeral root for 5 + hundreds classifier". Also, in both of these language groups, the serial counting numerals are similar but not identical to the quantifiers which appear in classifier expressions; the quantifiers are a set of numeral roots which remain largely invariant from one class to another. Thus, in Ponam, the first four serial counting numerals are *si, luof, talof,* and *faf,* while the corresponding numeral roots, to which a classifier is suffixed, are *sa-, lo-, tulu-,* and *fa-*. In fact these four numeral roots are the only ones which are suffixed. In Gele', however, classifiers are suffixed to each of the numeral roots for 1 to 9. It is not the case, then, that the AN classifier languages can be thought of as having a large number of different counting systems, one for each noun class; for most purposes, class-invariant counting numerals are used. Thus, there are no grounds for the sort of criticism, apparent in the nineteenth century literature, which suggests that those foundational societies which have numeral classification have been unable to develop an abstract concept of number, freed from the context in which counting takes place (Cassirer, 1953).

From examples given previously it is clear that, at least among the AN classifier languages, that there is a degree of similarity between them in the nature of the noun classes that they possess, and indeed we might expect this if their common POC ancestor also exhibited numeral classification. However, it is also apparent that differences exist between and within the various AN groups, particularly with regard to the numbers of classes each language possesses: within the Admiralties Cluster these vary between 20 to 40-odd while in the Micronesian languages they vary between 3 or 4 to 68. There is, in addition, a basic difference between groups in the word order used in quantifying expressions, the order QC being universal in the Admiralties and Micronesian languages while the order CQ appears to be common among the classifier languages of the Papuan Tip and Polynesian groups. Of the two exhibited word orders, it seems reasonable to suggest that one was present in POC and that the other is an innovation. Ross (1988) indicated that his view is that it is among the Admiralties and Micronesian languages that the innovation has occurred: "the possibly innovative feature is not the use of classifiers, which are reconstructible in POC, but the sequence of numeral + classifier, rather than the reverse" (p. 328). While this may be the case, it is relevant to note that, among the AN languages of the Indonesian region and south-east Asia, the most common form which occurs in classifier constructions is QC which tends to suggest that this may be the more archaic form. Other arguments from the peopling of the Pacific may differ in terms of possible later migrations.

Large Numbers in the NAN and AN Languages

We have seen, in the earlier sections of this chapter, that various societies within PNG and Oceania differ in significant ways according to the importance that they attribute to the enumeration of objects as well as to the types of objects counted and the circumstances in which counting is carried out. These societies also differ according to the extent to which they count: all have the resources enabling them to count at least to 10 and many have the capability of counting in an efficient way to higher orders of magnitude. In this section we consider:

1. the evidence for the existence of terms for large numbers in the counting systems of the NAN and the Oceanic AN languages;
2. whether these languages differ in significant ways in the resources available for the expression of large numbers;
3. whether there is any indication of the origin of terms for large numbers, and in the case of the AN languages, which terms were likely to have been present in Proto Oceanic; and, finally,
4. whether such terms are true numerals or whether they are largely descriptive or impressionistic in nature.

Large Numbers in Other Indigenous Cultures

In the nineteenth century literature on natural language numeral systems, it is not uncommon to find reference to societies which are such that their entire means of enumeration is encompassed by the terms "one, two, many". The numerical horizon of people possessing such "systemless" numeration, it was inferred, was very limited indeed: the scholars of the time, under the sway of social Darwinism, had little compunction in assigning such people to the very bottom rung of the hypothetical ladder of cultural evolution. It is worth noting here that with the data collected for this study, in particular the data relating to 2-cycle systems as summarised in Chapter 3, there is no evidence of the existence of systems which terminate at 2: even in the case of languages possessing "pure" 2-cycle systems, there exists the syntactic resources for the expression of numbers larger than 2. Setting aside, however, the question of the validity of claims that "one, two, many" numeration systems exists and whether such claims were an artifact of the way in which data were elicited, there still persists today, in popular texts on the history of number, the view that many indigenous peoples possessed, prior to the influence of European numeration, only the most rudimentary linguistic means of enumerating their world. There were, nevertheless, nineteenth century scholars such as Tylor (1871, pp. 240-272) and Conant (1896) who were aware that there were significant exceptions to this view and that there existed examples of indigenous cultures in Africa, the Americas, and the Pacific, which possessed terms for large numbers ranging from tens of thousands to millions. Conant, for example, notes that "the Tonga Islanders have numerals up to 100 000, and the Tembus, the Fingoes, the Pondos, and a dozen other South African tribes go as high as 1 000 000" (p. 33).

Among the Amerindians of North America, Schoolcraft (writing in 1851 and quoted by Closs, 1986) indicated that "the Dakota, Cherokee, Objibway, Winnebogo, Wyandot, and Micmac could all count into the millions, the Choctaw and Apache to the hundred thousands and many tribes to 1 000 or more" (p. 13). Similarly, in Central and South America, "the Aztec, Inca and Maya all counted into the millions". Evidence therefore exists which indicates that various indigenous cultures do have the linguistic resources for the expression of large numbers and we shall consider now whether this is also the case for the cultures of New Guinea and Oceania. In order to do this we shall consider first the evidence for the NAN language groups and, second, the evidence for various sub-groupings of the Oceanic AN languages.

Expressing Large Numbers in the NAN Languages

Relatively simple monomorphemic terms for the expression of numbers greater than or equal to 1 000 are rarely found in the NAN language of the New Guinea region. We have seen earlier, in Chapter 6, that the Yele of Rossel Island in the Milne Bay Province (PNG) have distinct terms for the numbers 1 000, 2 000, 3 000 and 4 000, and that the thousands from 5 000 to 8 000 are formed from these. Yele is a member of the East Papuan Phylum and it is largely among other members of this phylum that we find terms for 1 000 or more. In southern Bougainville, for example, there appears to be a unique local development among the East Papuan languages Nasioi, Buin and Siwai in which the term for domestic fowl is used to represent the number 1 000. This is also true for several neighbouring AN languages such as Banoni, Uruava and Torau. Immediately to the south of Bougainville, in the most northern of the Solomon Islands, the AN language Mono-Alu also represents the number 1 000 by the term for domestic fowl while, in the AN language Central-East Choiseul, the same term represents 10 000 rather than 1 000. The actual term used is *kokolei, (kokoree, kokorako*, and its various reflexes) and is AN in origin; thus, for the East Papuan Phylum languages mentioned above, the term for 1 000 is an AN loanword. In Nasioi, the reduplicated form *kokokokorei* is used to represent 1 000 000.

Seven NAN languages are spoken in the Solomon Islands and each of these is a member of the East Papuan Phylum. All languages possess a single term for 100 and in several cases it is clear that the term used is an AN loanword: Lavukaleve, for example, has *tangalu* and Nanngu has *te/ta lau*. There are, however, exceptions to this: Mbilua has *paizana* and Aiwo and Santa Cruz both have *tevesiki (tövisiki)*, neither of which appears to be AN in origin. All of the languages appear to have a single term for 1 000 and in several cases these are AN loanwords: Mbaniata, for example, has *tina* which is commonly used in the languages of the North-West Solomonic group. Savosavo has *toga* which is found in ten AN Solomons languages while Mbilua has *vuro* which appears to have been borrowed from the neighbouring AN Ghanongga and which, in *kia*, signifies the number 10 000. However not all terms appear to be borrowed: Santa Cruz has *siu (jiu)* and Lavukaleve has *lamukas*, for example.

While single terms for numbers of 1 000 and larger are rare among the NAN languages and indeed are largely confined to some members of the East Papuan Phylum (occasionally as AN loanwords), there is nevertheless a number of NAN languages which express numbers of the order of 1 000 by means of complex expressions. For example, Sulka, East New Britain Province, another member of the East Papuan Phylum, has complex expressions for the numbers 400, 800, and 1 600 which may be analysed as having the constructions respectively of 20x20, 20x40, and 20x80. The East Papuan Phylum language Pele-Ata has a distinct term for 100 and a complex expression for 1 000 which may be analysed as 100 x 10. Examples such as these indicate that the existence of single terms for large numbers is not a prerequisite for the expression of large numbers: this may be achieved by the use of complex expressions which may be analysed as having multiplicative constructions. Further complexities including the use of body parts and gestures for large numbers are indicated for Yu Wooi and Iqwaye in Chapter 3, while Kanum in Chapter 7 uses the repositioning of words among other linguistic ideas to represent large numbers.

Expressing Large Numbers in the AN Languages of Papua New Guinea

The existence of single terms which express large numbers varies considerably among the AN languages of PNG and between their various Clusters as identified by Ross (1988). Details of each of the languages mentioned below can be found in the appendices of Lean's (1992) thesis and in the GLEC database (GLEC, 2008) which are organised by Province. The North New Guinea Cluster languages, none of which now exhibits the "pure" 10-cycle numeral system of their POC ancestor, have a paucity of simple terms for numbers of the order of 1 000 or even of 100. In the Sandaun Province, two neighbouring languages, Ali and Tumleo, have a term for 100, *raput*. Kairiru, spoken in the East Sepik Province, has the term *wurol* for 100. The languages of this cluster which possess a single term for 100 are, however, largely located in the West New Britain Province, and one, Gitua, is in the Morobe Province. Each of the former, namely Bariai, Kilenge-Maleu, Kombe-Kove-Kaliai, Lamogai-Rauto-Ivanga, Arove, and Asengseng, has a term *vuno* (and its cognates) for 100. Gitua has *ai*. The remaining 68 languages of the North New Guinea Cluster show no evidence of possessing single terms for numbers above 20 although a proportion do possess complex expressions for 100 which have the form 5×20 or 20×5.

In the Papuan Tip Cluster, the 25 languages classified as "Nuclear" and for which we have data show no evidence of possessing single terms for the numbers 100 and above, although 100 is in some cases expressible as a complex multiplicative construction of the form 5×20. This is in contrast to most of the languages of the "Peripheral" Papuan Tip for which there is evidence of the existence of terms for 100, and in some cases, much larger numbers as well. Each of the languages Kilivila, Muyuw, Nimoa and Sud-Est possesses a single term for 100. In the Central Province, this is also true of the languages Nara, Roro, Mekeo, Gabadi, Doura, Motu, Sinagoro and Keapara. Motu has single terms for each of 1 000, 10 000 and 100 000, while Sinagoro and Keapara both have terms for 1 000. For each of the eight languages which possess a term for 100, the term used is *sinahu* (and its cognates: *sinavu, sinau, tinavuna*). This appears to be a local development without cognates in other Papuan Tip languages.

The 64 languages of the Meso-Melanesian Cluster, of which 23 belong to the North-West Solomonic group, are such that the majority possess single terms for 100. Some also possess single terms for 1 000: this is true for almost all of the North-West Solomonic group. Over the whole cluster, there is little uniformity regarding the terms used for 100. Certain terms are, however, uniformly found among regional groupings of languages. In at least nine of the New Ireland languages, the term for 100 is *(a) mar*: this is identical to the Tolai word for 100 and Lean suggested that it is likely that this term was borrowed by a number of New Ireland languages under the influence of Tolai missionaries working in New Ireland in the nineteenth century. In the North-West Solomonic group some 11 languages have the term *gogoto* for 100: this appears to be a local development and the term does not have reflexes in other members of the Meso-Melanesian Cluster. (The only language outside this group that has a related word *(vovoto)* is the NAN language Rotokas, in Bougainville). A further seven languages of the North-West Solomonic group have the term *gobi* for 100: this, too, appears to be a local development. One term, however, does appear to have an incidence which is not merely localised and is found in various forms in Nakanai in West New Britain (*salatu*), in seven languages of Bougainville (*latus, natus*), in Mono-Alu of the North-West Solomonic group (*latu*), and in several members of the South-East Solomonic group (for example Nggela has *hangalatu*). Significantly, this is also cognate with the Indonesian/Malaysian term for 100, *ratus*, and is thought to have been present in Proto Austronesian (PAN) as **Ratus* and in POC as **Ratu* (Harrison & Jackson, 1984, p. 69).

Apart from the three clusters of Ross's Western Oceanic we need also to consider the languages of the Admiralties Cluster, most of which as we have noted earlier, possess numeral classification. Indeed the terms for which the hundreds exist in these languages are classifier constructions so that, in Papitalai for example, 100 is *se-ngat*, 200 is *ru-ngat*, and 300 is *tulu-ngat* where *se-, ru-, tulu-* are numeral roots or quantifiers, and *-ngat* is the "hundreds" classifier. Similarly, there are in a number of the Admiralties languages terms for the thousands which also employ classifier constructions: Andra-Hus, for example, has *sa-po* for 1 000, *lu-po* for 2 000, and *tulu-po* for 3 000. The term for 10 000, however, is *pue-sih* which does not employ a classifier construction. Similarly, in Titan, while classifier constructions are used for the hundreds, they are not employed for the thousands so that 1 000 is *pue-si*, similar to the Andra-Hus term for 10 000. This discontinuity in the means of signalling quantification may be an example of what Hurford (1987) terms a "growth mark" of a numeral system, that is a point which at one time would have been a limit of counting but which now represents a transition point between one means of enumeration and another which extends the previous limit (pp. 81-85).

The discussion of terms for large numbers in the AN languages given above has concentrated mainly on those for the hundreds: terms for 1 000 and more are much less common. Indeed they appear to be virtually absent from the languages of the North New Guinea Cluster and largely absent from the Papuan Tip languages with the exception of Motu, Sinagoro and Keapara in the Central Province, for which the terms for 1 000 are, respectively, *daha, dagatana* and *ragana*. (See Appendix C for details of these three languages.)

Terms for 1 000 are found more commonly in the languages of the Meso-Melanesian Cluster and in particular among those of the North-West Solomonic group. In the latter, nine languages have *tina* for 1 000 (see Lean's (1992) thesis) while seven have *toga*. For the AN languages of Bougainville, reference has already been made to the use, in at least three cases, of the word for domestic fowl in representing 1 000; in four other languages the terms *piku, tapan*, and *tuku* are employed. In the East New Britain Province, both Tolai and Duke of York have *arip*. Each of the terms used in the various clusters appear to be localised occurrences, without cognates in languages outside a relatively small region (see Appendix C).

Expressing Large Numbers in Island Melanesia, Polynesia and Micronesia

In the Solomon Islands, the 56 AN languages spoken comprise 23 belonging to the North-West Solomonic group, 22 belonging to the South-East Solomonic group, 6 belonging to the Eastern Outer Islands group, and 5 Polynesian Outliers (Figure 8.3). The North-West Solomonic group belongs to Ross's (1988) Meso-Melanesian Cluster and was discussed above. With respect to the members of the South-East Solomonic group, all have single terms for the numbers 100 and 1 000; several have terms for larger numbers as well. The terms for 100 may be divided into three groups, each with its own variants: (a) *sangatu* (*thengetu, hathangatu*), (b) *talange* (*talanae*), and (c) *tangarau* (*tangalau*). Of the last, Codrington (1885) notes that "the most common word in use in Melanesia, as in Polynesia, is *rau* 'a branch or leaf'". He indicated that the explanation for the use of this term arises from the practice of using a frond from the *cycas* tree to tally days:

> beginning on one side of it a leaflet was counted each day, one being pinched down as a tally for every tenth. The frond when treated in this way on both sides furnished tallies for a hundred ... The same practice is found in the Solomon Islands, where ... not the simple *rau* but *tangarau* is the word in use. (p. 249)

In POC, the reconstructed form is **dau* which has the meanings "hundred, unit of hundreds" as well as "leaf" (Harrison & Jackson, 1984, p. 69).

Codrington's assertion that this term for 100 is the one most commonly used in Melanesia is not in fact correct: it is found only in eight languages of the South-East Solomonic group, in Fijian (*ndrau*) and Rotuman (*ta rau*). It is, however, found extensively in the Polynesian languages. A different term for 100, but one which still has the meaning of palm frond, is found in the languages of the northern islands of Vanuatu. This is *meldol (melnol, medol)* and was also recorded by Codrington (1885, p. 249). There is no one term in the South-East Solomonic group which is predominantly used to signify 1 000: *toga (toha)* is used in seven languages, *toni (to'ani)* is used in six, and several other terms *(sinora, meru)* are used in one or two languages. Finally, a term for 10 000, *mola,* is found in several languages while in Arosi the term used is *husia*.

In the languages of Vanuatu, the data available for terms for large numbers are less complete than those available for the Solomon Islands. It is clear, however, that in the northern islands at least most languages have terms for both 100 and 1 000: these are, respectively, *meldol, (melnol, medol),* mentioned above, and *tar (ter*). In the Sakao language the latter term is used for 100 rather than 1 000. In North and South Efate, the term *manu* is used for 1 000 and this is commonly found in the Polynesian languages; indeed, the presence of this term in the Efate languages may be due to the influence of the neighbouring Fila-Mele and Emai Polynesian Outliers.

The situation in New Caledonia appears to be not unlike that occurring in the North New Guinea Cluster of Western Oceanic, that is distinct terms for 100 and 1 000 are rare although more complex expressions for 100 exist which tend to have the form 5x20.

It is among the Polynesian and Micronesian languages that we most commonly find single terms for very large numbers. All Polynesian languages for which data exist possess terms for 100: in at least 24 languages the term is *rau* (and such variants as *lau, selau, 'au*, etc.). Similarly, all languages appear to have terms for 1 000, the most common being *mano*. At least six languages have, instead, *afe*, and for several of these the term *mano* is used not for 1 000 but for 10 000. Terms for numbers of the order of 100 000 and more are not uncommon. The American anthropologists P. and E. Beaglehole (1938) reported of the Polynesian Pukapuka that

> it was a favourite jest among informants that the Pukapuka could count to a higher power than we could; proof of this they argued was not only the presence of words indicating progressions to infinity, but also the ability of the culture hero Maui to find Pukapuka words

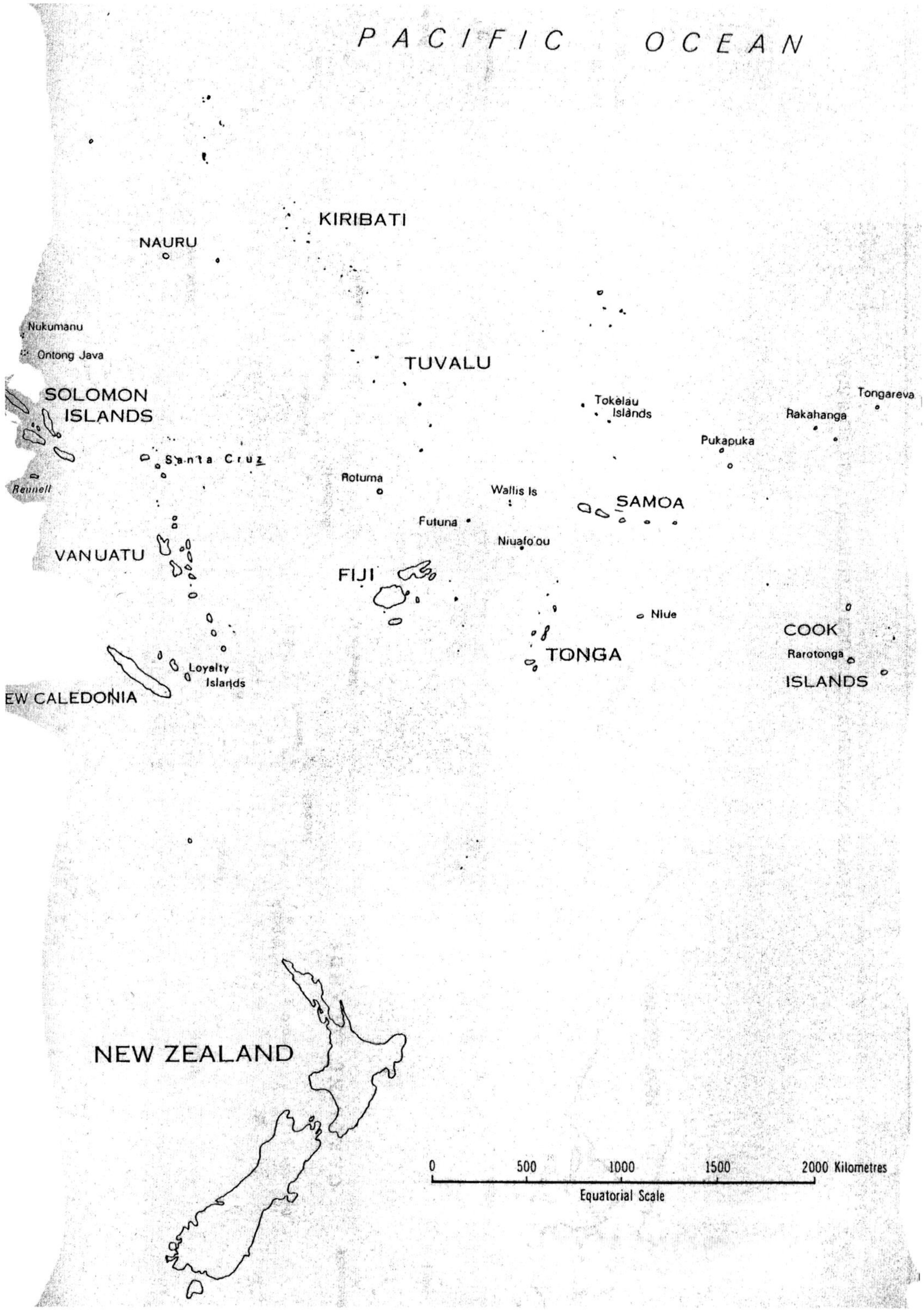

Figure 8.3. Location of societies cited in text (Island Melanesia and Polynesia).

which enabled him completely to enumerate the stars in the sky, the fish in the sea, the sands on the beach, and so forth. (p. 354)

Among the Mangareva, another Polynesian group, the terms used for large numbers when counting men, houses, boats, stars, and so on, are given in Table 8.4.

Table 8.4
Enumerating Large Numbers Among the Mangareva (Polynesian group)

Mangareva		Large Numbers
	rau	100
ten rau	*mano*	1 000
ten mano	*makiu*	10 000
ten makiu	*makiukiu*	100 000
ten makiukiu	*makorekore*	1 000 000
ten makorekore	*maeae*	10 000 000

Source. Buck (*Te Rangi Hiroa*) (1938, p. 417)

Terms for such very large numbers as those given in the latter part of the Table 8.4 are not uncommon among some of the Polynesian languages. Indeed, in Nukuoro, one of the Polynesian Outliers, there are terms for increasing orders of magnitude up to 10^{10} as shown in Table 8.5. Furthermore, Bender and Beller (2013) suggested that Mangarevan language has special words for large groups. But their special counting words are all decimal numbers multiplied by powers of two, which are 1, 2, 4, 8 … . Specifically, *takau* equals 10; *paua* equals 20; *tataua*, 40; and *varu*, 80. Those big numbers are useful for keeping track of collections of valuable items, such as coconuts, that come in large numbers. Bender and Beller realised that the Mangarevan counting system makes it possible to use binary arithmetic for calculations of large numbers.

Table 8.5
Large numbers in Nukuoro (Polynesian group)

Large Number	Nukuoro
10^6	*seloo*
10^7	*sengara*
10^8	*semuna*
10^9	*sebugi*
10^{10}	*sebaga*

Source. Harrison and Jackson (1984, p. 72)

Since Lean (1992) wrote on large numbers and classifiers providing most of the text for this chapter (as other chapters), Bender and Beller (2006), following Clark (1999) provided speculative terms for 10^5 **timi* and 10^6 **ki(l)u* for Proto Polynesian. For Proto Micronesian, they speculate for 10^4 **lopwa,* for 10^5 **sepu/sepi* or **depu/depi,* for 10^6 **nena.*

It should be noted, however, that terms for very large numbers are not restricted to the Polynesian languages only but have also been reported for Mbughotu, several languages of the South-East Solomonic group, for example Arosi and Fox (1931) reports that, for the Arosi, the term for 10^6 is *raurauni ha'aro* and that for 10^7 is *e ahusia* (p. 239).

Among the Micronesian (MC) languages, most languages for which we have data possess terms for hundreds and thousands. The Micronesian languages possess numeral classification and have both "hundreds" and "thousands" classifiers. In Puluwat, for example, the former is *-pwukuw* and the latter is *-ngeray*: each of these may prefixed by a numeral root from 1 to 9. Several languages have terms expressing higher orders of magnitude than these. Kiribatese, Ponapean and Woleaian, for example, have the terms given in Table 8.6.

Table 8.6
Showing Terms for Increasing Orders of Magnitude in Three Micronesian Languages.

	Kiribatese	Ponapean	Woleaian
10^4	*terebu*	*nen*	*sen*
10^5	*tekuri*	*lopw*	*selob*
10^6	*teea*	*rar*	*sepiy*
10^7	*tetano*	*dep*	*sengit*
10^8	*tetoki*	*sapw*	*sangerai*

Source. Harrison and Jackson (1984, p. 67)

According to Bender and Beller (2006), the Marquesas in Northwest preferred to count with $2 \times$ power of ten and in the Southwest with $4 \times$ power of ten. They counted very large numbers such as 4 million. Again certain items had specific counting words and similarly for Hawaiian as there is "some indication of the parallel usage of different systems, as specific objects were counted with specific terms, such as 40 fish referred to by *ka'au* or 40 tapa or canoes by *'iako* (Alexander, 1864, p. 14)" (Bender & Beller, 2006, p. 386).

The Origin and Development of Terms for Large Numbers

The data presented in the previous sections suggest that the degree to which the languages of New Guinea and Oceania possess the resources for the expression of large numbers varies both between and within the NAN and AN language groups. Generally speaking, the NAN languages rarely possess single terms for the expression of numbers of the order of 100 and there is no evidence to suggest that any NAN language possesses terms for the expression of numbers of high order of magnitude of the sort found in the Polynesian and the Micronesian languages. Those NAN languages which do have single terms for the order numbers of 100 and 1 000 belong, generally, to the East Papuan Phylum. Furthermore, not all East Papuan Phylum languages exhibit this property: only certain languages possessing a 10-cycle numeral system have the resources for the efficient expression of large numbers. In Chapter 6, those NAN languages which do possess 10-cycle numeral systems have developed these as a result of the influence of neighbouring AN speakers. Indeed, in some cases, though not all, the terms for large numbers are AN loan words. The borrowing of terms is only one manifestation of influence. A more fundamental aspect of influence is the adoption of the principle of denoting numbers, which might otherwise be expressed in complex multiplicative constructions, by a simple expression or a single word. Thus we have, for example, the replacement of a complex expression of the form 10×10 by a single term denoting 100 and, by extension, the replacement of expressions of the form $10 \times 10 \times 10$ or 10×100 by a single term for 1 000. It is this principle and not merely the possession of a 10-cycle system, which enables numeral systems to be extended indefinitely in an efficient way. It is important to point out, however, the replacement principle is most commonly a feature of 10-cycle systems and while, in theory, it could be applied in other systems, for example those with a (5, 20) cyclic pattern, this does not generally appear to be the case.

It is among the Oceanic AN languages that we most commonly find single terms for large numbers. Yet even among the AN languages there is a considerable variability: some groups such as certain Solomon Islands languages together with the Polynesian and Micronesian languages have resources for the expression of numbers of high orders of magnitude; other groups, such as the North New Guinea Cluster, do not have these resources and many indeed lack terms for numbers of the order of 100. For those AN languages which do have terms for 100 and 1 000, there is no obvious uniformity in the terms used. This raises the question whether terms for these numbers were present in POC. As indicated above, two different words for 100 have been reconstructed for POC: one is **Ratu*, present in PAN as **Ratus,* and the other is **dau*. The fact that two words are reconstructible, rather than one unique word, reflects the lack of uniformity among the daughter languages of POC in the terms used to denote 100. There is, among the Polynesian languages, a greater degree of uniformity in this respect: *lau,* and its cognates, are most commonly found. But even among certain of the Polynesian languages the word *lau* denotes not 100 discrete objects but refer, when counting collections, to 100 pairs or 100 quartets of objects, that is to 200 or 400 items. The Polynesian Outlier Rennellese (Solomon Islands) has the term *gau* for 100 which is used when counting various classes of objects; it is not used universally, however: *mano* is used when counting piles of bananas, *huata* is used in counting panels of thatch, and *kauhusi* is used in counting pairs of yams or breadfruit (Elbert, 1988, p. 190). This sort of variability in the terms used for numbers when enumerating various collections of objects is not uncommon among the Oceanic AN languages. In Arosi, in the Solomon Islands, for example, Fox (1931) indicated that *tangarau* (a reflex of POC **dau*) is seldom used except for counting men and coconuts. In counting houses, quartets of bats' teeth and pairs of yams the term for 100 items is ′arangi. In counting banana shoots the term used is *umuumu*; for pigs and dogs the term is *nahomera* (p. 236-239).

A similar situation applies when we consider the range of terms used to express 1 000 in a given AN language. In Rennellese, Elbert (1988) indicated that generally *noa* is used in the enumeration of various objects or collections of objects. However, in counting pairs of yams and breadfruit, *ahe* is used (p. 190). In Arosi, Fox (1931) reported that *meru* is normally used to denote 1 000 but that the term for 1 000 yams, taro, bananas, stones or mangos is wawaibe′o (the same term is used to denote 1 000 units of shell money where one unit comprises four fathoms.) Coconuts are counted in pairs and 1 000 pairs is *bwera*. The term for 1 000 pigs or dogs is *hagahaga* (pp. 227-240). Among the Polynesian languages, we have indicated earlier that the most common term used for 1 000 is *mano*. The same term may be used to denote 1 000 pairs or quartets: in Hawai'i, for example, *mano* is used for denoting $4 \times 1\,000$ objects while, in Tongareva, *mano* denotes $2 \times 1\,000$ objects (see Appendix B based on Lean's (1992) data). In at least four Polynesian languages, *mano* is used to denote not 1 000 but rather 10 000: this is the case in both Tongan and Tokelau, for example, in which the term *afe* is used to denote 1 000.

The foregoing discussion of the Oceanic AN terms used for 100 and 1 000 highlights the difficulties in attempting to reconstruct the historical development of the AN numeral systems and, in particular, the nature of the POC numeral system and the extent to which terms for large numbers were present in POC. We have, for example, the possibility that the practice of counting various categories of objects existed in which category-specific terms were used for enumeration rather than one unique set of numerals. This practice is widespread among the AN-speaking societies of Melanesia, Polynesia and Micronesia and it is not unreasonable to assume that it existed in the POC community. If this were the case then it is likely that there existed more than one term for the expression of each of the large numbers of the order of 100, 1 000 and more. Thus, for example, the term for 100 used in enumerating yams or taro may have been different from those used in enumerating units of shell money or pairs of breadfruit or quantities of eggs. The fact that at least two terms for 100 are reconstructible in POC, as indicated earlier, may well derive from this. Similarly, it is possible that more than one term for 1 000 existed in POC. The reconstruction of such terms, however, is complicated by at least two factors. The first is that it is difficult to nominate even a single candidate from the reflexes occurring in the POC daughter languages from which a term for 1 000 might be reconstructed. For example, while the term *mano* is common among the Polynesian languages it is not commonly found among the other Oceanic AN languages. The second factor arises from the uncertainty as to whether terms for numbers of the order of 1 000 and more actually denoted precise values or whether they were used to describe indefinitely large quantities. Elbert (1988), for example, in discussing the Rennellese, said that "the large numbers … used in food distributions have never been taken too literally but symbolise unfathomably large quantities, which are so admired in counting food" (p. 187). Two types of evidence suggest this possibility. First, in certain languages a given term is used to denote 1 000, while in other languages the same, or a similar, term is used to denote, say, 10 000. Thus in the Polynesian languages, *mano*, is commonly used to denote 1 000 but in several cases it denotes 10 000. The second type of evidence derives from the semantics of certain terms used for large numbers for which the general meaning is "countless" or "indefinitely large". Harrison and Jackson (1984), for example, provided instances from the Micronesian languages in which the nominal interpretation of terms for large numbers includes such meanings as "end", "limit", "sand", "soil", and so on (p. 69).

In summary, then, the languages of New Guinea and Oceania do not present a uniform picture with regard to the presence of simple terms for large numbers of the order of 100, 1 000, and so on. Among the NAN languages such terms are not generally found although there are a few exceptions: among members of the East Papuan Phylum, for example. These languages also possess 10-cycle numeral systems which thus have (10, 100, 1 000) cyclic patterns. Several of these languages have clearly borrowed their terms for large numbers from AN sources, however this is not true in all cases. Other NAN languages clearly have the resources for expressing numbers of the order of 100 but this is often done by the use of complex expressions rather than of single terms. The conclusion which might be drawn is that, generally speaking, the languages ancestral to those belonging to the various NAN phyla now found in the New Guinea region did not possess single terms for large numbers and that those languages which now possess such terms have acquired these by either direct or indirect AN influence.

It seems likely that POC possessed single terms for large numbers at least of the order of 100 and 1 000. Also, as is common among many of its daughter languages, POC possessed a variety of terms for counting collections of objects. The outcome of this is that there may exist not just one unique term denoting 100, say, but rather several terms, depending on the objects being counted. This possibility needs to be taken into account while attempting to reconstruct the POC terms for 100 and 1 000. Among the daughter languages of POC the evidence tends to suggest that some terms for large numbers are found only in certain regional groups: the inference which may be drawn is that such terms have been invented and have gained local currency rather than being derived from POC.

It is also apparent that there have been at least two other post-POC developments with respect to the expression of large numbers. One of these is that the languages of the North New Guinea and Nuclear Papuan Tip clusters have lost their terms for numbers of the order of 100 and 1 000 as indeed they have lost their POC-derived 10-cycle numeral systems. The second development appears to have occurred among certain Oceanic AN languages located largely outside the New Guinea region, most noticeably the Polynesian and Micronesian languages. This is the extension of their 10-cycle numeral systems to include single terms for numbers of the order of 10^4 and more. Harrison and Jackson (1984) reviewed data which

> suggest that the higher ten-power bases have to some degree a history distinct from that of the lower. In our view, they may have developed as numbers at a more recent historical period. … the systems found are the result of a number of independent innovations. (p. 73)

If this view is correct, then it seems possible that POC did not possess terms for numbers of this magnitude. The other, less parsimonious, possibility is that POC did possess such terms but that these were largely lost and that the Polynesian and Micronesian languages have subsequently invented new terms (Figure 8.4).

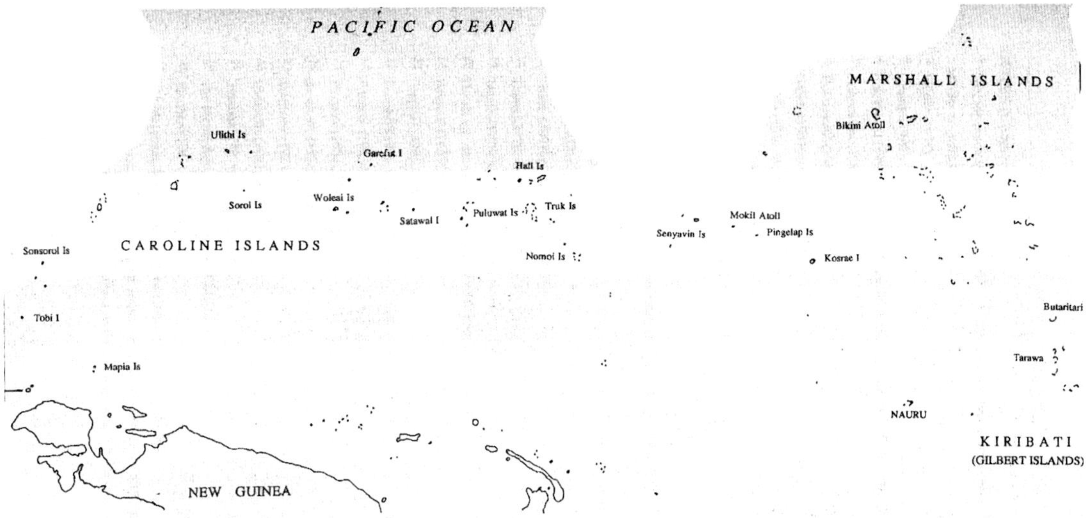

Figure 8.4. Location of societies cited in text (Micronesia).

It seems likely that certain terms now used for large numbers in the Oceanic AN languages did not originally denote numbers but rather had meanings which referred to imprecise amounts, collections of objects, or limitless quantities. Thus in the Micronesian languages we have the Kiribatese words for 10^7 and 10^8 have the meanings, respectively, "sand or soil" and "end" (Harrison & Jackson, 1984, p. 73). Elbert (1988), in discussing the Rennellese word *nimo*, indicated that this "theoretically means a million, but it is sometimes used for impossibly high numbers, such as the national debt. It ordinarily means 'to forget' or 'to disappear'" (p. 189). A number of languages possess a term for 100 *(lau, meldol)* which originally referred to a palm frond and had the meaning leaf: the implied quantity being the leaflets of the palm frond. Within a particular group of languages, a certain word which originally had the meaning of some large but unspecified quantity was eventually used to denote a specific number. The number denoted, however, was not necessarily the same across all languages in the group. Thus, in the Polynesian languages, *mano* is not commonly used to denote 1 000 but in several languages it denotes 10 000 instead. Similarly in several languages of the North Solomons Province (PNG), *kokolei* (and cognate terms) is used to denote 1 000 but in Central-East Choiseul (Solomon Islands) a similar term denotes 10 000.

It may be that such terms are used to denote numbers which represent the approximate practical limits of counting in these cultures: in some cultures this may be of the order of 1 000 while in others the limit may be larger. While it is certainly the case that some Oceanic AN societies have terms for much larger numbers than these, it seems unlikely that, even for ceremonial occasions in which large quantities of food or shell-money might be involved, counting up to the order of 10^5 or more would take place. As Beaglehole and Beaglehole (1938) observed about the presence of terms for large numbers among the Pukapuka:

> it is a little hard to see the function of high numerical concepts in an atoll culture … It is likely, however, that such words … indicate not so much a definite number as a progression of increasing greatness that is more sensed or felt than definitely apprehended. (p. 354)

References

Adams, K. (1989). Systems of numeral classification in the Mon-Khmer. *Nicobarese and Aslian sub-families of AustroAsiatic. Pacific Linguistics, B-101.*

Alexander, W. (1864). *A short synopsis of the most essential points in Hawaiian*. Rutland, VT: Charles E. Tuttle.

Alkire, W. (1970). Systems of measurement on Woleai Atoll, Caroline Islands. *Anthropos, 65*, 1–73.

Allan, K. (1977). Classifiers. *Language, 53*, 285–311.

Armstrong, W. (1924). Rossel Island money: A unique monetary system. *The Economic Journal, 34*, 423–429.

Armstrong, W. (1928). *Rossel Island: An ethnological study*. Cambridge, UK: Cambridge University Press.

Bathgate, M. A. (1985). Movement processes from precontact to contemporary time: The Ngi-Nggai, west Guadalcanal, Solomon Islands. In M. Chapman & R. Mansell Prothero (Eds.), *Circulation in population movement: Some concepts from the Melanesian case* (pp. 83–118). London, UK: Routledge & Kegan.

Beaglehole, E., & Beaglehole, P. (1938). Ethnology of Pukapuka. *Bernice P. Bishop Museum Bulletin, 150.*

Becker, A. (1975). A linguistic image of nature: The Burmese numerative classifier system. *Linguistics, 13*(165), 109–121.

Bender, A., & Beller, S. (2006). Numeral classifiers and counting systems in Polynesian and Micronesian languages: Common roots and cultural adaptations. *Oceanic Linguistics, 45*(2), 380–403.

Bender, A., & Beller, S. (2013). Mangarevan invention of binary steps for easier calculation. *Proceedings of the National Academy of Sciences of the United States of America, 111*(4), 1322–1327.

Berlin, B. (1968). *Tzeltal numeral classifiers: A study in ethnographic semantics*. The Hague, The Netherlands: Moulton.

Best, E. (1907). Maori numeration, the vigesimal system. *Journal of the Polynesian Society, 16*, 94–98.

Biggs, B. (1990). Extraordinary eight. In J. J. C. S. Davidson (Ed.), *Pacific Island languages: Essays in honour of G.B. Milner* (pp. 29–39). Honolulu, HI: University of Hawaii Press and School of Oriental and African Studies, University of London.

Bowers, N., & Lepi, P. (1975). Kaugel Valley systems of reckoning. *Journal of the Polynesian Society, 84*(3), 309–324.

Brainerd, B., & Peng, F. (1968). A syntactic comparison of Chinese and Japanese numerical expression. In H. Brandt Corstius (Ed.), *Grammars for number names*. Dordrecht, The Netherlands: Reidel.

Buck, P. (1938). Ethnology of Mangareva. *Bernice P. Bishop Museum Bulletin, 157.*

Burling, R. (1965). How to choose a Burmese numeral classifier. In M. Spiro (Ed.), *Context and meaning in cultural anthropology* (pp. 243–264). New York, NY: The Free Press.

Carrier, A. (1981). Counting and calculation on Ponam Island. *Journal of the Polynesian Society, 90*(4), 465–479.

Cassirer, E. (1953). *The philosophy of symbolic forms* (Vol. 1). New Haven, CT: Yale University Press.

Clark, R. (1999). Proto-Polynesian numerals. In E. Zeiton & P. J.-K. Li (Eds.), *Selected papers from the 8th International Conference on Austronesian linguistics* (pp. 195–204). Taipei, Taiwan: Academia Sinica.

Closs, M. (Ed.). (1986). *Native American mathematics*. Austin, TX: University of Texas Press.

Codrington, R. (1885). *The Melanesian languages*. Oxford, UK: Clarendon Press.

Conant, L. (1896). *The number concept*. New York, NY: Macmillan.

Crump, S. (1990). *The anthropology of numbers*. Cambridge, UK: Cambridge University Press.

Elbert, S. (1988). *Echo of a culture: A grammar of Rennell and Bellona, Oceanic Linguistics Special Publication* (Vol. 22). Honolulu, HI: University of Hawaii Press.

Elbert, S., & Pukui, M. K. (1979). *Hawaiian grammar*. Honolulu, HI: University of Hawai'i Press.
Fisher, J. (~2010). *Enriching students' learning through ethnomathematics in Kuruti elementary schools in Papua New Guinea*. Lae, Papua New Guinea: ResearchGate.
Foley, W. (1986). *The Papuan languages of New Guinea*. Cambridge, UK: Cambridge University Press.
Fortune, R. (1942). Arapesh. *American Ethnological Society, 19*, 1–60.
Fox, C. (1919). Social organization in San Cristoval, Solomon Islands. *The Journal of the Royal Anthropological Institute of Great Britain and Ireland, 49*, 94–179. doi:10.2307/2843437.
Fox, C. (1931). Arosi numerals and numeration. *Journal of the Polynesian Society, 40*, 236–243.
Franklin, K., & Franklin, J. (1962). The Kewa counting systems. *Journal of the Polynesian Society, 71*, 188–191.
Franklin, K., & Franklin, J. (1978). The Kewa dictionary, with supplementary grammatical and anthropological materials. *Pacific Linguistics, C-53*.
Frazer, J. (1923). *Taboo and perils of the soul*. London, UK: Macmillan.
Frege, G. (1950). *The foundations of arithmetic*. Oxford, UK: Blackwell.
Frege, G. (1967). *The basic laws of arithmetic*. Los Angeles, CA: University of California Press.
GLEC. (2008). *Glen Lean Ethnomathematics Centre*. Retrieved from http://www.uog.ac/glec/
Gwilliam, J. (1982). Some religious aspects of the *hiri*. In T. Dutton (Ed.), *The hiri in history: Further aspects of long distance trade in Central Papua* (pp. 35–63). Canberra, Australia: Australian National University.
Haas, M. (1942). The use of numeral classifiers in Thai. *Language, 18*, 201–205.
Haas, M. (1951). The use of numeral classifiers in Burmese. *Semitic and Oriental Studies, 11*, 191–200.
Harrison, S., & Jackson, F. (1984). Higher numerals in several Micronesian languages. In B. Bender (Ed.), *Studies in Micronesian linguistics* (pp. 61–79). Canberra, Australia: Pacific Linguistics.
Heider, K. (1991). *Grand Valley Dani: Peaceful warriors* (2nd ed.). Chicago, IL: Holt, Rinehart and Winston.
Hoa, N. (1957). Classifiers in Vietnamese. *Word, 13*(1), 124–152. doi:10.1080/00437959.1957.11659631.
Holzknecht, S. (1989). The Markham languages of Papua New Guinea. *Pacific Linguistics, C-115*.
Hurd, C. (1977). Nasioi projectives. *Oceanic Linguistics, 16*, 111–178.
Hurford, J. (1987). *Language and number: The emergence of a cognitive system*. Oxford, UK: Basil Blackwell.
Jacob, J. (1965). Notes on the numerals and numeral coefficients in old, middle and modern Kymer. *Lingua, 15*, 143–162.
Keller, K. (1955). The Chontal (Mayan) numeral system. *International Journal of American Linguistics, 21*, 258–275.
Lakoff, G. (1987). *Women, fire, and dangerous things: What categories reveal about the mind*. Chicago, IL: University of Chicago Press.
Lancy, D., & Strathern, A. (1981). "Making twos": Pairing as an alternative to the taxonomic mode of representation. *American Anthroplogist, 83*, 773–795.
Lawes, W. (1885). *Grammar and vocabulary of the language spoken by the Motu tribe*. Sydney, Australia: C. Potter, Government Printer.
Lean, G. (1992). *Counting systems of Papua New Guinea and Oceania* (Unpublished PhD thesis). PNG University of Technology, Lae, Papua New Guinea. Retrieved in 2008 from http://www.uog.ac.pg/glec/
Milner, G. (1966). *Samoan dictionary: Samoan-English, English-Samoan*. London, UK: Oxford University Press.
Mimica, J. (1988). *Intimations of infinity: The mythopoeia (cultural meanings) of the Iqwaye counting and number systems*. Oxford, UK: Berg Publishers.
Mosel, U., & Hovdhaugen, E. (1992). *Samoan reference grammar*. London, UK: Scandinavian University Press.

Muke, C. (2000). *Ethnomathematics: Mid-Wahgi counting practices in Papua New Guinea* (Unpublished MEd thesis). University of Waikato, Waikato, New Zealand.

Oliver, D. (1989). *Oceania: The native cultures of Australia and the Pacific islands* (2 Vols.). Honolulu, HI: University of Hawaii Press.

Omar, A. (1972). Numeral classifiers in Malay and Iban. *Anthropological linguistics, 14*, 87–96.

Panoff, M. (1969). The notion of time among the Maenge people of New Britain. *Ethnology, 8*, 153–166.

Panoff, M. (1970). Father arithmetic: Numeration and counting in New Britain. *Ethnnology, 9*(4), 358–365.

Pospisil, L. (1963). *The Kapauku Papuans of west New Guinea*. New York, NY: Holt, Rinehart, and Winston.

Ray, S. (1926). *A comparative study of the Melanesian island languages*. Cambridge, UK: Cambridge University Press.

Ross, M. (1988). Proto Oceanic and the Austronesian languages of western Melanesia. *Pacific Linguistics, C-98*.

Salisbury, R. (1970). *Vunamami: Economic transformations in a traditional society*. Berkeley, CA: University of California Press.

Schafer, E. H. (1948). Noun classifiers in classical Chinese. *Language, 24*(4), 408–413. doi:10.2307/410357.

Seidenberg, A. (1962). The ritual origin of counting. *Archives for History of Exact Sciences, 2*, 1–40.

Smith, G. (1978). Counting and classification on Kiwai Island. *Papua New Guinea Journal of Education, Special Edition the Indigenous Mathematics Project, 14*, 53–68.

Smith, G. (1981). *Traditional mathematics in Morobe* (Working Paper No. 21). Lae, PNG: Papua New Guinea University of Technology.

Smythe, W. E. (1970). Melanesian, Micronesian, and Indonesian features in languages of the Admiralty Islands. In S. Wurm & D. Laycock (Eds.), *Pacific linguistics studies in honour of Arthur Capell, Pacific Linguistics, C-15* (pp. 1209–1234). Canberra, Australia: Pacific Linguistics.

Smythe, W. E. (n.d.). *Admiralty Islands linguistics, Part 7: Gele' vocabulary*. Ukurumpa, Eastern Highlands Province, Papua New Guinea: Unpublished type script held at SIL library.

Song, J. J. (1997). The history of Micronesian possessive classifiers and benefactive marking in Oceanic languages. *Oceanic Linguistics, 36*(1), 29–64.

Strathern, A. (1971). *The rope of moka: Big-men and ceremonial exchange in Mount Hagen, New Guinea*. Cambridge, UK: University Press.

Strathern, A. (1977). Mathematics in the *moka*. *Papua New Guinea Journal of Education, Special Edition the Indigenous Mathematics Project, 13*(1), 16–20.

Terrill, J. (1986). Linguistics and the peopling of the Pacific Islands. *Journal of the Polynesian Society, 90*, 225–258.

Terrill, A. (2011). Languages in contact: An exploration of stability and change in the Solomon Islands. *Oceanic Linguistics, 50*(2), 312–337.

Thune, C. (1978). Numbers and counting in Loboda: An example of a non-numerically oriented culture. *Papua New Guinea Journal of Education, 14*, 69–80.

Tylor, E. (1871). *Primitive culture: Researches into the development of mythology, philosophy, religion, language, art and custom*. London, UK: John Murray.

van Heijenoort, J. (1967). *From Frege to Gödel: A source book in mathematical logic*. Cambridge, MA: Harvard University Press.

Wilder, R. (1974). *Evolution of mathematical concepts*. London, UK: Transworld.

Wilder, R. (1981). *Mathematics as a cultural system*. Oxford, UK: Pergamon Press.

Wolfers, E. (1972). *Counting and numbers* (Vol. 1). Melbourne, Australia: Melbourne University Press.

Zaslavsky, C. (1973). *Africa counts: Number and pattern in African culture*. Boston, MA: Prindle, Weber & Schmidt.

Chapter 9
Testing the Diffusion Theory

Kay Owens and Glen Lean

Abstract This chapter addresses the issue of how different counting systems occurred and in particular the theory of counting systems spreading from a centre. The most comprehensive theory of this kind before 1990 was that of Seidenberg. This theory is expounded and then several queries are raised. In general, the argument is put that the counting systems of Papua New Guinea and Oceania did not spread from the Middle East and the prominence of so-called neo-2 cycles and 10 cycles cannot be supported.

Keywords diffusion of counting systems • Seidenberg's diffusion theory • prehistory of number theories • innovation in counting systems

Introduction

The data presented in Chapters 3 to 8 and which summarise the material given in the appendices of Lean's (1992) thesis indicate the complexity of the counting system situation that exists in the traditional societies of New Guinea and Oceania. The main focus of this chapter is to consider a theory of how this situation may have come about. Do we take the view, for example, that each of the counting systems that are found today is a lineal descendant of a system which was invented by the ancestors of the present inhabitants of the region at some remote time in the past, the essential structural features of each system being retained despite the inevitable changes due to linguistic speciation over time. Such a view implies that once a particular society possessed a given counting system then the integrity of the system would be maintained through succeeding generations despite the possibility that the society may come into contact with another which possessed a different, perhaps more efficient, system.

An alternative view to this is that a society's counting system, far from being a stable and invariant feature of that society, is in fact very susceptible to external influence. If, for example, a society with a 2-cycle system comes into contact with another society which has a 10-cycle system then this view suggests that the most likely outcome of such contact is that the first society will abandon its 2-cycle system in favour of the second society's more efficient 10-cycle one. This transmission of a counting system from one society to another is an example of diffusion which, in a wider context, may also involve the transmission of other cultural institutions as well as artifacts and technologies. Generally speaking, a diffusionist interpretation of how the current counting system situation in New Guinea and Oceania came about would take the following outline. At some time in the past, the

K. Owens (✉)
Charles Sturt University, Dubbo, NSW, Australia
e-mail: kowens@csu.edu.au

G. Lean

K. Owens et al., *History of Number*, History of Mathematics Education,
DOI 10.1007/978-3-319-45483-2_9

ancestors of the current inhabitants possessed a particular type of counting system. Subsequently, new and different counting systems were introduced into the region in some sequence and, to use a tidal analogy, as each new system swept in it engulfed and overlaid certain systems already in place while by-passing others leaving them intact. The situation that is now apparent is the end result of a dynamic process of continual flux and change.

Of the two views outlined above, it is the diffusionist interpretation which has prevailed as the dominant explanatory theory of the prehistoric development of counting systems in human societies (see Chapter 1). In this chapter we will first outline the main conjectures of the most influential diffusionist theory of counting systems. Second, we will consider several aspects of the diffusion process itself which have been largely unaddressed in this theory. Third, the data available for this study will be used to elucidate the types of change which have apparently occurred to various counting systems in New Guinea and Oceania. Finally, a detailed evaluation and critique of the major diffusionist theory is provided together with an indication of the degree to which the data available support, or do not support, a diffusionist stance. This will then pave the way to make decisions on whether there is a third possible explanation of the diversity.

The Origin of Diffusion Theories

During the 19th century, as the amount of ethnographic data on the Indigenous cultures of the major continents increased, it became apparent that many societies, located in widely separated parts of the world, shared similar cultural traits, institutions, and artifacts. One view of why this should be the case, that of the independent inventionists, took the stance that each society invented its own cultural institutions and that the similarities which may be perceived in widely separated societies are the outcome of similar and spontaneous reactions of the human mind to the environment. Summarising the main points of this view, Raglan (1939) noted that

> The essence of this doctrine is that every human being is born with tendencies which lead him to make stone axes, bows and arrows, and dug-out canoes; to organise himself into totemic clans; and to believe in witchcraft, animism, and survival after death. These are assumed to be the mental and material equipment with which nature endowed primitive man, and which he proceeded to improve upon wherever local conditions allowed his innate progressiveness to develop. (p. 10)

It was also thought that Indigenous societies underwent separate but parallel development. It was apparent, however, that these societies, despite having similarities, were not identical and that there was considerable variation in the degree to which each society did develop and attain technological sophistication. We therefore find, particularly among 19th century scholars, that the proponents of the parallel development of cultural systems also adhered to a view of cultural evolution in which human societies could, in theory, be placed on a unilinear scale which ranged from the "primitive" to the "civilised". One major and influential representative of this view was Edward Tylor (1871) in his work *Primitive Culture.*

In the early part of the 20th century, at least two different theories were developed to counter that of the parallel and independent evolution of human societies. The first, developed by German and Austrian scholars was the Kulturkreise theory (Lowrie, 1937, pp. 123, 178-179) "that had cultures everywhere developing as a result of overlapping bundles or complexes of traits carried from some heartland in great waves or circles" (Riley, Kelley, Pennington, & Rands, 1971, p. xii). One adherent to this theory and, in particular how it applied to numeral systems, was the German linguist Fr. W. Schmidt (1926, 1929). The second theory was elaborated initially by G. Elliot Smith (1933) during the period 1910 to 1930 and is most clearly stated in his book *The Diffusion of Culture.*

After discussing a wide range of cultural practices, beliefs and myths shared by many societies around the world, Smith (1933) argued that these

> establish beyond question the unity of origin of civilization and the fact of an unbroken diffusion of culture for fifty centuries. When, however it is remembered that for hundreds of thousands of years before then men did none of these strange things nor conceived any of the fantastic ideas just enumerated, the conclusion is established that there is no innate impulse in man to create the material or the spiritual ingredients of civilization. (pp. 186-187)

Smith's view was that there was "one original source" from which

> the civilization of the whole world was derived (p. 232) [and that] the evidence which is now available justifies the inference that civilization originated in Egypt, perhaps as early as 4 000 B.C., but certainly before 3 500 B.C. (pp. 222-223)

A rival theory to that of Smith's was developed in the 1930s by Lord Raglan (1939). Raglan believed, like Smith, that the important inventions of human culture occurred once and once only in a centre of civilisation from which they were diffused all over the world. Unlike Smith, however, Raglan believed that these inventions originated not with the Egyptians but rather with the Sumerians of Mesopotamia.

The dichotomy between the independent inventionists and the diffusionists is apparent in the literature of the history of number. Writing in 19th century on "numerals as evidence of civilisation", Crawfurd (1863) noted that "the general conclusion, then, to which we must necessarily come is that each separate tribe invented its own numerals, as it did every other part of language" (p. 102). One hundred years later, Wilder (1974), in his influential book *Evolution of Mathematical Concepts*, stated that

> whether counting started in a single prehistoric culture and spread thereafter by diffusion or developed independently in various cultures (as seems most likely), is perhaps not too important for our purposes; interesting as it may be to speculate thereon … and since it seems impossible to find out when man developed counting … we may as well get on with what we know from the archaeological and historical records. (p. 35)

Abraham Seidenberg (1960), however, in the first comprehensive statement of the diffusionist view of counting, disagreed with this opinion: "on the contrary, there is a great amount of information scattered through the anthropological literature which can be brought to bear on the question of the origin and development of counting" (p. 215). Seidenberg's influential work was developed as a contribution to Lord Raglan's theory. Seidenberg's achievement has been such that he has effectively established what appears to be the prevailing view of the prehistory of number and one which finds its expression in more recent publications such as those by Flegg (1984, 1989), Barrow (1992), and Van der Waerden and Flegg (1975a, 1975b). It is perhaps no wonder that the diffusion view was strong given the monastic Catholic church's teaching on numbers, like other things, being created by God and unified by patterns (Book of Wisdom 11: 21, "Thou hast ordered all things by number, measure, and weight" that linked science, e.g. calendar histories, with religion (Brown, 2010). Lean investigated the validity of Seidenberg's theory, so it is necessary to provide here a summary of his views which will be the focus of subsequent discussion.

Seidenberg's Theory

The basic tenet of Seidenberg's (1960) theory is that several of the various methods of counting, which can now be discerned in the Indigenous cultures of the world, each had a single centre of origin: "it is known that ideas arose in the ancient civilisations; it is not known that ideas ever arose, anciently, anywhere else" (p. 218). The earliest method of counting to be diffused from its centre of origin (the pure "2-system", identical to the pure 2-cycle system discussed in Chapter 3) "spread out over the whole earth; later, other methods of counting arose and spread over almost all, but not quite all, of the world" (p. 218). The counting systems which we now see are "living documents of archaic civilisation", that is the survivors of counting systems which were diffused early and were not subsequently displaced by counting systems diffused at a later date. Each of these systems is discussed briefly below.

"The Pure 2-System"

Seidenberg (1960) argued that the oldest type of counting system and the first to be diffused throughout the world is the pure 2-cycle system.[1] The geographical distribution of this system among the indigeneous peoples of the world is such that it "appears now only at the edges, and seems ready to be wiped off the face of the globe" (p. 218). It might be argued that "systemless" counting, in which there are three or four cardinals that are not combined to form higher numbers might be considered as a candidate for the oldest type of counting. Seidenberg suggested, however, that systemless counting is only observed in regions which have predominantly 2-cycle systems and that it may be a degenerate form of the 2-cycle type rather than an earlier counting method. It is often observed that the 2-cycle system does not invariably occur in its "pure" form but has instead a secondary 5-cycle and, sometimes, a 20-cycle as well. Seidenberg suggested that the existence of such systems is the result of the hybridisation of two different types: the original pure 2-cycle system and the (5, 20) digit-tally system which was diffused at some time after the 2-cycle system.

In examining the way in which numbers were represented in ancient texts, Seidenberg considers that the arrangement of strokes to represent the numbers from 1 to 10 clearly indicate a paired arrangement indicative of a 2-cycle system. The earliest example of such representations is found in Sumerian pictographs which are dated at 3 500 BC. He thus takes the ancient centre of Sumerian civilisation to be the likely candidate for the origin and centre of dispersion of the 2-cycle system.

"The neo-2-System"

If the 2-system originated in an ancient centre of civilisation in the Middle East and is now only found in the margins of the world, there should nevertheless still be traces of it to be found between the centre of dispersion and its present locations. Traces of the 2-cycle in its pure form between the Middle East and, say, South America, where it is found, are not readily apparent. There is no evidence, for example, of pure 2-cycle systems existing anywhere in North America. Seidenberg argued that while traces of the 2-cycle system in its pure form are not apparent nevertheless there is another counting system which has an affinity with the 2-cycle system or was perhaps developed from it. This system he termed the neo-2-system which he classified into two types:

Type 1: The proto-neo-2-system which is characterised by having

6 = 2 x 3,
7 = (2 x 3) + 1,
8 = 2 x 4, and
9 = (2 x 4) + 1

[1] The terminology used throughout this book will be used to discuss Seidenberg's theory; it was not his terminology.

or, alternatively,

6 = 2 x 3,
7 = (2 x 4) - 1,
8 = 2 x 4, and
9 = 10 - 1.

Type 2: The method of representation of two equal or quasi-equal summands which is characterised by having

6 = 3 + 3,
7 = 4 + 3,
8 = 4 + 4,
9 = 5 + 4.

Seidenberg suggested that the origin of the neo-2 system, in either of its forms, is the 2-cycle system itself. If the latter is represented in tally form, we have (Figure 9.1):

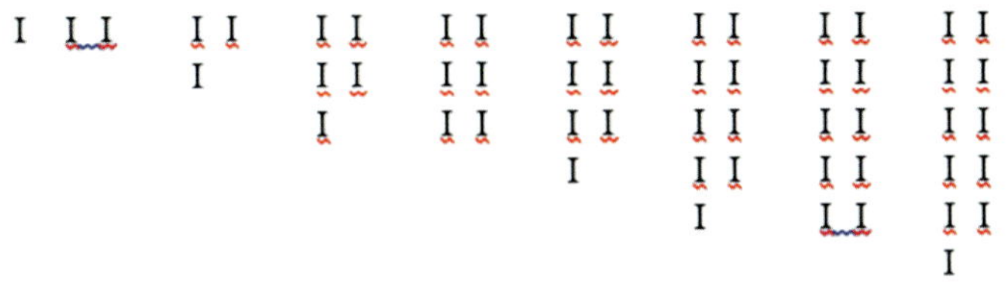

Figure 9.1. 2-cycle systems.

Depending on how such an arrangement is viewed, we can think of the number 6, for example, as being 2 + 2 + 2 or 2x3 or 3 + 3, each representation being, respectively, 2-cycle, neo-2 type 1, and neo-2 type 2.

Seidenberg traced the occurrence of this system in its two manifestations and indicated that they may be found in one form or another in North and South America, Africa, Australia and Papua New Guinea. He also observed that the neo-2 constructions as given above are often associated with 10-cycle systems and which can therefore be characterised as neo-2-10 systems. Seidenberg (1960) indicated that while 2-cycle systems and neo-2-systems have an affinity, they are nevertheless distinct: "the fact that they are distinct suggests to us not only that pure 2-counting came before neo-2-counting but that both of these came before, for example, 5-20-counting" (p. 237).

The 5-and 10-Cycle Systems: The Americas

Seidenberg distinguished two 5-cycle systems, the (5, 20) and the (5, 10). Using data accumulated by Kluge (in particular, 1939) and by reference to a map showing the distribution of counting systems throughout the world compiled by Schmidt (1926), Seidenberg first considered in some detail how the counting system distribution may have come about in the Americas paying particular attention to the 10-cycle system and the two types of 5-cycle system. The picture presented may be summarised as follows. The 2-cycle system was diffused first followed by the neo-2-system. The latter moved down the western side of South America and up the south-eastern side leaving intact the 2-cycle system in the central and north-eastern regions. This was followed by the diffusion of the neo-2-10 system. In South America this resulted in a distribution of neo-2-10 systems as well as a residue of intact neo-2- and 2-cycle systems. The next development was the diffusion of the (5, 20) digit tally system across the Bering Strait and down the western coast into Central America, the Innuit/Eskimos being the only main group in North America affected by this system. The subsequent diffusion of the (5, 20) system from Central America into both North and South America largely had the effect of introducing a 5-cycle into systems which had previously been neo-2-10, thereby producing (5, 10) systems: "the 10-counters took over finger counting from the 5-20 finger-toe counters. The reason they took over finger counting and not toe counting, was that they already had a consolidated 10-system" (p. 247).

There is, then, a fundamental difference between the (5, 20) and the (5, 10) systems: the former, Seidenberg suggested, is one of the primary types of systems, along with the 2-cycle and the neo-2-10 systems, which were diffused each from a single centre. The (5, 10) system, on the other hand, is a hybrid of the (5, 20) and the neo-2-10 system and it came into being in the regions where these two types of system met and interacted. One way in which the difference between these two types of system is reflected is in the construction of their second pentads. While with the (5, 20) system the second pentad numbers 6 to 9 have an explicit quinary construction, Seidenberg noted that the (5, 10) system rarely has an "intact" second pentad in which all the numbers 6 to 9 have an explicit quinary construction: "the non intact character of the 5-10 counting supports the idea that it is a cross" (p. 246).

Seidenberg argued that while the neo-2-10 system is a primary one which arose in a single centre and was diffused, this is not the case with the pure 10-cycle system. He surmised that when the (5, 20) system interacts with the neo-2-10 system, several variants may arise (such as the 2-10-5 system) in which the ideas underlying the way in which numbers are combined to form larger numbers become confused; this, together with linguistic change ("slurring"), results in numerals which appear to be distinct and in which the original components are no longer apparent. This process, Seidenberg suggested, "is enough to yield pure 10-cycle systems (also 10-20 and pure 20-systems)" (p. 247). He also noted that the 10-cycle system, having been "purified" in this way, could itself diffuse.

The 2-, neo-2, and 5-Cycle Systems in Africa

Seidenberg observed that each of the primary counting systems, i.e. the pure 2-cycle, the neo-2-10, and the (5, 20) system, existed in Africa and that these appeared there in the order given: this is inferred from their current geographical distribution in the continent. Furthermore he believed that, of the two types of neo-2 system discussed earlier, Type 1 entered Africa before Type 2. This is inferred because the former does not have a clear spread throughout the continent and that where it does appear it seems peripheral to Type 2. Seidenberg also indicated that his view that Type 1 is chronologically prior to Type 2 is supported by evidence from the Americas in that the former occurs in both North and South America but the latter in North America only.

There is ample evidence of the existence of the (5, 10) system in Africa. Seidenberg distinguished two types of this system: Type A which is such that "the numbers 6, 7, 8, 9 are built in a completely regular manner and use the conjunction 'and'. The word 'seven', for example, is literally 'five and two'" (p. 252). Type B is such that the second pentad numerals have the construction $x+n$ where x is not the word for 5, n takes the values 1 to 4 respectively, and no conjunction is used. We have, incidentally, noted earlier in Chapter 4, that there are three common ways in which the second pentad of 5-cycle systems is constructed: the two types distinguished by Seidenberg together with a third type which has a $5+n$ construction, that is the numerals 6 to 9 are explicitly constructed with 5 but no conjunction is used. Seidenberg's view is that his two types of (5, 10) system were diffused in Africa at different times: Type B, which is "a mixture of neo-2-10 and 5-20-counting" (p. 255), came first while Type A was diffused subsequently.

Digit Tallying and Body-Part Tallying

Seidenberg (1960) observed that body-part tallying, of the kind described in Chapter 4, is found in the Torres Strait islands and among certain groups of Australian Aboriginals. In both locations, 2-cycle numeral systems are also found. This, at first sight, seems paradoxical in that body-part tallying is rather more elaborate than the more common digit tallying and yet it occurs in regions where the simplest type of counting system exists: "the paradox consists in thinking that the simple is older than the complex" (p. 270). Seidenberg's view is that the body-part method is the older of the two and that its genesis had nothing to do with counting but rather with the practice of "parceling out various parts of the body to various gods" (p. 270). The practice of utilising body-parts to represent things evolved into using body-parts to maintain a tally and, in particular, as a calendrical device. The two practices, that of using body-parts and marking them off in a given order, and that of counting, came together and resulted in the use of body-parts for counting purposes. The reason for this fusion of the two practices was, Seidenberg suggested, the counting *tabu*, that is the practice in certain cultures in which the verbal counting of people or objects is expressly forbidden. Verbal counting, however, can be circumvented by the use of gestural, non-verbal counting:

> the gestures themselves had already existed, but became applied, under the circumstances of a *tabu*, to counting. The gestures having become standard symbols for numbers, the verbal descriptions of these gestures (possibly in the same effort to circumvent a *tabu*) began to stand for the numbers. (Seidenberg, 1960, pp. 270-271)

Thus we have the sequence in which:

1. names for numbers existed, but under the circumstances of a *tabu* being imposed on their use,
2. non-verbal body-part or digit tallying was used in which body-parts symbolised numbers; finally,
3. the verbal descriptions of the tallying process or the names of the body-parts themselves came to represent the numbers and displaced the proscribed original number names.

This view contradicted those of many earlier writers on this subject. Tylor (1871), for example, said that with regard to numbers "Word-language not only followed Gesture-language but actually grew out of it" (p. 246). Thus when a child learns to count by using his fingers, this reproduces "a process of the mental history of the human race; that in fact men counted upon their fingers before they found words for the numbers they thus expressed" (p. 246).

The practice of tallying or counting on the fingers is so widespread throughout the Indigenous cultures of the world that many scholars of the nineteenth century, Tylor (1871) and Conant (1896, pp. 7-17) among them, concluded that finger counting was an instinctual rather than a cultural phenomenon. Seidenberg (1960), on the other hand, took the view that the names for numbers and the various primary counting systems existed prior to the practice of using body-parts for counting. In addition, he argued that not only is the association of numbers with the marking off of fingers a cultural rather than an instinctual phenomenon but also that the order in which the fingers are enumerated has a definite significance and a cultural origin. Seidenberg's argument is essentially this: initially, the 2-cycle, neo-2-10, and the (5, 20), counting systems were not associated with body-part gestures. As a result of the operation of the counting *tabu* process outlined above, the (5, 20) counting system became associated with digit tallying and gave rise to two basic types of finger counting: that in which tallying began on the little finger and that in which tallying began on the thumb, the former being chronologically prior to the latter. Two further developments occurred when the (5, 20) system diffused and came into contact with the 2-cycle and the neo-2-10 systems. Seidenberg said that "the pure 2-counters were taught the gestures by the 5-20-counters, and, moreover, by 5-20-counters who began with the left little finger" (p. 263). When, however, the (5, 20) counters came in contact with the neo-2-10 counters, a different type of finger counting arose: "the neo-2-counters did not invent finger counting but were taught this practice by quinary finger counters, but, in acquiring the practice, modified it in accordance with their neo-2 habits" (p. 262). The modification was that counting began not with the little finger or the thumb but with the index finger.

A Genealogy of Counting Systems

The previous sections summarise briefly Seidenberg's views on those counting systems which he regarded as having originated in an ancient centre of civilisation and which were subsequently diffused all over the world. These systems and the order in which they were diffused are: a) the 2-cycle system, b) the neo-2 or neo-2-10 system of which, we have noted earlier, there are two types, and c) the (5, 20) system. The 2-cycle system is thought to have originated with the ancient Sumerians of Mesopotamia by 3 500 BC at the latest. No similar dates were, however, given for the origin of the other two systems. One other system which is commonly found among Indigenous societies is the (5, 10) system. As indicated earlier, this is regarded as a hybrid between the neo-2-10 system and the (5, 20) finger and toe counting in which the quinary nature of the latter was taken over by the former. This system, then, does not have the same primary status as the other three. One further system of considerable significance but which was also not granted primary status in Seidenberg's scheme is the 10-cycle system. It is clear that Seidenberg believed that this system existed prior to the (5, 20) system (p. 261 has a footnote indicating "10-counting preceded finger counting"). He suggested that it may have derived from neo-2-10 systems by a process of linguistic change in which the composite nature of the numerals less than 10 ceased to be apparent. How either type of neo-2 system acquired a 10-cycle in the first place is not explained: it is clear from the data quoted, however, that some, though not all, neo-2 systems do possess a superordinate 10-cycle (p. 271).

Seidenberg provided a diagram (Figure 9.2) which summarised "the genealogy of counting systems" (p. 271)

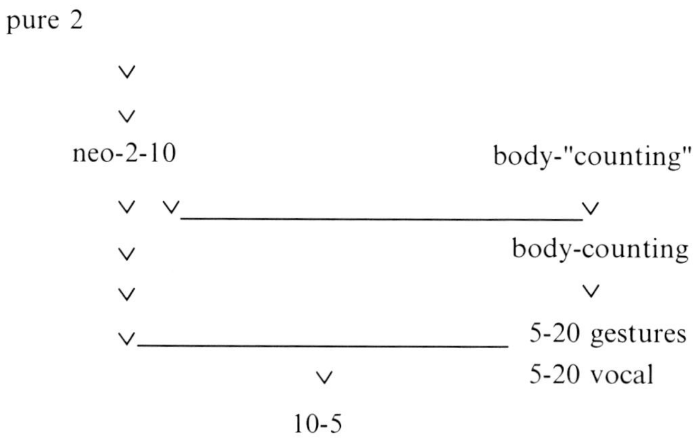

Figure 9.2. Seidenberg's counting system genealogy: A genealogy we dispute.
Source: Seidenberg (1960).

Additional Aspects of Seidenberg's Theory

We are concerned here with evaluating Seidenberg's (1960) views as they relate to the diffusion of various types of counting system and as set out in the previous sections. Several questions raised in Seidenberg's theory are not, however, addressed here largely for the reason that the database provided in Lean's (1992) Appendices will not provide answers to these. One question concerns the relation of numerals to grammatical number and whether the former preceded or followed the latter. Grammatical number, as it occurs in the languages of New Guinea and Oceania, does not merely involve the distinction between singular and plural but includes dual and often trial forms of the personal pronouns as well. Data on the form of personal pronouns have not been collected for this study and questions relating to this are not considered.

A further contribution which Seidenberg (1962) has made to the prehistory of number, although not dealing specifically with diffusion, is an essay on the ritual origin of counting which is an extension of Lord Raglan's (1939) ideas on the ritual origin of civilisation. Seidenberg's view is that "counting was invented in a civilised center, in elaboration of the Creation ritual, as a means of calling participants in ritual onto the ritual scene" (p. 37). In developing this idea, Seidenberg pointed to a belief held in many societies that numbers are sacred and are associated with deities, this belief being apparent in the myths of these societies. Myth and ritual are associated phenomena and the rite in which "deities", that is participants in the rite, are numbered is the Creation ritual or census in which the deities are called forward onto the ritual scene:

> the sacred character of numbers derives ... from the numbering of participants in a ritual. I go a step further, however, and see in the words used to call participants onto the ritual scene the very origin of number words and of numbers ... Counting is the secularization of the rite which called participants in ritual onto the ritual scene. (Seidenberg, 1962, p. 262)

Except to note these views, the details of this aspect of the theory are not addressed here nor, generally, will we be concerned with speculating about the genesis of numbers. We will, however, scrutinise several aspects of the diffusion theory as it relates to counting systems. We consider, first, whether there are any specific properties of counting systems which may affect their diffusibility and we also consider whether there are special circumstances which have to be met in order for diffusion to occur. Second, we list several types of change that have occurred to counting systems in the New Guinea region and consider whether these changes can be accounted for by diffusion. Finally, we provide a critique of Seidenberg's (1960) theory in the light of this discussion.

Counting Systems and Diffusion

In his book *The Anthropology of Numbers*, Crump (1990) concluded that "If there is one lesson to be learnt from the present study of traditional numeracy, it is that diffusion is the most common explanation of the emergence of numerical institutions in any local culture" (p. 147). In justifying his strong diffusionist stance, Crump pointed to several features which are characteristic of counting systems and which tend to make them easily diffusible relative to other cultural institutions or traits. The first of these characteristics is the abstract basis of counting systems which results in their being able to be diffused without making specific cultural demands: "these demands are in most cases linguistic", however, "once such an institution is understood in terms of the local culture, expressed in local language, it frees itself almost immediately from these cultural ties" (p. 147). The second feature "is that particular ways of using numbers ... relate to natural phenomena that know no cultural boundaries" (p. 148). Thirdly, Crump believed that among numerical institutions "there is a sort of 'survival of the fittest' ... The expression of numbers, beyond a certain low threshold, in terms of a polynomial with a single base (generally 10) is absolutely better than any alternative" (p. 150). The implication here is that a more efficient counting system will normally displace a less efficient one, a point which will be explored further when discussing the Oceanic AN counting systems below.

For these reasons, Crump maintained that "it is difficult for any society, however traditional, to defend itself against numerical institutions superior to those it already possesses, once they are knocking at the door" (p. 150). While the characteristics enumerated by Crump may well mean that the conditions under which the diffusion of counting systems might occur are favourable, there are nevertheless other factors which may influence the likelihood of diffusion between one group and another. Jett (1971) enumerated several such factors:

> the degree of friendliness, the intensity and length of contact; the degree of similarity of the values and technologies of the group involved; the degree of conservatism of the cultures, both in revealing and accepting ideas; the practical, prestige, luxury, or religious values of the traits, and the ease of learning them. (p. 21)

With particular reference to the diffusion of counting or tallying, whether or not the nature of the contact between two groups had an economic component, that is trade of some sort occurred, would also appear to be an important factor. Finally, it is possible that important changes can occur within a traditional society when it changes its environment by migration. If the migrating group moves into a region largely dominated by another culture, it may be to the advantage of the newcomers to adapt to certain aspects of the prevailing culture and this may include changes to its economy. Under such conditions it seems possible that a migrating group, which might otherwise be seen as the vehicle for diffusing new ideas, could well adapt its economy and numerical institutions to those of the culture already in place.

While the existence of similar cultural traits, including counting systems, among two traditional societies may provide evidence of diffusion, it may also be the case that the groups are related culturally and may share a common ancestor. Thus the situation which we now see among, say, certain traditional societies in Melanesia and Polynesia where there are obvious cultural similarities and clear linguistic affinities, is the result both of the migrations of these related societies and of the differentiation of their languages from a common ancestral proto language. Thus, Seidenberg (1960), in noting that the word for 5 in some Vanuatan languages is *lima*, added that "but it is known that this word is Indonesian" (p. 266) and made a basic error in assuming that somehow Indonesians contributed a word to the Vanuatans. This ignores the common Austronesian ancestry of the languages of Vanuatu and of Indonesian as can be determined by investigating the languages as a whole: isolating a counting system from its linguistic context can easily lead to this type of error. However, with this general caveat in mind, it is possible to find instances where one language possesses in common, in part or in whole, a numerical lexis with another language but in circumstances where the two languages are otherwise unrelated. In such cases the possession of loanwords does provide evidence that diffusion has occurred. The borrowing of numerals, however, is only one type of mechanism by which a counting system may change as a result of diffusion and we will now consider some of these changes as they occur in the languages of New Guinea and Oceania.

Changes Which Occur to Counting Systems

Crump (1990) had the view that

> no part of speech is more susceptible to linguistic borrowing and cultural diffusion than numerals. This in part explains not only why the lexical origins of numerals are so difficult to trace but also why numerals tend, intrinsically, to be so little related to other parts of the vocabulary. (p. 34)

That the counting systems of traditional societies do change is supported by abundant evidence from the data on New Guinea and Oceania. It is a not uncommon view among diffusionists that such changes occur as a result of the influence of one group on another and that the mechanism of change is that one group will borrow the other's numerical lexis, in whole or in part, resulting in an increase in the primary cycle of the borrowing group's counting system. Suppose, for example, that Group A possesses a (5, 20) digit-tally system and that it has some form of contact with Group B which possesses a pure 2-cycle system. The assumptions implicit in Seidenberg's theory is that, normally, influence is unidirectional with Group A affecting Group B so that the latter's 2-cycle system will be displaced or modified in some way, resulting in the adoption of a (5, 20) system. A likely mechanism which brings this change about is that Group B will augment its numerical lexis by borrowing from that of Group A.

In presenting the data that follow we shall investigate the validity of this view. In order to do this we will distinguish the various types of change which have evidently occurred in the counting systems of various traditional societies. We will consider, in passing, whether a system which has a larger primary cycle will always affect a system with a smaller primary cycle, and not vice-versa, and also whether change always results from influence rather than being spontaneously generated.

Borrowing and Displacement

Evidence that borrowing does occur can be seen by the presence of AN loanwords in the numerical lexis of several NAN languages. Three examples are given in Table 9.1 where we give the numerals 1 to 10 for three NAN languages, Yele (PNG), Ekagi (West Papua), and Mbilua (Solomon Islands), together with the corresponding numerals of Proto Oceanic (POC) for comparison.

Table 9.1
Showing the Numerals 1 to 10 for Three NAN Languages and POC

	POC	Yele	Ekagi	Mbilua
1	**kai, *sa, *tai*	*ngeme*	*ena*	*omandeu*
2	**rua*	*mio*	*wisa*	*omungga*
3	**tolu*	*pyile*	*wido*	*zouke*
4	**pat, *pati*	*paadi*	*wi*	*ariku*
5	**lima*	*limi*	*idibi*	*sike*
6	**onom*	*weni*	*benomi*	*varimunja*
7	**pitu*	*pyidu*	*pitiwo*	*sike-ura*
8	**walu*	*waali*	*waruwo*	*sio-tolu*
9	**siwa*	*tyu*	*isi*	*siak-ava*
10	**sangapulu*	*y:a*	*gati*	*Toni*

Note. POC is Proto-Oceanic. Yele (Milne Bay Province, PNG) data are from Henderson (1975) and supported by CSQs. Ekagi (West Papua, Indonesia) data are from de Solla Price and Pospisil (1966). Mbilua (Solomon Islands) from Tryon and Hackman (1983, pp. 123, 127, 131).

Yele appears to have borrowed at least the numerals 4 to 8 from an AN source while Ekagi has borrowed at least the numerals 6 to 8 and possibly 9. Mbilua, however, has borrowed the numerals 2, 3 and possibly 4 which appear in the second pentad and which are part of a 5-cycle construction for the numerals 7 to 9. In each case the AN numerals appear to augment an already existing numerical lexis and the result of this is to extend the original cyclic structure of each system. Lean suggested (see Chapter 5) that this is the means by which these East Papuan Phylum languages have acquired 10-cycle systems. The presence of AN loanwords can also be detected in a number of other NAN languages, for example Moi, Karon-Pantai, and Wodani (Galis, 1960; Lean, 1992, appendix on Oceania), each of which is located in West Papua. Generally speaking, however, it is clear from a survey of the NAN languages that the borrowing of AN numerals is not a common phenomenon even in the case where NAN languages are located in areas which are predominantly AN-speaking, for example Kuot in New Ireland (PNG) (Lean's field notes and Kluge (1941)) and Kovai on Umboi Island in the Morobe Province (PNG) (Lean's field notes and three CSQs).

We also need to consider whether any evidence exists to indicate that AN languages have borrowed NAN numerals. This is a more difficult task than the reverse situation in that whereas AN numerals are usually relatively easy to identify this is not the case with NAN numerals which exhibit enormous diversity. It is possible nevertheless to identify examples of AN counting systems which have a numeral lexis which is no longer recognisably AN in character and which we may infer is possibly due to NAN influence. Several instances of these are apparent: Sissano (and Sera) in Sandaun Province (PNG) and Maisin in the Oro Province (PNG) for example. The first five numerals of these are given in Table 9.2.

Table 9.2
Showing the Numerals 1 to 5 for Sissano, Maisin, and POC for Comparison

	POC	Sissano	Maisin
1	**kai, *sa, *tai*	*pontenen*	*sesei*
2	**rua*	*eltin*	*sandei*
3	**tolu*	*eltin pontenen*	*sinati*
4	**pat, *pati*	*eltin eltin*	*fusese*
5	**lima*	*eltin eltin pontenen*	*faketi tarosi*

Note. POC is Proto-Oceanic AN. Sissano data derive from 4 CSQs but are similar to records of Schmidt (1900) , Ray (1919), and Kluge (1938). Maisin data derive from Strong (1911) and is similar to 9 CSQs.

In neither set of numerals is the original AN character retained. Whereas in the case of the NAN systems we had an augmentation, by the AN loanwords, of an already existing sequence of NAN numerals, in the case of Sissano and Maisin the original numerals have been completely displaced by a new set of numerals quite unlike those we expect to find in AN languages.

We may therefore distinguish two types of change in the examples given above: borrowing and displacement. In the case of the former we have a sequence of several numerals borrowed from some source and grafted onto an already existing sequence. We can also find evidence of just a single numeral being borrowed. In Chapter 7, for example, we cited the case of several NAN languages having borrowed the AN word for domestic fowl used to denote the numeral 1 000. Indeed it often appears to be the case that when a single numeral is borrowed this will be a term for a large number. The second type of change, displacement, does not involve the adoption of a few additional numerals but rather the complete abandonment of the old system in favour of a new one. While it is possible to find isolated instances of displacement of counting systems having occurred in a number of languages in the New Guinea region during the pre-European period, it is now a much more commonly observed phenomenon as the English or Tok Pisin counting systems displace traditional systems. In PNG, in particular, this sort of displacement has occurred frequently among those language groups which possess 2-cycle counting systems. Among those groups that have 10-cycle or (5, 10) systems, it is often possible to observe the use of two parallel systems: the traditional systems which are used in traditional contexts, such as counting shell-money and bride wealth, and the introduced systems of either English or Tok Pisin which are used in non-traditional contexts. It also extends the systems where counting by 10s may occur with two counters with an existing (2, 4, 8) cycle system as in Hagen.

In the pre-European period it is clear that both borrowing and displacement occurred. Either of these types of change is relatively easy to detect when the interaction is between AN and NAN groups: detecting the interaction between two AN groups or two NAN groups, however, is not so easily done and thus it is probably impossible to estimate the degree to which borrowing or displacement may have occurred in such cases. From the data available on AN-NAN interactions, however, it is clear that neither borrowing nor displacement have occurred to any significant extent and that these types of change are probably not the principal means by which the diffusion of counting systems takes place and we need to examine the evidence further in order to establish what other types of change can occur.

Loss of the Numerical Lexis

The effect of borrowing numerals and increasing the basic numerical lexis is to increase the magnitude of the primary cycle of the borrower's counting system. Thus a language group which possesses a 2- or 5-cycle system could, by borrowing a sequence of numerals, acquire a 10-cycle system. The reverse situation is also possible: a language group with a 10-cycle system could, by the loss of part of its numerical lexis, acquire a system with a smaller primary cycle of, say, 5 or even 2. Such a change, however, might seem a retrograde step in terms of the reduction of counting efficiency. The assumptions of diffusionists like Seidenberg and Crump do not generally appear to encompass this type of change, yet, as we will see in the following discussion it is a common occurrence, particularly among the Oceanic AN languages of PNG and Island Melanesia.

A fundamental assumption on which the following analysis is based is that the speakers of POC possessed a 10-cycle numeral system with a basic numerical lexis comprising 10 monomorphemic numerals, as given earlier in Table 9.2: this is a well-established result of the historical linguistics of the Oceanic AN languages. The analysis to be presented below will be discussed in the context of the premise that any AN language which does not possess a "pure" 10-cycle numeral system has undergone a change and that this change has occurred as a post-POC development. In discussing each type of change we will consider whether it is likely that it has occurred as a result of direct or indirect external influence or whether, in the absence of any obvious influence, it has occurred as a "spontaneous" innovation.

(10) → (5, 10). The data presented in Chapter 5 indicate that, of the 420 Oceanic AN languages for which we have data, a total of 113, or 27%, possess counting systems with a (5, 10) cyclic pattern. The change which has occurred in such cases is the loss of numerals 6 to 9 in the second pentad. Three examples of this type, which occur in the languages of Tolai, Mota, and Kaliai, have been previously shown in Tables 5.8 and 5.9. The distribution of the (5, 10) system among the Oceanic AN languages may be seen in Figures 5.2 and 5.3 and this is such that it is found in PNG in 44 languages of the North New Guinea, Papuan Tip, and Meso-Melanesian Clusters, and in only one language, Nauna, of the Admiralties Cluster; it is also found in 68 languages of Vanuatu. Apart from these, the (5, 10) system is not found elsewhere in Island Melanesia, Polynesia, or Micronesia.

Most of the (5, 10) AN systems that are found in PNG occur in regions which are inhabited by NAN groups: along the north and north-east coast, in the New Britain, New Ireland, North Solomons, Central, and Milne Bay Provinces. In these cases it is possible that the AN groups were influenced by neighbouring NAN groups. In the case of Nauna, which is the easternmost language of the Admiralties Cluster, it seems unlikely, given the history of the Admiralties languages, that it was in contact with NAN groups. There is, perhaps, the possibility that the group or groups which left the POC homeland and eventually settled in the Manus region may have taken a route, say through New Ireland, which brought them into contact with NAN groups. Such a contact, though, would have needed to be sufficiently sustained in order to effect a change in the counting system of the migrating AN group. Similarly, with the languages of Vanuatu, there is no immediately obvious explanation of how they acquired a (5, 10) system. Was there, for example, a diffusion of AN groups carrying (5, 10) systems from New Britain, New Ireland, or Bougainville (but not the Solomon Islands where there is no evidence of this type of system)? Or is it possible that Vanuatu (and New Caledonia) sustained, at some time in the past, a NAN population which has since died out but has left, as its legacy, traces of its existence in the counting systems of the AN immigrants? Tryon (1984), in fact, noted with regard to the NAN languages in Melanesia that "there has been some speculation that they may have extended as far as southern Vanuatu and New Caledonia" (p. 152). If we are to assume that a change in the AN counting systems must be brought about by external influence then we need to allow for the mechanism by which this can occur. If we do not concede the possibilities mentioned above, we would have to conclude that the counting systems of Nauna and the Vanuatan languages, at least, have undergone spontaneous change and that their (5, 10) systems were innovations.

(10) → (5, 20). The (5, 20), or digit tally, system is found in 58, or about 14%, of the Oceanic AN languages, and in 5 non-Oceanic AN languages of West Papua. In PNG, the (5, 20) system is largely confined to the North New Guinea and Papuan Tip Clusters where it occurs in 30 languages. It is also found in southern Vanuatu (15 languages) and in New Caledonia (8 languages). The only Polynesian language not to possess a 10-cycle system, Faga-Uvea, is spoken in the Loyalty Islands of New Caledonia and this, too, has a (5, 20) system. The nature of the change which occurs in this system is that while the numerals of the first pentad are retained, the complete second pentad is lost. The original numeral 10 is replaced by an expression meaning 2x5 or "two hands", or something similar, as may be seen in Table 5.3. In some cases, there is reference in the second decade to "feet", as is common in digit tally systems. The original numeral 20 (2x10) is usually replaced by the word for "man", or in some cases, a construction involving the terms "hands" and "feet". (For further details, see Chapter 5, Lean's (1992) appendices or GLEC's (2008) EXCEL summary.)

In seeking to determine whether this type of system was induced in the AN languages by external influence and in particular by NAN languages possessing digit tally systems, the situation is somewhat similar to that discussed above with the (5, 10) systems. Those AN languages which are situated in the East Sepik, Madang, Morobe, New Britain, Oro, and Milne Bay Provinces (PNG), could all, conceivably, have been influenced by NAN neighbours. However, southern Vanuatu and New Caledonia are populated only by AN speakers and we would need to hypothesise either long-distance diffusion from PNG, in the north, or the prior existence of NAN languages in the region which have since become extinct. If neither of these were the case then we would probably have to ascribe the changes to the counting systems of this region to localised and spontaneous innovation.

(10) → (5, 10, 20). Less common than the other two 5-cycle systems, this type occurs in the North New Guinea and Papuan Tip Clusters in a total of 18 languages, and in New Caledonia where it is relatively common and is found in 19 languages. Elsewhere it is found only in West Papua in one Oceanic AN language and 8 non-Oceanic AN languages. The change which commonly occurs to produce this kind of system is the loss of the second pentad numerals 6 to 9: the numeral 10 is normally retained. The original numeral 20 (2x10) is normally replaced by the word for "man" as is common in digit tally systems. Three examples of the (5, 10, 20) system are given in Table 5.13.

In the case of the languages located in PNG, it is possible to attribute these changes in the AN counting systems to the influence of neighbouring NAN languages. In the case of the New Caledonian languages the situation is as set out above, that is we must allow for either long distance diffusion from PNG or for the prior existence of NAN languages which have since become extinct but which, at some time in the past, influenced the immigrant AN languages. If neither of these occurred then the changes most likely may be regarded as a localised innovation.

(10) → "Manus" type. The Manus type of 10-cycle numeral system was discussed in Chapter 6. The use of the term Manus to distinguish this type is because it is found very largely in the languages of the Admiralties Cluster located on or near Manus Island. The deviation from the usual 10-cycle system in this type is such that, normally, the numerals 7 to 9 are lost and are replaced by expressions implying subtraction from 10 or more precisely the number to reach the complete group of 10, although the numeral 10 does not explicitly appear in these. Three examples are given in Table 6.9. One dialect of Levei-Tulu (Manus) has subtractive constructions for the numerals 6 to 9. Elsewhere in New Guinea and Oceania, this type of system is extremely rare and the only instance found among the AN languages occurs in the Mioko dialect of Duke of York and Wuvulu-Aua west of Manus. It also occurs in several NAN languages of Bougainville and one NAN language, Nanggu, in the Solomon Islands. Further comment occurs in Chapter 6 and some details in Table 9.3.

Given the relative isolation of the languages of Manus Island and that the type of change discussed here is almost entirely restricted to them, it seems possible that we have here an example of a localised innovation. However we need to consider whether there is also the possibility that the Manus

Table 9.3 *Examples of 'Manus' Subtractive-Type AN and NAN Outside Manus Region*

	Mioko Dialect of Duke of York	Nanngu	Levei-Tulu
1	*ra*	*tate/šte, šte*	*eri*
2	*rua*	*lali, lšli*	*luweh*
3	*tul*	*latÿ, lštu*	*toloh*
4	*vat*	*lafo, lopo*	*ha-hup*
5	*lima*	*lamaf, lšmšp*	*limeh*
6	*nom*	*lšma, temo*	*choha-hup*
7	*talakatul*	*tumatu, tumtš*	*chotoloh*
8	*talakarua*	*tumali, temli*	*choluweh*
9	*tolotakai*	*tumate, tumšri*	*cho-eri*
10	*ra noina*	*napnu, nopnu*	*ronoh*
20	*rua noina*	*nopnu li*	*lunoh*
30	*tula noina*	*nopnu tu*	*sunuh*
100	*ra mar*	*telau šti*	*ranak*
1 000	*ra rip*		*ropop*

Note. Duke of York (Mioko) data are from Lean's 1986 field notes but reflect Parkinson's (1907, p. 745) and Kluge's (1941, p. 193) data. Nanggu data are from Tryon and Hackman (1983, pp. 123, 127, 131) and Cashmore (1972, p. 55). Levei-Tulu stretches across Manus and has two distinct dialects (if not languages) and the data here are from Lean's (1986) field notes and 2 CSQs but they reflect Smythe's earlier SIL data (prior to 1970) given in Z'graggen (1975).

system may have been induced by some sort of external influence. As was discussed above, in the case of Nauna, if such a change was induced by contact with NAN languages, we would have to assume that such contact occurred early in the history of the Admiralties languages, that is at a time when the speakers of the language(s) ancestral to the present-day languages were still in the POC homeland or during their voyage from the POC homeland to their present location. In the case of the latter, we would most likely have to assume that the voyage included visiting New Ireland where there was sufficiently sustained contact with NAN groups to induce some sort of change in the counting system of the AN travellers. This possibility will be explored further in the general commentary and summary below.

(10) → "Neo-2"-10. This type of change also has a very localised distribution and is found largely in the Central Province (PNG), east and west of Port Moresby. The system in question was discussed in Chapter 6 where it was distinguished by the term Motu type, however Seidenberg (1960) cited the various languages which possess this type of system as being among those which have his neo-2 system (p. 230). In fact the seven languages spoken in the Central Province and which have this system, exhibit Seidenberg's "Type 1 Proto Neo-2" features with 6=2x3, 8=2x4, and 9=(2x4)+1; four languages also have 7=(2x3)+1 while one dialect of Keapara appears to have 7=8-1 and 9=10-1. We thus have the loss of numerals 6, 8, and 9, and in several cases 7. Outside of the Central Province, only one other Oceanic AN language appears to have a counting system showing similar features: this is Wuvulu-Aua, the westernmost language in the Admiralties Cluster. Tables 6.11 and 6.13 show various examples of this type.

Seidenberg's interpretation of how such a system might come about is to hypothesise the existence of a primary type of counting system, the neo-2, which has the same status as the pure 2-system and the (5, 20) system, in that each of these, he believed, was invented and diffused. The data at our disposal, however, throw some doubt on this interpretation. Firstly, the Type 1 neo-2 system is not found at all among the NAN languages which, as was discussed in Chapter 1, were established in the New Guinea region prior to the advent of the AN languages: if the neo-2 system had been diffused some time after the 2-system we would expect to find it distributed among the NAN languages. Since the neo-2 features are found only among the counting systems of a small group of AN languages, all of which derive from an original proto-10 system, and since they could not have acquired them directly from a NAN source, then the likely interpretation is that this variant of the 10-cycle system is an innovation occurring within this group. This does not preclude the possibility that such an innovation was induced by some sort of NAN influence and this will be explored further in the commentary below.

(10) → (4). As was discussed in Chapter 7, only four AN languages have counting systems which exhibit 4-cycle features as can be seen in Tables 7.6 and 7.7. Two of these, Wogeo and Bam, have systems which show pure 4-cycles while the other two, Ormu and Yotafa, have systems which show both 4- and 5-cycle features. In the case of the former two, the only AN numerals to be retained are 1 to 3. It is possible that such systems can be regarded as localised inventions. It was noted in Chapter 7, however, that, along the northern coast of PNG, several 4-cycle systems occur among the NAN languages as well and it does not seem out of the question that there may have been some interaction between the AN and NAN groups. We therefore need to consider the possibility that this system may have occurred originally among some NAN languages and was diffused to the AN groups.

(10) → (2) or (2, 5) or (2″, 5). Perhaps the most surprising of the various transformations which the POC 10-cycle systems has undergone is where the primary 10-cycle has been reduced to a primary 2-cycle as was noted in Chapter 3. There are two examples among the Oceanic AN languages where there has been a loss of the numerals 3 to 10 resulting in pure 2-cycle systems (see Table 3.2). There are, in addition, 18 examples where the AN languages now possess systems with (2, 5) cyclic patterns. For the majority of these what we have is a 2-cycle system augmented by a (5, 20) digit tally system in which the word for 20 contains a "man" morpheme or both "hand" and "foot" morphemes: the full cyclic pattern in such cases is (2, 5, 20). Of the 18 groups which have this type of system, 13 belong to the Markham Family of which Adzera, discussed in Chapter 8, is a member. The remaining five, Sera (Sandaun Province), Roinji (border of Madang and Morobe), Dawawa, and Igora (Milne Bay), and Tomoip (East New Britain), are, like the Markham Family, all located in regions which have NAN languages with similar systems (see Table 3.6 and Appendices).

There are 12 AN languages which possess what Lean (1992) had termed a quasi-2-cycle system as discussed in Chapter 3. This type of system, denoted (2″), has a basic numeral set (1, 2, 3) with the numeral 4 having a 2+2 construction; the word for 5 usually contains a "hand" morpheme as is common with digit tally systems. Of the 12 languages having this system, seven belong to the North New Guinea Cluster and five belong to the Papuan Tip Cluster: all are located in regions which are inhabited predominantly by NAN language groups and it seems likely that these have influenced the AN groups.

Types of Change: Commentary and Summary

The various types of change, outlined above, which the counting systems of New Guinea and Oceania have undergone and which involve either the increase or decrease of their numerical lexis, do not provide an exhaustive list of the changes which have occurred. In particular, those changes which have obviously resulted from the introduction of the numerical institutions of the colonial languages (German, English, Dutch, Indonesian, and French) or of the Melanesian pidgins, have largely been ignored. These institutions now hold a predominant place in day-to-day commerce and in schools and are having an overwhelming impact on the numerical institutions of the traditional societies in the region (see, e.g., Saxe (2012)). These changes provide the most obvious and decisive evidence of the effects of the diffusion of counting systems which are an integral part of an introduced culture which has achieved political and economic dominance. The changes that we are concerned with here, however, are those which have occurred to the counting systems of the traditional cultures prior to the introduction of the colonial cultures.

In the cases where the borrowing of numerals has taken place we have clear evidence of language groups interacting in such a way as to bring about changes in the borrowers' counting systems. Such borrowing is most easily identifiable when it takes place between NAN and AN groups but is difficult to detect when it takes place between like groups. The degree to which borrowing appears to have occurred between unlike groups, however, is not particularly marked and this suggests that the borrowing of numerals cannot be regarded as the primary mechanism by which diffusion occurred in the pre-colonial, traditional context.

We have suggested that a substantial proportion of the Oceanic AN languages have had their counting systems changed by losing part of their numerical lexis. In some cases this has resulted in a change to the cyclic structure of the counting system; in others, the cyclic structure has remained essentially unchanged but there has been a change to the way in which the numerals of the second pentad have been constructed. For each of the changes to the AN counting systems discussed above, we have briefly indicated whether it seems possible that such changes may have come about as a result of influence by, say, NAN language groups, or, alternatively, whether they appear to be localised innovations. Unlike the case of borrowing where the presence of, for example, AN loanwords in an otherwise NAN numeral lexis provides direct evidence of influence, the evidence in the case where there is a loss of the numeral lexis must necessarily be circumstantial. Thus, when we observe that the Markham Family languages no longer have 10-cycle systems and have instead (2, 5) systems, we may inquire whether there appear to be any special circumstances which may have induced such a change. As was discussed in Chapter 8 with respect to the Adzera, in particular, and the Markham Family generally, it appears that the language groups ancestral to the present day groups left their predominantly maritime environment and eventually moved inland up the Markham Valley in the Morobe Province (PNG). In so doing, they moved into a region inhabited by NAN speakers and with whom they engaged in trade. The AN groups adapted themselves to the dominant NAN culture and modified aspects of their own cultural institutions, including their economy and their counting system, the latter becoming, like that of their NAN neighbours, a (2, 5) system.

Such an interpretation seems reasonably plausible and, indeed, a similar interpretation could be made in all cases where AN groups have acquired a (2, 5) system (or (2) or (2″, 5) systems as well). For many of the types of change discussed above it is possible to invoke similar mechanisms by which AN groups came into sustained contact with NAN groups and thereby acquired certain of their characteristics. This is particularly the case with those AN language groups belonging to the North New Guinea and Papuan Tip Clusters which moved onto the PNG mainland. There are a number of instances where change has occurred to the counting systems of AN languages which is less easily attributed to NAN influence. For example, the Admiralties Cluster languages, which now have the Manus type of counting system, would appear to have had little or no contact with NAN languages at all, at least according to their current distribution. In Chapter 1 we discussed Ross's reconstruction of the history of the Admiralties Cluster in which he envisaged the speakers of their ancestral language(s) leaving the POC homeland at a relatively early date and travelling to the previously uninhabited Manus Island. Whether these travellers had contact with NAN groups prior to their leaving their homeland, or on their way to Manus, perhaps in New Ireland, is unknown. The point is that we need to hypothesise that some sort of contact did take place if we assume that changes to counting systems are always induced and do not occur spontaneously. If we dismiss the possibility that such contact did take place then it seems difficult not to conclude that the Manus type of system is an example of an independently invented innovation.

We have suggested that the Motu type of system, which displays Seidenberg's neo-2 (Type 1) features in the second pentad, does not seem to have been acquired as a result of the diffusion of this system: where it does occur, it appears to be an innovation which has occurred among a small number of AN languages. As indicated earlier, nowhere in the New Guinea region does this type of system occur among the NAN languages: if the neo-2 system had been introduced into the region from outside, we would expect to find some trace of it among these. Since the AN languages did not acquire this system directly from NAN groups then it seems likely that we have here an example of a localised innovation. If this conclusion is valid, the implications for Seidenberg's theory, as far as the hypothesised neo-2 system is concerned, are considerable. We need, however, to consider, as we mentioned above with regard to the Manus type of system, whether such an innovation occurred spontaneously or whether it was induced by NAN influence.

The remaining systems that we need to consider are those in which a primary 10-cycle has been replaced by a primary 5-cycle and of which there are three different types. With regard to the (5, 10) system, Seidenberg's view was that this is a hybrid of a 10-cycle (perhaps a neo-2-10) system with a digit tally (5, 20) system. It also seems possible, although Seidenberg did not deal with these specifically, that the (5, 10, 20) and the (5, 20) systems are outcomes of interactions between AN 10-cycle systems and NAN digit tally systems. As indicated earlier, the AN 5-cycle systems found in the PNG region could have resulted from interactions with NAN groups in that the AN groups having these systems are all located in regions inhabited by NAN groups. This is not the case in Vanuatu and New Caledonia which are now inhabited only by speakers of AN languages but where we find, nevertheless, all three types of 5-cycle system. In order to account for this, we have suggested three possibilities: (1) long distance diffusion by other AN groups carrying 5-cycle systems with them from PNG, to the north, and incidentally by-passing the Solomon Islands and parts of central Vanuatu, (2) the prior existence of now extinct NAN groups in southern Vanuatu and New Caledonia which interacted with the immigrant AN groups, at some time in the past, to the extent of affecting their counting systems, or (3) that the AN groups in this region spontaneously changed their counting systems from 10-cycle to 5-cycle systems without external influence.

We have, then, a number of instances in which changes to the AN 10-cycle counting system could be accounted for by the direct influence of NAN language groups on AN language groups. In the cases, however, of (a) the Manus type, (b) the Motu type, together with (c) the 5-cycle systems of Vanuatu and New Caledonia, the reason why such systems occur requires a more complex explanation. First, with regard to (a) and (c), if these were outcomes of NAN influence, we have to allow for the languages having these systems to have come in contact with NAN groups at some time in the past because they exist in regions now uninhabited by such groups. Second, with regard to (a) and (b), neither of these systems is found to any extent in the NAN languages and therefore we cannot attribute their existence in the AN languages to direct diffusion from NAN groups. This, however, is not to discount the possibility that such changes to the AN systems were indirectly induced by contact with NAN groups and indeed Lean (1992) proposed a mechanism by which this may have occurred. Finally, we cannot discount the possibility that certain of these AN systems may have resulted from spontaneous change and were not induced, directly or indirectly, by NAN influence. In Chapter 10, we suggest a mechanism by which such spontaneous change may occur in certain special circumstances.

In considering the case of indirect NAN influence on AN counting systems, suppose that speakers of an AN language who possess an intact 10-cycle counting system come into a period of sustained contact with speakers of NAN languages who possess a digit tally system or, possibly, a 2-cycle numeral system augmented by a digit tally system. It is possible that a variety of outcomes could result from such a contact, the most likely being that the AN 10-cycle system is modified so as to acquire the 5-cycle feature characteristic of the digit tally system which, as we have seen, primarily affects the second pentad numerals 6 to 9 or 10. Thus we might expect the AN group to acquire any of the systems with (5, 10), (5, 10, 20), or (5, 20) cyclic patterns. However, we suggest that the effect on the AN system could be more generalised in that, while the construction of the second pentad numerals is affected, this does not necessarily involve the adoption of the additive 5-cycle construction of the form, say, $5+n$ but may involve instead other alternative constructions including the subtractive Manus type or the doubling Motu type. If we allow these possibilities we might expect to find in any related group of AN languages which had come in contact with NAN digit talliers, not only one resultant type of system but rather several different types, each with variant second pentads. Thus in the case of the Admiralties languages, for example, if the language(s) ancestral to these did come into contact at some time in the past with NAN groups possessing digit tally systems and this induced instability in the second pentad of the original AN 10-cycle system, then we might expect several different systems to be apparent in the daughter languages today. What we do find, in fact, is that in addition to the common Manus type of system, Nauna has a (5, 10) system, Seimat has a (5, 20) system, and Wuvulu-Aua has a system similar to the Motu type. Similarly, with the AN languages of the Central Province (PNG), which all belong to the (Peripheral) Papuan Tip Cluster, six possess the common Motu type of system, however both Kuni and Mekeo have (5, 10) systems while the Keapara system exhibits some subtractive constructions in its second pentad. We thus appear to have, within related groups of languages, a range of systems with variant second pentads which could be attributed to the interaction between the 10-cycle system and the (5, 20) digit tally. One such system includes Seidenberg's (1960) neo-2, or Motu, type which we suggest occurs as a result of such an interaction and cannot, as Seidenberg asserts, be regarded as a primary counting system in its own right.

The situation, then, in New Guinea and Oceania, regarding the way in which traditional cultures have interacted so as to produce changes in their counting systems is rather more complex and less predictable than that which we would expect from the views of diffusionists such as Seidenberg (1960) and Crump (1990). It is clear, for example, that the more efficient 10-cycle system of the AN speakers has not swept in and overwhelmed the less efficient systems of the NAN speakers and indeed in many cases the reverse situation has occurred. It is also clear, however, that diffusion of counting systems has occurred even if the way in which it has occurred is somewhat unexpected. There are few instances in the various types of change which were delineated above which cannot be attributed to direct or indirect influence of one language group on another. Even in the case of AN languages which are now located in regions which are uninhabited by NAN groups, we cannot rule out the possibility of past contact and hence NAN influence on the counting systems of the AN groups. We therefore have not yet found unequivocal evidence to support the idea that counting systems can be spontaneously and independently invented: the Manus and Motu type systems, together with the 5-cycle systems of Island Melanesia, may be examples which provide evidence that localised innovations do occur but it is also possible that these have not occurred spontaneously and may have been indirectly induced according to the mechanism described above. As will be discussed in the next section and in the next chapter with the views of recent research by Spriggs (2006, 2011), the only candidates which do appear to be genuine innovations are the 6-cycle systems and certain of the 4-cycle systems which occur among the NAN languages.

A Critique of Seidenberg's Theory

Seidenberg's theory is open to criticism on at least two levels. First, there are several shortcomings with regard to his methodology which consists largely in providing a big picture interpretation, at once historical and diffusionist, of the current geographical distribution of the various counting system types throughout the major continents. Second, it is possible to take issue with his interpretation of how this distribution came about with regard to the both the nature and the chronology of his various counting system types.

The Nature of Diffusion

Seidenberg (1960) did not consider in any detail the nature of the diffusion process. The two principles on which his theory was based are that 1) knowledge always arises in the centres of ancient civilisations and not anywhere else, and 2) "knowledge always passes from those who know to those who do not know, not the other way" (p. 218). No model was suggested in order to explain either the *vehicle* of diffusion, i.e the means by which a counting system is transmitted from one location to another, or the *mechanism* of diffusion by which features of a counting system are transmitted from one group to another. In both cases, there are certain necessary conditions that need to be met in order for diffusion to occur. If, for example, diffusion is thought to occur by a migration, or a sequence of migrations, from one location to another, then there must be a plausible geographical route which the migration could follow. Thus in considering how the 2-cycle system might have been diffused from its supposed origin in the Middle East to, say, the Australian continent, we need to take into account that the latter has been largely isolated since the last Ice Age, that is from 8 000 to 10 000 BP, and that there has probably been little human movement into the continent after that time. Furthermore, as we have noted above, when an immigrant group carries a counting system into new and inhabited territory, it is not necessarily the case that the introduced system will overwhelm and displace existing systems: the outcome may be quite the opposite, with the immigrant group accommodating itself to the existing culture and circumstances. Generally speaking, Seidenberg was not concerned with considering whether or not his theory is plausible in the light of the constraints placed upon it by the nature of the diffusion process itself.

Sources and the Lack of Collateral Evidence

The scholarly sources on which Seidenberg relied in order to develop his theory were drawn from anthropological, linguistics, and historical literatures, the data on Indigenous counting systems being derived from the first two. Much of this material was gathered in the nineteenth century and Seidenberg did not avail himself of the advances in either field in the period from 1930 to 1960. With regard to New Guinea and Oceania, Seidenberg's sources comprise about a dozen publications which deal with only a small proportion of the 400-odd AN languages and hardly any of the 700-odd NAN languages spoken in the region. Seidenberg's main source for the linguistic situation generally, and for the counting system situation in particular, both for this region and other parts of the world, was Schmidt's work published in 1926 but based on data gathered in a somewhat earlier period prior to 1910. Hence Seidenberg did not take into account, for example, the advances made by Dempwolff and others in the field of historical linguistics which established the Oceanic Hypothesis, mentioned in Chapter 1, and the essential unity of the Oceanic AN languages. While Seidenberg's failure to consult such material as this does not necessarily invalidate the main thrust of his arguments, it needs to be recognised that a wider review of pertinent literature would have provided a more secure basis for his theory.

A more serious omission concerns the lack of use of collateral evidence from fields outside the immediate purview of Seidenberg's sources: for example, the evidence from archaeology and prehistory places some important constraints on Seidenberg's speculations regarding the time scale that he allowed for the diffusion of his primary counting systems. As we have noted, Seidenberg located the origin of these in the ancient centres of civilisation of the Middle East - as though other equally ancient centres, such as those in India and China, need not be considered - and dates the genesis of the 2-cycle system at, or somewhat before, 5 500 BP. Archaeological data for the Americas, however, suggest that the major migrations from Asia occurred in the period 30 000 to 15 000 BP and that the ancestral Eskimo (Inuit) population was established by 4 000 BP (Zegura, 1985, p. 13). Similarly, the dates at which the main migrations into the New Guinea region, Australia, and Oceania occurred range from 50 000 or 60 000 BP to about 5 000 BP, as indicated in Chapter 1. These data suggest that the migrations which might have carried primary counting systems into these regions may have been largely completed before the time that Seidenberg suggested for the beginning of their diffusion. There is of course the possibility that such diffusion was not necessarily effected by the major migrations mentioned above but were rather brought about by a combination of the incursion of smaller groups and the subsequent transmission of systems by group-to-group interaction. This possibility seems less likely to bring about widespread diffusion in that the introduction of a new and relatively small group into an established culture which has political and economic dominance would probably result in the new group adapting to the existing culture rather than vice-versa.

Seidenberg's Theory: Data from Australia, New Guinea, and Oceania

Seidenberg's genealogy of counting systems was established very largely on data from the Americas and Africa and, to a lesser extent, on data from Australia and Asia. We consider here how his theory fits the data from New Guinea and Oceania by focusing on two main aspects of the theory: (1) the counting and tally sustems that are accorded primary status and those accorded secondary or hybrid status, and (2) the chronological order of the genealogical development. Two other aspects will also be addressed: Seidenberg's dating of the historical development of counting systems and the degree to which innovation and independent invention may have occurred.

With regard to the 2-cycle system, Seidenberg's view was that this would have been the first of his primary counting systems to have been introduced into the region and, generally speaking, the data from Australia, New Guinea, and Oceania do not controvert this view. As indicated in Chapter 3, various 2-cycle systems, including the pure 2-system, are commonly found throughout Australia and the Torres Strait islands. We also find, in the same locations, instances of pure 2-cycle counters also possessing body-part tallies. To a lesser extent, we also have evidence of the existence of (5, 20) digit tally systems in their pure form but also, in a larger number of cases, systems with (2, 5) or (2, 5, 20) cyclic patterns which could be interpreted as being hybrids of pure 2-cycle systems with digit tally systems. There is no firm evidence of other major types of counting systems being found in the Australian continent. While this situation does not controvert Seidenberg's views, there is no definite evidence to suggest that, of the three main types of counting system or tally that are present, one may have been introduced into the continent earlier than another. All we can infer is that it seems likely that they were introduced either with the early migrations into Australia or with the subsequent NAN migrations into the New Guinea region followed by diffusion southwards into Australia.

Wurm (1982) said of these migrations that

> the first wave of immigrants are believed to have been Australoids coming from the west who occupied first the northern part of what was then the single New Guinea-Australian continent, and gradually spread south through the entire continent. New Guinea and Australia became separated and isolated from each other by rising sea levels about 10 000 to 8 000 years ago … The first ancient Papuans entered New Guinea … and overlaid the Australoid population still present in New Guinea. Kirk … suggests that the immigration of the first ancient Papuans into New Guinea took place a short time after, or not too long before the isolation of New Guinea from Australia (through Torres Strait coming into being) and points out in support of this view that Papuans have not entered the Australian continent, at least not in significant numbers. (p. 226)

If we assume that counting systems were carried with the migrations into Australia then it seems likely, given Wurm's statement, that this diffusion was largely complete by 8 000 BP and that the systems were either carried by the original Australoid populations or were diffused from the early NAN immigrants into New Guinea. If this is the case, then the 2-cycle system, the (5, 20) system, and the body-part tally were not only present in Australia by 8000 BP but were present in the New Guinea region as well. We have noted previously that Seidenberg's two types of neo-2 system are not found in the NAN languages nor are they found anywhere in Australia. There are several implications of this interpretation for Seidenberg's theory. First, that in agreement with Seidenberg, we regard the 2-cycle system and the (5, 20) system as having primary status. Second, that, contrary to Seidenberg, we do not regard the neo-2 system, in either of its forms, as having primary status and that where it does occur, in its Type 1 form, among a few AN languages, this is possibly the result of an interaction of a 10-cycle system with a (5, 20) digit tally system. Third, Seidenberg's dating of the genesis of the 2-cycle system as being contemporaneous with the Sumerian civilisation, that is at about 6000 BP, seems unlikely in that the system was probably present in Australia at least 2000 years prior to that time but most likely tens of thousands of years earlier.

Chapter 1 provided an outline of the currently accepted picture of the various language migrations into New Guinea and Oceania. The relative chronological order of these is:

1. the Australoid migrations into Australia and New Guinea,
2. a sequence of two or three NAN migrations, and finally,
3. the introduction of AN-speaking groups carrying pre-POC.

Keeping this in mind, the current counting system situation for this region may be interpreted as follows. The 2-cycle system and its variants, the body-part tallies, and the (5, 20) digit tally system, are primarily associated with the NAN languages: these types were established in New Guinea prior to the advent of the AN languages. The AN-speakers brought with them a pure 10-cycle system and, after the breakup of POC, this system was carried throughout Island Melanesia, Polynesia, and Micronesia. Interactions between AN and NAN groups brought about changes to the counting systems of some members of both groups: certain NAN groups acquired systems with (5, 10), (5, 10, 20), (10), and (10, 20) cyclic patterns, while AN groups, as we have seen above, acquired systems with primary 2-cycles, 5-cycles, and even 4-cycles. We have also suggested that several variant 10-cycle systems, such as the Manus and Motu types, may have been induced as a result of an AN-NAN interaction. According to this interpretation, then, those counting systems which appear to be candidates for primary status are the pure 2-cycle system (and, associated with some 2-counters, the body-part tally), the (5, 20) system, and the 10-cycle system: the first two were introduced into New Guinea (and Australia) first, the 10-cycle system being introduced subsequently.

This interpretation of the counting system situation in New Guinea and Oceania has both points of agreement and disagreement with Seidenberg's theory. First, as we found in the case of Australia, the evidence suggests that the 2-cycle system and the (5, 20) cycle system both have primary status as suggested by Seidenberg. The only other counting system to which primary status can be accorded is the pure 10-cycle system, introduced by the AN immigrants. As we have discussed earlier, Seidenberg did not regard the 10-cycle system in its pure form as having primary status: he suggested that it was probably diffused initially as a neo-2-10 system, was introduced into the Americas prior to the (5, 20) system, and subsequently became a pure 10-cycle system by linguistic change. Our interpretation disagrees with this view in at least two respects. First, we do not regard the 10-cycle system as having been introduced into the New Guinea region as part of a neo-2 system but that it came in its pure form. Second, the 10-cycle system was introduced after both the 2-cycle and the (5, 20) cycle systems. There is a feature of the AN 10-cycle system which suggests that, at some earlier time, it may have evolved from a digit tally system: the POC word for 5, **lima*, is identical to the word for "hand". However that may be, the 10-cycle nature of the system was already established prior to its introduction into New Guinea as the reconstruction of the Proto Austronesian numerals shows.

We have suggested above that, wherever the (5, 10) system occurs, this had largely arisen as a hybrid of the 10-cycle system and the (5, 20) digit tally. Seidenberg, who has a similar interpretation of how the (5, 10) system came about and that this has a hybrid, secondary status rather than primary status. With regard to such systems, Seidenberg ascribed some significance to the way in which their second pentads are constructed and, in discussing the counting system situation in Africa, he distinguished two types of second pentad construction and suggested that these actually constitute two different types of counting system, each type being diffused into Africa at different times. The data available do not unequivocally confirm or deny such a view. It is true that the second pentad constructions of 5-cycle systems can be classified into a relatively small number of types, the most common being of the form $5+n$, $5+c+n$, and $x+n$, as described in Chapter 5. Summarising the data from Chapter 5 in a single table (Table 9.4).

Table 9.4
Showing the Numbers of NAN and AN Languages Having Particular Second Pentad Constructions for 3 Types of 5-Cycle System

	$5+n$	$5+c+n$	$x+n$	Type
NAN	14	17	26	(5, 20)
	2	5	6	(5, 10)
	4	2	6	(5, 10, 20)
AN	18	21	10	(5, 20)
	10	19	74	(5, 10)
	12	22	7	(5, 10, 20)

In PNG, an analysis of how these various types of second pentad construction are distributed geographically indicates that the distribution appears to be random. In the New Ireland Province, for example, the neighbouring and related languages Tigak, Tiang, and Nalik, each has a (5, 10) system with one of the three different second pentad constructions (Lean, 1992, appendix on New Ireland). In Vanuatu, however, the situation is far more uniform with some 60-odd AN languages having 5-cycle systems with the $x+n$ second pentad type (and thus accounting for most of the 74 languages in the table above).

In the foregoing discussion of the various types of counting system which are found in New Guinea, Australia, and Oceania, we have focused on those types which are given prominence in Seidenberg's theory. This does not include either of the two types of system discussed in Chapter 7, the 4-cycle and 6-cycle. The reason that these are given scant attention is, perhaps, that neither system is adequately encompassed by the diffusionist view. Unlike the 2-, 5-, and 10-cycle systems, they do not have a wide geographical distribution over most continents: outside of New Guinea they are found only in Africa and North America. In each of these locations their occurrence is sporadic and in the case of the 6-cycle system it is rarely found in its completely intact form and is usually such that only some of its 6-cycle features are displayed. The 4-cycle system, however, is found in its fully intact form in both New Guinea and North America but in both locations it is restricted to a relatively small number of groups.

The inference which we might draw from the pattern of distribution of the 4- and 6-cycle systems is that neither was diffused and dispersed over a wide region but rather that each is instead a localised development. We have suggested earlier that the 4-cycle system had its genesis in digit tallying and that, whenever this type of system is found, tallying is done by treating the hand as comprising the four fingers but not the thumb. It may also be the case that the 6-cycle system had a similar origin in that the fingers, including the thumb, of the hand were augmented by the thumb joint, although the evidence for this is less robust.

An interpretation which we may therefore put on this is that the (5, 20) digit tally system was diffused but that the 4- and 6-cycle systems were not: these arose from varying the way in which digit tallying is carried out. While the large majority of digit talliers retained the conventional method, a relatively small number of groups varied the tallying procedure according to their own needs giving rise to localised innovations. We thus have a situation, not encompassed by Seidenberg's theory, in which the occurrence of counting systems is plausibly accounted for by a combination of both diffusion and independent invention rather than by diffusion alone.

In conclusion, then, the main focus of this chapter has been to test whether the current distribution of counting system types among the NAN and AN languages and, to a lesser extent, the Australian languages, may be accounted for by a diffusionist interpretation of events, generally, and by Seidenberg's theory in particular. In the discussion above three categories of counting systems or tallies were distinguished. First, there are the primary systems which we may regard as the original or proto systems introduced into the region as part of the cultural baggage of the immigrants. The prehistory of these systems and their subsequent fate will be elaborated in Chapter 10. Lean (1992) suggested that the three candidates for primary status are the 2-cycle, the (5, 20) digit tally, and the 10-cycle systems: the body-part tally may also be included in this category and this will be pursued further in the next chapter. That the first two systems are accorded primary status is in agreement with Seidenberg's theory; according the pure 10-cycle system primary status is not. He also argued that the evidence does not support Seidenberg's idea of the primacy of the neo-2 system.

The second category comprises the secondary or hybrid systems which have their origin in the interaction of the primary systems and are thus the products of diffusion which has occurred within the region. Thus the (5, 10) and (5, 10, 20) systems may be regarded as outcomes of the interaction of the (5, 20) and 10-cycle systems. The hybridisation of systems is allowed for in Seidenberg's theory. The other possible outcome of the interaction of primary systems is the displacement of one system by another: we have argued that the evidence indicates that displacement does not appear to have occurred to any great extent although there is evidence to suggest that the borrowing of part of a numeral lexis to increase the primary cycle of a counting system has occurred in a number of instances.

The third category of counting systems encompasses the 4-cycle and 6-cycle systems, discussed above, which appear to be innovations that have occurred in the region and therefore cannot be regarded as the outcomes of diffusion. It may be that the Motu and Manus types of 10-cycle system may also belong to this category. With these exceptions, the contemporary counting system situation may be interpreted as having resulted from a degree of diffusion between language groups already in place rather than diffusion introduced from some external source. There are certain aspects of Seidenberg's theory that appear to be supported by the available evidence. There are, however, a number of details of Seidenberg's theory, for example the ritual genesis of counting, the origin of tally methods, and the status of the variant second pentad constructions of 5-cycle systems, which have not been addressed here in any detail. These matters are largely speculative and probably will not be resolved one way or another by the evidence available.

Finally, we have suggested that the evidence casts doubt on the validity of Seidenberg's conjectures regarding the place of origin of the 2-cycle system and the chronology for the diffusion of his primary counting systems: it seems likely that both the 2-cycle system and the (5, 20) digit tally system were present in Australia and New Guinea by at least 8 000 BP, that is 2 000 years before the time suggested by Seidenberg for the genesis of the 2-cycle system in Sumeria. The choice of Sumeria, or indeed any one of the ancient centres of culture in the Middle East, as the source of counting seems, in any case, a decidedly Eurocentric view which ignores the existence of equally ancient centres in, for example, India or China. As Barrow (1992) pointed out, however, "the most ancient evidence of human counting is to be found in the remains of smaller groups of hunters and gatherers who existed long before any of these great centres of civilisation" (p. 31). These remains comprise engraved bones found in Africa which are clearly tallying devices. One of these dates from 35 000 B.C.; another, the "Ishango" bone, dates from about 9 000 BC and an analysis of the groupings of notches on the bone suggest that it may have been used as a calendrical device (Barrow, 1992; Bogoshi, Naidoo, & Webb, 1987; Joseph, 1987). The main implication of such evidence is that the locus of the genesis of counting, one of the most momentous inventions in the intellectual history of the world, may be found not among any of the known centres of "civilisation" but rather in humbler "primitive" societies whose existence predates these by, perhaps, many thousands of years. However, Barrow (1992) disputed whether civilisations invented counting over and over to meet their living demands, resulting in all people counting. He claimed that even the few 2-cycle systems occurring in vestiges in Australia,

Africa and South America originated from Sumeria around 5 000 BP. The Persians around 2 500 BP used a decimal system. The Babylonian adaption of two marks in different positional relationships to record all 59 numbers for the 60 base system was regarded as too complex while base 5 was too small. However, the digit tally systems with base 20 were less easily discounted except to say they were too concrete. Similarly he noted body part tallies and classifier systems as too concrete. From his evidence in map form, he continued to explain the replacement by "civilisations" whom he noted interacted in one way or another to eventually reach what he claimed as the most efficient counting system using a zero and place value columns for recording and calculating. Barrow's (1992) argument, however, lacked the close grained analysis of Indigenous counting systems that Lean (1992) provided by a closer and more comprehensive analysis of the counting systems in one of the regions where multiple different kinds existed along with 2-cycle systems. Thus Barrow's analysis also needs scrutinising.

References

Barrow, J. (1992). *Pi in the sky: Counting, thinking, and being*. Oxford, UK: Clarendon Press.

Bogoshi, J., Naidoo, K., & Webb, J. (1987). The oldest mathematical artefact. *Mathematical Gazette, 71*, 294.

Brown, N. (2010). *The abacus and the cross*. New York, NY: Basic Books.

Cashmore, C. (1972). Vocabularies of the Santa Cruz Islands, British Solomon Islands Protectorate. *Working Papers in Linguistics, Department of Anthropology, University of Auckland, 17*, 1–79.

Conant, L. (1896). *The number concept*. New York, NY: Macmillan.

Crawfurd, J. (1863). On the numerals as evidence of the progress of civilization. *Transactions of the Ethnological Society of London, 2*, 84–111.

Crump, S. (1990). *The anthropology of numbers*. Cambridge, UK: Cambridge University Press.

de Solla Price, D., & Pospisil, L. (1966). A survival of Babylonian arithmetic in New Guinea? *Indian Journal of History of Science, 1*(1), 30–33.

Flegg, G. (1984). *Numbers: Their history and meaning*. New Orleans, LA: Pelican Books.

Flegg, G. (1989). *Numbers through the ages*. London, UK: Macmillan and the Open University.

Galis, K. (1960). Telsystemen in Nederlands-Nieuw-Guinea. *Nieuw Guinea Studien, 4*(2), 131–150.

GLEC. (2008). *EXCEL summary of counting systems in Papua New Guinea based on Lean's Appendices*. 2015, from http://www.uog.ac.pg/glec/counting_sys/t-coutning_sys/t-counting_sys.htm

Henderson, J. (1975). Yeletnye, the language of Rossel Island. In T. Dutton (Ed.), *Studies in languages of central and south-east Papua, Pacific Linguistics* (Vol. C-29, pp. 817–834). Canberra, Australia: Australian National Univerisity.

Jett, S. C. (1971). Diffusion versus independent development: The bases of controversy. In C. Riley, J. Kelley, C. Pennington, & R. Rands (Eds.), *Man across the sea: Problems of pre-Columbian contacts* (pp. 5–53). Austin, TX/London, UK: University of Texas Press.

Joseph, G. (1987). Foundations of Eurocentrism in mathematics. *Race and Class, 28*(3), 13–28.

Kelly, R. (1993). *Constructing inequality: The fabrication of a hierarchy of virtue among the Etoro*. Ann Arbor, MI: University of Michigan Press.

Kluge, T. (1938). *Die Zahlbegriffe der Australier, Papua und Bantuneger nebst einer Einleitung ueber die Zahl, ein Beitrag zur Geistesgeschichte des Menschen*. Berlin, Germany: Selbstverlag.

Kluge, T. (1939). *Zahlenbergriffe der Völker Americas, Nordeurasiens, der Munda, und der Palaioafricaner*. Unpublished manuscript. Berlin, Germany.

Kluge, T. (1941). *Die Zahlenbegriffe der Sprachen Central- und Südostasiens, Indonesiens, Micronesiens, Melanesiens und Polynesiens mit Nachträgen zu den Bänden 2-4*. Berlin, Germany. Ein fünfter Beitrag zur Geistesgeschichte des Menschen nebst einer principiellen Untersuchung über die Tonsprachen. Original unpublished manuscript held at Yale University, microfilm held at Australian National University, Canberra, Australia.

Lawes, W. (1895). *Grammar and vocabulary of the language spoken by Motu tribe*. Sydney, Australia: C. Potter (Government Printer).

Lean, G. (1992). *Counting systems of Papua New Guinea and Oceania* (Unpublished PhD thesis). PNG University of Technology, Lae, Papua New Guinea. Retrieved from http://www.uog.ac.pg/glec/

Lowrie, R. (1937). *The history of ethnological theory*. Berkeley, CA: University of California Press.

Parkinson, R. (1907). *Dreissig jahre in der Südsee*. Stuttgart, Germany: Strecker und Schröder.

Raglan (Lord). (1939). *How came civilization?* London, UK: Methuen & Co.

Ray, S. (1919). The languages of northern Papua. *Journal of the Royal Anthropological Institute of Great Britain and Ireland, 49*, 317–341.

Riley, C., Kelley, J., Pennington, C., & Rands, R. (Eds.). (1971). *Man across the sea: Problems of pre-Columbian contacts*. Austin, TX/London, UK: University of Texas Press.

Schmidt, M. (1900). Die sprachlichen Verhältnisse von Deutsch Neu-Guinea. *Zeitschrift für Afrikanische und Ozeanische Sprachen, 5*, 345–384.

Schmidt, W. (1926). *Die Sprachfamilien und Sprachenkreise der Erde*. Heidelberg, Germany: Carl Winter's Universitätsbuchhandlung.

Schmidt, W. (1929). Numeral systems. *In Encyclopedia Britannica* (14th ed., pp. 614–615). London, UK/New York, NY: Sears Roebuck.

Seidenberg, A. (1960). The diffusion of counting practices. *University of California Publications in Mathematics, 3*.

Seidenberg, A. (1962). The ritual origin of counting. *Archives for History of Exact Sciences, 2*, 1–40.

Smith, G. E. (1933). *The diffusion of culture*. London, UK: Watts and Co.

Spriggs, M. (2006). The Lapita culture and Austronesian prehistory in Oceania. In P. Bellwood, J. Fox, & D. Tyron (Eds.), *The Austronesians: The historical and comparative perspectives* (pp. 119–142). Canberra, Australia: ANU Press.

Spriggs, M. (2011). Archaeology and the Austronesian expansion: Where are we now? *Antiquity, 85*(328), 510–528.

Strong, W. (1911). The Maisin language. *Royal Anthropological Institute of Great Britain & Ireland, 41*, 381–396.

Tryon, D. (1984). The peopling of the Pacific: A linguistic appraisal. *Journal of Pacific History, 19*(3–4), 147–159.

Tryon, D., & Hackman, B. (1983). Solomon Islands languages: An internal classification. *Pacific Linguistics, C-72*, 509.

Tylor, E. (1871). *Primitive culture: Researches into the development of mythology, philosophy, religion, language, art and custom*. London, UK: John Murray.

Van der Waerden, B., & Flegg, G. (1975a). *Counting 1: Primitive and more developed counting systems*. Milton Keynes, UK: Open University.

Van der Waerden, B., & Flegg, G. (1975b). *Counting 2: Decimal counting words, tallies and knots*. Milton Keynes, UK: Open University.

Wilder, R. (1974). *Evolution of mathematical concepts*. London, UK: Transworld.

Wurm, S. (1982). *Papuan languages of Oceania*. Tübingen, Germany: Gunter Narr Verlag.

Z'graggen, J. (1975). Comparative wordlists of the Admiralty Islands languages collected by W. E. Smythe. *Workpapers in Papua New Guinea Languages, 14*, 117–216.

Zegura, S. (1985). The initial peopling of the Americas: An overview. In R. Kirk & E. Szathmary (Eds.), *Out of Asia: Peopling the Americas and the Pacific* (pp. 1–18). Canberra, Australia: Journal of Pacific History, Australian National University.

Chapter 10
Towards a Prehistory of Number

Kay Owens and Glen Lean

Abstract The previous chapters focused on the cycles of the languages, but this chapter summarises the similarities and differences of the various phyla to indicate how different systems influence others. This provides evidence of much earlier dates for people in this region and a possible timeline with the known movements for counting systems to have been modified especially from the Austronesian 10-cycle. As a result, it is possible to dispute Seidenberg's timeline and theory of the diffusion of number from civilisations in the Middle East. An alternate timeline is provided based on known events such as the end of the ice-age, uplifts and volcanic activity in island regions, the spread of Lapita pottery, and cultural developments. The variation in counting systems also provides evidence when considered in terms of the other evidence from linguistics, archaeology, geography, genetics and biogeography. Thus a possible timeline for the history of number is suggested indicating the probable antiquity of these counting systems from around 40 000 to 5 000 years ago in Melanesia.

Keywords Prehistory of number • Proto-Oceanic counting system • Lapita pottery spread • colonisation and adaption of number systems

Introduction

There are three main reasons for the importance of Seidenberg's theory and which justify its discussion and evaluation in the previous chapter. First, in the literature of the history of mathematics before 1990, it was the only coherent and comprehensive view to have been put forward of the prehistory of number and the appearance of various types of counting systems in human societies in the period prior to the advent of written records. Second, it established a means whereby a genealogy of counting system types can be inferred from an investigation of their nature and distribution in contemporary indigenous societies. Third, the establishment of a genealogy allows for the possibility of setting this ordered sequence within a chronological framework, thereby providing it with an historical dimension. It is clear, however, that the data accumulated by Lean may be interpreted in such a way as to be at variance with several of Seidenberg's conclusions while, at the same time, providing some agreement with a diffusionist perspective. In this chapter we will continue to develop in some detail an interpretation of the data available on the counting systems of both the Non-Austronesian (NAN), also called Papuan, and Austronesian (AN) languages, mostly Oceanic, the intention being to establish a revision of Seidenberg's genealogy. Findings in the fields of linguistics and archaeology assist to locate the genealogy within the known chronological framework of the migration of various

K. Owens (✉)
Charles Sturt University, Dubbo, NSW, Australia
e-mail: kowens@csu.edu.au

G. Lean

K. Owens et al., *History of Number*, History of Mathematics Education,
DOI 10.1007/978-3-319-45483-2_10

language groups into New Guinea and Oceania, thereby providing an outline of the prehistory of number for this region.

Foley (2010) discussed the convergence of languages within the region suggesting that there was convergence, his term, rather than diffusion between Austronesian and Non-Austronesian languages but also between Non-Austronesian families. He noted that the few speakers of some languages in the Sepik in the 1960s were bilingual with their neighbouring languages which had a thousand or more speakers, for example Yelogu being lost to Kwoma and Karawa to Bouye. A similar problem occurs with Binumarien whose speakers are also bilingual with neighbouring languages of the Markham valley (related Gadsup and Tairora and Austronesian Adzera) and whose population has declined through tribal fighting and malaria. Interestingly in some cases, one group may have learnt surrounding languages but this may not be reciprocated e.g., the Usarufa men in the Eastern Highlands speak the distantly related Kamano-Kanite and Fore languages but not vice-versa while the Siane men and women also speak Chuave. This is particularly common in border villages and with exogamous marriages.

> Within the Siane context there is a distinct prestige associated with multilingualism, which is regarded as a desirable accomplishment. There are many indications of this in the culture: the use of foreign languages in Siane songs and the ubiquity of translations from one language into the other on formal occasions as a means of showing off language proficiency. (Foley, 2010, p. 797)

Except for exogamous marriages, women did not tend to be multilingual but men who traded or travelled were more likely to be multilingual.

> The effects of this intense language contact over many millennia have been profound; the languages show borrowing and diffusion of traits at all levels: lexical items, phonological patterns, bound morphology, word order, syntactic constructions, discourse styles, genres. (Foley, 2010, p. 797)

This makes it difficult to classify some languages as NAN or AN with some calling them hybrid. Maisin (Table 9.2) is one language that is hard to classify. Mostly core vocabulary (nouns of concrete objects) and to a lesser extent bound morphemes are borrowed. In Chapter 4, we discussed identifiable NAN language families and the issue of their position in phyla.

The counting systems situation found in the NAN and AN languages will be dealt with separately. Many of the counting systems covered in this chapter are detailed in Appendix C. While the NAN languages display remarkable diversity, the counting systems and tallies which are found among them may be classified into a relatively small number of types. We will proceed by taking an overview of how these various types are distributed throughout the NAN languages according to their genetic classification, beginning with the phylum. For each phylum we will consider what the current distribution of counting and tally types is and, in particular, which types appear to predominate and have a general spread throughout the various stocks of the phylum and which types appear to be restricted to a relatively small class of languages. From such an analysis we will attempt to infer which types may have been present in the languages that existed in the early history of the phylum and which types may have arisen in the later history of the phylum as a result of invention or external influence. This analysis will provide the basis for reconstructing the prehistory of counting and tallying in the NAN languages. Much of this work is linked to the archaeological records and those of other disciplines as well as linguistic anthropology (Attenborough, 2010; Lincoln, 2010; Pawley, Attenborough, Golson, & Hide, 2005; Sheppard, Walter, & Roga, 2010; Swadling, 1997, 2010).

With regard to the AN languages, we provided in Chapter 1 a summary of the results of a study by Ross (1988, 1989) in which he used the methods of historical linguistics to unravel the complexities of the early history of the Oceanic AN languages in PNG. Using Ross's reconstruction, we will trace the fate of the counting systems of the various clusters and groups of the daughter languages of Proto Oceanic (POC) as they dispersed in PNG and throughout Island Melanesia and the Pacific. Developments in the archaeology of this region, in particular the discovery of the remains of a distinctive style of pottery in 1952 at Lapita in New Caledonia, have resulted in a large number of studies which trace the spread of the culture with which the Lapita style of pottery is associated, from Melanesia into the Pacific. Various scholars have identified the Lapita cultural complex as belonging to the speakers of the Oceanic AN languages: this means, essentially, that the estimated dates from various archaeological sites provides a means of establishing a chronological sequence of the post-POC migrations and, thus, the linguistic prehistory of the region and, by inference, the prehistory of the post-POC counting systems (Allen, 1996; Bellwood, 2010; Blust, 1976; Green, 2010; Kennedy, 1983; Kirch, 1982; Pawley & Green, 1985; Pawley & Ross, 2006; Spriggs, 2006, 2011; Tryon, 1984).

Counting Systems and Tallies of the NAN Languages

In the previous chapters it has been established that the counting systems and tallies which are found among the NAN languages of New Guinea, the islands lying to the east of the mainland, and the Solomon Islands, comprise mainly the various 2-cycle variants, the body-part tallies, the (5, 20) digit tally, several 4-cycle systems, and a small number of 6-cycle systems. There is also a number of languages which have counting systems possessing either a primary or secondary 10-cycle. In discussing each of these types individually we have indicated in previous chapters how they are distributed among the various phyla of NAN languages (Ross, 2005). This section covers Lean's (1992) synthesis to indicate how the counting systems and tallies are distributed among the NAN languages according to this broad linguistic classification. It seems likely that certain systems may have been present in the languages of a given phylum from an early date while other systems may have been introduced or have developed as localised innovations. We will then attempt to infer, for each phylum, which systems fall into each category and, in particular, which may have been present as proto systems in languages which existed in the early history of the phylum.

In order to infer which among the counting systems or tallies that are now found in a given phylum, may be possible candidates for such proto systems, certain assumptions are made. It will be assumed that as the original languages of the phylum differentiated they nevertheless retained the basic cyclic structure of their proto systems even though the numerical lexis of each daughter language would have diverged from the original language. We would therefore expect to find the descendants of a proto system to have a reasonably wide distribution throughout the major stocks or subphyla of the phylum. On the other hand, those systems which are found today to exist in only one or two families in a circumscribed geographical area can probably be rejected as candidates for proto systems and indeed it seems likely that such systems have developed in those languages as localised innovations. We also need to consider which systems in a given phylum may have resulted from influence external to that phylum. Each of these considerations will be taken into account as we discuss below the distribution of the major counting and tally types throughout the main phyla and their constituents.

Distribution According to Phyla

According to Laycock (1975) "number systems, at least in the New Guinea area, afford few indications of genetic relationship of languages; closely related languages may show widely-differing systems" (p. 219). While this is to some extent true, it is nevertheless possible to discern that certain types of systems are characteristic of broad linguistic groupings. The broadest level of linguistic grouping used to classify the NAN languages is the phylum. As discussed in Chapter 1, there are five major phyla: the West Papuan, East Papuan, Torricelli, Sepik-Ramu, and Trans New Guinea. There are also six minor phyla together with a total of seven phylum-level isolates (Wurm, 1982, pp. 18-19. See also Table 1.1 in Chapter 1). The distribution of the various counting system and tally types among the phyla is shown in Table 10.1. In this Table the minor phyla and isolates have been combined. The distributions in each phylum will be discussed below.

Table 10.1
Showing the Distribution of Counting System and Tally Types Among the Phyla of the NAN Languages

Types	West Papuan	East Papuan	Torricelli	Sepik-Ramu	Trans N. G.	Minor Phyla	Total
(2)	0	0	0	3	39	0	42
(2, 5)	0	1	16	5	86	1	109
(2′, 5)	0	1	3	5	17	1	27
(2″, 5)	0	0	5	3	31	1	40
(5, 20)	0	1	2	17	52	7	79
(4)	0	0	0	1	6	2	9
(6)	0	0	0	0	5	0	5+
Body-Parts	0	0	0	8	58	4?	70+
(5, 10)	2	12	0	3	4	0	22
(5, 10, 20)	5	0	0	0	4	3	13
(10)	1	8	0	1	2	0	13
(10, 20)	2	0	0	0	1	0	3

Note. These are numbers from collected data, which are most languages but not all languages in Papua New Guinea and Oceania. They exclude 11 West Papuan languages in North Halmahera. + recent work and language determination indicate more languages

The West Papuan Phylum

Ten of the 13 West Papua Phylum languages were collected (the 11 in North Halmahera outside New Guinea brings the total to 24).[1] All of these are located in the Bird's Head or Vogelkop of western West Papua. Referring to Table 10.1 it can be seen that the counting systems of these languages are such that their cyclic structures contain either a primary or secondary 10-cycle and that there is no trace of a 2-cycle. It has been indicated earlier, in Chapter 6 (Table 6.2), that several of these systems show traces of AN influence in that AN numerals have been borrowed into their numerical lexis. The islands peripheral to the Bird's Head are inhabited by speakers of non-Oceanic AN languages and it is from these that the considerable AN influence on the NAN languages of this region emanates (Wurm, 1982, p. 207). This influence is apparent in the counting systems of all the West Papuan Phylum languages, if not in the direct borrowing of numerals then in the presence of a 10-cycle. The inference that may be drawn from this is that, at some time in the past, the West Papuan Phylum languages possessed digit tally systems with a (5, 20) cyclic structure and that as a result of AN influence the present situation is such that two languages now have (5, 10) systems, five have (5, 10, 20) systems, and two have lost the original 5-cycle and have (10, 20) systems instead.

The East Papuan Phylum

The East Papuan Phylum consists of 23 languages located in New Ireland (1), New Britain (6), the North Solomons Province (8), the Milne Bay Province (1), all in PNG, and in the Solomon Islands (7) based on Lean's data given in the appendices of his thesis (Lean, 1992). A further language for which data exist, Kazukuru, was spoken in the Solomon Islands but is now extinct. At least 18,

[1] The 10 languages are: Tehit, Kalabra, Moi, Moraid, Karon-Pantai, Madik, Brat, Hattam, Borai, and Amberbaken. See Lean's (1992) Appendix D for more details; Figure E2 in Appendix E.

possibly 20, of these languages have counting systems with either a primary or secondary 10-cycle. One, Taulil, has a (5, 20) digit tally system while two, Sulka (Table 3.3) and Kol-Sui (Table 3.7), have 2-cycle variants with a secondary 5-cycle: these three are located in the East New Britain Province. At least four languages, Nasioi, Nagovisi, Siwai, and Buin (see Appendix B), all situated in the North Solomons Province, display numeral classification as discussed in Chapter 8.

All the languages of this phylum are located in regions which are inhabited predominantly by speakers of AN languages. Wurm (1982) noted that virtually all these NAN languages show some degree of AN influence, this being particularly so of the Reefs-Santa Cruz Family in the eastern Solomon Islands: "the languages of the family show evidence of overwhelming Austronesian influence on all levels ... As much as half of the vocabularies of the languages may be Austronesian in origin" (p. 240). It was indicated earlier, in Chapter 6, that Yele, a member of this phylum in the Milne Bay Province (PNG), has borrowed several AN numerals as has Mbilua in the Solomon Islands. While several other languages also have counting systems which have incorporated AN loanwords, the majority have not. However, the AN influence may be detected by the presence of a 10-cycle in the NAN counting systems and this has been achieved without direct borrowing.

The remaining three languages of the phylum which show no trace of a 10-cycle in their counting systems may perhaps provide an indication of the type of counting systems which existed in this phylum prior to the advent of the AN immigrants. The data suggest that candidates for such proto systems are the 2-cycle system and the digit tally with a (5, 20) cyclic pattern: while no pure 2-cycle systems are found, Sulka nevertheless has a 2-cycle system augmented by a digit tally system. There are, in addition, 12 languages which have systems with a secondary 10-cycle which also have a primary 5-cycle and it seems likely that these may have developed from (5, 20) systems. Whether these hypothesised proto systems were also associated with numeral classification is uncertain. While noun classification is not uncommon among the East Papuan Phylum languages, it is only among the languages of southern Bougainville that a well-developed type of numeral classification occurs.

The Torricelli Phylum

The languages of the Torricelli Phylum are located mainly in northwestern PNG in the Torricelli Mountains and in areas between these and the coast in the West and East Sepik Provinces. The Marienberg Family, comprising six languages, is situated in the coastal and inland region extending from the east of Wewak to the Murik Lakes in the East Sepik Province. Two languages of the Monumbo Family are situated near Bogia in the northern coast of the Madang Province. There is a total of 48 languages in the phylum of which data were obtained for 26 (Wurm, 1982, pp. 226-230).

Several of the Torricelli Phylum languages situated near the coast are adjacent to small AN enclaves. None of these, however, show any detectable signs of AN influence in their counting systems. Indeed, in some instances, there appears to be instead NAN influence on the systems of the AN languages: Sissano, for example, in the West Sepik Province, is an AN language spoken in a region adjacent to the NAN Olo language, a member of the Torricelli Phylum. As shown previously, in Table 3.2, Sissano now possesses a counting system with a (2, 5) cyclic pattern and with numerals which are no longer AN in character.

As can be seen from Table 10.1, 24 of the languages for which data exist possess a 2-cycle variant system. In most of these there is a secondary 5-cycle and we have essentially 2-cycle systems augmented by digit tally systems. Wurm (1982) noted that the morphological features of the Torricelli Phylum languages include the use of subject prefixes which "show number, gender, and (class) concordance. The concordance extends to numerals" (p. 229). This is true, for example, of the Southern, Mountain, and Bumbita Arapesh languages (East Sepik Province) (see Appendix B). In addition to possessing a complex system of noun classification, these three languages also have counting systems with unusual features. Mountain Arapesh, for example, has a main system with a (2′, 4, 24) cyclic pattern and a further system which shows some features of a rare 3-cycle. These systems appear to be unique to the Arapesh languages and may well be localised innovations. In addition to these, two languages of the Torricelli Phylum appear to have primary 5-cycles and may be digit tallies. There is no evidence that body-part tallies are found in this phylum. Thus, in attempting to determine likely candidates for the proto counting systems which existed in the languages ancestral to the current day languages of the Torricelli Phylum, we have either a single type with a (2, 5, 20) cyclic pattern or, perhaps, two types, one with a pure 2-cycle structure together with a digit tally having a (5, 20) cyclic structure.

The Sepik-Ramu Phylum

Larger in composition than the other three phyla just discussed, the Sepik-Ramu Phylum comprises 98 languages situated in northern PNG, mainly in the West Sepik, East Sepik, and Madang Provinces. Being larger, this phylum also displays a greater degree of heterogeneity in its counting system types than do the other three. The possibility of major AN influence on the languages of the Sepik-Ramu Phylum appears to be slight: their geographical distribution is such that the majority are located inland and away from the few areas of AN incursion. Only two languages, Boiken (Table 7.1) and Murik, are presently located in regions adjacent to AN languages and indeed it is possible that both of these may have interacted with their AN neighbours.

By reference to Table 10.1, there are 16 languages which have 2-cycle variant counting systems; 13 of these are augmented by a digit tally while three are pure 2-cycle systems that are associated with body-part tallies. There are 17 languages which have (5, 20) digit tallies with no trace of a 2-cycle; one of these, Boiken, has a dialect which is spoken in the islands lying off the north coast near Wewak and possesses a 4-cycle system (Table 7.1). At least eight language groups employ body-part tallies and the cycles of several of these range from 26 to 29. Although the cycles differ, five of these tallies have the first ten tally points in common with the tenth point being the shoulder. In addition, there are three languages which appear to have counting systems with (5, 10) cyclic patterns and another one, Wapi, which also has a system showing evidence of a 10-cycle. The latter has adopted the numerals of the neighbouring Enga language (see the Trans New Guinea section for more comment on Enga). Of those systems which appear to display a (5, 10) cyclic pattern, Murik, Kapriman, and Kopar (East Sepik Province), we have insufficient data for the latter two. Murik, however, may have originally possessed a (5, 20) digit tally system and, possibly through AN influence, now has a system with a secondary 10-cycle.

Except for the body-part tallies, the common element which is found in the counting systems of this phylum is the (5, 20) digit tally either in its pure form or with a primary 2-cycle. Thus, in attempting to identify those counting systems which may have been present in the languages ancestral to the contemporary Sepik-Ramu Phylum languages, we have at least two candidates, the 2-cycle system and the (5, 20) digit tally. The question as to whether body-part tallies were also present in these ancestral languages, possibly with those having pure 2-cycle numeral systems, is debatable. Several languages, such as Hewa, Abau, and Alamblak, which have such tallies, are located in regions which are adjacent to languages belonging to the Trans New Guinea Phylum and which also possess body-part tallies. Wurm (1982) noted that "the southernmost languages of the Sepik Hill Stock, especially Hewa, appear to have been subject to strong influence from the East New Guinea Highlands Stock languages and other Trans New Guinea Phylum languages" (p. 218). Lean (1992) suggested, therefore, that there is a possibility that body-part tallies may have been introduced into the Sepik-Ramu Phylum languages by such influence and that the tallies were not present as an original feature.

The Trans New Guinea Phylum (TNG)

The languages of the Trans New Guinea Phylum comprise about 70% of the total number of NAN languages spoken in the region under consideration. Wurm (1982, pp. 97-99) provided details of the complex classification of the phylum into its constituents. Rather than considering an analysis of the counting system situation by these constituents, we will proceed by considering the main counting and tally types and whether these have a general or specific distribution throughout the phylum. We will also consider the degree to which members of this phylum appear to have had their counting systems affected by AN influence.

With reference to Table 10.1 it can be seen that, taken as a group, the 2-cycle variants are the predominant type of system found in the TNG Phylum; as noted in Chapter 3, a total of 173 such systems occur in Lean's data set. Of these, 39 appear to be pure 2-cycle systems although it is possible that some of these may have secondary 5-cycles (in a few cases insufficient data were available to determine whether this was the case). About 25 of these 2-cycle systems are found in language groups which also possess body-part tally systems. The most common of the 2-cycle variants is the (2, 5) or (2, 5, 20) system which is found in 86 languages; the other two variants, the (2′, 5) and the (2″, 5), are together found in 48 languages. The distribution of the 2-cycle variants in the TNG Phylum is such that they are found in all the major stocks. In the Finisterre-Huon Stock, for example, of the 53 languages for which data exist, 30 possess a 2-cycle variant. Similarly, in the East New Guinea Highlands Stock, of the 49 languages and dialects for which we have data, 33 possess a 2-cycle variant.

The digit tally system in its pure form, that is without any trace of a 2-cycle, is found in 52 languages of the TNG Phylum, which is about 17% of these languages for which data exist. This type does not have a widespread distribution throughout the phylum and is most commonly found in the Finisterre-Huon Stock, in which there are 20 examples, and in the Madang-Adelbert Range Sub-Phylum which accounts for a further 17. Isolated examples are found in several other stocks. The geographical distribution of the (5, 20) digit tally is such that the majority are located in the north-eastern sector of PNG, inland from the coastal region which stretches from Finschhafen in the Morobe Province to Bogia in the Madang Province.

More than 80% of all the documented body-part tally systems which are found in New Guinea occur in languages belonging to the Trans New Guinea Phylum. Appendix D provides extensive detail. Their distribution throughout the main stocks, however, is not uniform. There is no confirmed evidence, for example, of the existence of body-part tallies in the Finisterre-Huon Stock nor in the Angan, Gogodala-Suki, Marind, Sentani, and various other smaller stocks. The exception for the Finisterre-Huon Stock is the work of Wassmann and Dasen (1994) who provided an example for men's counting from a Yupna language provided by a senior Elder. There appears to be only one example in the Madang-Adelbert Range Sub-Phylum. On the other hand, in the Central and South New Guinea Stock, there are 15, in the East New Guinea Highlands Stock there are 11, in the Teberan-Pawaian Sub-Phylum level Super-Stock there are 10, in the Trans-Fly Stock there are at least 8, and in the Border Stock, 4. The geographical distribution of the body-part tallies is given in Figure 4.2 in Chapter 4 and is such that the large majority are located in the central and southern part of the New Guinea mainland. The fact that there are two other languages spread along the northern border of the Phylum towards the East suggests possible earlier communication for ceremonial purposes or influence. However, it would seem to be localised innovations to overcome the limitations of 2-cycle systems. This may have been a recent example for the Yupna who would have been familiar with both Lean and Smith asking about whether there were body-part tallies and they may have seen this as an advantage or idea to pursue. However, the fact that only three old men had an idea of this system would suggest a possible ceremonial purpose and that the body-part tallies may have once been more widespread. As noted in Chapter 4, Wassman and Dasen (1990) reported one Elder included male genital body parts which would have been seen as inappropriate in other cultures.

Throughout the whole of the Trans New Guinea Phylum there is evidence of only six languages which exhibit examples of 4-cycle systems. Five of these belong to the East New Guinea Highlands Stock and are situated in the Western Highlands, Jiwaka, Southern Highlands, and Enga Provinces (PNG). In the Kewa and Wiru languages, the 4-cycle systems are a digit tally variant in which the four fingers but not the thumb comprise "one hand" as discussed previously in Chapter 7. Both of these languages have, in addition to these systems, body-part tallies. The Melpa dialect of Hagen has a counting system with a (2′, 4, 8) or (2′, 4, 8, 10) cyclic pattern and this too is essentially a variant digit tally in which 4 is "one hand". In addition to these, there is also the very unusual 4-cycle system which employs "cycle units", as described in Chapter 7, and which is found in the Mae dialect of Enga; the Gawigl or Kaugel dialect of Hagen; and the Angal-Heneng dialect of Mendi. The five languages of this region which possess 4-cycle systems all belong to the Central or West-Central Families (Kewa, Wiru, Hagen, Engal Mendi, data from Lean's (1992) appendix on Highland Provinces). We suggest in Chapter 7 that the occurrence of 4-cycle systems among these languages appears to be a localised innovation and an adaptation and extension of the standard digit tally. Strathern (1971) noted the importance of pairs in ceremonies and this may indicate a cultural reason for these developments.

Apart from these highlands languages, there is evidence of only one other member of the TNG Phylum possessing a 4-cycle system. This is Nafri which is located on the north coast of West Papua near the border with PNG (Galis, 1960, p. 144). Several of the languages spoken in the coastal areas to the east and west of the West Papua/PNG border have counting systems which show traces of a 4-cycle: Wutung and Vanimo, for example, both of which are NAN languages belonging to the Sko Phylum-level Stock, and Ormu and Yotafa (Galis, 1960, pp. 144-145 supported by older data for Yotafa, see Lean's, 1992 appendix on Oceania) which are both Oceanic AN languages. Nafri is spoken in a region adjacent to the Yotafa area which surrounds Jayapura. The inference which may be drawn is that Nafri has probably acquired its 4-cycle system from its AN neighbour.

There are five languages of the TNG Phylum which possess 6-cycle systems, as discussed in Chapter 7. Three of these, Kimaghama, Riantana, and Ndom, belong to the Kolopom Subphylum-level Family and are located on Kolopom Island in southern West Papua (Lean's appendix on West Papua but also see Galis (1960) and Boelaars (1950)). The other two, Kanum (Donahue, 2008) and Tonda (1 CSQ), belong to the Trans-Fly Stock and are located to the east of Kolopom Island and across the PNG side of the border. Figure 7.3 shows the location of these 6-cycle systems. There is no evidence of the existence of 6-cycle systems outside this region and we suggest in Chapter 7 that these systems appear to be a localised innovation.

The only remaining counting systems of the TNG Phylum to be considered are those possessing a 10-cycle. There is evidence of at least four languages which have systems with a (5, 10) cyclic pattern and a further four which have systems with a (5, 10, 20) cyclic pattern but others have insufficient data to confirm their patterns. Two and possibly the third of this latter group, Baham, Iha, and Semimi, all of which belong to the Mairasi-Tanah Merah Stock, are located in western West Papua in coastal regions which are also inhabited by non-Oceanic AN neighbours. A further three, Ulingan (5, 10), Pay (5, 10, 20), and Tani (5, 10 and possibly 20 cycle), which are members of the Madang-Adelbert Range Sub-Phylum, are located on the coast north of Madang near Malala Harbour in a region in which the AN language Medebur is situated. There is, in addition, Kovai (5, 10, 20), which belongs to the Finisterre-Huon Stock, whose speakers live on Umboi Island in the Morobe Province (PNG) and whose neighbours (Mangap, Mutu, Barim) are all AN. The remaining members of this group having counting systems with a secondary 10-cycle are Kwale and Magi, both belonging to the Central and South New Guinea Stock and which are situated in the Central Province (PNG). Kwale is located in a region adjacent to that of the AN Sinagoro while Magi is spoken along the south coast which is also inhabited by the speakers of the AN Magori, Ouma, and Yoba languages. Finally, there are three languages not included in this group of eight, Koita, Koiari, and Domu, which have counting systems that are 2-cycle variants with secondary 5-cycles but with tertiary 10-cycles. These are also members of the Central and South New Guinea Stock and are located in the Central Province. The first two are situated outside the Port Moresby region which is dominated by the AN Motu speakers while Domu is situated on the south coast near Cape Rodney and is adjacent to the region inhabited by the AN Keapara speakers. We thus have, for all of these TNG Phylum languages, circumstantial evidence that the presence of 10-cycles in their counting systems is due to the influence of proximate AN languages.

The remaining three languages which are shown in Table 10.1 as having counting systems with a primary 10-cycle are Enga, Lembena, and Ekagi. They are discussed in Chapter 6 and Table 6.6 shows the numerals of the Ekagi language which indicate clear evidence of borrowing from an AN source. Both Enga and Lembena are referred to in Chapter 6 and Table 6.5 where it is suggested that the original Enga 4-cycle system (Chapter 7, Table 7.5) has undergone change. Lean (1992) suggested that the modern Enga system has probably been affected by the Tok Pisin or English 10-cycle systems but given that Enga has loanwords from AN coastal languages (Swadling, 1997), it is possible that these ambitious people adopted a 10-cycle system earlier (see Appendix C for more details on Enga).

Comment on Innovation and Influence on TNG Phylum Language Changes

The foregoing survey of the counting system and tally situation in the TNG Phylum languages provides a means of inferring which counting systems and tally types may have been present in those TNG Phylum languages which are ancestral to the present-day languages. Of the various types discussed above, several may be eliminated as candidates for such proto systems on the basis that they appear to be innovations which have occurred in a relatively restricted group of languages and at a time when the TNG Phylum languages were already established in the New Guinea region. The 4-cycle and 6-cycle systems fall into this category. In addition, we have suggested that all systems which show evidence of a 10-cycle have acquired this feature as a result of the influence of language groups external to the TNG Phylum, that is mainly AN language groups.

The most obvious candidates which may be considered as possible proto systems of the TNG Phylum are the 2-cycle variants and the (5, 20) digit tally. In addition, it seems likely that some form of body-part tally was present in the languages ancestral to the present day ones. An interpretation of the data available suggests that there may have been at least two separate developments which have led to the distribution of counting and tally types that we now see in the TNG Phylum. First, it seems possible that the proto systems of the phylum were the 2-cycle numeral system and the body-part tally and that, second, the digit tally was introduced into the phylum subsequently, probably as a result of interaction with members of other phyla. Other aspects of this interpretation will be developed in later sections.

The Minor Phyla and Isolates

The six minor phyla account for some 29 languages of which data have been obtained for 21 (Wurm, 1982, pp. 247-251). The distribution of these among the phyla is as follows: the Sko phylum-level Stock (6), the Kwomtari phylum-level Stock (4), the Arai phylum-level Family (4), The Amto-Musian phylum-level Family (2), the East Bird's Head phylum-level Stock (2), and the Geelvink Bay Phylum (3). In addition to these there are seven phylum-level isolates and data have been obtained for six of them.[2]

In the Sko Stock, one language, Warapu, has a (2, 5) system; three languages, Vanimo, Wutung, and Rawo, have digit tally systems which show evidence of a 4-cycle and in which 4 is one "hand" (Chapter 7, Table 7.1, and Appendix C). Only the first language has a pure 4-cycle system, the latter

[2] The isolates were discussed in Wurm (1982, pp. 252-254). The six languages for which some data exist are: Warenbori, Taurap, Pauwi (Lean's Appendix D, West Papua); Busa , Yuri, Nagatman (Lean's Appendix C, Sandaun Province) and. Wurm included Maisin in his list of isolates: this is found in (Lean's Appendix A, Oro Province) where it has been taken to be an AN language which has been extensively influenced by NAN languages.

two showing evidence of a 5-cycle as well. The Sko languages are situated on or near the north coast of the New Guinea mainland near the border of West Papua and PNG. In this region there are several languages, both AN and NAN, which show evidence of a 4-cycle in their counting systems, and this appears to be a localised development peculiar to this northern coastline. Whether the innovation occurred with the NAN languages and was adopted by the AN languages, or vice versa, is uncertain. We have, in addition, two other languages, Sko and Sangke, which appear to have the standard (5, 20) digit tally. The common feature of all the languages is a form of digit tally either in this standard form, in its modified 4-cycle form, or as a 2-cycle system augmented by a digit tally (see Appendix C).

In the Kwomtari Stock, two languages, Fas and Baibai, possess 2-cycle variant systems. The speakers of the Kwomtari language appear to possess a body-part tally while the Biaka-speakers appear to have a digit tally. Kwomtari is spoken in a region adjacent to that of the TNG Phylum language Amanab. This and other languages of the TNG Phylum found in this general area all possess body-part tallies and there is a possibility that Kwomtari may have been influenced in this respect as this is the only member of the Kwomtari Stock to have such a tally. Apart from this, the common factor among the Stock appears to be the digit tally which, in two cases, augments a 2-cycle variant (see Appendix C).

The data for the Arai Family is such that there is somewhat incomplete information for four languages (Rocky Peak, Iteri, Bo and Nimo). None of these appear to possess any indication of a 2-cycle system nor, probably, a digit tally. It is possible, though not certain, that body-part tallies are used: these have not been counted as definite in Table 10.1. That such tallies may exist in these languages is also suggested by the prevalence of body-part tallies in the other neighbouring languages of this region, particularly Mianmin, a member of the TNG Phylum. Similarly, the data for Amto and Musian, the two constituents of the Amto-Musian Family, do not provide conclusive evidence regarding their counting or tally types except to indicate that there is no trace of a 2-cycle (see Appendix C).

Data exist for two languages of the East Bird's Head phylum-level Stock, Meax and Mantion, both of which have systems with a (5, 10, 20) cyclic pattern. As is the case with the languages of the West Papuan Phylum, which are located largely in the western half of the Bird's Head in western West Papua, there is evidence of marked AN influence on certain structural features of both Meax and Mantion. This influence is also apparent on their counting systems and is detectable by the presence of a 10-cycle which appears to have been incorporated into what was originally a (5, 20) digit tally (see Appendix C).

For the Geelvink Bay Phylum, data are available on three languages, Yava, Tarunggare, and Bauzi. The last has a (5, 20) digit tally while the Tarunggare system may have a primary 5-cycle. The Yava system, however, has a (5, 10, 20) cyclic pattern. The presence of a 10-cycle can most likely be attributed to AN influence as Yava is the only NAN language spoken on Yapen Island which is otherwise inhabited by the speakers of 12 different AN languages. The current system can be interpreted as being originally a (5, 20) digit tally which has subsequently incorporated a 10-cycle. The common feature shared by the counting systems of the members of the phylum is a primary 5-cycle and it seems likely that, at some time in the past, each possessed a digit tally.

Of the six isolates for which we have data, three are located in the West Sepik Province (PNG) and three are located in West Papua. The first three, Yuri, Nagatman, and Busa, each appear to possess a body-part tally. Yuri is spoken in a region adjacent to several TNG Phylum languages, including Anggor and Dera, both of which have body-part tallies. There is a possibility that the Yuri-speakers may have acquired their tally from these neighbours, and indeed the same may be true of Nagatman and Busa: both of these are located to the east of the Anggor region and are surrounded by language groups which have body-part tallies (see Appendix C).

The other three isolates, Warenbori, Taurap, and Pauwi, are such that the first two have systems which show traces of a 5-cycle while the last appears to have five distinct numerals although there is insufficient data to determine the full nature of the system. It seems likely, however, that at least the first two languages may have irregular digit tallies. No trace of a 2-cycle is apparent in any of the systems (see Appendix C).

The foregoing analysis of the counting system and tally situation as it pertains in each phylum and its constituents has attempted to identify which types are possible candidates for the proto system(s) may have existed in the early history of the phylum. In addition, an attempt has been made to identify which types may have originated in the later history of the phylum either as a result of influence external to the phylum or as a result of innovations. The results of this analysis will now be used in an attempt to reconstruct the prehistory of number and counting in the NAN languages.

Reconstructing the Prehistory of Number in the NAN Languages

The Papua New Guinea Context

Archaeological discussions on the prehistory of Sahul (New Guinea-Australia-Tasmania before the end of the last ice-age in the late Pleistocene era) indicate occupation of valleys such as the Ivane valley in the Owen Stanley Range in the south eastern mainland of New Guinea. It included the edges of rainforest areas and a recognition of the rock-types used for forming flints (Ford, 2011) dating inhabitants to 52 - 47 000 BP. Also indicative in this area are several different occasions when inhabitants came and occupied the landscape. Later dates indicate at least consumption of crops such as taro. In the highlands region, early occupation from around 42 000 BP have been found but not consistently across the mountain valleys. For example it is less likely in Simbu Province than Wahgi valley during the waning of the last ice-age in the late Pleistocene period or early Holocene. In the Wahgi valley it seems that people both foraged and assisted, e.g., by burning, to have an increase in food that allowed less travel when foraging (Terrell et al., 2003). This last study suggested movements that might be sporadic, reduced in area for foraging but at other times more widespread but adapting to both the ice-age and what the rainforest had to offer. It was the sporadic and adaptive husbandry of forest, swamp edges, and higher ground that was particularly noticed by Terrell and colleagues. Much of the diversity and innovation is perhaps contrary to beliefs that agriculture came with the Austronesians. In fact, it is clear that forest plants like taro, sugarcane, pandanus, and sago were largely recognised and encouraged from the mountain valleys to the coast and that there was some travel and trading occurring between people as early as 40 000 years BP.

In the Holocene era, according to Spriggs (2006, 2011) who put much of the variation and influence on agriculture and use of plants down to the Austronesians, there was extensive trading especially of obsidian rock and as indicated by pottery pieces. Swadling (1997) provided considerable evidence for the changes in languages resulting from interactions of different languages facilitated by movement across the inland seas on the northern side of PNG mainland. She suggested this ease of movement was a more likely explanation for language interaction than the impact of a later AN Oceanic colonisation. Swadling (2010) also emphasised that the environment led to two distinct trade routes – one to the east up the Ramu, Lower Sepik and Yaut rivers into the highland regions while the one to the west continued across the mountains into the middle Sepik then into the highlands and across to the other side of the mainland. Evidence from shells and mortars indicates this trade. The counting system data collected by Lean does not show a fine distinction in terms of systems but there is evidence of similarities between these languages. This was particularly evident with the use of body-part tally systems discussed in Chapter 4 and summarised in Appendix D. Furthermore, Swadling (2010) noted that a couple of tribes had considerable influence and spread. Gammage (1998) discussed alliances of tribal groups and demarcations of land by drains during the first European contact incursions into the western regions of Papua New Guinea beyond Mt Hagen. The Chambri were chased out in 1905 by the Iatmut and only recently returned. The Murik had been the dominant traders early in the 1900s both along the trade routes inland and along the coast.

There are a number of coastal regions where related language groups can be found with large gaps between relatives such as the Siassi Family from the North coast area of Madang to Siassi

Islands in Morobe or the Bukawa influence well south of the intervening Lae region. The Austronesian languages spread right up the Markham valley in Morobe. Languages also seemed to be diversified by warfare that occurred in the Markham valley and probable friendships and negotiations to avoid warfare.

Although relatively recent changes in the 19th century, Holzknecht (2001a) provided a detailed account of the movements of the Lahe or Aribwatsa, as the last remaining speaker in 1985 called her language, which was one of the four Busu subgroup languages of the Lower Markham Group of the Austronesian Markham Family. These language groups had spread up the Markham valley but the settlements of the other Lower Markham language the Wampar, who had also taken up residency in the mountains by a tributary of the Markham, chased them back down the Markham spreading the speakers to be refugees with friendly neighbours. The last Aribwatsa speaker could read and write seven languages although she heard her own mother tongue as a child but she also lived among another language group, Kawa, which became the dominant language for the refugee children. It was evident that certain items in the early 1900s were not prevalent including sweet potato for which the language appeared to have a transliteration from another language. The counting system had a 2-cycle as well as a digit tally system that had developed in the much more distant past.

Musom held on to its position in the Markham Valley but like Adzera and other languages had adopted a 2-cycle system (0 *mak*=negative verb, 3 *siruk da wen* is 2+a definite singular article, 4=2+2) within its digit tally system so that 6 was literally "one hand altogether and take one from the other side". The use of words to replace taboo words led to some interesting modifications in the language while plurals were identified by reduplication (Holzknecht, 2001b). So if recent modifications have occurred during the last 150 years, it is likely that there was plenty of earlier tribal interchanges both friendly and otherwise.

Archaeological, Linguistic and Genetic Studies

The archaeological evidence by which we may judge the minimum timespan of human habitation in the New Guinea region indicates that people were living in the Owens Stanley Range of Central Province probably 50 000 years ago, in the Huon Peninsula (Groube, Chappell, Muke, & Price, 1986) in the Morobe Province (PNG) at least 40 000 years ago, and in New Ireland and Bougainville at least 30 000 years ago (Allen, Gosden, Jones, & White, 1988; Wickler, 2001). Similar dates from archaeological sites in Australia such as those at Keilor in Victoria (40 000 BP) and at Lake Mungo in New South Wales (41 000 to 39 000 BP), provide additional evidence in that it is assumed that the migratory routes into Australia passed through New Guinea (Thorne & Raymond, 1989). The dating of materials from the highlands region of New Guinea indicate the presence of a pre-agricultural people at least 25 000 years ago (Bulmer, 1975). Given this vast time-scale, it is difficult with current methods to reconstruct the nature of the languages spoken by the various early immigrants into the New Guinea region: while the methods of historical linguistics have been used to reconstruct features of such proto languages as Proto Oceanic and Proto Austronesian, the time-scale for these is of the order of 5 000 to 7 000 years. It is more problematic to reconstruct, say, the numerical lexis of a language spoken in New Guinea 10 000 years or more ago and thus know the precise nature of its counting system. However, some of the linguistic work associated with the archeological work does indicate some interesting findings from around this time. This work is further supported by genetic statistical analyses (Reesink et al. 2009). For example, the language families within the Trans New Guinea (TNG) Phylum showed an 88% consistency for males in the one society and language (Attenborough, 2010). The antiquity of the proto TNG language, rather than the spread of something like agriculture to allow advances and dominance in the culture, seems to account for the high degree of variation but similarities (Foley, 2000).

Deep and Surface Features of Counting

A counting system, however, can be thought of as having both surface and deep features. The surface feature is the numerical lexis; this is subject to the normal and inevitable linguistic change over time. The deep feature is the counting system's cyclic structure and the evidence suggests that this is the one single feature of a system which is likely to remain relatively stable as the daughter languages of some proto languages differentiate and diverge. The history of the counting systems of the AN languages, for example, indicates the remarkable stability of the systems' 10-cycle structure over thousands of years despite the divergence in the numerical lexis of one language from another. However, it is also evident from Chapters 6 and 7, that the cyclic structures of the counting systems of certain of the AN languages, particularly those in the PNG region, have undergone change. Such change, Lean (1992) suggested, is not the norm but occurs only under certain special circumstances. It is largely in those AN groups which moved into regions occupied and dominated by NAN groups and were influenced by these to the extent that changes occurred to their traditional cultures and economies, that a significant and concomitant change also occurred to the cyclic structure of their counting systems. The act of counting is embedded in the culture of a society and is an integral part of ceremonial occasions and economic transactions; to change the nature of a society's counting system requires changes to its institutions.

Evidence of Early Counting Systems

As discussed in Chapter 1, Wurm and his co-workers have attempted a reconstruction of the possible prehistoric linguistic migrations into New Guinea using evidence from both linguistics and other disciplines. Those aspects of the reconstruction which concern us here are that, first, the Australoid migrations into the single New Guinea-Australian continent occurred 50 000 BP or earlier and, second, that at a much later date, perhaps around 40 000 BP, the first migrations of the speakers of the ancient NAN languages occurred. Wurm (1982, pp. 258-275) suggested this occurred by 15 000 BP (now thought to be earlier 40 000 to 30 000 BP) and that these languages may still have direct descendants today in the West Papuan Phylum and East Papuan Phylum languages, as well as in some of the minor phyla and isolates, all of which he believes to be archaic. The members of the Torricelli Phylum are also thought to be archaic and pre-date the spread of the Sepik-Ramu Phylum languages (Swadling, 1997). Third, Wurm indicated that the main NAN migration was that of the TNG Phylum languages which entered the New Guinea mainland in the southern Bird's Head-Bomberai Peninsula area of West Papua, subsequently spreading west to east. Wurm gave the date of the beginning of this west-to-east migration as about 5 000 BP but it now seems to be much earlier. Wurm based this argument on the presence of AN loanwords in TNG Phylum languages which, Wurm suggested, occurred as a result of the interaction of speakers of ancestral forms of TNG Phylum languages with AN groups in western West Papua prior to the main migrations to the east. Given this postulated late date of entry into New Guinea, the timescale does not allow for the degree of linguistic diversity now apparent in the TNG Phylum languages and thus Wurm believed that their ancestral form started to split up 9 000 years or more ago, considerably earlier than their entry into New Guinea. While this brief summary of the NAN migrations into New Guinea omits further details, given by Wurm, of subsequent migrations of the language groups of various phyla within the New Guinea area, it is sufficient for our purposes here but we need to recognise more recent interpretations. So, for example, as mentioned earlier, Swadling (1997) suggested these loanwords, for example in Enga, and the loss of 10-cycle in the AN languages may be a result of the large inland lakes in the Ramu-Sepik region based on archeological findings (e.g. Dongan site) between the minor ice-ages when navigation across the lake made it possible for AN and NAN languages to more easily communicate.

However, there is now further evidence of even earlier habitation to the east. For example, on Buka in the north of Bougainville Province, Kilu Cave dates back to 29000 BP (Wickler, 2001). Thus the NAN languages were likely to spread down through the Solomon Islands and to Island Melanesia thousands of years before the AN Oceanic travellers came through. The common ancestral language was likely some 10000 BP when people could more freely travel across the region. Sheppard et al. (2010) suggested that for the Solomon Islands, Austronesian speakers pre-dated the Lapita pottery period. The two-way travel between places and the obsidian trade suggest a date prior to 2300 years ago.

In considering the results of the analysis of the counting system variants within different phyla, the distribution of the suggested proto systems and tallies in the major and minor phyla is remarkably restricted: in all phyla the candidates comprise only 2-cycle variants and/or the (5, 20) digit tally. In the large majority of cases, where such systems occur in the languages spoken today, they take the combined form of a 2-cycle variant augmented by a digit tally. While it is possible that this combined form is itself a proto system, however, in most phyla, there is evidence of the two types existing independently and in the West Papuan Phylum languages there is no trace of a 2-cycle. We may recall, in addition, that the 2-cycle variant systems and the digit tally are both present in the Australian languages. The joint existence of both types in New Guinea and Australia does not provide any compelling evidence that one type of system has historical priority over the other. On the other hand, their joint existence in Australia suggests the possibility that they were present prior to the separation of New Guinea and Australia 8000 years or more ago. If this were the case, did the two types of systems enter with the original Australoid populations or were they introduced subsequently as a result of NAN migrations into New Guinea and the diffusion of the systems southwards into Australia? Furthermore, did such diffusion necessarily occur prior to the separation of New Guinea and Australia or did it occur after separation with the diffusion occurring via the Torres Strait islands?

With the current evidence available, earlier dates are much more likely. As Wurm (1975) indicated, "it has long been believed that Torres Strait constituted a clear linguistic boundary between the Australian and Papuan language areas" (p. 915). Subsequent work, however, has revealed a degree of interaction between the two; Wurm noted that "connections between Papuan and Australian languages across Torres Strait can be attributed to mutual linguistic influence and the adoption of loanwords. It seems the influences have gone in both directions, with northward Australian linguistic influence antedating a scattered southward Papuan influence" (p. 922). Thorne and Raymond (1989) indicated that cross-Torres Strait interaction was not just linguistic: "there was considerable contact across this narrow water gap … and many elements of Melanesian culture were taken up by people across the north of Australia: outrigger canoes, skin drums, smoking pipes, funeral posts, and certain initiation ceremonies and hero cults" (p. 146). Ancient pottery has been found on a Torres Strait island (Butler & Dean, 2013).

It seems that with considerations such as these we cannot eliminate the possibility of the cross-Torres Strait diffusion - in either direction - of counting systems and tally methods. However, while such presumed trading contact between southern New Guinea and northern Australia is probably a necessary condition for such diffusion to occur, the question of whether it is also a sufficient condition is debatable. The circumstances in which, for example, the AN groups on the east coast of PNG have had substantial changes induced in their traditional counting systems have involved more than just the establishment of trading relations with their NAN neighbours. The changes to numerical institutions are outcomes of more fundamental changes which occurred to the traditional politico-economic institutions of the AN societies as they established sustained and geographically proximate contact with dominant NAN groups. The Australian-New Guinea connection, on the other hand, appears to be more tenuous than this and probably does not satisfy the rather more stringent conditions necessary to induce such basic change. If this is the case, then it seems likely that proto counting systems were present in Australia prior to the separation of New Guinea and Australia and, moreover, that these were already present in the languages of the Australians or were introduced by the early NAN migrations that began prior to separation.

The picture that we have of the distribution of proto systems in the various NAN phyla does not strongly suggest that the languages of one phylum rather than another may have been the carriers of each of the two proto systems into New Guinea. The most economical interpretation of events, rather, suggests that the 2-cycle numeral system and the digit tally were present from an early date in the ancient languages ancestral to each of the major and minor phyla. It may be that the digit tally entered New Guinea independently of the 2-cycle numeral system and that they were brought by different language groups. However subsequent interactions between the NAN groups over many millennia have resulted in both types being present, either separately or in hybridised form, in each of the phylic groups. In addition, other developments within New Guinea due to localised innovations and the introduction of AN influence have introduced a degree of complexity into a situation which, Lean (1992) suggested, was not originally there.

There is an important exception to the idea that it is difficult to associate the proto counting and tally types with a particular phylum, given the present-day distribution of languages. We have suggested above that the body-part tally appears to be associated specifically with members of the Trans New Guinea (TNG) Phylum and that, where this is found sporadically in other New Guinea languages, this may probably be attributed to influence by neighbouring TNG Phylum languages. However, we need to recall that, as indicated earlier, body-part tallies very similar to those found in New Guinea are also found in southern Australia and in the islands of Torres Strait. In attempting to interpret this situation we may consider, first, the possibility that the body-part tally was not present in New Guinea prior to the advent of the languages ancestral to the TNG Phylum and that it was the speakers of these who introduced it. If this was the case, then the presence of the tally in Australia suggests that it was diffused from New Guinea southwards into Australia.

We have the same difficulty here as was discussed above regarding whether the diffusion took place before or after the separation of New Guinea and Australia. If the body-part tally was introduced to Australia prior to separation then this requires the TNG Phylum languages to have been established in the region 8 000 to 10 000 years ago. This date conflicts with Wurm's suggestion that these languages began their west to east migration into New Guinea about 5 000 BP. However, this date is based on the observation that AN loanwords are found in certain of the TNG Phylum languages and that the contact with AN groups probably occurred in the western part of New Guinea when the AN-speakers arrived in that region perhaps about 5 500 years ago. If the TNG Phylum languages were, on the other hand, already established in New Guinea long before this date then it seems likely that the AN contact occurred not in the west but in the east, perhaps in the Markham Valley of the Morobe Province (PNG), after the establishment of POC and the migration of AN-speakers onto the mainland (in a personal communication with Lean, Lynch suggested this and also noted Wurm and colleagues indicated this probability).

An alternative scenario, which we might note, for explaining the joint presence of the body-part tally in New Guinea and Australia is that this tally was already present in Australia prior to the advent of the NAN languages in New Guinea. Sometime after the establishment of the TNG Phylum languages in the region there was a northward diffusion of the tally as a result of Australian influence on these NAN groups, as noted by Wurm (1975, p. 922), with subsequent further diffusion into the New Guinea highlands regions. Interesting though this conjecture is, on balance, it seems less likely than the alternative southwards diffusion in that, while the body-part tally has only a sporadic appearance in Australia, it has a substantial occurrence in New Guinea suggesting that the latter was its original homeland.

It was observed earlier that the body-part tally is not uniformly distributed throughout the TNG Phylum languages and that it tends to occur in certain stocks located in the central and southern highlands, in the Sepik Provinces (PNG), and in parts of West Papua. One possible explanation of this is that the ancestors of the TNG Phylum languages may have comprised at least two different strands, one strand bringing with it the 2-cycle numeral system and the body-part tally, and another strand bringing a 2-cycle numeral system and a digit tally. Alternatively, if the ancestors of the TNG Phylum entered New Guinea bringing only the 2-cycle numeral system and the body-part tally then we must assume that those language groups which have lost the body-part tally have done so within New Guinea as a result of contact with languages from other phyla. Generally speaking, according to this interpretation, the change has been such that the body-part tally was displaced by the digit tally. In certain cases, particularly for languages in the Southern Highlands Province, a modified 4-cycle digit tally was adopted and used for counting while the body-part tally was retained for calendrical purposes. For a number of languages in the Finisterre-Huon Stock and the Madang-Adelbert Range Subphylum, both the 2-cycle system and the body-part tally have been displaced by the digit tally. This is in line with the findings of an easterly occurrence of a body-part tally in Yupno in the Finisterre Ranges (Wassmann & Dasen, 1994).

If Seidenberg's view of the diffusion of counting practices, as dealt with in the previous chapter, are applied to the situation in New Guinea and Australia, the events that we have discussed above would have developed in the following way. The ancestors of the Australian languages and those of all the major and minor phyla of the NAN languages would not have possessed a means of enumeration and they would remain in this state until the 2-cycle system had diffused across the world from the Middle East, with the diffusion of other systems following subsequently. On the other hand, the view of the prehistory of counting and number which has been developed above in this chapter presents a different picture in which the possession of numerals and a means of tallying is seen as an archaic feature of human societies. In this view, the 2-cycle numeral system, the digit tally, and the body-part tally, were in Australia and New Guinea as part of the cultural knowledge of the ancient immigrants whose descendants now speak the languages which exist today. Similarly, as discussed further below, the 10-cycle system entered the New Guinea region with the original AN immigrants. We therefore have that the essential features of the counting and tallying situation which we see in New Guinea and Australia today had already been laid down by the time that the city-states of the Middle East were being established and that, therefore, the origins of counting cannot be sought in these but in the managed agricultural and neolithic societies of a much earlier historical period.

Reconstructing the Prehistory of Number in the AN Languages

The current situation regarding the nature of the counting systems distributed throughout the AN languages of New Guinea and Oceania is delineated in the previous chapters. In Chapter 1 we discuss Ross's classification of the AN languages of PNG and the northwest Solomon Islands into two first-order subgroups, namely the Admiralties Cluster and Western Oceanic, the latter comprising three clusters termed the North New Guinea, the Papuan Tip, and the Meso-Melanesian (Ross, 1988). (See Ross' diagram in Appendix E (Ross, Pawley, & Osmond, 2003).) The remainder of the Oceanic AN languages of Island Melanesia and the Pacific comprise a further subgroup, Central/Eastern Oceanic, delineated by Lynch and Tryon (1985). In Chapters 3 to 8, the distribution of the various counting system types among the PNG clusters and the subdivisions of Central/Eastern Oceanic are discussed. In Chapters 8 and 11, a number of brief case studies of several societies are given and these deal with certain matters relating to the place of counting and number in those societies. Other matters relating to the existence and nature of numeral classification were dealt with, as was the extent to which large numbers are found in various languages. Finally, in Chapter 9, the changes which appear to have occurred to the cyclic structure of certain of the AN counting systems were summarised together with some discussion of how these changes may have come about.

In this section we will begin by summarising the material given in Chapters 3 to 8 as it relates to the AN languages. In Table 10.2, the distribution of the various counting system types now apparent among the AN languages of New Guinea is given. The subdivisions of the languages are Ross's four clusters together with a group of five Oceanic AN languages which are located in West Papua.

Table 10.2
Showing the Distribution of Counting System Types Among the Four Clusters of the PNG Languages and the AN Languages of West Papua

	Admiralties/ St Matthias	North New Guinea	Papuan Tip	Meso-Melanesian	West Papua	Total
(2)	0	2	0	0	0	2
(2, 5)	0	15	2	1	0	18
(2″, 5)	0	7	5	0	0	12
(5, 20)	1	19	11	1	2	34
(5, 10)	1	20	7	17	0	45
(5, 10, 20)	0	13	5	0	1	19
(10, 100)	23	0	9	41	0	73
(10, 20)	0	0	0	3	0	3
(4)	0	2	0	0	2	4
Totals	25	78	39	63	5	210

Table 10.3 provides a summary of how the counting system types are distributed among the remaining AN languages of Island Melanesia, Polynesia, and Micronesia.

Table 10.3
Showing the Distribution of Counting System Types Among the Languages of Island Melanesia, Polynesia and Micronesia

	S.E. Solomonic	Vanuatu	New Caledonia	Fiji/ Rotuma	Polynesia/ Micronesia	Total
(5, 20)	0	15	8	0	1	24
(5, 10)	0	68	0	0	0	68
(5, 10, 20)	0	0	19	0	0	19
(10)	28	19	0	3	39	89
(10, 20)	0	0	0	0	6	6
Totals	28	102	27	3	46	206

The discussion which follows uses the language subdivisions employed in each table and for each subdivision an attempt is made to reconstruct the prehistory of the counting systems of its constituent languages. This task is approached in a somewhat different manner from that used in the previous discussion of the prehistory of the NAN languages. Instead of attempting to infer the nature of the proto system of the AN languages we assume that this is known. The methods of historical linguistics have enabled linguists to reconstruct the numerals of Proto Oceanic (POC) language, given previously in Table 6.7, and the ten distinct reconstructed numerals indicate that the POC system was a 10-cycle one. The reconstruction of the prehistory of the AN counting systems involves, essentially, tracing the fate of this system as the daughter languages of POC developed and dispersed throughout New Guinea and Oceania and, where possible, placing this dispersal in a chronological framework.

The POC Language Community and the Lapita Cultural Complex

The reconstruction of the nature of the POC language community, the dating of its establishment and of its breakup and dispersal into Island Melanesia and the remainder of the Pacific, has been a joint enterprise between linguists and archaeologists. The lexical reconstruction of POC now extends to about 2 000 items. Pawley and Green (1985) noted that these "represent only a fraction of the total vocabulary of the language community, but tell us a good deal about the culture" (p. 129). They added that

> these reconstructions indicate that POC speakers had an economy based jointly on gardening and fishing. The major root and tree crops of contemporary Oceanic societies, other than the sweet potato and cassava, are represented: yam, taro, breadfruit, coconut, etc. A variety of fishing techniques were exploited, including nets, lines, basketry traps, and plant poisons. (Pawley & Green, 1985, p. 130)

The reconstruction of certain kinship terms suggests the possibility "that POC society had descent groups in which land rights were invested" and a further set of terms suggests that the society also had hereditary chiefs (p. 132).

It was indicated in Chapter 1 that the beginning of the POC community is thought to be New Britain, in or near the Willaumez Peninsula (Ross, 1989, p. 143). Ross noted that if

> it remains the case that the earliest settlement date from the Admiralties lies around 1850 B.C., then the latest possible date of arrival in New Britain for the ancestors of the Proto Oceanic speech community will be around 2 100 B.C. (p. 149)

There are two points to note regarding these dates. The first is that they are based on work by Kennedy in the Admiralties, the results of which were published in 1981. Subsequent studies, also by Kennedy (1983, p. 118), have however resulted in the establishment of earlier dates for settlement of the Admiralties at about 4 500–5 000 BP and this implies that the POC community must have been in place at some time prior to this period. The second point to note is that the dating of these events is based on archaeological evidence and on an assumption that there is a connection between the POC community and the cultural complex associated with the Lapita-style ceramic tradition found throughout Island Melanesia and western Polynesia (Allen, 1996; Pawley & Green, 1985; Spriggs, 2011). This connection provides a means of establishing a chronological sequence for the spread of the descendants of the POC community from their putative homeland onto the PNG coastal areas and southwards into the remainder of Melanesia. However, Ross and others suggest there were probably at least two separate migrations to the south from this area.

Tracing the Fate of the POC Counting System in PNG

According to Ross (1988), "at least two groups of people - some of whose descendants spoke Proto Admiralty and Proto South-East Solomonic respectively - departed from the (POC) homeland area before the occurrence of the innovations which characterise Western Oceanic languages" (pp. 382-383). The most probable migratory route of the first group from the POC homeland to the Admiralties is, as also noted by Ross, through New Ireland and Mussau in the St Matthias group. While these travellers have left no trace on the languages spoken in New Ireland, Ross indicates that the language spoken on Mussau, Emira-Mussau, may be related to those in the Admiralties Cluster. Once the separation from the New Britain-New Ireland occurred, it seems likely that "Proto Admiralty and its descendants would have developed without much external linguistic influence" (p. 383). As indicated above, the earliest date established for the settlement of the AN immigrants in the Admiralties is the period 4 500-5 000 BP.

From Table 10.2 we can see that 23 languages in the Admiralties/St Matthias group have 10-cycle systems while one, Nauna, has a (5, 10) system and a further one, Seimat, has a (5, 20) system. Of those languages possessing 10-cycle systems, only one, the Mussau dialect of Emira-Mussau, has a pure 10-cycle system with a completely intact second pentad. The remainder are such that 21 systems belong to the subtractive Manus type and one, Wuvulu-Aua, has features of the Motu type (Lean's terminology). We should also note that with regard to Emira-Mussau, there is some evidence, published by

Friederici in 1913, that the Emira dialect had at one time a (5, 10) system but this no longer appears to be in use. Thus, all the members of the Admiralties Cluster now have counting systems which deviate in some way from the system of their POC ancestor and the changes are such that they occur in the numerals of the second pentad. As discussed in Chapter 7, there are two ways of viewing these changes. The first is that the changes in the second pentad are innovations which were not induced by external influence but occurred spontaneously. The second view is that the second pentad instability may have resulted from the interaction of the original 10-cycle system with the (5, 20) digit tally system perhaps by means of the AN migrants coming in contact with NAN groups, conceivably in New Ireland, for a sufficiently sustained period for changes to be induced in the AN numerical institutions. Whatever the mechanism, the subtractive construction used in most of the counting systems of the Admiralties Cluster appears to be a post-POC innovation that, with one exception, is not shared with other AN languages.[3]

One further aspect of number in the Admiralties languages, discussed in Chapter 6, is the occurrence of numeral classification and, in particular, the use of classifier constructions in which numeral roots, the quantifiers (Q), are suffixed by classifiers (C), that is a QC order is used in noun phrases, as indeed we also find in the Micronesian languages. While Ross (1988, p. 328) indicated that classifiers are reconstructible for POC, he has suggested that this QC sequence, rather than the reverse, is an innovation. On the basis that the QC sequence is commonly found in the non-Oceanic AN languages of South-East Asia, perhaps this is the more archaic form (as suggested in Chapter 6). Be that as it may, the phenomenon of numeral classification does not now generally appear in the majority of the Oceanic AN languages with the exception of some of the Papuan Tip Cluster and a number of the Polynesian languages. It may be the case that the ancestors of the Admiralties Cluster, having left the POC homeland relatively early, retained this original feature of POC.

After the departure of the ancestors of the speakers of Proto Admiralty and Proto South-East Solomonic from their homeland, the innovations unique to the Western Oceanic group occurred. This group, according to Ross (1988, p. 351), comprised the remaining three clusters of the AN languages located in PNG and the north-west Solomon Islands. We now consider the fate of the POC counting system in each of the clusters.

Referring to Table 10.2, of the 78 languages of the North New Guinea Cluster for which we have data, none now possesses a counting system with a primary 10-cycle. In reconstructing the history of the cluster, Ross (1988, p. 188) argued that the current languages are descended from an early Oceanic communalect which developed in the POC homeland. Two groups subsequently separated from the homeland: the first, the ancestors of the Proto Schouten group, moved onto the PNG mainland and moved along the north-west coast and adjacent islands, occupying Manam Island and dispersing from there to the islands of Wogeo and Bam, and also along the north coast to the Sepik Provinces. Ross (1988, p. 124) noted that in Proto Schouten the POC numeral 10, **sangapulu*, has been replaced by **kulemwa*. Other notable changes which occurred in this group are the adoption, by the Wogeo and Bam (Table 7.6) language groups, of a 4-cycle system (see Chapter 7) and the adoption of a 2-cycle system by the Sera and Sissano (Tables 3.2 and 9.2) language groups, most likely as a result of the influence of NAN languages. With these exceptions, all the other languages of this group now have systems with a primary 5-cycle.

The second group of the North New Guinea Cluster to depart from the POC homeland were the speakers of Proto Huon Gulf who settled first on the coast of the Huon Gulf in the Morobe Province with subsequent migrations inland. One such group which moved inland was the ancestors of the Markham Family, discussed in Chapters 6 and 7, and whose member languages now mainly possess 2-cycle counting systems. The exception to this is Labu (see Appendix C) which has a (5, 10, 20) system and which shows evidence of having possessed numeral classification such that the classifier preceded the numeral, that is it had a classifier-quantifier order (CQ) sequence. The other group that migrated inland gave rise to the languages now found in the Mumeng region of the Morobe Province, most of which now possess 2-cycle variant systems (see Appendix C: Mumeng, Piu, Mapos Buang, Manga Buang).

[3] The exception is the Mioko dialect of the Duke of York language.

The North New Guinea Cluster group remaining in the vicinity of the POC homeland and who spoke, according to Ross, Proto Ngero/Vitiaz, now has descendants located in the East and West New Britain Provinces as well as in the Madang and Morobe Provinces. With the three exceptions of Matukar, Roinji, and Nenaya, which have 2-cycle variant systems, the remaining descendants of Proto Ngero/Vitiaz all possess systems with a primary 5-cycle.

The inference that can be drawn from this analysis is that very early in the history of the North New Guinea Cluster, the ancestral communalect spoken in the region of the POC homeland had its counting system affected in such a way that it ceased to have a primary 10-cycle and by loss of the numerals 6 to 9 became instead either a (5, 10) or (5, 10, 20) system. This may have occurred as a result of interaction with NAN groups located in the West New Britain area and who possessed (5, 20) digit tally systems. The migrants who departed from the homeland and moved onto the PNG coastal regions carried their 5-cycle system with them. Further interaction with NAN groups on the mainland resulted in some of the AN languages losing more of their numeral lexis, thus acquiring 2-cycle variant systems. In the case of Wogeo and Bam, the adoption of a 4-cycle system may have been a localised innovation although, as noted in Chapters 5 and 7, 4-cycle systems have a sporadic appearance along the northern seaboard of PNG and West Papua which suggests the possibility of diffusion having occurred. There are no firm archaeological dates resulting from Lapita sites on the PNG coast which might give an indication of when the North New Guinea Cluster languages may have entered this area; however this must have occurred after the departure of the ancestors of the Admiralties Cluster who were settled in the Manus region by 4 500 BP.

Ross said that "Proto Papuan Tip seems to have separated from an early Oceanic dialect chain … and, to judge from the exclusively shared innovations of most member languages, to have remained separate from it" (Ross, 1988, p. 193). Ross also indicates that a subsequent split in the original group that migrated away from the POC homeland gave rise to two daughter networks which he terms the Nuclear Papuan Tip and the Peripheral Papuan Tip (Ross, 1988, p. 191). The languages of the first network are now largely located in the Milne Bay and Oro Provinces on the mainland and the D'Entrecasteaux Islands adjacent to it. None of the languages for which we have data now possesses a 10-cycle system: 7 have a 2-cycle variant and the remaining 18 have systems with a primary 5-cycle. The languages of the Peripheral Papuan Tip network are located in two main areas: the eastern islands of the Milne Bay Province, and along the southern coastal and inland region of the Central Province.[4] At least two languages situated in the Milne Bay Province, Nimowa and Sud-est, have 10-cycle systems with numerals very similar to those of their POC ancestor. The other four languages of this network and which are situated in this Province each has a system with a (5, 10) cyclic pattern. This Milne Bay group is also distinguished by the possession of numeral classification which is such that the classifier precedes the numeral, that is a classifier-quantifier (CQ) sequence operates. The remaining 10 languages of this network and which are situated in the Central Province are such that three have (5, 10) systems and the remainder have the Motu type, discussed in Chapter 6.

The foregoing discussion suggests the following interpretation. The speakers of Proto Papuan Tip left the POC homeland with an intact 10-cycle counting system. After the split into the Nuclear and Peripheral Papuan Tip networks, the languages of the former network lost at least the numerals 6 to 9 thus changing the primary cycle of their counting system from 10 to 5. Some of these languages had a further loss of their numeral lexis in the first pentad giving rise to systems which are 2-cycle variants. These changes to the counting systems of the Nuclear group may be most likely attributed to the influence of NAN groups which have a strong presence in the Milne Bay and Oro Provinces. With regard to the Peripheral Papuan Tip languages, only those situated in the easternmost islands of the

[4] Trobriands, Marshall Bennet, and islands of the Louisiade Archipelago.

Louisiade Archipelago retained a 10-cycle system while Kilivila, Muyuw, and Budibud developed systems with a (5, 10, 20) cyclic pattern. The remaining part of this network migrated to the south coast of the Central Province, possibly settling inland where the Sinagoro language is now spoken (Ross, 1988, p. 145). The development of the Motu type of system may have developed here. Several languages of this group have systems with a primary 5-cycle and this may have occurred prior to migrating from the Milne Bay Province. Kirk (1982) noted that the archaeological evidence indicates that the migration of AN groups along the southern coastline of the Central Province appears to have happened at about 2 000 BP.

The 63 languages of the Meso-Melanesian Cluster for which we have data are now located in West New Britain, New Ireland, parts of East New Britain, the North Solomons Province, and the north-western half of the Solomon Islands. Ross (1989) indicated that

> the linguistic evidence is ambiguous as to whether there were one or several migrations (that is by speakers of different dialects at different times) from the Meso-Melanesian homeland to New Ireland and its offshore islands, but it is clear that the portion of New Ireland from the present day township of Namatanai southwards became a centre of dispersal, from which Proto Northwest Solomonic originated. (p. 145)

In addition to this southward migration, some groups also migrated back across St George's Channel and into East New Britain.

Two languages in West New Britain, descended from the stay-at-home group of the Meso-Melanesian Cluster, possess 10-cycle systems and a further four have systems with a (5, 10) cyclic pattern.[5] The languages of southern New Ireland, which formerly served as the centre of dispersal, uniformly have 10-cycle systems, however as we proceed north the majority of languages have systems with a (5, 10) cyclic pattern. Lihir, spoken on an island off the east coast, has a (5, 20) digit tally system while Tomoip, located in East New Britain, has a system with a (2, 5, 20) cyclic pattern. Proceeding south to Nissan Island and Buka, the languages have 10-cycle systems while, on Bougainville, several languages have 10-cycle systems and a number have (5, 10) systems instead. South of Bougainville, in the northwest Solomon Islands, the 23 Meso-Melanesian languages of this region uniformly possess 10-cycle systems, three of these having secondary 20-cycles.

This discussion of the current distribution of counting systems in the Meso-Melanesian Cluster enables the following interpretation to be made. The speakers of Proto Meso-Melanesian had a 10-cycle system which they carried with them to New Ireland. While the date of their first arrival in New Ireland is unknown, the earliest Lapita site, at Ambitle Island in the Feni Group to the east of southern New Ireland, has yielded an estimate of 3 200 BP (Allen, 1984, p. 197), and this suggests that New Ireland proper would have been settled sometime prior to this time. After this settlement, the 10-cycle system was carried southwards into Nissan and Buka. On Bougainville, certain languages lost part of their numeral lexis and acquired (5, 10) systems, probably as a result of the strong presence of East Papuan Phylum languages which influenced the immigrant AN groups. In New Ireland, as the AN migrants moved northwards they may have also come into contact with established NAN groups (although today only one such group, the Kuot-speakers, still survives) and as a result acquired (5, 10) systems with a loss of their numeral lexis in the second pentad. A subsequent migration brought the ancestors of the Tolai-speakers to East New Britain, carrying their (5, 10) system with them.

The foregoing discussion deals with the vicissitudes of the POC 10-cycle system as the descendants of POC dispersed from their homeland and into the coastal and island regions of PNG. We consider, finally, the fate of the POC system as the AN migrants moved southwards into Island Melanesia and beyond.

[5] The two languages with 10-cycle systems are Bola and Nakanai. The four languages with (5, 10) systems are: Bali-Vitu, Bola, Meramera, and Harua.

Tracing the Fate of the POC Counting System in Island Melanesia, Polynesia, and Micronesia

The two groups which left the POC homeland early before the innovations unique to Western Oceanic occurred were the ancestors of the speakers of Proto Admiralty, discussed above, and of Proto South-East Solomonic. This latter group moved southwards into the Solomon Islands, and possibly further, and may be the source of the Central/Eastern Oceanic group of languages which, as noted in Chapter 1, comprises "all the languages of north and central Vanuatu, Fiji, Polynesia, and Micronesia … together with the southeast Solomon Islands, Utupua and Vanikoro, the south Vanuatu, and possibly the Loyalties and New Caledonia groups" (Ross, 1989, p. 136).

Referring to Table 10.3, there are 28 languages comprising the South-East Solomonic group and the Eastern Outer Islands group and all of these have systems with a primary 10-cycle. Ross indicated that various features of the South-East Solomonic languages imply that they have been little disturbed by external linguistic influence, including NAN influence, since their separation from POC (Ross, 1988, p. 384). The counting systems of this group, therefore, are probably direct descendants from the POC 10-cycle system and they have retained the original 10-cycle nature unlike the systems of many of the Western Oceanic languages. If the migration away from the POC homeland occurred relatively early, as Ross suggested, then we might expect that the arrival in the Solomon Islands may have happened at a time not too different from that when the other early immigrants from the homeland arrived in the Admiralties, that is sometime before 4 500 BP. The earliest date available from archaeological evidence, however, is that from an excavation in Santa Ana which yields an estimate of 3 500 BP (Allen, 1996).

Similarities and variations in the languages and counting systems are likely to have developed as a result of biogeographical considerations. For example, it is known that people sailed from the Outer Eastern Islands in the Solomons along the zenith star lines to the Santa Cruz island and back maintaining links between the two in terms of the long-standing and newer settlements with archaeological evidence of occupation 3 000 years ago. The divergence of some of the languages in the Eastern Outer Islands of the Solomons seems to have occurred well before 1 200 AD when Polynesian travellers came to the islands and their languages dominated. Variations on where languages were found were also influenced by land upheavals and the covering of coral atolls and higher areas with volcanic ash thus giving knowledge that these languages were existing between 3 150 and 3 000 BP. Links were maintained for cultural and language reasons but also fertile land space for the population was also significant for the occupation of different language groups (Green, 2010). Nevertheless, Green cited a workshop paper by Pawley, to support his argument that one reason for the Polynesian Oceanic languages tending to be spread across the oceans and occupying more language space than each of the Oceanic languages of Island Melanesia is most likely due to culture and values. As the chief families and commitments dominated, the valuing of the language of communication seemed to be strengthened. It is also likely that there was then a need to count large quantities of tributes and hence the development of the counting systems involving powers of 10 or other numbers.

Similarly, the archaeological evidence from Western Solomons especially from Roviana lagoon and surrounding land suggests a later spread of Lapita pottery after 2 600 BP. This is supported by the agricultural evidence. Given that in the ice-age this area was still separated from the chain of other Bougainville islands, this is not surprising (Sheppard et al., 2010). Oral histories also indicate the interaction of NAN language groups, especially Kazukuru with the AN groups in this area. Initially this appears to have been through friendly marriage developments but perhaps later around 1 200 AD this became more akin to dominating coastal areas and to headhunting rituals. The presence of matrilineal societies such as on Vella Lavella speaking NAN languages indicates some connections with the NAN matrilineal societies of Bougainville suggesting that cultural developments would precipitate changes in counting associated with customs such as bride price and tributes to chiefly families. More violent interactions between groups seem to be very recent in the 18th and 19th centuries reducing numbers speaking languages and languages themselves. Importantly, the connection between archaeology and language shows that variation can occur slowly even with the more rapid expansion of Oceanic Lapita cultures to the rest of Island Melanesia and further east (Sheppard et al., 2010).

In Vanuatu, the linguistic situation is now such that there are 105 AN languages spoken, three of which are Polynesian Outliers. No NAN languages are present. Referring to Table 10.3, of the 102 non-Polynesian languages, 19 have systems with a primary 10-cycle, 68 have (5, 10) systems, and 15 have systems with a (5, 20) cyclic pattern. The distribution of these is not uniform throughout the islands of Vanuatu. The northern Torres and Banks Islands account for 13 languages, all of which have (5, 10) systems. The remainder of the (5, 10) systems are located on the central islands of Espiritu Santo (19), Maewo (3), Pentecost (4), Malekula (21), Ambryn (4), and Efate (3). The 10-cycle systems are mainly found in parts of these central islands, including Malo Island to the south of Espiritu Santo. It is on Malo that the earliest Lapita sites in Vanuatu are located and these have yielded dates that indicate occupation at about 3 000 BP (Green, 1979). The languages with (5, 20) systems are located in the central and southern islands, particularly Tanna and Aneityum. Archaeological sites on each of these islands have yielded dates indicating occupation by 2 370 BP and 2 900 BP respectively (Spriggs, 2006).

However, more recent research is suggesting that there were Melanesians in the Solomons and Vanuatu before the Lapita culture and that it is likely that there were on-going movements to the west as well as to the south and east over a long period of time. For example obsidian and phalanges were found in Indonesia suggesting a western movement. Furthermore, there is evidence from nuts, stones and shell implements both across the north of New Guinea and inland as well as through island Melanesia dating well before 3 500 BP, probably as early as 5 000 BP (Allen, 1996). The consideration is that there have been considerable interrelationships over time that neither a power perspective of invasion nor an innovation perspective could explain. Furthermore, the mix of Southeast Asia with Melanesian over a considerable period of time could explain the differences in appearance commented upon by early explorers. However, it is much more feasible that the easterly movements were also return journeys explaining the mix of languages in Micronesia, Vanuatu, the Solomons, Fiji and other islands (Allen, 1996; Spriggs, 2011; Terrell et al., 2003; Thomas et al., 1989). After all, the sailing distances to the south and east (Vanuatu, Nauru, Kiribati, Tuvalu, Marshall Islands, Fiji, Samoa and Tonga) were shorter from the southern coast of the Solomon Islands or the south east of Papua New Guinea than from South-East Asia.

The existence of 5-cycle systems in Vanuatu has been discussed previously in Chapter 9. In that discussion, three possibilities for their existence were suggested:

1. that the 5-cycle systems were introduced by long distance diffusion from PNG, by-passing the Solomon Islands;
2. interaction of the ancestors of the Vanuatu languages in situ with now-extinct NAN languages; and
3. uninfluenced, spontaneous change.

There is no compelling evidence, archaeological or linguistic, for the second option. With regard to the first, the internal linguistic evidence of the Vanuatu languages does not appear to suggest that there was an original AN settlement followed later by one or more other intrusive AN migrations which may have brought a (5, 10) cycle system, as well as other innovations, with them. On the other hand, archaeological evidence from central and northern Vanuatu does suggest that several unrelated pottery traditions, only one of which is Lapita, overlap (Pawley & Green, 1973). This may indicate that there were several different movements of AN groups into the islands over a period of two or three thousand years.

With regard to the third option, it has not been generally necessary to invoke the mechanism of spontaneous change in the counting systems of PNG because of the strong and ubiquitous presence of the NAN languages and the likelihood of their influence on AN groups. In Vanuatu, however, if as seems likely we cannot attribute counting system change to NAN influence in situ and if, for the moment, we discount long-distance diffusion, then uninfluenced and spontaneous change seems to be the default mechanism. The circumstances which may bring about such change need to be considered.

Returning to Bougainville in PNG, Lincoln (2010) concluded "the notion that peculiarities in Banoni or closely related Piva are due to Papuanisation is undermined" (p. 225) by his careful analysis supporting his earlier claim (Lincoln, 1976) and Lean's (1992). His argument began by revealing the multiple patterns of prefixes in 13 languages from Bougainville and Lauru. Most have *ka-* for many words but only three have it consistently for all numbers 1 to 9. Others mainly use *to-* but some languages use *e-* for 2 to 5, another uses *tau-* for 5 to 9. Lincoln (2010) also showed the system is based on counting on one's fingers so that a word like 3 is readily replaced by *pisa* or similar word by most of the AN Oceanic Bougainville and Lauru languages but not in other Pacific languages. Metonymic use of a word for an object such as a finger when counting has been shown in several studies to result in spontaneous change (Blust, 2010). Ten is frequently similar to *manoγa* suggesting some protolanguage spread. Finally, the possible influence of the NAN languages if the lexical data is considered must have been quite complex. Nevertheless, the connection between 7, 8, and 9 and the pattern of $x+n$ is still strikingly like counting patterns found in many NAN languages. However, the neighbouring NAN languages also have classificatory suffixes that adds to the difficulties of identifying any direct adoption of NAN systems. Interestingly, NAN languages in the south of Bougainville have different ways of expressing 6 to 9. Namely, Nagovisi and Motuna count up to 10 with Buin (Uisai) representing 7 and 8 as how many "less" than the complete group (Manus-type language) and Nasioi counting up from 5.

First, we have suggested above that counting system change is an outcome of changes to a society's politico-economic institutions. In Vanuatu, such changes might conceivably have occurred as groups settled on some of the larger islands and moved into the interior regions of these, thus abandoning their traditional maritime economies. Second, the counting system change is in the direction of the introduction of a primary 5-cycle. It may be that finger tallying, and probably toe tallying as well, may have been an accompaniment to the POC method of counting: this is suggested by the use of POC **lima*, that is "hand", for the numeral 5. Thus, in serial counting to 10, two alternative methods may have customarily been used: either all ten numerals were enunciated or, on reaching 5, the units were repeated until 10 was reached; this essentially introduces a primary 5-cycle. This is a somewhat radical proposal because it suggests that, while the reconstructed POC system is a 10-cycle one, nevertheless the exigencies of everyday counting mean that the system has an inherent instability which, under certain circumstances, may cause it to become a 5-cycle one and that this may occur without external influence, for example by NAN languages. With certain groups, this method of enumeration could become the norm, resulting in the atrophy of the original 10-cycle system in favour of a (5, 10), (5, 20), or (5, 10, 20) system.

The situation in Vanuatu, then, may have developed as follows. An AN migration from the north brought settlers - who were probably the initial settlers of the region - carrying with them a 10-cycle counting system. These occupied the northern/central region of the islands. Over a period of time there was probably a further filtering through the islands southwards to Tanna and Aneityum and then to New Caledonia. Tryon noted that an early migration from Vanuatu appears to have reached Fiji by 3 500 BP (Tryon, 1984). He added that

> at about this time, a set of migrations apparently began in the north/central Vanuatu region, one moving north, spreading the Austronesian languages through Micronesia (for which there is evidence of an east to west spread), another moving southeast to the Fiji group. From there, after a period of consolidation, the Polynesian languages evolved, moving out from the Tonga-Niue area sometime around 1 000 BC. (p. 155)

The counting system situation in Fiji, Rotuma, Polynesia, and Micronesiais such that all the languages, with one exception, have 10-cycle systems (see Table 10.3). The exception, Faga-Uvea, is a Polynesian Outlier in the Loyalty Islands (see Appendix C): it has a (5, 20) system and is situated in a region where 5-cycle systems are universal. The inference which can be drawn is that the AN migrants to Fiji and Micronesia carried 10-cycle systems with them and that this system was retained in Proto Polynesian and was subsequently carried with the Polynesian dispersal throughout Triangle Polynesia which was largely complete with their arrival in New Zealand by 900 AD. The situation in Micronesia is distinguished by the marked presence of numeral classification which, like the

Admiralties group, has a QC order (Quantifier-Classifier). In the Polynesian languages which still exhibit numeral classification, a CQ order (Classifier-Quantifier) is the norm. Even though the two sequences differ, the presence of numeral classification in both the Micronesian and Polynesian languages suggests that it may have been present in the languages ancestral to those groups and which were spoken in the north/central Vanuatu region.

Referring to Table 10.3, the counting system situation in New Caledonia, including the Loyalty Islands, discussed in Chapter 5, is such that 19 languages have systems with a (5, 10, 20) cyclic pattern and 8 have (5, 20) digit tally systems. The geographical distribution of these is given in Figure 5.4. The uniform occurrence of systems with a primary 5-cycle present similar problems of interpretation as in the case of Vanuatu, discussed above. There is archaeological evidence of a ceramic tradition in New Caledonia dated at 5 000 BP (Kirch, 2000; Pawley & Green, 1973). Furthermore, the DNA information suggests that there is considerable intermingling of the Melanesians with Austronesians over a considerable period of time as they moved across or back and forth from the islands of South East Asia and the north coast and islands of Papua New Guinea (Gibbons, 2001). As Gibbons noted:

> All this means that the synthesis of genetics and archaeology is slowly coming together. But the linguistics still point to Taiwan as the source for Austronesian languages, whereas genetics and archaeology rule it out as the place of origin for genes and Lapita culture. "I don't see a problem at all with that," says linguist Blust. "Languages can spread without preserving the genetic makeup of the [original] speakers. Think Latin. (p. 1737)

Whether this culture was an AN or NAN settlement is debatable (Schutler, 1978) but it seems unlikely to be NAN especially considering that the developments in navigation craft were AN (Thorne & Raymond, 1989) and bacterial investigations suggest links to Taiwan (Moodley et al., 2009). While there is evidence of a migration prior to Lapita, it seems unlikely that the change to the AN counting systems can be attributed to the influence of NAN groups in situ. One possibility is that once the 5-cycle system was established in Vanuatu, AN migrants moved south into New Caledonia carrying the system with them or there was on-going movements in the region.

Summary

Of the 420 Oceanic AN languages for which we have data, only 171 or 41% have counting systems with a primary 10-cycle while 209 or 50% have systems with a primary 5-cycle. The original AN immigrants who settled in the POC homeland, probably about 5 000 years ago, brought a 10-cycle system with them, as judged from the reconstructed Proto Austronesian numerals (Dahl, 1976, cited in Smith, 1988, p. 45). The essential features of this system have survived in certain of the daughter languages of POC which are now found in PNG, Island Melanesia, Micronesia, and Polynesia, and thereby displaying remarkable stability over thousands of years. And yet, as the discussion above has indicated, various changes have occurred over time to the way in which these early AN immigrants enumerated their world. If, as seems likely, the inhabitants of the POC homeland used numeral classification, this survives today in a relatively small number of languages. Also, the majority of the Oceanic AN languages now have counting systems with a primary 5-cycle, although many of these still have a secondary 10-cycle. For many of these languages in PNG, the change to a 5-cycle system and, in some cases, 2-cycle variants, can be attributed to the influence of NAN languages. However, for some languages, this attribution can be somewhat tenuous and it seems that, under certain conditions, the 5-cycle systems may have developed from the original POC system through the practice of accompanying serial counting with finger tallying, a mechanism which induces change without the requirement of an external stimulus such as the sort of diffusion suggested by Seidenberg. New Guinea and Oceania, once the NAN and AN languages were in place, may be regarded as a virtually closed system: the initial conditions having been set, the situation evolved over time under its own internal dynamics to produce the complex picture that we see today. Within this system we may trace the genealogy of the counting systems and tally methods that exist in the contemporary Indigenous societies of the region and this reconstruction may provide an indication of how enumeration developed in human societies generally.

Genealogies of Enumeration in New Guinea and Oceania

The prehistoric epoch which begins with the introduction of various means of enumeration into New Guinea and Oceania may be divided into two periods. First, that in which the NAN languages arrived and then subsequently spread throughout New Guinea, undergoing speciation beginning about 40 000 years ago. This occurred after the much earlier Australoid migrations through New Guinea and into Australia: the NAN migrations which followed overlaid the remnants of whatever Australoid groups remained in New Guinea. The second and later period begins at about 5 500 BP when AN immigrants reached the western part of West Papua. A further migration to the east brought AN settlers to the POC homeland by about 5 000 BP after which a series of movements out of the homeland resulted in the occupation of the coastal and island regions of PNG and eventually the remainder of Oceania. In PNG and Island Melanesia, this brought about a degree of interaction between AN and NAN language groups. Figures 10.1 and 10.2 summarise the results of the above discussion and they respectively provide outlines of the genealogical development of enumeration during each of these two periods.

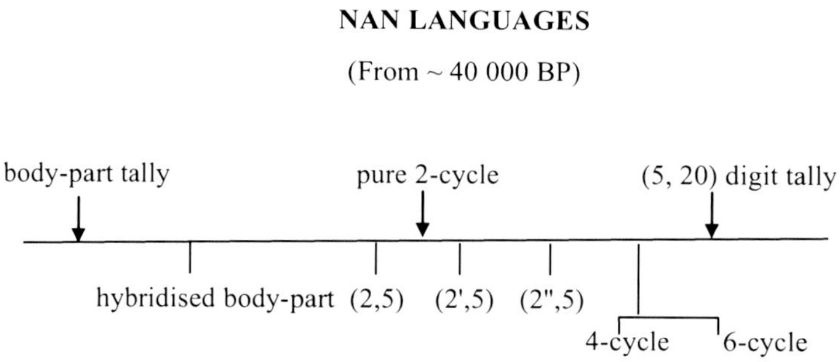

Figure 10.1. Genealogy of the NAN counting systems and tally methods prior to the AN migrations.

The date of 40 000 BP given for the entry of the NAN languages into New Guinea is not meant to indicate the date of origin of the 2-cycle system or the tallies. By 35 000 BP, people had moved through New Guinea across the sea to New Britain and New Ireland, Manus by 21 000, Buka-Bougainville and Solomon Islands by 28 000 BP when most islands were joined (Ross et al., 2003). Similarly, the date of 7 000 BP assigned to the existence of the 10-cycle system in Proto Austronesian is not meant to indicate the date of origin of that system. In both cases the dates provide the maximum time spans for their existence. In Lean's (1992) thesis he conservatively mentioned 10 000 – 15 000 BP for NAN but there is new archaeological evidence that suggests a much longer time period for occupation and hence of some developments. He also noted that the 10 000 – 15 000 BP was possibly the latest dates for the Trans New Guinea Phylum as a later group entering PNG but it is so widespread it is not likely to have replaced earlier ones which are mainly around the coast. In fact, despite the mountainous areas, sometimes very cold during that period, the large upland valleys would be fertile along with the forest for supporting populations, they may have pushed to the coast for warmth but much of the coast may also have been inhospitable as there is much swampland today.

It is worth repeating that the Non-Austronesian languages (about 800) in the region belong to "a dozen genetic stocks and isolates which are on present evidence unrelated to each other or to any languages outside of this region" (Ross et al., 2003, p. 21). Most of the 480 Austronesian languages in the region strikingly belong to one Oceanic group; the exceptions are about 30 in the Bird's Head of New Guinea and two in Micronesia despite the enormous distances across the oceans. Interestingly, while the Austronesian languages in New Guinea remain on the fringes, they dominated the islands of Near Oceania (Island Melanesia) but it seems were the only travellers to Remote Oceania across the Pacific. Distinctive and somewhat revolutionary (there were other pottery sites) Lapita pottery appeared in the Bismarck Archipelago (New Britain, New Ireland) by 3 200 BP or earlier and spread throughout Island Melanesia and the Pacific and onto the mainland with considerable evidence accumulating (Green, 2003). They reached Fiji and Tonga by 3 000 BP as a result of navigational skills and the single-outrigger canoe. The culture appeared to be significant in terms of their canoes, sails, chisels, knives, fishing hooks, and stone adzes as well as domesticated animals but more significantly the cultural organisations associated with language. Thus there was intrusion on the existing communities although there was also integration (through marriages but cultural evidence such as use of earth ovens and foods, for example) and innovation such as the pottery and stone axe and adze styles. With fewer variety and number of animals and plants compared to mainland New Guinea, coastal waters were very important to the Oceanic dwellers and explorers. The sites give evidence of extensive trading across long distances and several variations of the pottery. The results suggest that there is not just diffusion from neighbours but also from trading partners further away, even across the seas. There may be some change as a result of invasion and power or negotiated settlement. Negotiated settlement is not unusual in PNG. Some people on the Manus coast moved to and from the north coast of New Guinea at least 2 000 years ago. On the coast of Manus, there are recent new settlements from 150 years ago (Thorne & Raymond, 1989); near Aitape in Sandaun Province, the village hit by the Tsunamai in 1998 was a migration from West Papua; the Yu Wooi (Mid-Waghi) in Jiwaka have an immigrant clan incorporated with only some variation in language (Muke, 2000) while the arrivals from about the 1940s into a valley in the Chuave area of Simbu have been adopted but currently keep their own language (Susie Daino, personal communication).

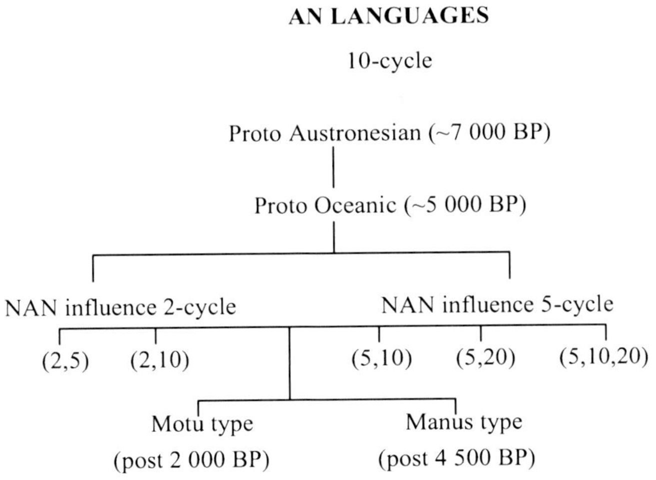

Figure 10.2. Genealogy of the AN counting systems.

The acceptance of multiple language skills, indeed prowess, also suggests that multiple counting systems may have been encouraged in some places. The need to value exchanges in places where 2-cycle systems existed seems to have led to a range of approaches including tallying with body parts or digits. The nature of the language, noun or verb oriented, also seems to have been relevant in the usage of different systems including classifiers.

References

Allen, J. (1984). In search of the Lapita homeland. *Journal of Pacific History, 19*(3–4), 186–201.

Allen, J. (1996). The Pre-Austronesian settlement of Island Melanesia: Implications for Lapita archaeology. *Transactions of the American Philosophical Society, 86*(5), 11–27. doi:10.2307/1006618.

Allen, J., Gosden, C., Jones, R., & White, P. (1988). Pleistocene dates for the human occupation of New Ireland, northern Melanesia. *Nature, 331*, 707–709.

Attenborough, R. (2010). Genes languages and agriculture in New Guinea population prehistory. In J. Bowden, N. Himmelmann, & M. Ross (Eds.), *A journey through Austronesian and Papuan linguistic and cultural space: Papers in honour of Andrew Pawley* (Vol. 615, pp. 59–78). Canberra, Australia: Pacific Linguistics, ANU.

Bellwood, P. (2010). Language families and the history of human migration. In J. Bowden, N. Himmelmann, & M. Ross (Eds.), *A journey through Austronesian and Papuan linguistic and cultural space: Papers in honour of Andrew Pawley* (Vol. 615, pp. 79–94). Canberra, Australia: Pacific Linguistics, ANU.

Blust, R. (1976). Austronesian culture history: Some linguistic inferences and their relation to the archaeological record. *World Archaeology, 8*, 19–43.

Blust, R. (2010). Five patterns of semantic change in Austronesian languages. In J. Bowden, N. Himmelmann, & M. Ross (Eds.), *A journey through Austronesian and Papuan linguistic and cultural space: Papers in honour of Andrew Pawley* (Vol. 615, pp. 525–546). Canberra, Australia: Pacific Linguistics, ANU.

Boelaars, J. (1950). *The linguistic position of south-western New Guinea*. Leiden, The Netherlands: Brill.

Bulmer, S. (1975). Settlement and economy in prehistoric Papua New Guinea: A review of the archaeological evidence. *Journal de la Société des Océanistes, 31*, 7–75.

Butler, M., & Dean, B. (Writers). (2013). *First footprints* (4 episodes). Australia: ABC Broadcasting.

Dahl, O. (1976). Proto Austronesian. In *Scandinavian Institute of Asian Studies, Monograph Series* (Vol. 15). Lund, Sweden: Scandinavian Institute of Asian Studies, Studentlitteratur.

Donahue, M. (2008). Complexities with restricted numeral systems. *Linguistic Typology, 12*, 423–429. doi:10.1515/LITY.2008.043.

Foley, W. (2000). The languages of New Guinea. *Annual Review of Anthropology, 29*, 357.

Foley, W. (2010). Language contact in the New Guinea region. In R. Hickey (Ed.), *The handbook of language contact* (pp. 795–813). Hoboken, NJ: Wiley & Sons.

Ford, A. (2011). Learning the lithic landscape: Using raw material sources to investigate pleistocene colonisation in the Ivane Valley, Papua New Guinea. *Archaeology in Oceania, 46*, 42–53.

Galis, K. (1960). Telsystemen in Nederlands-Nieuw-Guinea. *Nieuw Guinea Studien, 4*(2), 131–150.

Gammage, B. (1998). *The sky travellers: Journeys in New Guinea 1938–1939*. Melbourne, Australia: Miegunyah Press, Melbourne University Press.

Gibbons, A. (2001). The peopling of the Pacific. *Science, 291*(5509), 1735.

Green, R. (1979). Early Lapita art from Polynesia and island Melanesia: Continuities in ceramic, barkcloth, and tattoo decoration. In S. Mead (Ed.), *Exploring the visual art of Oceania: Australia, Melanesia, Micronesia, and Polynesia* (pp. 1–31). Honolulu, HI: University Press of Hawaii.

Green, R. (2003). The Lapita horizon and traditons: Signature for one set of oceanic migrations. In C. Sand (Ed.), *Pacific archaeology: Assessments and prospects. Proceedings of the First International Conference for the 50th anniversary of the first Lapita conference* (Vol. 15). Noumea: Les Cahiers de L'Archelogie en Nouvelle-Caledonie, New Caledonia Museum.

Green, R. (2010). The outer eastern islands of the Solomons: A puzzle for the holistic approach to the anthropology of history. In J. Bowden, N. Himmelmann, & M. Ross (Eds.), *A journey through Austronesian and Papuan linguistic and cultural space: Papers in honour of Andrew Pawley* (Vol. 615, pp. 207–223). Canberra, Australia: Pacific Linguistics, ANU.

Groube, L., Chappell, J., Muke, J., & Price, D. (1986). A 40 000-year-old human occupation site at Huon Peninsula, Papua New Guinea. *Nature, 324*, 453–455.

Holzknecht, S. (2001a). Musom morphology and grammar sketch and Musom word list. In S. Wurm (Ed.), *Pacific linguistics D-89: Materials on languages in danger of disappearing in the Asia-Pacific region No. 1: Some endangered languages of Papua New Guinea: Kaki Ae, Musom, and Aribwatsa* (Vol. D-89, pp. 69–142). Canberra, Australia: Research School of Pacific and Asian Studies, Australian National University.

Holzknecht, S. (2001b). Aribwatsa: A "lost" language of the Markham Family, Papua New Guinea? In S. Wurm (Ed.), *Pacific Linguistics D-89: Materials on languages in danger of disappearing in the Asia-Pacific region No. 1: Some endangered languages of Papua New Guinea: Kaki Ae, Musom, and Aribwatsa* (Vol. D-89, pp. 145–164). Canberra, Australia: Research School of Pacific and Asian Studies, Australian National University.

Kennedy, J. (1983). On the prehistory of western Melanesia. *Australian Archaeology, 16*, 115–122.

Kirch, P. (1982). Advances in Polynesian prehistory: Three decades in review. *Advances in World Archaeology, 1*, 51–97.

Kirch, P. (2000). *On the road of the winds: An archaeological history of the Pacific Islands before European contact*. Berkeley, CA: University of California Press.

Kirk, R. (1982). Linguistic, ecological, and genetic differentiation in New Guinea and the western Pacific. In M. Crawford & J. Mielke (Eds.), *Current developments in anthropological genetics* (Vol. 2, pp. 229–253). New York, NY: Plenum Press.

Laycock, D. (1975). Observations on number systems and semantics. In S. Wurm (Ed.), *New Guinea area languages and language study. Vol. 1: Papuan languages and the New Guinea linguistic scene* (pp. 219–233). Canberra, Australia: Pacific Linguistics.

Lean, G. (1992). *Counting systems of Papua New Guinea and Oceania* (Unpublished PhD thesis). PNG University of Technology, Lae, Papua New Guinea. Retrieved from http://www.uog.ac.pg/glec/

Lincoln, P. (1976). Austronesian languages: Bougainville Province. In S. Wurm (Ed.), *New Guinea area languages and language study, Vol. 2: Austronesian languages* (pp. 419–440). Canberra, Australia: ANU, Pacific Linguistics.

Lincoln, P. (2010). Count not on substrata. In J. Bowden, N. Himmelmann, & M. Ross (Eds.), *A journey through Austronesian and Papuan linguistic and cultural space: Papers in honour of Andrew Pawley* (Vol. 615, pp. 225–244). Canberra, Australia: Pacific Linguistics, ANU.

Lynch, J., & Tryon, D. (1985) Central-Eastern Oceanic: A subgrouping hypothesis. *Austronesian linguistics at the 15th Pacific Science Congress (C-88)*. Canberra, Australia: Pacific Linguistics, ANU.

Moodley, Y., Linz, B., Yamaoka, Y., Windsor, H. M., Breurec, S., Wu, J.-Y., et al. (2009). Peopling of the Pacific from a bacterial perspective. *Science, 323*(5913), 527–530.

Muke, C. (2000). *Ethnomathematics: Mid-Wahgi counting practices in Papua New Guinea* (Unpublished MEd thesis). University of Waikato, Waikato, New Zealand.

Pawley, A., Attenborough, R., Golson, J., & Hide, R. (Eds.). (2005). *Papuan pasts*. Canberra, Australia: Australia: Pacific Linguistics, ANU.

Pawley, A., & Green, R. (1973). Dating the dispersal of the Oceanic languages. *Oceanic Linguistics, 12*, 1–67.

Pawley, A., & Green, R. (1985). The Proto-Oceanic language community. In R. Kirk & E. Szathmary (Eds.), *Out of Asia: Peopling the Americas and the Pacific* (pp. 161–184). Canberra, Australia: Journal of Pacific History, Australian National University

Pawley, A., & Ross, M. (2006). The prehistory of Oceanic languages: Current view. In P. Bellwood, J. Fox, & D. Tyron (Eds.), *The Austronesians: Historic and cultural perspectives*. Canberra, Australia: Pacific Linguistics, ANU.

Reesink, G., Singer, R., & Dunn, M. (2009). Explaining the linguistic diversity of Sahul using population models. *PLoS Biology*, 7(11), e1000241. doi:10.1371/journal.pbio.1000241.

Ross, M. (1988). Proto Oceanic and the Austronesian languages of western Melanesia. *Pacific Linguistics, C-98*.

Ross, M. (1989). Early Oceanic linguistic prehistory. A reassessment. *Journal of Pacific History, 24*(2), 135–149.

Ross, M. (2005). Pronouns as a preliminary diagnostic for grouping Papuan languages. In A. Pawley, R. Attenborough, J. Golson, & R. Hide (Eds.), *Papuan pasts: Cultural, linguistic and biological histories of Papuan-speaking peoples* (pp. 15–65). Canberra, Australia: Pacific Linguistics.

Ross, M., Pawley, A., & Osmond, M. (2003). *The lexicon of Proto Oceanic: The culture and environment of ancestral Oceanic society. 2: The physical environment*. Canberra, Australia: Pacific Linguistics, Australian National University.

Schutler, R. (1978). Radiocarbon dating and Oceanic prehistory. *Archaeology and Physical Anthropology in Oceania, 18*, 122.

Sheppard, P., Walter, R., & Roga, K. (2010). Friends, relatives and enemies: The archaeology and history of interaction among Austronesian and non-Austronesian speakers in the western Solomons. In J. Bowden, N. Himmelmann, & M. Ross (Eds.), *A journey through Austronesian and Papuan linguistic and cultural space: Papers in honour of Andrew Pawley* (Vol. 615, pp. 95–112). Canberra, Australia: Pacific Linguistics, ANU.

Smith, G. (1988). Morobe counting systems. In *Papers in New Guinea Linguistics* (Vol. A-76, pp. 1–132). Canberra, Australia: Pacific Linguistics, Australian National University.

Spriggs, M. (2006). The Lapita culture and Austronesian prehistory in Oceania. In P. Bellwood, J. Fox, & D. Tyron (Eds.), *The Austronesians: The historical and comparative perspectives* (pp. 119–142). Canberra, Australia: ANU Press.

Spriggs, M. (2011). Archaeology and the Austronesian expansion: Where are we now? *Antiquity, 85*(328), 510–528.

Strathern, A. (1971). *The rope of moka: Big-men and ceremonial exchange in Mount Hagen, New Guinea*. Cambridge, UK: University Press.

Swadling, P. (1997). Changing shorelines and cultural orientations in the Sepik-Ramu, Papua New Guinea: Implications for Pacific prehistory. *World Archaeology, 29*(1), 1–14. doi:10.1080/00438243.1997.9980360.

Swadling, P. (2010). The impact of a dynamic environmental past on trade routes and language distributions in the lower-middle Sepik. In J. Bowden, N. Himmelmann, & M. Ross (Eds.), *A journey through Austronesian and Papuan linguistic and cultural space: Papers in honour of Andrew Pawley* (Vol. 615, pp. 141–159). Canberra, Australia: Pacific Linguistics, Australian National University.

Terrell, J., Hart, J., Barut, S., Cellinese, N., Curet, A., Denham, T., et al. (2003). Domesticated landscapes: The subsistence ecology of plant and animal domestication. *Journal of Archaeological Method and Theory, 10*(4), 323–368.

Thomas, N., Abramson, A., Brady, I., Green, R. C., Sahlins, M., Stephenson, R. A., et al. (1989). The force of ethnology: Origins and significance of the Melanesia/Polynesia division [and Comments and Replies]. *Current Anthropology, 30*(1), 27–41.

Thorne, A., & Raymond, R. (1989). *Man on the rim: The peopling of the Pacific. Sydney*. Sydney, Australia: Angus and Robinson.

Tryon, D. (1984). The peopling of the Pacific: A linguistic appraisal. *Journal of Pacific History, 19*(3–4), 147–159.

Wassmann, J., & Dasen, P. (1994). Yupno number system and counting. *Journal of Cross-Cultural Psychology, 25*, 78–94.

Wickler, S. (2001). *The prehistory of Buka: A stepping stone island in the northern Solomons*. Canberra, Australia: Department of Archaeology and Natural History and Centre for Archaeology Research.

Wurm, S. (1975). Possible wider connections of Papuan languages: Torres Strait and North Australia. In S. Wurm (Ed.), *New Guinea area languages and language study. Vol. 1: Papuan languages and the New Guinea linguistic scene (C-38)* (pp. 915–924). Canberra, Australia: Pacific Linguistics.

Wurm, S. (1982). *Papuan languages of Oceania*. Tübingen, Germany: Gunter Narr Verlag.

Chapter 11
Indigenous and Western Knowledge

Patricia Paraide

Abstract The argument of this book is then extended to recognise how Indigenous number knowledge in fact is knowledge important in the global understanding of number. By taking an in-depth study of the Tolai mathematical activities, Paraide argues for the richness of Indigenous number knowledge particularly in context and for cultural preservation. Tolai use of groups of 2s, 4s, 6s, 10s, 12s, and 20s is particularly rich in terms of cultural activities but also school learning if incorporated into the school mathematics. These cultural practices extend beyond the use of the morpheme and word patterns that can reinforce school mathematics learning in the early years and which are ubiquitous opportunities with most Indigenous counting systems.

Keywords Tolai counting and measuring • counting systems with groups of 4s, 6s, and 12s • importance of pairs • representations of large numbers • teachers' perceptions of ethnomathematics and Indigenous knowledge • teacher education for Indigenous societies

Strengthening an Understanding of Number Through Indigenous Knowledge

Ascher (2002) provided considerable detailed study of Indigenous peoples' mathematics, such as the Incas of South America and their remarkable way of organising data structure; the Mayans of South America and their calendar system with a variety of cycles; the Amprym in Vanuatu and their kinship system; and the Iqwaye people in Papua New Guinea (PNG) (Mimica, 1988) and their use of body parts to represent numbers. Ascher clarified the value of mathematics in other Indigenous peoples' lives and how this knowledge is applied in their everyday activities. Ascher also discussed the similarities between those Indigenous people's knowledge for farming, trading, artwork, fishing, weaving, and games, and Western mathematical knowledge. Similarly Jannok Nutti (2007, 2013) in Sweden documented the Sámi mathematics from her experience as a Sámi and her participation in a Sámi school. In this chapter, we outline how a PNG counting system, namely Tinatatuna, spoken by Tolais in East New Britain, PNG, understand number in its cultural context.

Many school children and adults assume that counting is and always has been a base 10 system, especially in the West. A few societies talk of vestiges of the past like the French use of twenty in numbers like 90 *quatre-vingt-dix* (4 x 20 + 10) or the English talk of 12 as a dozen or 20 as a score. One aspect of schooling that is often not well taught or grasped by children is the importance of composite units. The rituals of counting in groups such as 4, 8, 12, 16 is not always associated with the grouping

P. Paraide (✉)
Divine Word University, Madang, Papua New Guinea
e-mail: pparaide@dwu.ac.pg

K. Owens et al., *History of Number*, History of Mathematics Education,
DOI 10.1007/978-3-319-45483-2_11

of four although some curriculum such as the Singaporean and Japanese curriculum encourage the use of rectangles or boxes to encourage the splitting and re-grouping of numbers and for composite units. However, recognition of the counting systems of the world that use five as a group helps with not only the notion of 10 as a composite unit in the base 10 place value system but also for multiplication. This goes beyond the use of five as a stepping stone to 10 as in some Asian languages for splitting numbers before remaking them. Furthermore, recognition of groups of 4s, 6s and 20s in existing languages or of alternate emphases in the relationships of numbers can provide a stronger mathematical concept of counting than is currently available to school children around the world.

Transitions in Education on Number for Indigenous Children

There are other reasons for Indigenous children who recognise they have a family culture and language to know and use their counting systems. First, it assists in the valuing and maintenance of culture and the child's identity. Second, the child may be fluent in the home language and hence learn the mathematical concepts better than in a new relatively unfamiliar language. Third, many of the languages include some arithmetic permitting the child learning bilingually to learn the concepts and facts of arithmetic more easily. For example, if the language has 7 as a numeral plus 2, it can be associated with 5 + 2, and then this number fact will be more quickly learnt by the child. Furthermore, the child will see the pattern for other numbers up to 9. For example, in Tinatatuna spoken by the Tolai of East New Britain (Table 11.1) addition patterns for numbers between 5 and 10 and for two digit numbers is evident as well as the multiplication structure for tens and hundreds. Thus local languages can provide key features of numeracy systems.

Table 11.1
Tinatatuna Number Structure

Number	Tinatatuna	Meaning, Pattern
7	*lavurua*	$x+2$
8	*lavutul*	$x+3$
9	*lavuvat*	$x+4$
and larger numbers also have patterns		
11	*a vinun ma tikai*	10 and 1
15	*a vinun ma a ilima*	10 and 5
20	*a ura vinun*	2 tens
30	*a utul a vinun*	3 tens
50	*a ilima na vinun*	5 tens
100	*a mar*	
200	*a ura mar*	2 hundreds
1 000	*arip*	
2 000	*a ura arip*	2 thousands

Introduction to Paraide's Study

The main purpose of Paraide's (2010) study was to understand:

1. In what ways teachers integrate Indigenous and Western mathematical knowledge in formal learning environments?
2. In what ways parents and students integrate Indigenous and Western mathematical knowledge in the home learning environments?

3. How teachers, students, and parents view Tinatatuna (vernacular language which is also known as Kuanua and Tolai) as the formal language of instruction within the bilingual or vernacular education policy together with their views on the integration of Indigenous and Western mathematical knowledge in the formal learning environments?

In addressing the three major research questions, Paraide further explored

- How is Indigenous number and measurement knowledge positioned and/or constructed, compared to Western number and measurement?
- How do such positioning or constructed practices affect the implementation processes?
- How can the current constructed positions be reconstructed?

To do this Paraide recorded and analysed her own Tolai number and measurement knowledge in a cultural context as a result of her own upbringing, participating in the everyday cultural activities, and discussions with family especially her mother.

Methodology

Paraide conducted a two-year case study in one of the Tolai people's communities, namely her own, whose dialect of the language Tinatatuna she speaks well. This assisted her to understand her people's attitudes and practices regarding the implementation of the education reform initiatives, the resistance to vernacular and bilingual instruction, and the integration of Indigenous and Western knowledge in formal schooling. As Reinharz (1983) stated:

> 'case study' refers to research that focuses on a single case or single issue, in contrast with studies that seek generalisations through comparative analysis or the compilation of large number of instances. (p. 164)

However Paraide was able to compare and contrast issues relating to the findings of this case study with other data collected in some of her other studies. This enabled some degree of generalisation about educational reforms in PNG but all PNG communities differ in many aspects especially language and culture. Case studies can also be used to evaluate theories in practice in the real world. In her study, the theory that students learn better in their first language in the early years of schooling was examined.

In addition to generating and evaluating theory, Reinharz discussed three other major purposes of case studies. These are to:

- analyse the change in a phenomenon, over time;
- analyse the significance of a phenomenon for future events; and
- analyse the relations among parts of a phenomenon (Reinharz, 1983, p. 164)

Paraide's sources of data were classroom and community observations, artefacts of classroom work, interviews with teachers, community members, and parents, and focus group discussions with students, field notes, informal discussions, participation in the research participants' community activities, and her familiarity with her own Indigenous number and measurement knowledge. The qualitative method of research is usually used to capture attitudes that govern practices in particular social settings. Henn, Weinstein, & Foard (2006) explained:

> Critical social researchers argue that the purpose of social research ought to be to uncover the fundamental nature of social reality by revealing these underlying mechanisms and structures with which capitalism has successfully persuaded the vast majority of people to act (mistakenly) against their own best interests. (p. 298)

Paraide focused on understanding the meanings and values of their particular practices and the representations of themselves as teachers, students, and parents. She further explored the ways in which their gender, cultural group, and social class can shape their identities, and the ways in which their identities are enacted through what they do in the classroom and home. Her study also intended to provide some insights into the cultural divide and power struggles that exists in current educational practices, especially the teaching strategies that are used when implementing the elementary and lower primary mathematics syllabi, and the integration of Indigenous and Western number and measurement knowledge in formal education. Currently, teaching strategies used at these levels of education have a strong Western linear organisation and rote repetition of number facts, which are not effective in teaching the integration of Indigenous and Western knowledge in some cases together with a considerable teacher-talk approach. The application of the qualitative method of research captured an in-depth insight into this issue. An elementary school and lower primary school were selected for the study because much of the Tinatatuna number and measurement knowledge is still used in the community and the dialect was known by Paraide. The teachers and all the parents gave their overwhelming support, and wanted to participate in the study.

The formal mathematics curriculum prescribed cultural mathematics for elementary level and the integration of Indigenous and Western knowledge in both the elementary (Grades Pre-Elementary – 2) and lower primary level (Grades 3 – 5). The students in this community speak Tinatatuna, which was also the language of instruction in the elementary school and bilingual instruction in the lower primary school site. All the lower primary teachers speak Tinatatuna. The participation of these two school sites allowed Paraide to pose provocative questions regarding the implementation of vernacular and bilingual instruction and the integration of Indigenous and Western mathematical knowledge.

One elementary Grade 2 class was the focus of the study. Within this class, ten students were selected for the study because this enabled close interaction with them. It allowed for easy tracking into the next year of school and a closer focus on their understandings of number and measurement and how this knowledge is applied in their home and formal learning environments. The following criteria were used to select the ten students — five females and five males:

- they speak Tinatatuna as a first language;
- both parents are first language speakers of Tinatatuna;
- they live in the Tinatatuna speaking community around the school; and
- would enrol in the lower primary school near the elementary school site the following year.

The Grade 2 elementary and Grade 3 primary school teachers were interviewed (and audiotaped), four mathematics lessons were observed in each classroom, and four student focus group discussions were conducted during the study. Ten parents were also interviewed during the study. The interviews and focus group discussions were audiotaped, transcribed and later analysed. Field notes were also taken regarding the environments of the schools, actual practices in schools, and other teachers', parents', and students' attitudes towards the education reform initiatives and normal practices in the school and home environments. Interviews with parents and other adults were in Tinatatuna. Tinatatuna, Tok Pisin and English were used interchangeably during interviews and discussions with the lower primary teachers. Tinatatuna was used in discussions and other communication with the student participants. All of the interviews, lesson observation notes, and recorded transcripts were translated from the Tinatatuna and Tok Pisin languages to English.

The ten students, teachers, and parents were interviewed in order to:

- gauge their views and general attitudes on how vernacular instruction is used;
- determine whether they know of, or acknowledge, the similarities in the Indigenous and Western mathematical knowledge;

- assess the current strength of the integration of Indigenous and Western mathematical knowledge in formal and informal teaching and learning situations; and
- try to understand any influences that may affect the implementation of the language policy and mathematics curricula in elementary and lower primary levels of education.

In 2004, two two-week and one one-week visits (five weeks) were made to the elementary school site. At this school, the three teachers and most of the students in this school spoke Tinatatuna fluently, so all teaching was done in Tinatatuna. A few students were from the cocoa and coconut plantations around the school and were fluent in Tok Pisin. The teachers occasionally used Tok Pisin to assist these students to understand the lessons. Intriguingly, these students understood well the lessons taught in Tinatatuna and gave correct answers in Tok Pisin when they were asked to do so. Tok Pisin was rarely used during the time with the students in and out of the classroom.

The case study continued in 2005 to explore how the Grade 3 teachers in the lower primary school coped with the elementary graduates, their attitudes towards bilingual instruction, and the integration of Indigenous and Western mathematical knowledge at the lower primary level. The lower primary school site had eleven classes — four classes in Grades 3 and 4, and three in Grade 5. The total number of teaching staff was eleven for a school population of 240 students coming from three elementary schools. It was government policy that bilingual teaching be implemented in all grades at this level of education (Department of Education Papua New Guinea, 1998, 1999a, 1999b, 2000, 2004). At the school site, Tinatatuna and English were supposed to be used in lower primary as the languages of instruction. Two two-week and one one-week visits (five weeks) were also made to this school during the second year of the study. At other times during the two years, Paraide was able to visit engendering greater social interaction and trust. At the teachers' request she also took a bilingual lesson within an inservice on bilingual education.

The researcher focused on:

- the language used to teach number and measurement in the classroom;
- teaching methods used to teach number and measurement lessons;
- use of Indigenous number and measurement knowledge as a stepping stone when teaching;
- how Indigenous number and measurement knowledge is integrated in formal and informal teaching and learning situations;
- how students, teachers, and parents view vernacular instruction, Indigenous knowledge versus Western knowledge;
- any power play that existed in the various classroom situations and home environment; and
- teachers' views regarding the language of instruction.

As Bordo (1999) and Henn et al. (2006) discussed about interacting with research participants, Paraide's interaction with the research participants was of considerable assistance in the data analysis. Furthermore, the interactions provided an in-depth understanding of the use of mathematical skills and knowledge, inside and outside the classroom, and how these skills are linked to formal mathematical learning. Paraide was a participant during community activities, such as death and bride-price ceremonies in the participants' home environments, and in staff development. She was a non-participant during lesson observations and some community activities. As Bordo (1999) and Henn et al. (2006) stress, using several methods for data collection enables triangulation and validation, and helps establish the reliability of the data. The classroom observations enriched the data collected on issues being addressed in the study.

During the study, Paraide was always conscious that what she observed in the classroom was influenced to a certain extent by how she positioned herself in relation to the issues being discussed, by her understanding of the PNG reform curriculum, and by how she valued mathematics. As a

Papua New Guinean who is a mathematics and language specialist, she may have observed and noted different aspects in this lesson from an outside researcher or non-specialist Papua New Guinean (Scheurich, 1997, p. 31).

Semi-structured interviews were used primarily because this technique allows for the "core issues to be covered, while at the same time, leaving the sequence and the relevance of the interview free to vary, around and out from that core" (Freebody, 2003, p. 133). At first, if Paraide noted that interviewees were giving answers they expected she wanted, but she was able to ask further questions to elicit other parts of the "story". She listened to the stories and avoided taboo subjects. She noted that this type of interaction with the participants introduced an informal flavour into the interview and strengthened her working relationship with the participants. Consequently, a comfortable sharing of information between the participants and herself developed during the interview sessions. She was able to follow particular lines of discussion, using ad hoc follow-up questions, and rephrasing if participants were unsure of the question (Paraide, 2010). Selected sections which showed some kind of pattern were analysed in-depth.

Artefacts were also collected during the study to supplement the data that were collected from interviews and classroom observations. Such artefacts included the students' mathematical exercises to provide an in-depth insight into issues that were encountered during the observation and interview sessions. The teachers collected artefacts during the study and gave them to the researcher. Teachers' involvement enhanced their active participation in the actual evaluation of their students' mathematical learning during the study. For example, if the students' work showed that they can add up one-digit and two-digit numbers, then these data strengthened the view that the students have understood addition concepts and can now apply them to more advanced addition concepts. Alternatively, if the students' work showed continual mistakes in simple and advanced addition, this indicated that they have not really understood the addition concept, and cannot apply it to other situations. Such information formed the basis of discussions concerning the students' learning of mathematical concepts with the selected students and their teachers. These data also captured the students' activities that were completed in the absence of the researcher. Also, these data showed how the teachers gave feedback on students' strengths and weaknesses during their learning.

Results on Tolai Understanding of Number and Measurement

Paraide's (2010) study found that Tolai number and measurement knowledge are still used in everyday practical activities. For example, they are used when making gardens, building canoes and houses, trading, fishing, navigating canoe trips, studying the weather, gathering coconuts, bananas, eggs, and taro, processing *tabu*, and other activities. *Tabu* (shell-money) is a currency still used by the Tolais for trading, bride-price, fines, school fees (elementary schools) and labour payments.

Smith (1981) found that tools were available in foundational societies for conducting the business of gardening, trading, feasting, and wealth exchanges. Paraide's study found that the Tolais also had traditional tools which they used for conducting business in their communities, before outside influences arrived. According to Paraide, Elders had counting and measuring systems that were used in gardening, trading, feasting, and wealth exchanges, and which are still used today. For example, Tolais use groups of four, five, six, ten, 12, 20 and 40 when counting specific items, primarily for recording and trading purposes. The Tolais also use adult arms and steps to measure length of larger values of *tabu* (shell money), mark out appropriate spacing between various crops, and measure the length and width of new houses and canoes to be built.

In many PNG and Melanesian societies, status was achieved by the distribution of wealth. Often, these distributions implied a reciprocal obligation on the part of the recipients. Hence, it was important to know how much was received (Smith, 1981). Tolais' distribution of wealth shows clan or family wealth status. Contributions to feasts from others, who assist to make displays impressive, are counted and recorded for the purpose of the favour being reciprocated and also to secure continuous social support.

Smith (1981) noted that counting the number of items which were distributed is not as important as other forms of measurement. Impressive public visual displays are also very important when measuring the quantity and quality of items in ceremonies. The Tolais execute displays during death ceremonies. On these occasions, the rich and powerful Tolai clans display huge *loloi* (rings) (see Figure 3.11n) of *tabu* (shell money) on erected bamboo structures. A lot of thick *loloi* measure the strength and wealth of a clans or families. The size and value of the *loloi na tabu* for such ceremonies vary in sizes. The smallest *loloi* usually is valued at one hundred fathoms of *tabu*, and the thickest and heaviest are usually valued at five hundred fathoms of *tabu*. Generally, the *loloi na tabu* that are displayed during the ceremonies are not distributed. They are displayed purposely to advertise the clans' power, strength, and wealth. Other *tabu* is distributed to the people. This display sets apart the rich and powerful people from the others in this particular society. The rich are labelled the *uviana*, and those who own less *tabu* as *malari*.

In addition to the *tabu* display, bananas and pigs are displayed in a certain way. A specific species of banana is used to build up the base of a pile. A variety of other specific types of bananas are piled on top, and lastly the pigs' carcasses. The wider the base of the pile, the taller it is, and the bigger the pigs, the more valuable.

A further addition to all this food and *tabu* are dances that are performed during death ceremonies. Sometimes, the *tubuan* and *dukduk* (usually the clan's) perform as well. Skills on headdress creation, application of body paints and skills in dancing are assessed during performances. This is also when the skilled people are set apart from the less-skilled in these particular areas. The skilled people secure commissioned work for the creation of headdresses, body paints, the choreography of new dances, and teaching dances for future ceremonies. They are paid in *tabu* for their expertise. In present-day society, modern currency is also used for payments. The choreographers and individual dancers and *kundu* drum beaters are also paid in *tabu* when they perform.

During ceremonies the sponsors of the feast distribute *tabu* and uncooked food to the people. All contributions from individuals, towards the feast, are itemised, counted, and recorded because the recipients have to reciprocate such contributions with the same quantity or more. Any contributions to marriage, death, and other activities are always recorded. For recording purposes, bananas are counted in sets of fours (*ainangava*), taros in sets of sixes (*kurene*), coconuts in twelves *(tanguvani)*, and *tabu* for distribution in tens of fathoms (*arivu*). For pigs and the introduced items, the bales, boxes, or individual items are counted individually.

Smith (1981) also noted that other PNG Indigenous peoples used special terms for small groups of objects, for example, three bunches of bananas tied together, or groups of four or five clay pots. This could facilitate the recording of small numbers of items that are given or received. This phenomenon is widespread in PNG, and frequently the counting of different objects proceeds in different ways. It is common for taro, fish, or coconut, to be counted in named groups of four or five.

Often, a certain number in these groups constitutes a larger named unit. As Smith (1981) found the Siassi people in the Morobe Province group count taro in a specific way. A unit of five taros is known as *ndir*, and 20 such groups are known as *ndingnding*. A larger unit, also known as *ndingnding*, is made up of 20 smaller *ndingnding*, and is considered to be a sufficient quantity for a feast.

The Tolais also group taro and count them in a similar way. However, the Tolais group taro in sixes and call one set of six, *a kurene na pa* or *tikana kurene na pa*, two sets of six are called *a ura kurene na pa*, three sets are called *a utula kurene na pa*, and so on. *Pa* is taro. Ten sets of sixes then become *a/tikana pakaruati na kurene pa*. In Tolai communities, items were, and still are, grouped and counted for feasting and trading purposes.

Smith (1981) found that counting was an important activity in trading among Indigenous PNG. He noted that, apart from exchanges for purposes of gaining status, there was also a good deal of trading between the groups that were connected by trading ties. There were also trade routes between coastal and inland people. A variety of goods were exchanged, and as Harding (1987) suggested, there may have been problems with establishing equivalent values for different commodities. As there was no accepted unit of currency, and given the many different types of commodities, a lot of ratios would have to be established if accurate exchange values were to be maintained.

Trading did proceed without these exact equivalences, although there were undoubtedly rough agreements as to the relative value of different items (Smith, 1981). Often, exchange partners were regarded as special kinds of relatives, and exchanges would be accompanied by hospitality and would be on more favourable terms than with strangers. The object of trade was not to make a profit, but to meet certain specific needs and maintain social relationships, so strict accounting procedures were not necessary.

According to Paraide's Elders, the Tolais were also traders. They traded with inland people, and neighbouring places such as West New Britain, the Duke of York Islands, and New Ireland. Brown (1908) and Mennis (1972) recorded these trading routes. The recounts of Paraide's Elders, along with the accounts from Brown and Mennis, show that the Tolais did barter trading with each other and also with non-Tolais. Trading transactions among the Tolais were in *tabu*. Some of the main reasons for the Tolais' sea voyages to other neighbouring lands were to search for more *tabu* shells and other items, and the general exploration for new trading partners and land.

Smith (1981) also found that some of the Austronesian languages of the coastal area around the Huon Gulf and Vitiaz Strait in Morobe Province had counting systems which appeared to be better suited to counting and ascribing quantity than hand and leg tallies. He noted that the Mutu language—an Austronesian language—has words for 10 and 20, and the terms 30 and 40 can easily be formed when counting large quantity of goods for feasting or trading. He noted that Austronesian languages have a very widespread distribution throughout the Pacific and Asia, and there are common links between number words in some Morobean Austronesian languages and those of other Pacific areas.

Tinatatuna has words for the numbers one to a thousand. Words for more than a thousand can easily be formed. For example, *vinunu/vinun* is used for tens, *mari/mar* for hundreds, and *arivu/arip* for thousands, with markers like *tikai*, and *aurua autul* to mark the exact figures. For example, *aura vinun* (20), *aura mari/mar* (200), *aura arivu/arip* (2 000), *aura vinun ma urua* (22), *aura mar/mari ma urua* (202), and *aura arip/arivu ma urua* (2 002).

Linda Smith (2005) noted that Indigenous knowledge was claimed, recorded as new discoveries, and interpreted from Western and other perspectives:

> Indigenous Asian, American, Pacific, and African forms of knowledge, systems of classification, technologies, and codes of social life, which began to be recorded in some detail by the Seventeenth Century, were regarded as 'new discoveries' by Western science. These discoveries were commodified as property belonging to the cultural archive and body of knowledge of the West. (p. 61)

Indigenous contributions in their own knowledge, and how they were explained and used were generally ignored or regarded as not 'scientific'. Much of this data also suggested a demeaning of the intelligence, organisation, and creativity of these Indigenous knowledge owners.

Comparing Tolai and Western Number and Measurement Knowledge in Tolai Activities

The following excerpt from an interview with an Elder shows how these areas of knowledge are applied and the superficial similarities between Indigenous and Western knowledge. The interview reveals how number and measurement knowledge is applied in everyday activities:

Elder: The coconut tree has a lot of uses- all parts of the coconuts is used by our people. The flesh of both the young and dried nuts are used for cooking. The leaves are used for mats, fans, baskets, brooms, and walling. The matured trunks for posts, and most importantly, we use all the parts of coconuts for fuel. We use them in barter trading with those who do not have them in exchange for other things. It is now also one of our important cash crops. We count the nuts as we gather them. We always start with two, then four, then one set of six, and two sets

> of six, which becomes one set of twelve. Then we count in twelves and then in sets of ten for recording purposes. We call the sets of ten twelves- *a pakaruati na tanguvani*. Now we have wheel barrows and trucks to transport them to our homes or copra drier so we just heap them into vehicles without counting them. But we still count like this, if we have to carry them on our backs or shoulders or head.

Figure 11.1 illustrates how the coconuts are tied and counted during the gathering process, as explained by the above quote from Paraide's Elder. This interview and the photographs show that this type of counting is still used in the research participants' lives because coconuts are still an important crop. Some writers have claimed that, in this particular case that the Tolais are counting in base six and twelve. However, this does not explain the counting in twos, fours and tens that is also applied simultaneously during the coconut gathering process. They do not fall neatly into any Western base counting. Paraide's observation is supported by Wolfers (1971) who argued that, although the mathematical distinction between a modulus and a base is quite clear-cut, it is not always equally clear just how particular counting systems operate.

a: *a evutu* (two)   b: *a varivi* (four)   c: *a kurene* (six)

d: *a kurene ma evutu* (six plus two)   e: *a kurene ma varivi* (six plus four)

f: *a/tikana tanguvani/nanguvani* (one set of twelve)   g: *a/tikana tanguvani/nanguvani* (one set of twelve) stacked

Figure 11.1. Groupings of coconuts related to number understanding.

When gathering coconuts, Tolais count them, knowing exactly how to apply this entire group counting. They do not even stop to analyse or even explain precisely how they do it. If the Tolai villagers were asked by a pure mathematician or an "outsider" what base counting is applied in this counting process, they may not be able to identify them. This counting comes naturally for them. Whatever base is used to count in this particular activity is not important. However, what is important is that, the coconuts are recorded and organised for easy manual transportation to the home or to the copra drier. The coconuts are usually organised, so that, an average healthy adult can easily carry four sets of twelve coconuts. Some may ask, why coconuts are always tied in twos. The answer is very simple. It is easier, for a younger child to carry two coconuts in one hand, or an older child to carry two groups of two in each hand, during transportation to a destination. An adult, can carry six coconuts in each hand easily, when the nuts are woven into each other. It is also easier to weave them into sixes, and twelves, when they are tied in twos, as shown in Figure 11.1 f and g.

The Elder's interview and the photographs show that the Tolais apply multiples of 2, 4, 6, 12 and 10 concepts simultaneously when gathering coconuts. This shows the unique application of this particular counting system. This also illustrates that multiplication, division, addition, and subtraction concepts are applied in this activity. As Smith (2005) stated:

> One of the supposed characteristics of primitive peoples was that we could not use our minds or intellects. We could not even invent things, we could not produce anything of value, we did not know how to use the land and other resources from the natural world, and we did not practise 'the arts' of civilisation. By lacking such virtues, we disqualified ourselves, not just from civilisation, but from humanity itself. (p. 25)

From these data, it can be suggested that the Tolai number knowledge is just as valuable as the Western counting system and, thus it is a vital component in Tolai children's education. Furthermore, this suggests that, with better teacher training and professional support in the preparation for the integration of Indigenous and Western mathematical knowledge in formal learning environments, many more Tolais can achieve equivalence with Western mathematical capabilities. The Tolais and other PNG Indigenous peoples have intellectual capacities and capabilities which can equal those of the Western population. The integration of Indigenous and Western addition, subtraction, multiplication, and division concepts is possible in the formal teaching environment.

According to Souviney (1981) and Saxe (1981), the question of the influence of the traditional counting system on cognitive development and mathematics achievement remained unresolved. However, later Saxe (2012) provided a model integrating how change occurred as social and personal ontologies interacted through micro changes. His study of the Oksapmin using a methodology of interviews at specific points in time with different age/education-experience groups illustrated and explained the model. Paraide's (2010) study found that with the current integration of Indigenous and Western number learning in schools, the cognitive development and mathematical achievement issue is beginning to be resolved. For example, the lesson observation data in this study showed that, by the time the research participants were in Grade 3, they were able to count competently in groups of two, three, four, five, six, seven, eight, nine, ten, and twelve. They were able to apply this knowledge to more complex problems. This was particularly evident when the students were given multiplication problems to solve during a bilingual teaching strategy demonstration.

Paraide was requested by the lower primary teachers to conduct this demonstration. The demonstration participants used stones and sticks to solve the multiplication problems in this particular exercise. The solving process applied by the students to reach the answer to 12 x 14, for example, intrigued the teachers. The students had already worked out the answer for 12 x 12 = 144. To get the answer for 12 x 14, they simply added two more sets of twelve to the 12 x 12 that they had grouped earlier, to reach the answer 168. This astounded the lower primary teachers, because the students were not just memorising the times table, but they were actually thinking, and applying counting in sets, to a slightly more advanced problem.

Paraide noted that the twelve times table chart in Rose's classroom ended at 12 x 12 = 144. The English word "multiplication" was not used at all during this demonstration. The Tinatatuna words used were *tinatogo* (the name for sorting-noun), *toga* (sort/group-verb), and *totogo* (description of the activity). The students were asked to; *totogo avarivarivi* "sort in groups of fours", *totogo a kurekirene* "sort in groups of sixes", *tatogo a pakapakaruat(i)* "sort in groups of tens", and *tatogo a tangutanguavan(i)/nagunaguvan(i)* "sort in groups of twelves".

In another multiplication problem, one of the female student participants disputed the answer that Paraide had written on the board for 9 x 6. Paraide wrote the answer as 56. She pointed out that the answer should be 54 and demonstrated this to her. The other students supported her claim. Her courage to dispute Paraide's answer also astounded the teachers. It is not normal for students to dispute their teachers' answers in this particular community. The teachers are generally viewed as "experts". Furthermore, it is not normal for female students to dispute authority, even if errors are noted. Her courage to speak out may have been the result of her being used to Paraide in the formal learning environment that had activities for children to learn from through investigation. Thus the girl had more ownership of her mathematics. This incident illustrated that she was actively processing the problems with which she was presented. She was using her intellectual capacity to solve a mathematical problem.

The data also showed that the students in this particular group were not working just from memory. They were applying their counting-in-sets knowledge to work out advanced multiplication problems and test their answers. Paraide built on the students' existing Indigenous number knowledge during this demonstration. The teachers viewed this teaching strategy as strengthening the students' understanding of basic multiplication concepts. These particular students already knew the practical application of counting in groups.

During discussions after the demonstration, the teachers seemed convinced that using Indigenous number knowledge as a stepping stone when introducing similar or new Western number knowledge during formal learning had value. It appeared that some of the teachers were now prepared to place Indigenous and Western number knowledge on the same level. Their established view that Western number knowledge was superior to Indigenous knowledge was beginning to "crack" during this discussion. Such a view is the consequence of the exclusion of Indigenous knowledge in formal learning during the colonial era and after Independence.

The exclusion of Indigenous knowledge in the young peoples' education was a conscious strategy used by the colonial powers to wipe out Indigenous knowledge as a form of discipline for "savages and uncivilised peoples":

> The most obvious forms of discipline were through exclusion, marginalisation, and denial. Indigenous ways of knowing were excluded and marginalised. (L. T. T. R. Smith, 2005, p. 68)

The demonstration gave rise to the teachers' acknowledgement, that students can understand formal number concepts better, and advance faster in their number learning, when their Indigenous knowledge and language are used as 'stepping stones' while teaching similar or new knowledge in formal learning environments.

These findings support what learning experts (e.g., Bialystock, 1991; Collier & Thomas, 2007; Garcia, 1994; Pica, 2003; Thomas & Collier, 2003) claim, namely that students have learning difficulties when they learn content in a language that they do not understand well, but that they learn well and faster when they learn in a language in which they are already competent, and can relate to the content they are learning. The evidence from this study indicates that mathematical knowledge is embedded in PNG languages. Therefore, the integration of Indigenous and Western number knowledge is possible, and can assist students to advance faster in their learning.

However, caution must prevail during the integration of Indigenous and Western knowledge (Battiste, 2002), and in this case, number and measurement knowledge. Elementary and lower primary teachers must understand how the students in various cultures view their world, how they use number and measurement knowledge in their particular communities, and the values, meaning, and importance of such knowledge in their lives. For example, according to Thune (1978), the Loboda people from Milne Bay Province of PNG have a different way of seeing their world to others. In place of an abstractly oriented numerical, or mathematical system, Loboda people make use of a whole series of other ways of organising the world which, by their very nature, preclude the use and expansion of the counting system that is present.

Thune (1978) argued that this indicated that teaching Lobada children to use, in this case, Western mathematical systems is not simply a question of the substitution, through the direct translation of English numbers for Loboda numbers. Rather, it is about introducing a different way of thinking and organising the world. In a sense, it is a problem of introducing the concept of number, and encouraging the view of the world to be compatible with it. Thune suggested that, this may be accomplished without destroying the Loboda worldview with all the values, meaning, and importance that it has for the Loboda people.

Thune (1978) further stressed that, in many systems, counting is carried out using what are essentially abstract, and more importantly, absolute categories and units of measure, which exist independently of the things which they measure. In contrast, for Loboda people, categories and measures of counting tend to be taken as relative, as the size, length, or quantity of one object or collection of objects is compared to another. For example (see Chapter 8 and Appendix B for more details, under Dobu), Thune (1978) noted:

> stating that a necklace is fifteen inches long in contrast does not relate it to any other length, but only to an absolute standard. Through always thinking in terms of relative (but uncountable) measures, Loboda people assure that the unit of counting or measure remains intimately linked to that which is counted. The two in effect (and frequently in fact) lie side by side as counting or measurement takes place, the one informing the other of its quantity. In other words, Loboda people tend to be more interested in the relative differences between two things or collections of things than in their absolute size. Hence, a measurement or indication of quantity, size, or age in Loboda is not made through the use of an independent "sign"; that is, through using an abstraction disassociates from the thing to which it refers, or measures, or counts. Rather, measurement is performed using as a scale or measure an "index" which is tied to the thing which it counts or describes, rather than being abstract and free from it. (p. 73)

Other groups of people in PNG, including the Tolais, apply similar principles to measure quantity and length. Links are made to what is available that can be related to and known by the listener, and not to abstract units of counting or measurement.

In formal mathematical learning, students generally are learning the abstract representation of number and measurement which can be difficult to comprehend when they are unable to connect the basic concepts with familiar things and activities for clarification. For example, a study by De Abreu, Cline, and Shamsi (2002) concerning Pakistani migrants and white students enrolled in primary schools in England found that the students usually do not connect formal mathematical learning with the mathematics which they come with, or use, in the home environment. Therefore, they have difficulties understanding similar basic concepts in formal learning situations. Consequently, students do poorly in formal mathematics learning. However, these migrant students were able to apply the same concepts competently in the practical activities in which they participated, in their home environments. However, they did not seem to know or realise that they were applying mathematical knowledge. These students sold newspapers, traded at markets, and assisted in their families' business bookkeeping, yet they did not link these activities with formal mathematics. They tended to view formal mathematics or classroom mathematics as different to the mathematics that they use in their home environments.

This finding is similar to those in Paraide's study concerning Tolai students' learning of formal mathematics. Even though the Tolai culture has mathematical concepts which have similarities to Western mathematics, Tolai students could perform better in national examinations. Only a few perform well, yet the Elder and parent interviews and lesson observation data show that Tolai students in this particular study apply similar concepts competently in their everyday activities back in their communities. This suggests that the Tolai students may also have difficulties in linking the mathematics they apply in their home environments to the formal mathematics learned in the formal learning environments. This dilemma may continue, if they are not made aware of the similarities, as Jude[1] did during his number lesson, where he used the counting of *tip na laptikai* to stress the similarities between counting this value of *tabu* with the formal x 12 multiplication concept. Teachers have to be made aware of similarities in the mathematics that is present in the two knowledge systems at all levels of education. Indigenous mathematical knowledge is actually used in the peoples' lives and teachers must be trained to link this to the type of mathematics taught in the formal learning environment as the two knowledge systems can be used to complement each other. In a conference in 2008, a speaker for one of the secondary schools in East New Britain claimed that the students understand formal mathematics better because they are linked to the carpentry, trading and other practical subjects taught in the school. It was claimed that the students' academic performances in mathematics and sciences had improved considerably since the introduction of this teaching strategy. This school caters for Grades 9 to 12 classes. This teaching strategy can be used to link Indigenous and Western mathematical knowledge in formal education.

Diminishing Indigenous Number Knowledge

Some of the vital Indigenous knowledge used by the Tolais has been lost primarily as a consequence of students being taken away from their home learning environments to learn Western knowledge in introduced learning institutions in English. L. T. T. R. Smith (2005) referred to Fanon's (1952) and later writers such as Nandy's (1983) discussions which suggested that imperialism and colonialism brought complete disorder to colonised peoples, disconnecting them from their histories, their landscapes, their languages, their social relations and their own ways of thinking, feeling and interacting with the world.

Paraide found that, as a consequence of the "push" for economic development and progress, and the dominant use of English and Western knowledge, some types of counting are no longer used. For example, the type of counting discussed by her Elder is not used when easy transportation of coconuts is readily available in certain areas of the Tolai communities. Consequently, this unique counting of coconuts is now diminishing in such areas because the lifestyle and economic environments have changed. The Tolais in such communities do not see the value and logic of using this particular counting system anymore.

These data suggest that this could be the beginning of the loss of such Indigenous knowledge because other counting systems now dominate current counting activities. In the case of coconuts, copra bags are counted individually, not in sets of twelve. With the continuous use of individual accounting, the Tolai method of counting coconuts could eventually be forgotten by the younger generations.

Smith (1981) found a similar trend when he studied the loss of mathematical knowledge as a result of influences from other non-Austronesian languages. He noted that there are no instances in Morobe Province of complete decimal systems with separate words for one to ten, as found in Austronesian languages in many other areas. It seems that such a decimal system has been lost as a result of the influence of neighbouring non-Austronesian languages. An extreme example is the Roinji language, an Austronesian language of the North Coast of the Huon Peninsula, which has numeral words only for "one" and "two". The Adzera family languages—Austronesian languages spoken further inland than is usual—also appear to have been modified in this respect (see Chapters 3 and 8).

[1] Pseudonyms used throughout the discussion.

Integration of Tolai and Other PNG Indigenous Peoples' Measurement Knowledge

Contact with others influenced the way our people designed and built houses after first contact with outsiders. Figure 11.2 shows photographs taken by early missionaries and show changes in Tolai houses. The two photographs were taken 27 years apart in a village which is about a 30-minute drive from the study site. The first was taken by Rev George Brown and the second by Rev. R. H. Rickard (Gash & Whittaker, 1975).

a: Early 1900s b: Early 1930s

Figure 11.2. Changes in houses over a 27 year period.
Source. Gash & Whittaker, 1975.

The first photograph in Figure 11.2 shows a Tolai house before the missionary era while the second shows a more permanent type built during a peaceful period. The Rev. George Brown had earlier predicted that such changes were inevitable, given the influences of the new settlers, and in particular, the missionaries (Gash & Whittaker, 1975). Introduced designs were viewed by missionaries as being better. Tolai houses were labelled as poorly designed and constructed. Tolais began to integrate new designs and materials when building houses. This shows that the Tolais have been integrating other peoples' building knowledge with their own, since the first contact with missionaries. Brown (1908) described the Tolai people's houses as:

> very small and very poorly constructed. They are generally oblong in shape and very low. In the olden days there was often a rattle suspended in the doorway at night, against which one attempting to enter in the night would hit his head and so arouse the inmates. This was done as some protection against surprise and treachery. (pp. 23-24)

They also built their houses close to the beach and had minimal possessions as warfare would result in too much loss. It is important to stress that the Western yardstick was used to measure the quality of workmanship on the Tolais 'tent-like' houses. What was not sufficiently stressed was that the houses were constructed this way for particular reasons such as fear of warfare, as Paraide's Elders informed her. Their own standards existed such as keeping these communities clean (Brown, 1908). With a decrease in warfare and influence of missionaries, people then began to build more permanent houses and began to adapt and integrate other designs such as sewn sago leaves *morata* for the roof introduced by Pacific Island entrepreneurs and missionaries. Tolais showed ingenuity in adapting and selecting new designs for new conditions, learning new skills for building. This desire to absorb the outside ideas that were often viewed as progressive also appears to have applied to number and measurement knowledge.

The research showed that the Tolai people measured length, width, height, straightness, thickness, and weight when building houses. For example, the posts that were used to support houses had to be of a certain thickness to withstand the weight of the materials used for the roof. The thatched roofs had to be a certain thickness to keep out the sun, and the rain. The walls had to be strong enough to hold-up the house. These data show that the Tolais used measurements when building their simple houses. Tolai house construction should not be compared with Western others and labelled as poorly constructed houses. They were constructed the way they were for a temporary purpose.

Paraide's (2010) study also found that geometry is applied during the process of weaving walls for houses. The patterns that her Elder talked about are square, diamond, rectangles, and triangles which are woven into bamboo walls called *balaen/balaeni*. The same size shapes are usually duplicated and joined in straight lines horizontally and vertically on large pieces of walls. The woven patterns may have been a later acquisition from the missionaries or others, when the structure of the houses changed. This shows that the Tolais understood geometrical concepts which they applied in this practical activity, and have passed on to later generations.

Weavers create patterns in mats and baskets by actually counting as they weave, and then shapes begin to appear. On the basis of the interview data it is suggested that Indigenous practical knowledge can be used in formal learning environments when teaching basic measurement and geometry. This helps students to link practical local applications to classroom geometry and measurements. This should strengthen and enrich the students' basic knowledge of measurement and geometry and enhance a more compatible application to more advanced counting, measurement, and geometrical problems.

Integration of Measurement in the Tolais' Gardening and Trade

Paraide's (2010) study found that people's Indigenous measurement knowledge was applied in gardening activities. The following excerpt from an interview with her Elder shows this:

Paraide: Talk about how our people garden…what they actually do…when gardening.

Elder: For gardening, we select a piece of land that we like and we know is suitable for gardening. According to our Indigenous knowledge, when the leaves at the base of the trees are really thick, then we know that, that piece of land is now suitable for gardening. It shows that the soil is rich and crops will grow well in this particular area. The trees are cleared and burnt off. Then planting begins.

When planting bananas, they are planted *a pokono ma a iki* (one and a half adult steps) apart. When planting *kaukau* (sweet potatoes), half adult arm length spacing is used between each *kaukau* mound. For *pa* (taro) the spacing is not so wide, for *tigapu/sigapu* (Chinese taro) the spacing is wider than taro. We use our eyes and experience a lot to estimate the spaces between the crops and when preparing stakes for the yam and banana crops on many occasions. Banana crops are staked in preparation for the windy seasons. …We planted yam crops because we could keep them for a long time, after they were harvested in preparation for the *e na vitoloni/mulmulum* (hunger season or dry season). Our people prepared for such times so they lived on yams when other crops failed during the extreme dry seasons.

The research showed that the participants also apply measurement in their gardening activities. Adult arm lengths and steps are used to measure plots of gardening land and spacing between the various crops. They also estimated the measurement of garden plots, the spacing between crops during the planting process, and the appropriate lengths of bamboo or other wood stakes for the yam and banana crops. Yams were cultivated in the past as a crop that was planted in preparation for the dry season or *a e na vitoloni/mulmulum* (hunger season) because of its durability.

Paraide's (2010) study found that number and measurement knowledge are also applied when processing *tabu* — the Indigenous Tolai currency. The next interview excerpt from her Elder shows how they are applied in this activity:

Paraide: How do we count *tabu*?

Elder: With *tabu*, we start with the value of five *tabu* shells. These are counted individually. For example, I can buy a banana from you with five *tabu* shells. Now we cannot buy anything with five *tabu* shells anymore and very few items are sold for ten *tabu* shells. Today, most people start with twelve *tabu* shells…some still begin with ten *tabu* shells and this may be able to buy one banana…a bundle of peanuts…greens…beans… then the next value is *a tip na rivu* which is twenty *tabu* shells. The next value — two sets of twenty *tabu* shells (forty) are called a *vartuku*. That is when we stop counting the individual *tabu* shells. Then *tabu* is now threaded in thin strips of cane and the value is measured by lengths. The length is measured with an adult's arm length. *A viloai, a bongabongo, a tabara, a leke, a papari,* then *a pokono*. After the *pokoko* (an adult's arm length), we measure the value of the lengths… so a bit more than ten fathoms is called *tikana a rivu* (10 x 1 fathoms), ten lots of *arivi* is *tikana marai* (10 x 10 = one hundred fathoms), then we count in hundreds, if we are counting large quantities of *tabu*. *A tutunana* is when we count in five hundreds. The *tabu* is rolled into large *loloi* (rings) of five hundred fathoms each and wrapped with dried banana leaves, or paper. There is an identification mark in each ring for each individual owner and the amount of *tabu* in each ring. Owning *a tutunana* is a sign of wealth, prestige, and strength of individuals and clans. The death ceremony of a rich clan leader usually has many *tutunana* on display. Some of these are hired from other clan members. *Tabu* is used to hire the *loloi*.

Figure 11.3 illustrates how *tabu* is counted, threaded, and measured for value.

The value of *tabu* that is approximately one fathom is counted as 1, Figure 11.3 (i) and (j). The measurement of *tabu* is generally not exact as illustrated in the pictures. This is done purposely to show generosity and to secure trading and social relationships. The use of exact measurement of *tabu* when trading and giving can be interpreted as people being greedy or selfish. The value of *tabu* Figure 11.3 (k) is approximately 10 fathoms, which is counted as 10 x 1 while Figure 11.3 (l) is approximately 20 fathoms or 10 x 2. It is important to note the use of the grouping unit of one fathom. This allows for further large numbers. In Figure 11.3 (m) the value of the *tabu* is 10 x 10 x 3 or 300.

This *tabu* was a bride-price payment, which took place during this study. Large amounts of *tabu* are rolled up and put away as a saving for death ceremonies. These rings are called *loloi*. The two *loloi* shown in Figure 11.3 (n) are valued at *tikana mari ma ilima* and *a mari* — 100 and 150 fathoms. *A tutunana* that Paraide's Elder talked about is five times thicker and heavier — approximately 500 fathoms.

Paraide's Elder made the following observations about the current "inflation" of *tabu*:

Elder: Today, measuring *tabu* is done differently from the past. In the past, we measured exactly a fathom and we accepted that. We did not ask for more. Now we want the *tabu* to be longer than a fathom. We want more for bride-price. In the past, the value of bride-price was lower. Now we are influenced by other Tolais who ask for so much more for bride-price.

This illustrates that the counting and measurement of *tabu* is still widely used in this particular community. The use of *tabu* for trading, bride-price payment, labour payment, and other purposes is still valued highly. For example, during this study, an electrician was paid in *tabu* for labour to do electrical wiring in my tradestore building. He preferred to be paid in *tabu* instead of the modern PNG currency, *kina* and *toea*. Paraide paid him three *arivu na tabu* (3 x 10 fathoms). Paraide exchanged K150 for the three fathoms of *tabu* with another member in this community. He wanted to exchange *tabu* for kina in order to pay his children's school fees. The electrician's request to be paid in *tabu* was unexpected because, in these current times, Paraide expected him to accept the modern form of payment.

a: *a utula ilima na palatabu* (5 x 3 = 15)

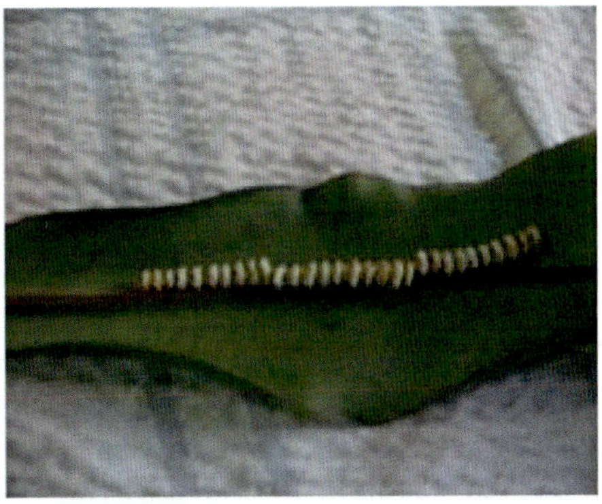

b: *a utula tip na ilima* (10 x 3 = 30)

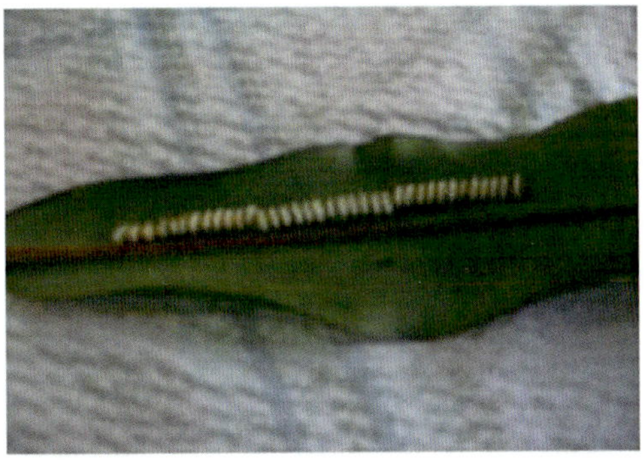

c: *a utula tip na lapikai* (12 x 3 = 36)

d: *a utula tip na rivu* (20 x 3 = 60)

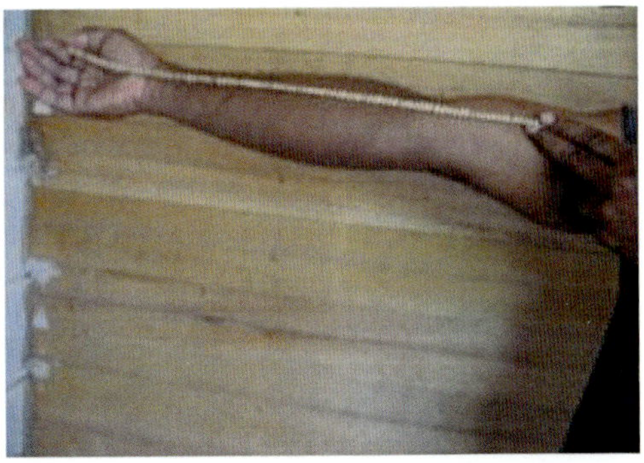

e: *a tura malimalikunu* (to the elbow)

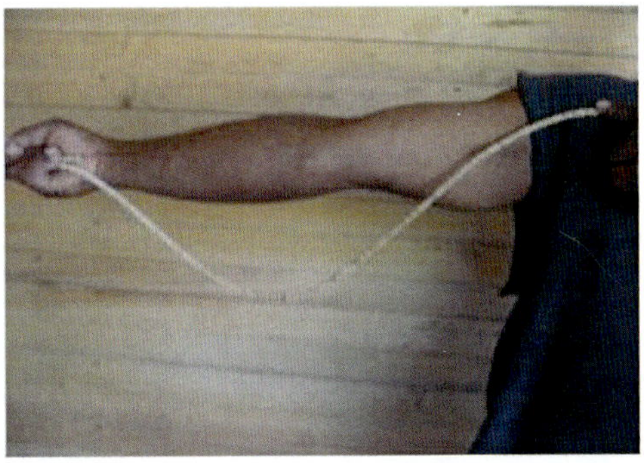

f: *a viloai* (to the shoulder)

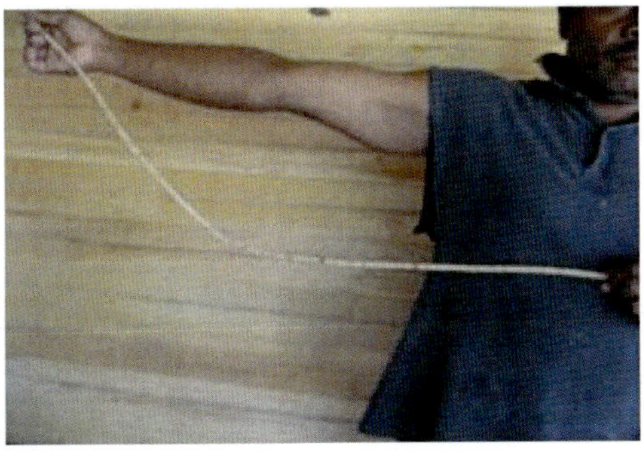

g: *a bongabongo* (to the chest)

h: *a leke* (to the other shoulder)

Figure 11.3. *Tabu* made, measured and counted.

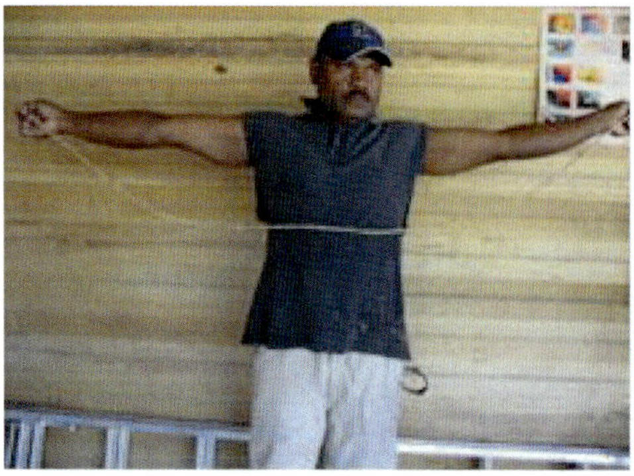

i: *a pokono* or *tikana pokono* (an adult's two-arm-lengths)

j: *a pokono* or *tikana pokono* (an adult's two-arm-lengths)

k: *tikana arivu(arip)/arivu(arip)* (10 x 1)

l: *aura arivu* (10 x 2 = 20)

m: *a utula mari na tabu* (10 x 10 x 3 = 300)

n: *loloi* ring *tikana mari ma ilima* and *a mari* — 100 and 150 fathoms

Figure 11.3. (*continued*)

This shows that the use of *tabu* has the same value as the modern currency in this particular community. This study also found that other Tolai communities in East New Britain Province place *tabu* and *kina* and t*oea* at the same level of importance. Given such a prominent standing, and the greater demand for *tabu* use, it further strengthens the notion that a successful marriage between this particular number and measurement knowledge and Western knowledge is possible in formal learning environments.

It was found that number and measurement are integrated with all the other Indigenous mathematical knowledge, and most importantly, with the other Indigenous knowledge such as building houses, gardening, collecting coconuts, and processing *tabu*. Indigenous measurement and number are still widely used and applied in practical ways in this community. There were degrees of integration of Indigenous and Western knowledge in the practical activities.

References

Ascher, M. (2002). *Mathematics elsewhere: An exploration of ideas across cultures*. Princeton, NJ: Princeton University Press.

Battiste, M. (2002). *Indigenous knowledge and pedagogy in First Nations education: A literature review with recommendations* (Working paper). Ottawa, Canada: Apamuwek Institute.

Bialystock, E. (1991). *Language processing in bilingual children*. Cambridge, UK: Cambridge University Press.

Bordo, S. (1999). Feminist skepticism 'Maleness' of philosophy. In S. Hesse-Biber, C. Gilmartin, & R. Lyndenberg (Eds.), *Feminist approach to theory and methodology* (pp. 29–44). New York, NY: Oxford University Press.

Brown, G. (1908). *Pioneer-missionary and explorer: An autobiography*. London, UK: Hodder and Stoughton.

Collier, V., & Thomas, W. (2007). Predicting second language academic success in English using the prism model. In J. Cummins & C. Davison (Eds.), *International Handbook of English language teaching, Part1* (pp. 333–348). New York, NY: Springer.

De Abreu, G., Cline, T., & Shamsi, T. (2002). Exploring ways parents participate in their children's school mathematical learning: Case studies in multiethnic primary schools. In G. De Abreu, A. Bishop, & N. Presmeg (Eds.), *Transitions between contexts in mathematical practices*. Dortrecht, The Netherlands: Kluwer.

Department of Education Papua New Guinea. (1998). *Elementary curriculum statement*. Waigani, PNG: Department of Education Papua New Guinea.

Department of Education Papua New Guinea. (1999a). *Cultural resource book for mathematics (Trial version)*. Waigani, PNG: Department of Education Papua New Guinea.

Department of Education Papua New Guinea. (1999b). *Elementary education: Making your own community curriculum. A guide for elementary trainters and teachers*. Waigani, PNG: Department of Education Papua New Guinea.

Department of Education Papua New Guinea. (2000). *Lower primary mathematics: Bridging to English resource book*. Port Moresby, PNG: Department of Education.

Department of Education Papua New Guinea. (2004). *Lower primary mathematics syllabus*. Port Moresby, PNG: Department of Education, Papua New Guinea.

Fanon, F. (1952). *Black skin, white masks*. London, UK: Paladin.

Freebody, P. (2003). *Qualitative research in education*. London, UK: Sage.

Garcia, E. (1994). *Understanding and meeting the challenges of student cultural diversity*. Boston, MA: Houghton Muffin.

Gash, N., & Whittaker, J. (1975). *A pictorial history of New Guinea*. Melbourne, Australia: Melbourne University Press.

Harding, S. (1987). *Feminism and methodology: Social science issues*. Milton Keynes, UK: Open University Press.

Henn, M., Weinstein, M., & Foard, N. (2006). *A short introduction to social research*. London, UK: Sage.

Jannok Nutti, Y. (2007). *Matematiskt tankesätt inom den samiska kulturen – Utifrån samiska slöjdares och renskötares berättelser, licentiate thesis [Mathematical thinking within the Sámi culture – On basis of Sámi handcrafters' and reindeer herders' stories]* (Masters). Luleå University of Technology, Luleå, Sweden.

Jannok Nutti, Y. (2013). Indigenous teachers' experiences of the implementation of culture-based mathematics activities in Sámi school. *Mathematics Education Research Journal, 25*(1), 57–72. doi:10.1007/s13394-013-0067-6.

Mennis, M. (1972). *They came to Matupit.* Vunapope, Rabaul, East New Britain Province: Mission Press.

Mimica, J. (1988). *Intimations of infinity: The mythopoeia (cultural meanings) of the Iqwaye counting and number systems*. Oxford, UK: Berg Publishers.

Nandy, A. (1983). *The intimate enemy: Loss and recovery of self under colonialism*. New Delhi, India: Oxford University Press.

Paraide, P. (2010). *Integrating Indigenous and Western mathematical knowledge in PNG early schooling* (PhD). Deakin University, Geelong, Victoria, Australia.

Pica, T. (2003). Second language acquisition: Research and applied linguistics. *Working Papers in Educational Linguistics, 18*(2), 1–26.

Reinharz, S. (1983). Experiential analysis: A contribution to feminist research. In G. Bowles & D. Klein (Eds.), *Theories of women's studies* (pp. 192–209). London, UK: Routledge & Kegan Paul.

Saxe, G. (1981). Body parts as numerals: A developmental analysis of numeration among the Oksapmin in Papua New Guinea. *Child Development, 52*, 306–316.

Saxe, G. (2012). *Cultural development of mathematical ideas: Papua New Guinea studies*. New York, NY: Cambridge University Press.

Scheurich, J. (1997). *Start your research*. London, UK: Falmer Press.

Smith, G. (1981). *Traditional mathematics in Morobe* (Working Paper no. 21). Lae, PNG: Papua New Guinea University of Technology.

Smith, L. T. T. R. (2005). *Decolonizing methodologies: Research and indigenous peoples*. New York, NY: Zed Books.

Souviney, R. (1981). Teaching and learning mathematics in the community schools of Papua New Guinea (Education, Trans.) *Indigenous Mathematics Project Working Paper Series* (Vol. 20). Port Moresby, PNG: Department of Education.

Thomas, W., & Collier, V. (2003). The multiple benefits of dual language. *Educational Leadership, 61*(2), 61–64.

Thune, C. (1978). Numbers and counting in Loboda: An example of a non-numerically oriented culture. *Papua New Guinea Journal of Education, 14*, 69–80.

Wolfers, E. (1971). The original counting systems of Papua and New Guinea. *Arithmetic Teacher, 18*(1), 77–83.

Chapter 12
Integration of Indigenous Knowledge in Formal Learning Environments

Patricia Paraide and Kay Owens

Abstract Building on Paraide's feminist, postcolonial study of the impacts of Western schooling on cultural knowledge, Paraide uses her data from Tolais in East New Britain to argue for bilingual education. This is supported by Matang's quantitative study of the use of vernacular languages in early education in other regions of Papua New Guinea. In particular, the chapter looks at the impact of social, postcolonial attitudes, changing education policies, and the importance of quality teacher education that addresses these issues.

Keywords Tolai education • Kâte education • ethnomathematics • bilingual education • mathematics and language • changes in mathematics • Papua New Guinea education

Limited Integration of Indigenous and Western Number Knowledges

The data presented in Chapter 11 from Paraide's (2010) study show that the integration of Indigenous and Western number and measurement knowledge is being practised in the research participants' community. This shows that this integration is already a norm and established practice in this particular community. Therefore, the integration of Indigenous and Western knowledge in the formal learning environments is possible, if the teachers are better prepared professionally for such integration and are encouraged to take a more positive view on the integration of the two knowledge systems.

The following excerpts suggest the limited integration of Indigenous and Western number and measurement in the lower primary mathematics curriculum documents. In the lower primary teachers' guide there is only one general statement that encourages the integration of Indigenous and Western number. Out of the fifty-six indicators for the number learning area, only one refers to the integration of Indigenous and Western number which is "compare traditional number systems with Arabic number systems" (Department of Education Papua New Guinea, 2004, p. 14). Out of the eighty indicators for measurement only one relates to the integration of Indigenous and Western measurement which is "apply the idea of measuring length in real life practical activities such as gardening" (Department of Education Papua New Guinea, 2004, p. 18).

P. Paraide (✉)
Divine Word University, Madang, Papua New Guinea
e-mail: pparaide@dwu.ac.pg

K. Owens
Charles Sturt University, Dubbo, NSW, Australia
e-mail: kowens@csu.edu.au

K. Owens et al., *History of Number*, History of Mathematics Education,
DOI 10.1007/978-3-319-45483-2_12

These statements show that the integration of Indigenous and Western number is limited in formal learning at the lower primary level. The one indicator each on number and measurement suggest that the actual integration of Indigenous and Western number and measurement in formal learning was given low priority during the development of the curriculum documents. The two indicators could not possibly bring out much Indigenous knowledge that could be used as a basis on which to build when introducing similar Indigenous and Western number and measurement knowledge when teaching.

Paraide's (2010) study found many examples of the practical application of number and measurement in the research participants' community. However, these documents are primarily developed to guide and assist the lower primary teachers to implement the current lower primary curriculum, which includes mathematics, but it has limitations concerning guidelines for the integration of Indigenous and Western number and measurement.

Paraide's study found that lower primary teachers in this particular school are of the view that they have received inadequate preparation for bilingual teaching, the bridging of Tok Ples (vernacular) and English, and the integration of Indigenous and Western knowledge in formal learning environments. Furthermore, they believe that they have forgotten some of their Indigenous number and measurement knowledge, and therefore, encounter difficulties when integrating these two knowledge systems.

Such a situation could be rectified if more indicators are developed for the integration of Indigenous and Western number and measurement in the curriculum mathematical documents. Given the difficulties that this particular group of teachers seem to be encountering, examples of a variety of Indigenous counting and measurement activities could be included in the lower primary mathematics curriculum documents. This could assist those teachers, who may have lost some or all of their Indigenous mathematical knowledge, to relearn or research their own Indigenous numbers and measurement knowledge for the purpose of integration with the Western others. This could assist them to regain their in-depth Indigenous knowledge in order to develop an achievable integration with Western number and measurement knowledge.

The elementary teachers', parents', and Paraide's Elders' interviews, and classroom observation data that were presented in other sections of her doctoral thesis show that the integration of Indigenous and Western number and measurement knowledge is already a strength in the elementary school site. Her other research experience also found that the integration of Indigenous and Western knowledge is a strength in other elementary schools in this particular province, and other provinces as well (Evans et al., 2006, p. 83). Therefore, it is reasonable to suggest that the integration of Indigenous and Western mathematical knowledge in formal learning situations is possible in the lower primary level of education, if the teachers are adequately supported and monitored during the implementation stages.

Diminishing Indigenous Mathematical Knowledge in the Communities

Number and measurement are also applied in fishing and when building canoes. However, outside influences have limited their use and application in these areas because of the use of modern transportation. Modern sea, land, and air transport are now used for travelling long distances, and canned fish is easily obtained.

The following excerpt is from an interview with Paraide's Elder which shows the current limited use of number and measurement in canoe building and fishing:

Paraide: How else is measurement and number used?

Elder: We use measurement when building canoes. We usually harvest *aiting* trees for canoes. If we want to make a large canoe then we harvest a tall and straight *aiting* tree that has a wide trunk. Smaller trees are used for smaller canoes. The tree is chopped down and the desired length of the canoe is measured and cut out. The log is then hollowed by first burning, then hollowed by a stone axe, and then shaped into a canoe. Stone axes were used in the past to further hollow out the canoes and shaped them into appropriate shapes. Now white man's tools are used when building canoes. Bamboo and smaller pieces of a specific wood are measured out and assembled for the outrigger. Specific strong wood is also measured out and shaped into paddles.

This shows that the community cultivated specific trees for canoes and that the length and diameter of the tree determines the size and space and weight capacity of the canoes. Straightness was an important factor in canoe building because a canoe that was not curved the right way could prove difficult to steer in the water.

The next excerpt from an interview with the Elder, shows the different sizes of canoes that were built:

Paraide: What kinds of canoes were built by our people?

Elder: Canoes built for fishing on the reefs and just beyond could accommodate two fishermen but they usually fish alone. Canoes built for ocean fishing were bigger and longer and could usually take up to four to six people. Fishermen went in groups for ocean fishing because of the heavy materials that were used for this type of fishing. Canoes built for long voyages and trading were longer and wider and could fit up to twenty people including cargo for trading and a lot of food and water for the traders. Canoes had to be curved well in the front and back and the outrigger had to be strong to keep the canoes afloat and steer well in water. Now only small canoes are built for fishing. Now people use boats, trucks, and planes for transport. They also just go to the shops to buy canned fish and other food.

This interview excerpt shows that Indigenous number and measurement knowledge were used to determine the size of the canoes and their carrying capacity. They show that smaller, medium, and large canoes were built for different and specific purposes. The estimated weight of passengers and cargo also influenced the size of the canoes that were built for specific activities.

The use of some number systems in PNG is also diminishing because of the influence of other languages. For example, Smith (1984) found that number systems diminished in some Austronesian languages in Morobe Province. He found that, in the Roinji language, there are no complete decimal systems with separate words for one to ten, as found in Austronesian languages of many other areas. It seems that such a decimal system has been lost as a result of the influence of neighbouring non-Austronesian languages. The Roinji language now has numeral words only for one and two although, as mentioned earlier in Chapter 3, there seems to be a pairing $2+2+2+1=7$ and $10=5\text{x}2$ although 5 is not clearly "hand".

Paraide's research found that, in the Tinatatuna language, the people still have words for the numbers one to ten, and still use them. However, there is now an increasing use by the younger generations, of English words for numbers when they are counting. If this trend continues, the Tinatatuna words for the numbers one to ten may diminish, as is the case of the Roinji language in the Morobe Province.

Indigenous number and measurement words in the Tinatatuna language have been gradually replaced by English words and counting in groups is now gradually diminishing, given the dominant use of individual counting. Some of the Tolai Indigenous counting systems are not used as extensively as in the past. The research suggested that the integration of Indigenous and Western mathematical knowledge is now being applied when building canoes, fishing, and trading.

The following excerpt shows some more of the changes that have taken place in this particular community:

Paraide: Tell me more about building canoes.

Elder: In the past, a man's canoe was also his coffin. The fishermen were buried in their canoes, and therefore they made sure that they were well-maintained and replaced when necessary. Others were buried wrapped in mats. Modern coffins have replaced canoes and mats.

This excerpt shows that the people in this particular community are adapting other practices to suit the current changes in the communities. Narokobi (1980) also noted the changes in the Tolai and other PNG Indigenous peoples' lives. He stressed the issue of preserving Indigenous knowledge which he observed as being lost in museums:

> Today, many traditional items are being substituted by buying aluminum pots, plastic belts, calico, Hong Kong umbrellas, trousers, plastic helmets, and airline bags from tradestores. It is important that, while we still have with us our old men and women who can remember the old ways, their hands can be used to make items belonging to the past, such as clay pots, bark belts, grass skirts, pandanus rain caps, tapa cloth caps, wigs, and bush material containers. Such people should be stimulated to provide material for their museum so that their children and future grandchildren of the community can learn about the old ways. (p. 52)

Changes in Indigenous practices are taking place in this particular community because of contact with other Papua New Guineans, and other nationalities. With such interaction, the other practices, beliefs, and attitudes have influenced the way that people practise their Indigenous activities. This also shows that, in this particular community, any new knowledge is refined, and applied to suit the current times. Narokobi (1980) made these observations regarding the loss of Indigenous knowledge:

> In certain areas of PNG, it is unfortunate that, through the narrow minds of some church leaders, who forbade most activities related to the past, the younger generation has been deprived of their own social laws and knowledge, which was normally conveyed from father to son as part of the initiation cycle preceding the reaching of manhood. Their fathers have become ashamed of their ancestral wisdom, their ceremonies shunned, and the society has become nearly sterile. The degree of articulation does not exist in all communities…To overcome this major problem it is essential that village wise men and women are given the opportunity to convey to their children, with confidence, their knowledge. (pp. 52–53)

Similar trends were found in this particular Tolai community. For example, canoes are now built only for small-scale fishing. Western dinghies, commonly known in this area as "banana boats", have replaced ocean fishing and long voyage canoes. Smith (1981) stated that:

> Apart from exchanges for purposes of gaining status, there was also a good deal of trading between the groups….they were connected by trading ties and there were also trade routes between coastal and inland people. (p. 8)

Paraide's Elders informed her that her people build canoes for fishing and trading purposes along the coast and neighbouring islands. History books show that the Tolais had ocean fishing and voyages in canoes that no longer exist today (Gash & Whittaker, 1975). Some use of Indigenous number and measurement in navigation, trading, and ocean fishing skills could diminish further, given their reduced application in these activities.

Teachers' Diminishing Indigenous Mathematical Knowledge

Most of the teachers in the school site claimed that they did not know very much of their Indigenous knowledge anymore. However, as other data in the study suggested, these teachers actually knew their Indigenous knowledge, but it was dormant, and they did not know how to incorporate it in their teaching. These teachers had never applied the integration of two knowledge systems in the formal learning environment in their past teaching experiences because of the English instruction and Western knowledge dominance in formal learning environments. Furthermore, these particular teachers have not been well-prepared to incorporate Indigenous and Western mathematical knowledge in the formal learning environment. The following excerpt from a teacher's interview shows this:

Paraide: Talk about your knowledge of Indigenous mathematics and your teaching…Do you know our Indigenous knowledge … counting…measuring?

Tina: I teach number and measurement and other mathematical knowledge. If I do not understand some things, I go to our Elders in the community to gain some more knowledge so that I can pass it on to these students. I ask for knowledge like counting, measuring a house, our Indigenous knowledge…education. I go back there to ask what this is or what that is, so I can pass this on to the students. They explain how we were able to work with time, about counting… measuring…some we do not understand…because new ones (counting and measuring) are now coming up … used…so if I am not clear about some of this mathematical knowledge I go back to my old mother…to get clarification.

These comments suggest that Tina seeks a lot of assistance from her Elders regarding Indigenous mathematical knowledge. Furthermore, there are outside influences to number and measurement knowledge, as reflected in the statement, *"some we do not understand because new ones …are coming up"*. This shows that Tina does not know some of her Indigenous knowledge, which could be why she goes to her Elders for assistance. However, when asked if she did gardening, processing of *tabu*, gathering coconuts or making more *tabu,* Tina stated that she did. The next excerpt from an interview with her shows this:

Paraide: Do you participate in gardening, gathering coconuts, processing *tabu*, and marriage and death ceremonies?

Tina: Most definitely. We have to make gardens in order to feed our families. We gather coconuts for copra and I have to make *tabu* in preparation for my mother's death ceremony.

Paraide: Do you think you are using number and measurement in these activities?

Tina: I think so, but I have not really thought about it…Yes…I think so.

Tina is involved in activities in her community which involve Indigenous number and measurement knowledge. She does not seem to make the link between Indigenous number and measurement which are applied in these activities with the number and measurement that is prescribed in the elementary cultural mathematics syllabus. This is shown by her comment "I think so, but I have not really thought about it…Yes…I think so" in response to the question, "Do you think you are using number and measurement in these activities?" This illustrates that Tina had some difficulty integrating her Indigenous number and measurement with the Western concepts that are prescribed in the syllabus.

Tina had not forgotten her Indigenous number and measurement knowledge. She was not aware of the similarities in the Indigenous and Western number and measurement knowledge, and therefore, found the integration of the two knowledge systems difficult to implement. Paraide also noted that Tina was not really as passionate about her Indigenous knowledge as Jude. Tina was of the view that English is a more effective language of instruction at elementary level than Tinatatuna, and that Western content knowledge was better than her own Indigenous knowledge. Consequently, she was reluctant to acknowledge the existence of her Indigenous number and to implement the integration of the two knowledge systems.

A further hindrance for Tina was that she did not know how to teach the Indigenous number and measurement knowledge within the Western education context, yet she did not seek Jude's assistance because of the power play that existed between them. Friction was evident between Tina and Jude with Jude inferring he would make a better Head Teacher than Tina as he thought men were better in this role than women. Nevertheless, Tina tried as shown in the following classroom excerpt:

> Counting coconuts… This is the way we count coconuts …*Tanguvani*… "One set of twelve coconuts" (a word used exclusively for sets of twelve coconuts).

The dualistic thinking of counting in twos, fours, sixes, twelves, and tens the Western and Indigenous ways was evident in Tina's teaching strategy in this case. However, she was then getting children to rote count.

> Counting coconuts... This is the way we count coconuts. We will start counting in twos to one set of twelve... look this way...one set of two is two... two sets of twos... is four... up to...twenty-four. One set of two is two...Ok, one, two, three ...one set of two is two...

This is not how the counting of coconuts is taught in the communities. Furthermore, this type of counting is only taught when gathering coconuts, and most importantly, the children learn the skills of counting them only during the gathering process.

The following lower primary teachers' focus group interview excerpt also shows that the teachers know their Indigenous number and measurement. However, they have not used this knowledge very much in formal teaching environments. The teachers were convinced they have received inadequate professional support on the integration of Indigenous and Western mathematical knowledge:

Mike: We know about counting *tabu*, coconuts, eggs, and bananas, but we just do not think about them as mathematics. ...Not when we are teaching here.
Julie: Yes...The concepts used in them are like the multiplication, division, addition, and subtraction we teach, but we just do not think about it this way here.
Joyce: Now that you have made us aware of it, we will now think about it more and link it with what we teach in mathematics.
John: We have noted that the students who graduated from elementary School here know more about them.
Mable: mmm... and we also noted from your teaching demonstration that these students already know more advanced mathematics.
Rose: They know about grouping....
Marg: We need to find out more about what they already know when they enrol here and also work closely with the elementary teachers in the elementary schools.

This demonstrates that Mike, John, Margaret, Rose, Joyce, Mable and Julie use Indigenous number when processing *tabu,* and when counting eggs, coconuts, and bananas, but did not seem to be aware of the fact that counting these items is similar to the formal Western number operations which they teach in the classrooms. During this discussion, some of them also claimed that they had lost some of their Indigenous counting. This shows that their number knowledge was diminishing. These teachers acknowledged that the elementary graduates know Indigenous number operations and were of the view, that their mathematical knowledge was advanced. The research indicated that the lower primary teachers were willing to work closely with their elementary school counterparts on a professional basis, to assist the students in their learning.

This loss of knowledge could in fact be an on-going process over a long period of time whenever cultural groups interact. Paraide's findings of the diminishing Indigenous number knowledge are similar to Smith's (1981) commentary regarding the Austronesian speakers in the Morobe Province. Smith (1981) found that the Austronesian speaking people in the inland of Morobe Province had lost some of their counting knowledge concerning the numbers one to ten that are still present in other Austronesian languages elsewhere.

Paraide's study found that the younger generations in the research participants' communities count items individually, and in many cases English numbers are now used when counting and recording. The following excerpt from an interview with Paraide's Elder shows traits of some Indigenous number knowledge being lost because of the dominant use of single item counting:

Paraide: Can you talk about *kiau na ngeoki* (wild fowl eggs)?
Elder: We dig for wild fowl eggs, mainly to sell. We always leave one or two in the ground so they can hatch again so that our birds do not die out. When we have collected four eggs,

> we say *tikana kevai* (one set of four), when we have eight we say *a ura kevai* (two sets of four), and twelve *a utula kevai* (three sets of four). We count in fours. It is the same with counting bunches of bananas. We also count in fours. *Tikana inangava* (one set of four), *a ura inangava* (two sets of four). It is the same, but we use different words. This kind of counting is rarely used these days to count bananas and wild fowl eggs. They are now counted individually. Many people in the younger generation do not know this type of counting anymore.

The non-usage of some specific counting systems seems to have led to the loss of some Indigenous counting knowledge. The older generation still knows and uses this counting system, but this knowledge does not seem to have been successfully passed on to younger generations. This could be seen as the diminishing of some areas of Indigenous number, because individual counting now dominates the counting of wild fowl eggs and bananas — an influence from Western individual counting.

A further contributing factor to such loss is that now people can record information in print. Therefore, they do not need to count in sets for the purpose of easy memorisation of information, as their ancestors did, given the people's oral history. Yet at the same time, they were also conserving their scientific knowledge. This particular group of people practised conservation as reflected in this statement: "We always leave one or two in the ground so they can hatch again, so that our birds do not die out".

The following excerpt is from a teacher's interview and provides more information on the reasons for the diminishing Indigenous mathematical knowledge:

Paraide: Talk about your students...what they already know when they begin elementary school.

Jude: Some of them can count coconuts and *tabu* and measure *tabu*. Those students who are taught by their parents about our knowledge, know them. Some students do not know this knowledge because their parents do not involve them much in these activities.

Some parents do not know our Indigenous knowledge anymore. So, when I teach cultural mathematics, here it is new for them, but they learn very quickly. Some of them go back home and teach their parents about our Indigenous knowledge. Our students know how to count *tabu* and measure them because we use *tabu* a lot in this community. Many of us do not have much of the modern currency so we use *tabu* to buy food and other stuff during times of need.

This excerpt shows that, in this particular community, the younger generation parents who were taught to count coconuts and *tabu* and how to measure *tabu* do pass on the information to their children. This suggests that the actual usage and application of Indigenous knowledge in the current times can enhance the maintenance and restoration of Indigenous knowledge and practices. The evidence shows that some young generation parents no longer recall some of their Indigenous knowledge and so their children do not use it.

This next excerpt from a parental interview supports vernacular instruction:

> It is the same. Like we count coconuts and *tabu* at home. The counting is the same in English. So one set of six coconuts is six coconuts, it is just the same as what they learn in school. They count other things at home like betel nut. They count two betel nuts for so many *tabu* shells, but they know this. They count those themselves. Just like building houses — when we measure out spaces for houses, we count in Tinatatuna. The men use steps or arm lengths to measure timber and other materials that need to be measured. This is similar to what they do in English — how they measure, but they use a tape measure. We use that, too, these days. When the children are playing, they gather stones and count them. They add and then they take away some. Sometimes, they line up coconuts and coconut husks and count them individually. They orgainse them into shapes like they are building houses. During this activity, they count both in English and Tinatatuna, just to make sure the number is correct.

> Using Tinatatuna is good because what they learn in elementary school is not difficult for them to build on when they go to lower primary school. It is easy for them because many things are similar. They already know how to count at home and in elementary school, so when they move up to Grade 3 they just learn more. So it is easy for them and they learn better. You see, for you and me when we learned to count in English, we did not really know or understand what we were saying. But now, because the children already know how to count in Tinatatuna they understand what the numbers are and so when they introduce them in English it is easy for them because they already know what the numbers are. They learn very quickly because they already know a lot from elementary school and from home. It is not so difficult for them to count or measure in school because they already know and understand what they are. So their learning is faster. In the past, this was difficult for us because there was no link. We just counted in English, we did not know the meanings of what we were saying in school, but now they do know what they are learning when the teachers talk and explain knowledge to them in Tinatatuna. So learning is better and students learn quickly.

This parent pointed out the advantages in learning in a language that her child knew best. She compared her own learning experience when she was her son's age. She stated that when she was in school, she repeated English words that were meaningless to her during her early years of formal schooling, and consequently, learning was difficult for her. She compares her education, where English was the dominant language of instruction, with her son's education.

She was of the view that her son had a better understanding of school subjects. She uses her own experience to evaluate her lack of understanding of formal subjects when English was used as the language of instruction and her son's level of understanding of subjects now that Tinatatuna is the language of instruction. Her evaluation seems to be that when she learned in a language that she did not know well, she had difficulties in understanding the subject content.

This parent's view, like the teachers' views given in Chapter 11, is supported by learning experts, (e.g., Bialystock, 1991; Collier & Thomas, 2007; Garcia, 1994; Pica, 2003; Thomas & Collier, 2003). These experts have found that cognitive development and academic development in one's first language have positive effects on second language learning. Academic skills, literacy development, concept formation, subject knowledge, and strategy development that are learned in the first language can be transferred to the second language being learned.

This interview data show this interviewee's views on the integration of Indigenous and Western number and measurement and vernacular instruction:

> See, we have mathematical knowledge at home, too. So, if you want to buy items with *tabu* then, if someone is selling food for, say, sets of twenty *tabu* shells, we count shells in twos until we reach twenty. Then we use that to pay for the food. If they ask for more for the items, then *tabu* is measured using the arms to measure off different values of *tabu*.
>
> We also count coconuts, so if you want to send a child to collect coconuts, you either ask for two, or one set of four, one set of six, or one set of twelve. At school, what the children learn at home is the same in school; we count in twos, fours, sixes, and twelves. Also, some other school knowledge about counting, like counting, in twos, threes, or fours, is similar to how we count coconuts. It is like the times tables.
>
> In school, we also do addition problems, which is the same as what they do at home. You just keep adding more. If you give away something then this is similar to take away problems in school. With gardening, when you harvest peanuts, and if you want to sell some, then you tie them in small bundles and you count them in tens. So two lots of ten is twenty, and three lots of ten is thirty. When planting our garden, the crops are spaced out according to the type of crop being planted. If you plant and space out particular crops well, then, they will grow well and you will be able to get a good harvest.

Sometimes the cultural mathematical knowledge is new to the students because the parents do not use the Indigenous counting anymore. So when we teach our Indigenous counting here, it is new knowledge for some of them. Once they have learned it here, they teach their parents about our Indigenous counting and measuring. A parent-teacher said:

> I think learning first in Tinatatuna is better than learning first in the white man's language. Using the language that they know best is good because that is the language they speak at home and they can communicate with that. So when they come here and we teach the school subject content in Tinatatuna, they understand and learn faster because they already know the language. So they learn first to read and write in Tinatatuna and then later they will be able to transfer that when learning English and also translate English words into Tinatatuna.
>
> For me, learning first in Tinatatuna is much better. The students' learning is effective because I use the Tinatatuna language and so they understand well the content knowledge that I teach them. Some children have Tok Pisin as their first language, so when they come here and the language of instruction is Tinatatuna, I see that the young parents learn Tinatatuna from their children. This is why I think Tinatatuna as the language of instruction is very good.
>
> While it is true that I did not learn to read and write in Tinatatuna when I went to school, my view changed when I began training for elementary teaching. I began to relearn a lot of words that had been replaced by the white men's or others' languages, like pawpaw. Now I know our own name for it. I began to understand a lot more and appreciated and valued learning in Tinatatuna because I know more about our people and what they already had before the white men arrived. So I really believe that learning in Tinatatuna is very good because we learn about what we already know in school and build on from that and then learn about things we do not know.

Like the female parent, this elementary teacher believes that using the students' best known language for formal instruction is important because the students can communicate well with him. This enhances a better understanding of subject content being taught, and as a result, the students are able to learn faster, as shown in this statement:

> Using the language that they know best is good because that is the language they speak at home and they can communicate with that, so when they come here and we teach the school subject content in Tinatatuna they understand and learn faster because they already know the language.

From a teaching perspective, Jude believes that building on what the students already know enables faster learning. This is reflected in this statement:

> For me, learning first in Tinatatuna is much better. The students' learning is effective because I use the Tinatatuna language, and so they understand well the knowledge that I teach them.

His assessment is supported by August and Hakuta (1997) who claim that research strongly supports the idea that acquisition of vernacular language literacy skills and use is of a great advantage in English language acquisition. They further claim that second language students make sense of the second language by using many of the same strategies that worked so well in acquiring their first language. What is different, however, is that second language students already have an understanding of the meanings, uses, and purposes of language. They now must go on to learn how the second language uses oral and print modes to express those purposes, uses, and meanings (Lindfors, 1987).

This particular elementary teacher, Jude, seemed to be convinced that the students will be able to transfer their Tinatatuna reading and writing skills to English when they learn to read and write in it in lower primary school. His view is supported by language experts mentioned above who claim that being literate in the first language makes it easier for learners to acquire a second language because they are able to transfer the reading and writing skills when learning a new language. Jude is of the view that English, as the formal language of instruction hinders, the students' learning because they may not understand the basic concepts being taught, and therefore, have difficulties when applying them in more advance learning.

This elementary teacher did his formal schooling in English and he values that. However, since he received his elementary teacher training, his attitude towards Indigenous knowledge, changed because he realised that his people had valuable knowledge which they applied in their everyday lives, before outside influences. Furthermore, he realised that much of his Indigenous knowledge is similar to other knowledge. This suggests that he went through a school system where English was the dominant language of instruction, which he seems to value. However, when he received his elementary teacher training, he was made aware of his own Indigenous knowledge and took the initiative to learn more about it from his own self-discovery. His training seems to have enabled him to identify the similarities between Indigenous and Western mathematical knowledge — in this particular case, number— and therefore, was able to integrate the two knowledge systems successfully in his teaching.

Resistance to Bilingual Instruction

Bilingual teaching was prescribed in all the lower primary subjects' syllabi (Department of Education Papua New Guinea, 1998, 2000, 2004) at the time of Paraide's (2010) study. This study found that the dominant language of communication between teachers and their colleagues, and between teachers and students in the lower primary school site was still English. The study also found that these teachers have been trained to teach formal subjects in English, and have been teaching in English since they graduated from their teachers' colleges. Furthermore, most of them had been educated in a school system where English was the only language of instruction. Literature also shows that the language policy for decades has been that English should be the only language of instruction and communication in PNG school environments (Department of Education Papua New Guinea, 1974, 1976; Dickson, 1976; Geoffery Smith, 1975; Tololo, 1976). Severe punishments, such as caning or cutting grass or other forms of manual labour were administered, if students did not adhere to the former language policy. This changed, when the Reform of Education language policy was introduced. Now the policy has returned to English as the language of instruction but policy also says that culture and vernacular language should be used to assist learning.

Unfortunately, the promotion of English only in the school environment has engineered the beginning of the loss and lack of respect for Indigenous languages, culture and knowledge. As L. T. T. R. Smith (2012) observed about Indigenous peoples who had been colonised:

> Imperialism still hurts, still destroys, and is reforming itself constantly. Indigenous peoples as an international group have had to challenge, understand, and have a shared language for talking about history, the sociology, the psychology, and the politics of imperialism and colonialism as an epic story of telling huge devastation, painful struggle and persistent survival. (p. 19)

The English language and Western knowledge have been promoted in formal schooling in the past, therefore, students began to value these more than their own languages and knowledge.

Battiste (2002) observed that Indigenous knowledge and teaching strategies were viewed by some academics as inferior to Western ones. Recent attempts have been made to include Indigenous knowledge in school curricula in order to place it at the same level as Western knowledge. In the same, way the Tolai Indigenous knowledge was viewed as inferior, and the similarities between Indigenous and Western mathematical knowledge were ignored in formal education. Western knowledge dominated students' formal learning. Consequently, Indigenous number and measuring, Indigenous teaching strategies and students Indigenous backgrounds were suppressed, and in some cases, forgotten because of Western and other dominations.

Geoffrey Smith (1975) discussed such domination as the deliberate attempt to Westernise Indigenous people with limited Western knowledge so that they would obey their colonial masters, but also to set them apart from the Western others. As a result, students who attended formal schooling no longer participated in the practical application of number and measurement, which was the common practice before the influence of formal schooling.

The former PNG school curricula were developed from Western perspectives, which ignored the similarities between Indigenous and Western number and measurement and teaching strategies. The data from the elementary lesson observations show that Indigenous and Western number and measurement and teaching strategies complement each other. The data show that the use of Indigenous number and measurement is really the practical application of these two areas of knowledge. The Reform elementary cultural mathematics curriculum aimed at correcting this attitude, and place Indigenous and Western number and measurement and teaching strategies at the same level of importance as the Western ones in formal learning environments. However, there is a general reluctance from lower primary teachers to change established practices in the school environment.

However, as the data in this study show, the success of such implementation is dependent on how the teachers view their own Indigenous knowledge, if they still know it, and if they value it sufficiently to have the courage to explore ways to successfully marry the two knowledge systems. The data indicate that those people who are most resistant to the current language policy and lower primary curriculum are the current lower primary teachers in the school site. This is primarily because they were trained to teach in English, and are not yet confident to conduct bilingual teaching, even though they have had some inservice sessions on bilingual teaching. Consequently, they continue to use teaching strategies with which they are most familiar, and resist the current changes.

In fact, because of the prevalence of this attitude and lack of quality training on bilingual education and the integration of cultural and Western number and measurement in school, the government has undertaken a significant modification of the reform. Along with this came a much needed detail of what outcomes were in terms of classroom expectations. They went back to "objectives" but more importantly they provided teachers with more knowledge about what to teach to achieve what is now called "Standards". Grappling with this with the diversity of teachers' and children's backgrounds has led to current broad standards that are yet to be clarified.

In Summary

The findings show that Western influences, including languages, and the adaptation of their way of doing things have contributed to some of the diminishing Indigenous knowledge in the research participants' lives. For example, the influence of the missionary settlers enhanced peaceful living, and therefore, the Tolai communities including the research participants' communities began building more permanent houses, which were structured like the missionary settlers' houses. This also influenced the use of introduced tools and materials for houses. The findings show that Indigenous houses were built on the ground, but this style of house no longer exists in the research participants' community. Current houses are built on stilts and introduced materials are generally used. Introduced dinghies have also replaced most of the canoes for travel between islands. Only smaller canoes are now built, specifically for fishing purposes.

The study found that the research participants have been integrating Indigenous and Western number and measurement in their community since the first contacts. The integration of different knowledge systems has now become normal in this community. This finding shows that with better professional support for elementary and lower primary teachers regarding vernacular and bilingual education and the integration of Indigenous and Western knowledge, a successful implementation of these programs is possible.

Bilingual Education for Maintenance of Knowledge

There are several arguments made in the literature already referred to in Chapter 11 that bilingual education encourages maintenance of culture and cultural knowledge. In addition, the cognitive processing required for learning in a first language is more efficient for a child to learn basic mathematical concepts. Understanding is then enhanced when transitioning to a second language. This premise will be further discussed in this section together with the issues faced by Papua New Guinea education.

Non-Acceptance of Tok Ples Instruction and Integration of Knowledge

This study found that the Reform language policy, which encourages Tok Ples or vernacular and bilingual instruction at the lower levels of education, has met with resistance from some of the PNG's educated population since it was first implemented in many communities around PNG, as a result of established practices created during the colonial era. As Young (2001) stated:

> post-colonialism's central preoccupation is with the politics …The problem is compounded by the fact that, at independence, power often passed to a native bourgeois elite produced during the time of colonialism that took on board many Western presuppositions; for example, the idea of the nation-state itself. Power passed to those who identified themselves nationally rather than to those with international or local identities and allegiances…The geographic boundaries of the State, and the legal and political structures that are the legacy of colonialism, exist in a continual state of contestation by Indigenous ethnic… Elsewhere, the colonialism of the past has given way to societies whose make-up still reflects the disjunctions of their specific colonial history. (pp. 59-60)

This study found that, as a consequence of the implementation of English as the main language of instruction at all levels of formal education during the colonial era, it is now an established practice. As a further consequence, Indigenous languages are now viewed as less-effective languages of formal instruction by some of PNG's Indigenous population. This is reflected in this statement from a teachers' focus group discussion:

> We were taught in English when we went to school and we learned well so this was okay. We were also trained to teach in English we are used to this, teaching subjects in English. Teaching in vernacular is new to us and may be it is okay, but we are not sure. However, the students must learn English well so they can learn better. They must understand English well because they have to participate in the Grade 8 National Examination, which is in English. Now with vernacular and bilingual education we really do not know if they can learn English well to understand this exam.

The tone of the teachers' voices, and the expressions on their faces indicated an inner conflict regarding the implementation of vernacular and bilingual education. These data suggest that English is now viewed by these teachers as the normal. It also illustrates, that they are not convinced that the vernacular and bilingual language teaching strategies can generate the mastery of the required English literacy skills, needed by the students in order to perform well in the Grade 8 National Examinations. Consequently, they resist vernacular languages as the formal languages of instruction. A male teacher commented:

> When we use vernacular, it is like we are going backwards, not forward, like when we use English, we progress, we learn well.

Similar comments such as, "vernacular instruction means PNG is going backwards or not progressing" have been made by other educated Papua New Guineans in relation to the implementation of vernacular instruction in elementary schools, and bilingual instruction in lower primary. Matsuura (2008) countered these views in this statement:

> Languages are indeed essential to the identity of groups and individuals and their peaceful coexistence. They constitute a strategic factor of progress towards sustainable development and a harmonious relationship between the global and the local context. They are of utmost importance in achieving the six millennium goals of *Educational for All* and the *Millennium Developmental Goals* on which the United Nations agreed in 2000. (p. 1)

The six millennium goals are eradication of extreme poverty and hunger; achieve universal primary education; promote gender equity and empower women; reduce child mortality and improve maternal health; combat HIV/AIDS, malaria and other diseases and ensure environmental sustainability; and develop a global partnership for development. These data infer that the resistance to vernacular and bilingual instruction is a consequence of English being the dominant language of instruction during the colonial era. Vernacular languages were forbidden in government schools. In the 1960s, mission schools also made English the dominant language of formal instruction, when a policy was passed that only schools which used English as the medium of instruction would receive funding from the colonial administration (Barrington-Thomas, 1976; Geoffery Smith, 1975). (See P. Smith (1987) and Owens, Muke, Owens, & Clarkson (in press) for further details on the history of education in PNG.)

The colonial administration declared English as the only formal language of instruction and made it, in Said's (1993) and Pennycook's (1998) terms, the language of power. Although vernacular languages were equally effective languages of instruction in Indigenous knowledge, this was overruled when the policy was passed promoting English as the only language of formal instruction. Since this policy was passed, English as the formal language of formal instruction became the normal practice in formal learning environments. It is now also viewed as the only effective language of formal instruction in PNG.

Only a few of the female lower primary teachers had the courage to state that Tinatatuna was also an effective language of instruction. However, their facial expressions indicated that they did not feel comfortable with the bilingual teaching strategy. This could have been because Paraide was female and doing PhD research in this area. Therefore, they did not have the courage to voice their uncertainty, because she had a positive view of the bilingual teaching strategy. Another possibility could be that they did not dare to express negative views on this teaching strategy because she had shared with them earlier that she was taught to read and write in Tinatatuna during her first three years of formal education and still did very well in the Grade 6 National Examination. Also learning in Tinatatuna during the first three years of her formal schooling did not seem to have any negative effect on her learning to speak, read, and write in English and progress in her life.

Despite an element of resistance to vernacular and bilingual teaching, it was found that changes were beginning to emerge in some of the participants' views regarding vernacular languages as the formal languages of instruction and the integration of Indigenous and Western knowledge. This confirms much anecdotal evidence that some Indigenous people in PNG who have had positive experiences with vernacular education have positive views concerning the integration of Indigenous and Western knowledge.

Teachers' Perceptions of Indigenous Number and Measurement Knowledge

The following excerpt from an interview reveals some of these changes. It shows the ease in which vernacular instruction is used by a male teacher to integrate Indigenous and Western number and measurement knowledge.

> Mathematical knowledge is applied in activities in the community. Counting … counting and measuring when processing and using *tabu*, building houses making gardens building canoes. Estimation of spaces between plants is done when planting peanuts, taro, bananas and other crops. Estimation is also done on the quantity of peanuts to buy and prepare for a marked space in the garden for the peanuts. Women use mathematics when preparing their goods to sell. They calculate the amount of money or *tabu* they will make from one sale. Some of the mathematical knowledge taught in the classroom is just like some of the mathematical knowledge used in the community. I take the mathematical knowledge that our people use like the types of counting, and build on that when teaching. Sometimes, students make connection with the counting they do in the classroom and those they do at home like when counting *tabu*, coconuts, when they count *kaukau* (sweet potato), bananas to cook.

Paraide's (2009) research found that coconuts, goods for trading, and *tabu* are counted in groups, but in different ways. The x 2, x 4, x 6, x 12, and x 10 multiplication concepts are applied when counting coconuts. The x 2 and x 10 multiplication concepts are applied when counting goods for trading. The processing of *tabu* integrates the number and measurement knowledge. This application involves the use of x 5, x 10, x 12, x 20, and x 40 multiplication concept and measurement. *Tabu* is measured using the various distances between an adult's two arm lengths. Appropriate lengths of various *tabu* values are measured out for purchasing goods, paying fines, paying bride-price and labour. The process of measuring the different values of *tabu,* using the arms, was demonstrated during a number lesson. For example, the value of *tabu* measured from the tip of the finger to the elbow is called *a tura malimalikun/u,* and *tabu* the length of two adult arms is called *a pokono.*

The male elementary teacher knew his Indigenous mathematical knowledge well. He pointed out the similarities between the Indigenous and Western number and measuring knowledge. He also built on his Indigenous measurement and number knowledge when teaching these knowledge areas in the formal learning environment. He used the elementary cultural mathematics syllabus as a guide when teaching.

The male elementary teacher also stated during the interview that his training to teach formal subjects in Tinatatuna and the integration of Indigenous and Western subjects enhanced his understanding and appreciation of his Indigenous knowledge. This prompted him to research his Indigenous knowledge to gain more insight into the knowledge and how they are applied in everyday living. The research enhanced his understanding of how and why his people practise and use Indigenous knowledge. He also stated that, occasionally, he was able to assist the students to link their classroom and home mathematics. On the other hand, this elementary teacher's colleague, a female, who had been teaching the Grade 2 class at the beginning of the study, occasionally approached her Elders for Indigenous knowledge. She was not as passionate about teaching in vernacular. This illustrates that the male teacher took the initiative to research his own Indigenous knowledge. Through his research, he identified the similarities between Indigenous and Western number and measurement and was able to integrate the two knowledge systems. It further suggests that, because he knew his Indigenous mathematical knowledge well and most importantly understood it, he was able to bridge these two knowledge systems with ease.

The male elementary teacher also seemed to succeed in assisting the students to note the similarities in the classroom mathematical exercises and the home activities. This was evident in his number lessons where he integrated the x 12 concept with the counting of a particular *tabu* value – *a tip na laptikai*. This value applies the x 12 concept. For example; *a tip na lapita* (12 x 1 = 12 *tabu* shells), *a ura tip na laptikai* (12 x 2 = 24 *tabu* shells), *a lavutulu na tip na laptikai* (12 x 8 = 96 *tabu* shells). Once the students understood the similarities, they applied this concept in solving the x 12 problems quickly and accurately.

The problems were represented in Western number symbols on the blackboard. The students used *tabu* shells to work out the x 12 problems during this particular lesson. These students' have an oral culture, therefore they do not have symbols to represent numbers. However, they know the different values of *tabu* for purchasing various goods. They already knew how to count in a variety of groups to make up the various values of *tabu*. The *tip na laptikai* value was an example of this skill. This elementary teacher tapped into this existing knowledge when the lesson's focus was on counting in sets of twelve.

Most of the lower primary school teachers claimed that they do not really know their Indigenous mathematical knowledge anymore. Most importantly, they had received inadequate professional training and support in bilingual teaching strategies and the integration of Indigenous and Western mathematical knowledge. The following excerpt from a teachers' focus group discussion shows this:

> We need more in-services on bilingual teaching so we can assist the students better…Today is the first time for us to become aware of our Indigenous counting and measurement and to think of them as part of formal learning or mathematics. We use them all the time, but we have never linked them to formal schooling or mathematics.
>
> …..
>
> Yes… and we must know more about how to bridge Tok Ples to English.
>
> Oh yeah, and I thought bilingual teaching was only done in language lessons.
>
> …
>
> In the past, we did not think about Indigenous number and measurement when teaching mathematics.
>
> …
>
> Some of us have not received any workshops on bridging from vernacular to English and bilingual teaching.
>
> …
>
> Some teachers have attended these workshops, but they do not run workshops for us after the completion of their workshops.

The above comments indicate that the teachers lacked effective professional training and support for the implementation of the bilingual teaching strategy. They had never linked their Indigenous mathematical knowledge to the formal learning environment because they had not been trained to do so. Additionally, there was also an absence of such integration in past school curricula. All Indigenous knowledge was suppressed during the colonial era. Such exclusion was a conscious strategy used by the colonial administration to alienate students from their Indigenous ways of life. Consequently, this generation of teachers places lesser value on their Indigenous knowledge than English and Western knowledge. As a further consequence, some have forgotten a proportion of their Indigenous knowledge. L. T. T. R. Smith (2012) made similar observations in his study of various literatures that have presented Indigenous peoples' ways of knowing. She argued:

> Imperialism, history, writing and theory…tend to provoke a whole array of feelings, attitudes, and values. They are words of emotion which draw attention to the thousands of ways in which Indigenous languages, knowledge, and cultures have been silenced, misrepresented, ridiculed, or condemned in academic and popular discourses. (p. 20)

On the other hand, Jude's passion about his Indigenous knowledge is attributed to the type of teacher training he received and his attitude towards his own Indigenous knowledge, as shown in these statements:

> My teacher training which assisted me to teach formal subjects in Tinatatuna enabled me to appreciate my Indigenous knowledge more. We were able to use our own knowledge before the white men's influence. The two knowledge systems are similar in some ways.

The training seemed to have facilitated a smoother integration of Indigenous and Western mathematical knowledge in Jude's teaching.

Tina's resistance to Tinatatuna as the language of instruction, and the integration of Indigenous and Western number knowledge is shown in these statements:

> I think English should be introduced earlier in elementary School. The students need English to learn better. I do not know some of our Indigenous counting so I go and ask the Elders to explain them to me. It is hard to teach number in Tinatatuna. Sometimes, it is hard to relate them to the syllabus content, but I try to teach the content.

Despite Tina's elementary teacher training, which prepared her to teach formal subject content in Tinatatuna, she was not convinced that it was an effective teaching strategy. Tina's attitude on the current language policy and reform curriculum seemed to influence her classroom practice. For example, she used the Western teaching strategy to teach the gathering of coconuts. The multiplication concepts of x 2, x 4, x 6, x 12, and x 10 are used simultaneously when gathering coconuts. Tina isolated these and used rote to teach them. "The students and I were bored and very restless during this particular lesson."

Tina did not encourage the students to use stones or any concrete material to assist them to count in sets, in the absence of coconuts. Jude did this when insufficient *tabu* shells were available during his number lessons. Had Tina done what Jude did and focused only on one set of counting, for example, the x 6, perhaps the students would have better understood the application of counting in groups. Furthermore, they could not link this particular counting with the gathering of coconuts because coconuts were not used. The students understood the x 12 better in Jude's number lesson because he focused only on one value of *tabu*. Tina focused on too many group counting applications in one lesson. Furthermore, the recitals went on for half an hour, with no practical activities to assist the students to understand counting in groups. This seemed to be the major contributing factor to the students' restlessness and lack of response. Jude asked the students why they count out the value of *a tip na laptikai na tabu* in his number lesson. The students answered, "to buy a bundle of peanuts", and "to buy fried flour", and "to buy two ripe bananas". The *tabu* used for purchasing items such as peanuts are broken off from the stringed *tabu*. What is usually done is that, if there are five children in a family or extended family who wish to buy peanuts, the adult or the older children measure out the sets, but do not break off each set. Instead, the sets are twisted into the opposite direction to clearly mark the five sets of twelve. This amount is then broken off and used to purchase the five bundles of peanuts.

> Today we are going to learn how to count *tabu*. Counting of *a tip na lapika* (value of twelve *tabu* shells) is done like this. You hold two shells and count them as one. Ok, let us do it together. Remember you hold two and count them as one. One, two, three, four, five, six... Ok, twist that set of twelve shells in the opposite direction. How many is that? Raise your hand. Don't shout. Twelve. Twelve what? *A tip na laptikai* (one set of twelve *tabu* shells). Good. Ok, let us add another set of twelve. One, two, three, four, five, six. How many do we have altogether now? *Aura tip na lapikai ba aura vinun ma ivati na palatabu* (Two sets of twelve *tabu* shells or twenty-four *tabu* shells). Good. Who can work out the answer for *a lavuvati na tipu na lapikai* (12 x 9)? You add twelve more shells to that... How many Cory... James... how many, Apex? One hundred and...Apex say it louder... How many? *A lavuvati na tip na lapikai* or one hundred and eight *tabu* shells. Good. Let us give Apex a clap.

The students in this particular case told Paraide that they enjoyed this particular lesson because the activity was something with which they were familiar, and therefore they understood what they were doing.

This illustrates that the integration of Indigenous and Western number was applied with some success. Students were able to understand the concept of counting in sets of twelve faster, and were able to apply the concept to the more advanced problems with which they were presented. They added seven more sets of twelve *tabu* shells quickly to work out the answer to the allocated problem: 9 x 12 = 108.

Jude was able to integrate the Indigenous and Western number counting in sets of twelve. He used the Tinatatuna language to teach these concepts. The students understood the concepts well because they were already competent in Tinatatuna. In a recent study on the implementation of the elementary curriculum in PNG, Evans et al. (2006, p. 83) also found that Tok Ples vernacular instruction bridged community knowledge better in formal learning situations than Tok Pisin, which is a widely spoken lingua franca in PNG. It was also found that the students participated actively in the classroom activities. They were able to communicate well with their teachers in their vernacular languages and were able to understand the subject contents better.

Jude had the courage to use the mathematical knowledge with which the students came to school, and built on that when teaching number and measurement which is prescribed in the elementary cultural mathematics syllabus. However, most of the interviewed teachers preferred to use the teaching strategies that they knew and with which they were more familiar, but as they began to realise the value of the current prescribed teaching strategies, they gradually began to apply them. For example, Tina also attempted to link Indigenous and Western mathematical knowledge in her lessons. She taught the Indigenous counting of coconuts (x 2, x 4, x 6, x 12, and x 10) using the Western teaching strategy. Tina used the conventional Western mathematics teaching strategy to teach the counting of coconuts possibly because she did not know how to integrate the two knowledge systems in formal learning situations. Therefore, she used the one with which she was most familiar with — in this particular case, rote learning. The teaching strategy she elected to use did not show how coconuts are grouped and counted during the gathering process, or why they are grouped and counted in that particular way.

On the other hand, the parents did not seem to find the integration of Indigenous and Western mathematical knowledge an issue. They did not seem to need any training to assist them to integrate the two knowledge systems in their everyday activity. They did it according to how they have always integrated new knowledge that they acquire through contact with other people. In this particular case, it seemed that the teachers' attitude, coupled with the lack of knowledge about the bilingual teaching, bridging from vernacular to English, and the integration of Indigenous and Western mathematical knowledge hindered the language policy implementation.

This finding is similar to Kaleva's (2001) study that focused on secondary school mathematics teachers. Kaleva found that teachers are willing to integrate Indigenous and Western mathematical knowledge, if they have sufficient Indigenous mathematical knowledge to begin with, and if they value their own Indigenous knowledge.

Furthermore, these findings suggest that the training conducted in preparation for the vernacular and bilingual instruction implementation, and the success of the implementation is dependent on how well teachers are prepared to implement the language policy, their view of the language policy, and how well-informed they are about the current teaching strategies. Their views and perceptions of this policy, awareness of the rationale for the change in the language of instruction policy, how well they have been trained, and the number of inservice training sessions received have a large bearing on their willingness to implement the policy and to change their established classroom and teaching strategies and biases.

The research found that the parental participants in the community site still view Indigenous knowledge positively. This next excerpt of interview data from a parental interview illustrates this:

> We use mathematics in many things. We count coconuts, *tabu* shells, measure out plots for gardens, measure out spacing between crops, measure out materials for building houses and spaces when building houses, and count goods that are to be sold. We count coconuts in twos, fours, sixes, twelves, and tens. We count taro and pitpit in sets of sixes as well. We also count *tabu* shells. We start with just five *tabu* shells; that is, the base value, followed by sets of ten *tabu* shells, then sets of twelve *tabu* shells, and lastly sets of twenty *tabu* shells. The value increases as the number of shells increases. After forty *tabu* shells, we start to measure out *tabu* like in pokono.
>
> We measure them out using our arms. We also use measurement when we build houses. In the current time, when we build houses, we use tape measures as well to measure the

> length and width of the buildings and materials to be used. We use steps and arm lengths, too. We also use both our own materials and the modern ones to build our houses. Now we also use the white men's way of counting a lot of things.

During the interview, interviewee also talked about how his son, understands mathematical subject content, and how he attempts to link that to the activities he participates in at home. He also stated that the use of vernacular as the language of instruction is effective because his son was able to better understand subject content. He stated that the integration of Indigenous and Western mathematical concepts is similar to the type of education his children and his siblings receive at home. Ludwick was of the view that this teaching strategy assists students to progress faster in their learning. This is shown in these statements:

> The counting that he learns in school is the same as what we use at home. He enjoys learning because he understands the subject content which is similar to what he learns at home.

In summary, these data support Goodman (Goodman, 1967; Goodman & Fleming, 1969; Goodman & Goodman, 2004), Clarke (1980), Cummins (1979), Alderson and Urquhart (1984), Wallace (1988) and Lopez and Greenfield (2005) who claim that children learn better in a language that they know best. One issue is the demotivation of learning in a second language (Zareian & Jodaei, 2015) while another is the loss of culture and its subsequent social issues (Reyhner & Tennant, 1995) thus making the maintenance of first language of high importance. In addition, there are cognitive advantages in learning basic concepts in your home cultural language and when languages are both strong to strengthen conceptual understanding. Code-switching and other language tools are important resources in multilingual classrooms especially when the teacher is also able to make use of both languages (Miranda & Adler, 2010; Setati, Adler, Reed, & Bapoo, 2002). The first language supports through communication and coherence as the second language, especially the mathematical technical terms are being consumed (Venkat & Adler, 2012). It also shows that Ludwick still values his Indigenous knowledge. His passion for this subject was shown by the smile on his face when relating his own Indigenous number and measurement knowledge to me and the rise of his voice when he pointed out the similarities with Western concepts.

Ludwick's in-depth knowledge of his own Indigenous number and measurement related with confidence how the integration of Indigenous and Western number and measurement is done in his community. The general acceptance in which the two knowledge systems were integrated and new knowledge adapted to suit the changing environments in the community. There were no signs of resistance on the integration of knowledge systems during the discussion. This illustrates the success and acceptance in some elements of adaptation of Western number and measurement knowledge in this community. This is reflected in this statement:

> We use tape measures as well to measure the length and width of the buildings… Now we use the white men's way of counting a lot of things.

This shows that the integration of Indigenous and Western number and measurement is already a normal practice in this particular community. The next excerpt from a students' focus group interview also shows that young children are taught to use number and measurement in their community activities.

The Students' Home Environments

The following summary of excerpts from students' comments show how they apply number and measurement in their home activities:

Activities at Home

Mixina: We count stones… count coconuts…
Apex: Weed at home…
Conxie: Weed our garden… take care of the crops in the garden…
Grace: We scrape coconuts…
Kiroro: Collect leaves for cooking and to wrap food in…
Cory: We gather food from the garden…
Kiroro: Collect firewood and coconuts for cooking…

Building houses

Conxie: They ask us to collect nails, hammer… timber… corrugated iron…
Apex: We hold things while they nail them down…
Kiroro: Cut them with a saw…

Gardening

Cory: Yes we work in the garden….
Pauline: Sometimes we do….
Conxie: No we just play in the garden…
Apex: We make small gardens…
Kiroro: We help our parents…

Activies in the garden

Kiroro: We use our bush knives to cut the grass…
Conxie: We make new gardens…
Mixina: Weed our garden…
Cory: We grow peanuts… taro…
Joy: Cassava…bananas…
Conxie: Corn… cucumbers…

Marketing

Pauline: Yes we sell some of them …
Conxie: Eat some…
Mixina: Yes we eat them…

Earnings

Apex: We sell them …beans… peanuts…
Mixina: Cucumbers for 20 t…
Pauline: 50 t… cooking bananas…
Conxie: We sell a hand for K1.00…
Kiroro: K2.00… a whole bunch…
Joy: Small…K5.00 …
Cory: A big bunch… K10.00…

During discussions, the students stated that they assisted their parents when at home. They did not seem to be aware that they were using number and measurement in the activities.

Paraide had the privilege of visiting some of the students' home environments. They participated in collecting firewood for fuel, cooking meals, gathering coconuts, making small gardens, maintaining the gardens, selling the produce from the gardens, processing *tabu,* and playing games. Her access to some of the students' homes was quite easy because she was related to all the student research participants. It was expected that she would visit the close relatives and their families.

During such visits, Conxie and Mixina used both Indigenous and Western number and measurement in their normal activities. For example, they counted, measured, and estimated the quantity of food needed when cooking, and the spaces between the crops they planted in their gardens. The students observed their parents during these activities, while working alongside them. They were encouraged to immediately apply any new skills that they had learned. For example, when Conxie and Mixina wanted to buy a bundle of peanut or fried flour, the parents measured out their sets of *tip na lapitaka* in order to buy them. They stood close by and watched how this was done. This learning process is similar to Jude's *tip na laptikai na tabu* value lesson, where students learned to count this particular value of *tabu*.

When Conxie and Mixina were learning new skills they were closely supervised. For example, during the study, Conxie and Mixina had to learn how to process *tabu* in preparation for a family death ceremony. They were encouraged to watch the whole process of preparing *tabu*. The adults measured out the appropriate lengths of strips of cane—about half a metre long—dressed the cane strips, chose the right size *tabu* shells to string together, and set appropriate spacing between the *tabu* shells during threading. The shells had to be one centimetre apart, and tight fitting on the strips of cane. Conxie and Mixina were encouraged to participate. Their younger siblings also stood by and watched, or participated by holding the dressed cane for stringing the *tabu* and choosing the right size of *tabu* shells to string. This shows that Conxie's and Mixina's home learning—especially in the processing of *tabu* —is similar to some of their Grade 2 number lessons that were observed in the elementary school site. In this particular case, they were applying both Indigenous and Western number and measurement knowledge when working with the *tabu*.

After completing Grade 2, Mixina, Apex, Conxie, Grace, Cory, Kiroro, Pauline and Joy progressed to Grade 3 in the lower primary School site. (Joe transferred to another province and bright Matis left school.) Their Grade 3 teacher was Rose. During Paraide's first visit, Rose's group-counting lesson was similar to Tina's lesson. Rose did not use Tinatatuna as the dominant language of instruction during my first visits, as prescribed in the lower primary curriculum documents. She used English as the dominant language of instruction. When Paraide asked about Matis, Rose told me that he had dropped out of school because he did not understand English. Rose made these statements about him:

> He never said anything in class and I thought he was a slow learner. When I explained the exercises to him in Tinatatuna he was happy and he did his mathematics exercises really fast and accurately. He is a very bright boy, and he understood mathematical concepts well when I explained them in Tinatatuna. He just dropped out of school because he did not understand English.

Rose also used the rote learning strategy to teach multiplication. The students recited the times tables during two of her number lessons. Paraide noted that Mixina, Apex, Conxie, Grace, Cory, Kiroro, Pauline, and Joy were restless during these lessons. They completed their multiplication faster than most of the students, and then sat waiting expectantly for the next tasks.

There were differences between Tina's and Rose's number teaching strategy. For example, Tina attempted to tap into the students Indigenous number knowledge during her number lesson and attempted to link Indigenous and Western number knowledge. On the other hand, Rose got the students to recite the x 2, x 3, x 4, x 5, x 6, x 7, x 8, x 9, x 10, x 11, x 12 multiplication tables. Mixina, Apex, Conxie, Grace, Cory, Kiroro, Pauline, and Joy had learned the x 2, x 4, x 6, x 10, and x 12 in elementary school. They did not have any problems with x 3, x 5, x 7, x 8, x 9, and x 11 concepts during this number lesson. They applied their existing knowledge of counting in groups to work out the problems using stones and sticks.

Rose did not link Indigenous mathematics with her number lesson. Linking the known knowledge to the unknown is prescribed in the lower primary mathematics syllabus. However, during the second and third visits, Rose's teaching strategy had changed. During these particular visits, she made the effort to use what the students already knew as a stepping stone when introducing new number and measurement knowledge.

Matang's Study of Children's Arithmetic Conceptualisation and Language of Instruction

Matang began his doctoral study with his own language group, Kâte. This is a non-Austronesian language in the mountains of the Huon Peninsula. Like many other non-Austronesian languages of PNG, this language is likely to be at least 20 000 years old. The counting words are made up of the main frame words of 1, 2, 5 and 20. As a digit-tally system, it has the secondary cycle of 5 with a tertiary cycle of 20 and involves links to the fingers and toes which are often used in counting. The counting word examples are given in Table 12.1. Matang commented:

Table 12.1
Kâte Counting System Illustrating Embedded English Arithmetic

English numeral in figures	Equivalent *Kâte* number word	*Kâte* operative pattern for each counting number words
1	*moc*	1
2	*jajahec*	2
3	*jahec-â-moc*	3 = 2 + 1
4	*jahec-â-jahec*	4 = 2 + 2
5	*me-moc*	5
8	*me-moc â jahec-â-moc*	8 = 5 + (2 + 1)
13	*me-jajahec â jahec-â-moc*	13 = 10 + (2 + 1)
15	*me-jajahec â kike-moc*	15 = 10 + 5 or 15 = 5 + 5 + 5
20	*ŋic-moc (ngi moc)*	20 (or 20 = 4x5)
23	*ŋic-moc â jahec-â-moc*	23 = 20 + (2 + 1)
40	*ŋic-jajahec*	40 = 20 + 20 (or 40 = 2x20)

[Furthermore, from Table 12.1] it can be noted that unlike the disjoint nature of the individual number relationships found within the English (Hindu-Arabic) numeration system, the counting structure of the Kâte numeration system is such that the use of each number word automatically provides the important number relationships between the individual counting numbers in terms of their order of occurrence in any counting tasks … For example, the Kâte number word for 8 is *me-moc â jahec-â-moc* translated into its operative pattern is 8 = 5 + (2 + 1) hence apart from emphasizing the relative sizes of the counting numbers 8, 5,

> 2 and 1, it also reinforces three important mathematical concepts associated with the operation of elementary addition These are firstly the concept of addition as an operation quantifying the counting numbers 5, 2, and 1, observing 8 as the resulting sum representing the total quantity of all the addends Secondly, the order of operation whereby the operation inside the grouping symbol "( )" indicates that this operation has been performed first. In everyday counting tasks, when one counts up to 8 in Kâte, the emphasis is placed on the counting associations namely, *me-moc* and *jahec-â- moc* where the connecting letter "*â*" represents the idea of a plus sign (+) used in the English (Hindu-Arabic) system. Thirdly, the idea of 5 as a composite unit upon which numbers between 5 and 20 are built, likewise for 20, which is a composite unit for every Kâte numeral beyond 20 (Matang, 2005, p. 507)

Thus it is reasonable to expect that students using their foundational counting system will learn arithmetic concepts well. In fact, Matang did find this. In his area were schools that taught only in Kâte (two were used, one near a road and one three-hours walk from the road so the children had virtually no contact with non-Kâte speakers or life outside a village) and one in which the teaching was in the lingua franca Tok Pisin with a little English even though 90 % of the children spoke Kâte in a settlement area. He assessed each student on a schedule of early arithmetic covering counting, forward and backward number sequences, subitising small numbers, addition, subtraction and multiplication/division. There were three items per concept/skill totalling 21. Children in Schools A and B were interviewed first in Kâte and then in English while children in School C were only interviewed in English. Results are presented in Table 12.2.

Table 12.2
Comparison of Means for Seven Arithmetic Tasks Between Children With Instruction in the Lingua Franca or the Children's Foundational Language

Numerical Tasks	School A- Kâte, near road				School B – Kâte, more remote				School C – Tok Pisin, some English			
	Mean	Stdev	N	%	Mean	Stdev	N	%	Mean	Stdev	N	%
FNWS	2.82	0.53	17	94.0	2.29	0.77	17	76.3	2.72	0.57	18	90.6
BNWS	1.06	0.90	17	53.0	1.35	0.79	17	67.5	2.11	0.90	18	70.3
Subitising	0.94	0.24	17	94.0	0.94	0.24	17	94.0	0.89	0.32	18	89.0
Counting	2.59	0.80	17	86.3	2.88	0.33	17	96.0	2.67	0.59	18	89.0
Addition	1.06	0.83	17	53.0	1.88	0.93	17	62.7	2.06	1.06	18	68.7
Subtraction	1.71	1.05	17	57.0	1.47	1.01	17	49.0	1.17	1.04	18	39.0
Mult/Div	0.65	0.49	17	65.0	0.47	0.51	17	47.0	0.61	0.50	18	61.0
Overall Mean	1.55	0.69		71.7	1.61	0.65		70.4	1.75	0.71		72.5

Note. FNWS – forward number word sequence; BNWS – backward number word sequence; % percent correct on the items; Mult/Div – multiply/divide.

It is interesting to note that the students in school C were slightly more able to count backwards and add but to conceptualise ideas, as in subtraction, they were not as able as the other students. Overall, none of the students did well in subtracting which is not unusual even in other countries but for these students' context, they are rarely like to do more than add or count in culture. The paired sample correlation in results for children who were asked in both Kâte and English was highly significant (r = 0.991 for School A and 0.921 for School B, p ≈ 0).

Matang then went on and looked at schools in different language groups of PNG from four different Provinces (Matang & Owens, 2014). The schools were categorised into six groups depending on the language or languages of instruction. These were Tok Ples (vernacular) only, Tok Ples and English, Tok Ples and Tok Pisin (lingua franca), Tok Pisin and English, Tok Pisin only, and Tok English only. In this study, children could choose the language for answering. He compared the number of errors made by children. To obtain a more equitable comparative group, he selected only the top third of students from each school to be involved in the study as the bottom third especially varied considerably from school to school. The children chose the language to respond in and Matang was able to use that language as he knew enough of each language having taught in those areas. It should be noted that the EO group were likely to come from more educated families and that those in EP were likely to be in towns. There were 272 children in total. Matang showed that for students in the monolingual situations (VO, EO, PO), those in vernacular only classes had fewer incorrect response rates as shown by the z-scores and slightly better than those in VE, the Tok Ples – English combination (Table 12.3).

Table 12.3
Students Incorrect Response (ICR) on Items of Matang's Assessment of Early Numeracy

Language of Formal Instruction School Group	ICR mean scores (%) on TENA task groups								Task Group Mean	Final z-score
	TG1	TG2	TG3	TG4	TG5	TG6	TG7	TG8		
PE *(N=32; S=4)*	3.47	2.23	19.64	8.18	11.46	36.46	38.54	28.13	18.51	-0.99
VO *(N=94; S=7)*	4.49	6.84	23.25	10.16	7.45	26.24	45.74	27.66	18.98	-0.86
VE *(N=27; S=1)*	3.70	3.17	21.69	12.35	8.64	23.46	49.38	29.63	19.00	-0.85
EO *(N=18; S=1)*	3.70	7.14	23.81	9.13	14.81	38.89	66.67	30.56	24.34	0.64
PO *(N=49; S=5)*	8.62	5.83	28.28	16.13	9.52	32.65	56.46	42.86	25.04	0.84
VP *(N=52; S=4)*	5.13	8.52	30.77	16.03	9.62	38.46	64.10	38.46	26.39	1.21
Mean of all groups	4.85	5.62	24.57	12.00	10.25	32.69	53.48	32.88	22.04	0.00
SD of all groups	1.95	2.44	4.17	3.45	2.59	6.52	10.91	6.27	3.58	1.00
TENA z-score	-4.80	-4.59	0.71	-2.81	-3.29	2.98	8.78	3.03	0.00	

Notes. TG1=Numeral identification; TG2=Forward number word sequence; TG3=Backward number word sequence; TG4=Subitising; TG5=Counting; TG6=Addition; TG7=Subtraction; TG8=Multiply/Divide; ICR= Incorrect responses as percentage of total number of questions in task – lower % means higher mathematical competency. PE – Tok Pisin-English; VO – Tok Ples (vernacular) only; VE – Tok Ples – English; EO – English only; PO – Tok Pisin only; VP – Tok Ples, Tok Pisin.

On average, children in VO schools made fewer incorrect responses in contrast to children in PO schools who scored the highest number of Incorrect Responses (ICRs) out of the three school groups. These results suggest that children in VO schools have greater educational advantage in terms of competency skills in mathematics when taught in their own mother tongue compared to PO and EO children. However, the Tok Pisin-English group, situated in towns with facilities like books and working parents, did perform best but it needs to be considered with the other groups using Tok Pisin who did not do well at all. Then in order to assess the value of Tok Ples in education, Matang made a comparison of groups with some Tok Ples compared to those with none. As a result, he showed that Tok Ples did have a significant positive effect on learning (Table 12.4).

Table 12.4
Language of Formal Instruction Influence on Incorrect Response Rate on Mathematical Assessment of Children's Mathematics Competency Level– Tok Ples Grouping versus non Tok Ples Grouping

Tok Ples Grouping		Non Tok Ples Grouping	
VO schools	18.98	PO schools	25.04
VP schools	26.39	PE schools	18.51
VE schools	19.00	EO schools	24.34
Group mean	21.46	Group mean	22.63
Stand. deviation	4.27	Stand. deviation	3.59

Note. Lower scores indicate higher mathematics competency level as the measure was measured by Incorrect Responses (ICR).

Tok Ples grouping, on average, displayed a higher level of competency skills in early school mathematics based on a lower mean score of 21.46 of incorrect responses compared to 22.63 for children in the non Tok Ples grouping who performed at a lower mathematics competency level. A one-way ANOVA indicated a considerably large F-test ratio of 6313.95 for the between-groups variance compared to 0.001 for the within-groups variance. These results imply that the differences between the means for ICRs for children in the Tok Ples and non Tok Ples groupings are unlikely to be due to chance (Bryman, 2016). Interestingly, the Grade 3 teachers' questionnaire also showed that teachers felt children who have learnt basic mathematics concepts in Tok Ples do well and have a strong conceptual basis for school mathematics.

Matang also carried out a comparison of reaction times to respond and again he found those in the Tok Ples schools performed faster indicating a better conceptual understanding. These were used to suggest higher cognitive efficiency (Ericsson, Chase, & Faloon, 1980). Children in the Tok Ples – English (VE) and Tok Ples only (VO) groups showed highest efficiency overall compared especially to Tok Ples – Tok Pisin (VP) group as their correct reaction time was lower (Table 12.5).

Table 12.5
Effect of Language of Formal Instruction on Children's Cognitive Efficiency Levels (CEL) across eight TENA Task Groups

Language of formal instruction School Group	Main TENA task groups/CRTs mean scores										
	TG1	TG2	TG3	TG4	TG5	TG6	TG7	TG8	LFI Mean	LFI z-score	Rank
VE *(N=27; S=1)*	1.65	6.00	9.31	16.10	8.22	9.36	13.61	23.38	10.95	-1.13	1
PE *(N=32; S=4)*	1.40	5.38	9.05	13.23	6.24	8.96	16.01	29.23	11.19	-1.00	2
VO *(N=94;S=7)*	1.64	6.44	11.43	15.20	8.05	9.03	20.42	24.95	12.15	-0.44	3
PO *(N=49; S=5)*	1.89	6.55	11.58	17.46	9.39	8.45	25.16	28.72	13.65	0.44	4
VP *(N=52; S=4)*	1.78	7.27	11.28	15.72	8.71	10.96	20.98	38.25	14.37	0.86	5
EO *(N=18; S=1)*	1.65	5.58	9.06	19.32	9.47	11.55	35.32	28.42	15.05	1.26	6
TENA Mean	1.67	6.20	10.29	16.17	8.35	9.72	21.92	28.83	12.89	0.00	-
Stand. Deviation	0.16	0.70	1.26	2.07	1.19	1.24	7.71	5.17	1.71	1.00	-
TENA z-score	-6.56	-3.91	-1.52	1.92	-2.66	-1.85	5.28	9.32	0.00	-	-
TENA Effect	-11.22	-6.69	-2.61	3.28	-4.54	-3.17	9.03	15.94	0.00	-	-

Notes. TENA=Task-based early number assessment; TG=Task group; S=No. of Schools; LFI=Language of formal instruction; CRT=Correct reaction time; Measure of efficiency is Final Group Means for Correct Reaction Times (CRT) (in seconds). Lower scores suggest great cognitive efficiency levels. *Source*: Matang & Owens (2014).

When Matang compared the groups who had some Tok Ples with those with none, there were significant difference in the time taken to provide correct responses (Table 12.6). A one-way ANOVA indicated a large *F*-test ratio of 1648.90 for the between-groups variance compared to 0.001 for the within-groups variance. This result implies that the differences between the means for response time for correct responses are unlikely to be due to chance (Bryman, 2016).

The use of Tok Ples as a language of instruction increases children's cognitive efficiency as indicated by correct reaction times as well as reducing incorrect responses to arithmetic items (Matang & Owens, 2014, p. 547).

In summary, Paraide's (2010) study found that the integration of Indigenous and Western number and measurement in formal learning is now an established practice in the elementary School site. A recent study on the implementation of the elementary curriculum (Evans et al., 2006, p. 83) also found that the integration of formal and community knowledge is now also a strength in other elementary schools. Matang's (2005) study on one language group indicated the value of learning in Tok Ples or home language and his study across a number of language groups showed that the specific sample using English did well, other students who had access to Tok Ples to some extent did show better conceptual understanding than those who had none. Matang's (Matang & Owens, 2014) and Paraide's studies found that the bilingual teaching strategy is yet to be fully accepted by the lower primary teachers. Attempts are now being made to embrace this teaching strategy. However, the use of Tok Pisin is limiting the use of Tok Ples in many areas reducing the value of using Tok Ples as a language of instruction (Muke, 2012). Furthermore, with the change in government policy on English as a language of instruction and standards beginning in 2015, there are some issues around the general curriculum principle of integration of knowledge and the value of bilingual education.

Table 12.6
Language of Formal Instruction Influence on Cognitive Efficiency Levels– Tok Ples Grouping versus Non Tok Ples Grouping

Tok Ples Grouping		Non Tok Ples Grouping	
VO schools	12.15	PO schools	13.65
VP schools	14.37	PE schools	11.19
VE schools	10.95	EO schools	15.05
Group mean	12.49	Group mean	13.30
Standard deviation	1.74	Standard deviation	1.95

Note. Cognitive Efficiency is measured by CRT (Correct Reaction Time) mean scores (in seconds). Lower scores means greater cognitive efficiency. *Source*: Matang & Owens (2014).

These studies bring into account, together with those of Saxe (2012), the impact of schooling on mathematics as a system and in a cultural context. We see the changes that occur through the possible loss of knowledge or the integration of knowledge. These modern occurrences on the diffusion and supplanting of systems, local modification, and dualities of counting within cultures to cater for contact, enhance our understanding of earlier historical situations prior to Western schooling. These studies also highlight the importance of teachers being well trained to bring Indigenous and Western knowledges together in the classroom. Saxe's (2012) and Paraide's (2010) studies indicated the effect of a teacher's position in society on the possible loss of Indigenous knowledges if they are not well versed in their cultural knowledge. It is therefore more urgent that the systems of Papua New Guinea are maintained as they can provide excellent comparisons for enhancing school mathematics concepts, uses in cultural situations, and a richness to a global understanding of mathematics and the history of mathematics.

References

Alderson, C., & Urquhart, A. (1984). *Reading in a foreign language*. London, UK: Longman.

August, D., & Hakuta, K. (1997). *Improving schooling for language minority children: A research agenda*. Washington, DC: National Academy Press.

Barrington-Thomas, E. (1976). Problems of educational provision in Papua New Guinea. In E. Barrington-Thomas (Ed.), *Papua New Guinea Education* (pp. 3–16). Melbourne, Australia: Oxford University Press.

Battiste, M. (2002). *Indigenous knowledge and pedagogy in First Nations education: A literature review with recommendations* (Working paper). Ottawa, Canada: Apamuwek Institute.

Bialystock, E. (1991). *Language processing in bilingual children*. Cambridge, UK: Cambridge University Press.

Bryman, A. (2016). *Social research methods* (5th ed.). Oxford, UK: Oxford University Press.
Clarke, M. (1980). The short circuit hypothesis of ESL reading or when language competence interferes with reading performance. *Modern Language Journal, 64*, 203–209.
Collier, V., & Thomas, W. (2007). Predicting second language academic success in English using the prism model. In J. Cummins & C. Davison (Eds.), *International Handbook of English language teaching, Part1* (pp. 333–348). New York, NY: Springer.
Cummins, J. (1979). Cognitive/academic language proficiency, Linguistic interdependence, the optimum age question and some other matters. *Working Papers in Bilingualism, 19*, 197–205.
Cummins, J. (1996). *Negotiating identities: Education for empowerment in a diverse society*. Ontario, CA: California Association for Bilingual Education.
Department of Education Papua New Guinea. (1974). *Report of the Five-Year Education Plan Committee, September, 1974 (The Tololo Report)*. Port Moresby, PNG: Department of Education.
Department of Education Papua New Guinea. (1976). *Education Plan, 1976–1980*. Port Moresby, PNG: Department of Education.
Department of Education Papua New Guinea. (1998). *Elementary curriculum statement*. Waigani, PNG: Department of Education Papua New Guinea.
Department of Education Papua New Guinea. (2000). *Lower primary mathematics: Bridging to English resource book*. Port Moresby, PNG: Department of Education.
Department of Education Papua New Guinea. (2004). *Lower primary mathematics syllabus*. Port Moresby, PNG: Author.
Dickson, D. (1976). Murray and education: Policy in Papua, 1906–1941. In E. Barrington-Thomas (Ed.), *Papua New Guinea education* (pp. 21–45). Melbourne, Australia: Oxford University Press.
Ericsson, K., Chase, W., & Faloon, S. (1980). Acquisition of a memory skill. *Science, 208*, 1181–1182.
Evans, T., Guy, R., Honan, E., Kippel, L. M., Muspratt, S., Paraide, P., & Tawaiyole, P. (2006). *PNG Curriculum Reform Implementation Project - Impact Study 6*. Faculty of Education Deakin University and National Research Institute PNG.
Garcia, E. (1994). *Understanding and meeting the challenges of student cultural diversity*. Boston, MA: Houghton Muffin.
Gash, N., & Whittaker, J. (1975). *A pictorial history of New Guinea*. Melbourne, Australia: Melbourne University Press.
Goodman, K. (1967). Reading: A psycholinguistic guessing game. *Journal of the Reading Specialist, 6*, 126–135.
Goodman, K., & Fleming, J. (1969). *Psycholinguistics and the teaching of reading*. Newark, NJ: International Reading Association.
Goodman, K., & Goodman, Y. (2004). To err is human: Learning about language processes by analyzing miscues. In R. Ruddell & N. Unrau (Eds.), *Theoretical models and processes of reading* (pp. 620–639). Newark, NJ: International Reading Association.
Jablonka, E., Wagner, D., & Walshaw, M. (2013). Theories for studying social, political and cultural dimensions of mathematics education In M. Clements, A. Bishop, C. Keitel, J. Kilpatrick, & F. Leung (Eds.), *Third International Handbook of Mathematics Education* (pp. 41–68). New York, NY: Springer.
Kaleva, W. (2001). *Linkages between Ethnomathematics and Language: A case for collaborative research*, Paper presented in the 11th World Congress on Comparative Education Societies Conference, Korea, July 2001.
Lindfors, J. (1987). *Children's language and learning*. Englewood Cliffs, NJ: Prentice Hall.
Lopez, L., & Greenfield, D. (2005). The cross language transfer of phonological skills of Hispanic Head Start children. *Bilingual Research Journal, 28*(1), 1–18.
Matang, R. (2005). Formalising the role of Indigenous counting systems in teaching the formal English arithmetic strategies through local vernaculars: An example from Papua New Guinea. In P. Clarkson et al. (Eds.), *28th Conference of Mathematics Education Research Group of Australasia, Melbourne* (pp. 505–512). Adelaide: MERGA. https://www.merga.net.au/documents/RP562005.pdf.

Matang, R., & Owens, K. (2014). The role of Indigenous traditional counting systems in children's development of numerical cognition: Results from a study in Papua New Guinea. *Mathematics Education Research Journal, 26*(3), 531–553. doi:10.1007/s13394-013-0115-2.
Matsuura, K. (2008). *Why languages matter: Meeting Millennium Development Goals through local languages*. Dallas, TX: SIL International.
Miranda, H., & Adler, J. (2010). Re-sourcing Mathematics teaching through professional development. *Pythagoras, 72*, 14–26.
Muke, C. (2012). *Role of local language in teaching mathematics in PNG* (PhD). Australian Catholic University, Melbourne, Australia.
Narokobi, B. (1980). *The Melanesian way-Total cosmic vision of life*. Port Moresby, PNG: Institute of PNG Studies.
Owens, K., Muke, C., Owens, C., & Clarkson, P. (in press). Change and continuity in mathematics education in Papua New Guinea. In J. Mack & B. Vogel (Eds.), *Mathematics and its teaching in the Asian/Pacific region*. Hackensack, NJ: World Scientific Press.
Paraide, P. (2009). Number in the Tolai culture—Occasional paper. ResearchGate.
Paraide, P. (2010). *Integrating Indigenous and Western mathematical knowledge in PNG early schooling* (PhD). Deakin University, Geelong, Victoria, Australia.
Pennycook, A. (1998). *English and the discourse of colonialism*. London, UK: Routledge.
Pica, T. (2003). Second language acquisition: Research and applied linguistics. *Working Papers in Educational Linguistics, 18*(2), 1–26.
Reyhner, J., & Tennant, E. (1995). Maintaining and renewing Native languages. *Bilingual Research Journal, 19*(2), 279–304. doi:10.1080/15235882.1995.10668606.
Said, E. (1993). *Culture and imperialism*. London, UK: Chatto & Windus.
Saxe, G. (2012). *Cultural development of mathematical ideas: Papua New Guinea studies*. New York, NY: Cambridge University Press.
Setati, M., Adler, J., Reed, Y., & Bapoo, A. (2002). Incomplete journey: Code-switching and other language practices in mathematics, science and english language classroom in South Africa. *Language and Education, 16*(2), 128–149.
Smith, Geoffrey. (1975). *Education in Papua New Guinea*. Melbourne: Melbourne University Press.
Smith, G. (1981). *Traditional mathematics in Morobe* (Working Paper no. 21). Lae, PNG: Papua New Guinea University of Technology.
Smith, G. (1984). *Morobe counting systems: An investigation into the numerals of the Morobe Province, Papua New Guinea* (MPhil). Papua New Guinea University of Technology, Lae, Papua New Guinea.
Smith, L. T. T. R. (2012). *Decolonizing methodologies: Research and indigenous peoples* (2nd ed., ebook). New York, NY: Zed Books.
Smith, P. (1987). *Education and colonial control in Papua New Guinea: A documentary history*. Melbourne, Australia: Longman Chesire.
Thomas, W., & Collier, V. (2003). The multiple benefits of dual language. *Educational Leadership, 61*(2), 61–64.
Tololo, A. (1976). A consideration of some likely future trends in education in Papua New Guinea. In E. Barrington-Thomas (Ed.), *Papua New Guinea education* (pp. 221–225). Melbourne, Australia: Oxford University Press.
Venkat, H., & Adler, J. (2012). Coherence and connections in teachers' mathematical discourses in instruction: original research. *Pythagoras, 33*(3), 1–8.
Wallace, C. (1988). *Learning to read in a multicultural society: The social context of second language literacy*. New York, NY: Prentice Hall.
Young, R. (2001). *Postcolonialism: An historical introduction*. Oxford, UK: Blackwell.
Zareian, G., & Jodaei, H. (2015). Motivation in second language acquisition: A state of the art article. *International Journal of Social Sciences & Education, 5*(2), 295.

Chapter 13
Rewriting the History of Number from Papua New Guinea and Oceania Evidence

Kay Owens and Glen Lean

Abstract In this chapter, we bring to a conclusion the studies of counting systems given in the previous chapters together with the arguments presented on context and prehistory discussed earlier. We justify this study in terms of its quality for historical studies. This study draws on multiple disciplines including mathematics, mathematics education, linguistics, anthropology and archaeology. We use recent studies on contact and change, hegemony and loss, to expand on Lean's study of the history of number. In particular we provide ways in which Papua New Guinea and Oceania Indigenous knowledges can expand understandings of a sense of number, patterns of number systems, groupings and place value, parts and wholes, multiplicative thinking, representations of number, and large number lores based on our knowledge of the history of number from a PNG and Oceania perspective. This chapter also considers implications for viewing and valuing early numeracy and numerosity differently and hence ways of catering for diversity in mathematics education.

Keywords History of number • diffusion of number systems • number system patterns • numerosity • large numbers • diversity of learning in number

The Data Used as Evidence for this History of Number

In 1970 Glen Lean began to collect data from his students at the Papua New Guinea (PNG) University of Technology on their counting systems using the Counting System Questionnaire (CSQ). This was a fascinating journey that occupied 22 years of his life. He collected information on two-thirds of the 850 (then seen as 750+) languages of New Guinea and also Oceania (West Papua, Island Melanesia, Micronesia, and Polynesia). He sourced written documents that were often in obscure records and in many parts of the world – from Papua New Guinea, Australia, Hawai'i, other USA States, The Netherlands, United Kingdom and Germany to name most of them. He made numerous field trips in PNG. For example, he particularly travelled to Enga to check out the Mae dialect information as he was getting conflicting information from younger men. He collated this information in four appendices (initially 16 volumes for the Provinces of PNG). Examples of this extraordinary work can be found in Appendices B and C of this book which contain data relevant to our arguments on the history of number presented in Chapters 8 and 10. Appendix D provides data on the diversity of body-part tally systems and Appendix E has maps adding to those given in Chapter 8. Significant is the fact that much of Lean's data from the field trips and Counting System Questionnaires were given by Indigenous speakers of the languages while others were recorded by linguists, especially from SIL,

K. Owens (✉)
Charles Sturt University, Dubbo, NSW, Australia
e-mail: kowens@csu.edu.au

G. Lean

K. Owens et al., *History of Number*, History of Mathematics Education,
DOI 10.1007/978-3-319-45483-2_13

who spent years studying the languages in situ. Although he used Wurm and Hattori's (1981) language atlas, he collated in one or more tables, the data that he had collected for each language. In some cases, considerable variations occurred from earlier and later sources, but mainly from different places speaking the language which often indicated different dialects, many of which are now considered languages. Lean had to make judgements about what to record as the commonly recognised words and spelling (remembering that these were oral languages with only some having standard orthographies even today). One significant aspect of his work is the collation of recordings of Indigenous knowledge not long after colonisation and often from first contact. The fact that he collected data in the 1970s from students who were among the first to complete a school education and who still spoke their home languages is also important. In quite a few languages, there are few people in these language areas today (less than 50 years later) still using these counting systems on a regular basis whereas in other places significant modifications have occurred. However, in many places people do know these systems and use them in everyday conversation. Earlier Reform government policies (1990s to 2013) on elementary schools strengthened the use of the vernacular counting systems. For people wanting to maintain or revitalise their cultural knowledge and language Lean's data will prove to be invaluable in the future.

Lean's analysis used Salzmann's (1950, 2006) cycle system simplifying the number of groups compared to Smith (1984) but also separating the use of digit tallies into two categories—cycles of (2, 5, 20) and (5, 20). Details of languages for each of his categories (2-cycle, 5-cycle, 10-cycle, 4- and 6-cycle, and body-part tally) exemplify Lean's extraordinary analysis on which to argue for antiquity as well as diffusion, invention, and integration. For example, by taking the variety of 2-cycle systems together, it is possible to notice how they cluster in certain areas but also in different pockets (Chapter 3). Thus all the processes that are likely to generate variation in counting systems are evident. The body-part tally systems are discussed in a similar vein (Chapter 4). Dwyer and Minnegal (2016) brought new data on body-part tally systems shedding light on diversity and neighbourhood influences within this kind of system. Saxe's (2012) work provided a detailed analysis of conceptual development within one language group by discussing change over time. By drawing on this evidence, it is possible to show that systems did diffuse especially within language families but also changed from integration due to relationships with other groups or inventions occurred seemingly for cultural reasons such as an emphasis on counting exchanges, valuing multilingualism, building trade and alliances, or even geographic changes.

Without separating the 5-cycle systems some of the differences between the systems may have been overlooked (Chapter 5). Most of the (5), (5, 10, 20), and (5, 20) systems are digit-tally systems which are very common in Non-Austronesian (NAN) languages but can also be found among Austronesian languages which provides a significant argument for integration of counting systems rather than diffusion of counting systems. The 10-cycle system spread with the Oceanic languages but there appears to be influence and invention for NAN 10-cycle systems (Chapter 6). Chapter 7 details the unusual 4- and 6-cycle systems that have developed intuitively and creatively in some areas and often found in neighbouring languages suggesting the importance of counting in cultural activities especially those involving two language groups. Throughout the chapters, examples of mathematical thinking to create, for example, systems for large numbers are presented.

These chapters set the scene for the critique of the diffusionist theory of counting systems (Chapter 9). A time-frame for these ancient counting systems is established based on linguistic, archeological, genetic, and biogeographic evidence together with that provided by the deep structures of the various counting systems themselves (Chapter 10). These systems were established well before the often touted spread of counting systems, one after the other, from the Middle-East region (Seidenberg, 1960) and not in the same way as Seidenberg suggested. Lean's (1992) thesis set the scene for alternative approaches to explain diversity.

A number of more recent studies have provided in-depth studies of counting, including counting in context while other analyses have suggested a stronger emphasis on numerosity than counting. This is particularly the case for Chapters 11 and 12 by Paraide on the Tinatatuna (Kuanua) language of the Tolais of East New Britain. Examples of counting systems for large numbers and paired counting in

Yu Wooi are provided by Muke (2000) from his study in Mid-Wahgi in which he gained responses from 73 speakers from different tribal groups who use one of the four dialects. Body parts represented the number of hundreds but this could be extended. Examples of paired counting are given also in Austronesian languages. Not only large numbers in the NAN Iqwaye language developed but the importance of counting in cultural contexts and the valuing of counting was established by Mimica (1988) who immersed himself in their cultural thinking.

Recent studies of the 6-cycle systems on Kolopom Island off the New Guinea coast and neighbouring languages by Hammarström (2009), Donohue (2008), and Evans (2009) assist in extending thinking about how counting systems spontaneously developed and spread and their role in cultural systems and values. At least one Kanum system (south coast area of New Guinea) has exponentiation as a result of cultural thinking. The use of the fist and fingers of one hand for groups of 6 while the other hand presents units is another example of body parts, their position, and gestures being used for higher numbers in a non-written language. Furthermore, as Donohue (2008) noted and Leslie, Gelman, and Gallistel (2008) explained, one-to-one counting of small numbers is restrictive but people are able to use visual measure comparisons in a successive sense of size associated often with some form of ratio perspective to compare and know large numbers. Perhaps this is the innate means by which counting systems were generated in different societies creating complex and different systems. Although these systems seemed to lack the numeric language words, these systems are nonetheless powerful ways of thinking mathematically.

Furthermore, studies of paired counting in Oceania indicate significant ways of ensuring numerosity and large numbers in their systems providing a different backdrop to early number concept development compared to one-to-one counting. Having oral rather than written traditions resulted in different approaches to number and hence a significantly different history and perspective on number as espoused in this book, especially in Chapters 8, 9 and 10. Culture, language, and number interplay and influence each other resulting in innovations. The high intellectual cognitive abilities of these Indigenous communities are evident in the efficient use of language patterns that were culturally developed, and resulted in diverse counting systems. Such examples are evident in the classificatory systems that incorporate objects, features of objects, and quantities such as powers of 10 (Chapter 8).

This book presents the findings of these studies in order to establish an appreciation of the numeracy of Indigenous communities across the southern Oceanic regions from communities that have continued for tens of thousands of years, and have variously moved across rugged landscapes and vast oceans. It is anticipated that this study would lead to a new appreciation of the history of number that extends beyond that developed from studies of Europe and Asia and written documents only. In particular, due to the recent contact of these Indigenous communities, oral historic data are valued in explaining the history of number in this region. Nevertheless, a multidisciplinary approach enables the analysis and history to be developed without written records.

Meeting Quality Research Criteria in Historical Studies

One of the critical aspects of a study of the history of number is the trustworthiness of the argument and data upon which the arguments are based (Schoenfeld, 2007). This book is largely built on Lean's (1992) study which met this particular criteria well. He undertook an exhaustive study of historic records and collected data from speakers of the languages through the Counting System Questionnaire and field visits. By using a cross-disciplinary approach he was able to take not only a quantitative argument in the sense of sourcing and where possible triangulating his tabulated data of counting systems but he took note of context and changes over time. His arguments related particularly to the possibilities of innovation and of neighbouring influences. He analysed the systems largely in terms of their cycles or as body-part tallies. However, the frame words and neo-2 cycle systems were carefully documented and compared. While the analysis in terms of cycles assisted in justifying Lean's argument against the diffusionist theory of Seidenberg (1960), he was well aware of other cultural impacts on counting systems and their developments and use. Paraide's (2010) critical analysis in Chapter 12 adds to our understanding of the importance of culture in establishing a history of number and a valuing of different number systems.

There is no doubt that in terms of numbers in his sample of languages, Lean's work has strong grounds from a quantitative perspective (Karp, 2014). Nevertheless,

> quantitative methods in historical research, of course, are quite feasible and useful. Yet one ought not to contrast them with other kinds of methods and assume that quantitative methods are more reliable. Without qualitative analysis and comparisons with other facts, computations are unlikely to be meaningful. (Karp, 2014, p. 19)

However, this book also draws authenticity from the fact that the cultural aspect is not missing, as Karp (2014, p. 17) notes as important. Culture is at the forefront of Chapter 8 that provides case studies of languages across the regions of Papua New Guinea, from West Papua, Island Melanesia, and across Oceania. These studies focused on the link of culture and counting systems and across languages. Paraide's (2010) work discussed in Chapters 11 and 12 and Muke's discussed in Chapter 3. Both undertook in-depth studies of their cultural counting systems knowing both their cultures and western mathematics well. It is their works and other studies such as that by Mimica (1988) that have provided a further dimension to Lean's theory of innovation and neighbours' influence on the history of number (see Chapters 8 and 9). Paraide illustrates how changes occur and how loss of knowledge can be arrested.

A western only view of mathematics and counting is limiting the numeracy knowledge of people in community and globally. Paraide's (2010) study led to our further point that there may be a diminishing knowledge of foundational practices for people in a cultural group. However, by recognising the links and changes, stronger theoretical, educational, and practical bases can be established for communities as shown by Paraide and Matang (Chapter 11) as they discussed the influence of schooling. This book does meet the demand of "identifying and elucidating connections with the general course of the development of society …[as] a crucial problem for the historian of mathematics education" (Karp, 2014, p. 10). Similar affects of school are discussed by Saxe (2012) who also showed the influence of monetary systems and movement beyond the village on development.

We might also ask whether the argument put forward on the history of number is relevant to Papua New Guinea today and whether it is relevant as a body of knowledge for others in the global world (Schoenfeld, 2007). We argue that both is the case. It is significant for Papua New Guineans to know their cultural roots but also to appreciate that change has occurred and to be able to tease out the significance of mathematics in culture, in school, in the modern world, and in the interaction spaces between these worlds (Bhabha, 1995). Such could be the argument given in the summary from the Aboriginal and Torres Trait Islanders Mathematics Association conference (ATSIMA, 2015) where the third space of relationships was key to having not a pipeline affect between community, school and business with blame for failure dragging down students. Preference was given to a model that is circular with two-way relationships between the three areas. Thus community knowledge is in a two-way relationship with school and business. Community knowledge is seen as mathematical.

Another notion that this study fulfils in terms of Karp's (2014) expectation of historical studies is providing a strong argument countering any notion of "primitive", or that base 10 counting was embedded in exotic ritualistic processes (Seidenberg, 1962). Mimica (1988) previously showed that Hallpike's (1979) discussion linking Piagetian notions of sensorimotor and preoperational thought to be similar to Indigenous cultures, especially those using a binary system or digit tally system or stick representations of number could be discounted. He argued in terms of the incorporation of number in culture and the subsequent development of Iqwaye counting to a level above 800 as a power of 20 and with a notion of infinity thus attained. Whether this cultural knowledge is called a worldview, attitude or spirituality, it has a rich sophisticated layer of culture that resulted in a different perspective and an innovative extension of counting beyond one and two or even twenty (see Chapter 3). Cycles and recursions in kinship systems and over time are particularly strong in Indigenous cultures and closely link with concepts that the West may attribute to number. This is evident in the Yolngu language from Northern Australia as documented for this community in book and video by Thornton and Watson-Verran (1996).

The data and comparisons of data are supported by anthropological linguistics, environmental pasts (e.g., Pawley, Attenborough, Golson, & Hide, 2005; Ross, 1988, 1989), and archeological findings (e.g., Denham, 2011; Swadling, 1997) often in recent times. By taking these studies and analysing and comparing the counting systems, an argument can be presented about the nature and spread of mathematics in the region, predating the spread of counting systems in the northern hemisphere. Indeed, it is noted that there are a range of sophisticated systems including base 10 from around 7 000 BP. New analyses such as those by Hammarström (2009) and Donohue (2008) also highlight new understandings of the diffusion of number. In their studies, they point out the complexities of the systems and the internal variations. The rarity of systems assists in the analysis upon which the history of number can be established rather than the similarities (Hammarström, 2010).

Systematic Analysis and Variance

Lean began his analysis by recognising the different phyla of the Non-Austronesian languages and the clusters of the Austronesian languages. Each counting system was identified on a map giving neighbouring systems and its Family and phyla. Then he classified each set of counting words into various 2-cycle systems, or those with primary 5- or 10-cycle systems or body-tally systems. Interestingly these were the only systems that he found in over 1 200 systems from across PNG and Oceania although each of these primary systems often developed secondary systems such as the 4- and 6-cycle systems or the (5, 10, 20) systems. In some languages, there are more than one system. In one language there may be many variations in terms of actual words although the cycle system would be consistent. There are however some interesting recombinations of words and morphemes by members of the community for their own purpose or understanding. A fixed, strict set of words that form the counting order does not necessarily hold. In many instances, it is not the difference between small numbers that counts as much as large numbers and their display representations.

The 2-cycle systems have a great variation and Lean carefully distinguished between different systems by the terms used for 3 and 4. He also emphasised that there could be a secondary cycle and even a third. So, for example, many systems were (2, 5, 20) cycle systems and in most cases these were digit tally systems in the sense that 5 was a hand and 10 two hands or complete hands, and then a foot morpheme was used before the whole man was reached for 20. Some systems used a word (not hand) plus 1, 2, 3, and 4 for the numbers 6 to 9. Together with the cycles, Lean provided the frame words, that is the words that formed the minimum number words from which all other numbers could be identified. In addition, he gave the operative pattern(s) that used addition, subtraction and multiplication and showed how words were combined to make new number representations. Examples of these can be found in Appendices B and C of this book.

However, there was more to his analysis as he found two common variations, particularly among the Austronesian languages on the northern area around Manus and in the southern coast of PNG around Port Moresby. The former involved number 6 being "four up to 10" to form the complete group, 7 as "3 up to 10", "8 as 2 up to 10", and "9 as 1 up to 10" while the latter variation involved pairs so 7 was 2 x 3 + 1 and so on. It seems that pair counting is quite widespread especially among Austronesian languages but possibly others.

There were a few pockets of languages that had either a cycle of 4 or of 6. These seemed to relate to cultural activities and to use of hands for counting. It is also evidence of local innovations with some adaptation or diffusion by groups with whom the groups may have traded or intermarried as there was some proximity. However, there were also pockets in far distant areas suggesting innovation.

Another important variation among systems involves the identification of quantities for different classes of objects, qualities of objects, and quantities. In some cases, words were significantly different, for example, when a ten of an object was named as "complete group of that object" and it differed from 10 of another kind of object. These were generally important and/or common objects in cultural activities. In other cases, a suffix or occasionally prefix is used but there are recognisable morphemes for different numbers although those morphemes would not be used alone as English uses, for example, seven as an abstract noun. The number may be before or after the category and this identifying feature was used to discuss possible relationships between languages that might indicate a diffusion of counting systems. This is particularly noticeable in Oceanic Austronesian languages (Bender & Beller, 2006).

The quantifying categories are also used to represent powers of 10 so that the Oceanic languages frequently provided a means for people to represent in an efficient word pattern very large numbers. This use of large numbers and the keeping of reciprocity knowledge also involved various operations on large numbers even though the terms add, subtract, multiply and divide were not explicitly used. It is an interesting contrast to languages in which speakers could not answer questions with numbers in the thirties (Wassmann & Dasen, 1994).

One final category of systems are the body-part tally systems. Some authors have suggested that these are not calculating systems and are therefore not arithmetic systems. We beg to differ on this point. First, children in Saxe's (2012) study were noticed using points of the body to represent and compare and even to add on beyond ten, sometimes using the traditional system and sometimes 10 as a place marker using the other side of the body. However, the interesting aspects about the body-tally systems which seem to occur only in New Guinea and Australia, is that they exist mainly in Trans New Guinea Phylum, Sepik-Ramu Phylum and Isolates in the Sandaun, East Sepik, Southern Highlands, Hela, Western and Gulf Provinces. The more northerly coastal connections may have occurred at the time of the inland sea allowing closer access for diffusion. However, there is the easterly example of Yupno reported by Wassmann and Dasen (1994) although this could be disputed. There is one possible example, Mangarevan, especially used with large numbers in Oceania (Bender & Beller, 2013). It should be remembered that body-parts were used for representing in numbers in many parts of the world in order to designate large numbers e.g., Bede's system and some Asian systems (Bender & Beller, 2012). These authors particularly note the interconnection between language, embodiment, and cognitive processing. This book also showed their use for representation of powers and multiples such as the Yu Wooi (Muke, 2000) and Iqwaye (Mimica, 1988) and in Oceania.

These gross variabilities were further modified by individual languages and within languages. Lean catered for some diversity in his table formats (see Appendices B and C for examples) but he also reduced differences especially in collating multiple Counting System Questionnaires. It turns out that part of the issue was that there are now far more identified languages (many previously called dialects) than Wurm and Hattori (1981) provided as indicated by the SIL ethnologue (nd), Dwyer and Minnegal (2016), Pawley et al. (2005) and others.

Exemplar Findings of Significance for Our Final Arguments

Contrary to Seidenberg's claim that 2-cycle systems are found only at the extremities of the world in South America, Africa and Australia as other cycles displaced them elsewhere, they were also in PNG. Of the 430 Non-Austronesian (NAN) languages in Lean's data set, only 42 (10%) were of the pure 2-cycle system with 2 Austronesian (AN) languages (1% of the AN data set). In most cases, it is not possible to tell if there is a higher than 5 secondary cycle due to lack of information but in 18 of the 42 NAN systems, they occur together with body-part tally systems as discussed in Chapter 4. Interestingly the two AN systems were reductions in the magnitude of the primary cycle expected from Proto Oceanic but there are other examples with less than 10 counting words and primary 10-cycle systems.

Of his data set, Lean found 109 NAN languages had (2, 5) or (2, 5, 20) cycle systems, mostly of the latter but he has not separated these in discussion. In this category, Lean's meticulous analysis indicated a variety of different combinations for 3 such "2 and another" and one case of 1 + 1 + 1 with 4 being 2 + 2. Many of these systems are 2-cycle augmented by the digit tally system with 5 and 10 having a "hand" morpheme and 20 as "man" or "being" although some have "hand" and "foot" morphemes for 20. There are 18 AN languages also of this nature, mostly in the Markham Valley of Morobe Province indicating considerable NAN influence, often associated with friendship building but may be a result of war and power struggles. Interestingly no AN language of this type has a distinct word for 4 while 27 of the NAN languages do. However, there are 12 AN languages that have a distinct word for 3 but 4 is 2 + 2 with another 40 NAN languages of this nature.

Among the systems with 2-cycles are those that proceed to 4- and 8-cycles such as the Hagen language. Some counting in this system is linked to counting pairs of fingers — two pairs on one hand, two on the other and then the two thumbs giving ten. The tens are tallied often by another person in ritualistic style. However, the 8-cycle system dominates in larger numbers with two, three and larger sets of 8. In summary, 58% of the Trans New Guinea Phylum languages possess a 2-cycle variant, 93% of the Torricelli Phylum and 38% of the Sepik-Ramu Phylum collated by Lean (1992). All 32 Oceanic (AN) languages are found in PNG, mainly in the North New Guinea Cluster (24) but there are seven in the Papuan Tip Cluster and one in the Meso-Melanesian Cluster. Not all currently recognised languages are included in these numbers but the quantities given in these last few paragraphs are indicative of diversity, frequency and frequency of each type of system.

Counting in a Cultural Context

Throughout the cultures mentioned in Lean's and the other studies referred to in this book, exchange and relationship building were the main reasons for counting to be used and developed. There were different emphases and values placed on the counting and any scheme that may develop. Whether counted or not, in many cases the display of the exchange was also important together with the distribution associated with recognition of relationships. In small "thank you" situations, a money note was significant and hence the basic unit for larger numbers. Thus a "20 shilling" or K2 note or these days a K20 note is expected.[1] Changes in culture exchanges may have resulted in changes in counting systems but certainly many developed ways of denoting large numbers. For example, the Mid-Waghi people used body parts to identify the number of 100s and then 1 000s involved in exchanges (Muke, 2000) (see Chapter 3). Philosophical approaches and worldviews would influence the emphasis placed on counting and several recent studies such as those of Muke (2000), Paraide (2010), and Mimica (1988) have highlighted this point.

Such an in-depth study of the number systems of the Iqwaye by Mimica (1988) indicates a very complex use of digits for representing number and for thinking about infinity. This produced an innovative way of representing large numbers through powers of 20. Numbers maintain relationships with the essence of their mythology as well as other people (see Chapter 3). Another study of neighbouring languages that had 1, 2, 2 + 1, 2 + 2 as the start of their counting system indicated that powers of 6 began to emerge for large numbers (Donohue, 2008) (see Chapter 7). Among the Tolai and many other groups with shell traditional money, people counted large quantities, again with display and generally in terms of strings of shell money using powers of 10 but they also built on 12 (Paraide, 2010) (see Chapter 11). In the Oceanic languages, the need for tribute probably generated many of the develop-

[1] During colonial times 20 shillings and the pound were used and old people still often refer to the 10 toea in the current decimal currency as a shilling. The smallest note is two kina (K2).

ments of powers of 10 or of 2s (see Chapter 6). Pairs were commonly valued and hence systems with 2-, 4-, 6-, 8-, and 10- cycles developed. Often counting words were used for the pair or for a group. These studies have counteracted the myth that people with only 1 and 2 or may be 5 counting words had unsophisticated counting or number systems. Even when there is not sufficient evidence for a language that the counting system could be extended, there would still be ways of thinking mathematically to large numbers in cultural contexts (Hammarström, 2010; Leslie et al., 2008).

As an historical study of mathematics, this book extends currently accepted notions of counting portrayed in school mathematics systems around the world. It also fills an important gap about building on the diversity of counting systems in schools. These two key points are discussed in the following sections.

Indigenous Knowledge Elucidating Number Knowledge in Western School Mathematics

The argument presented in Chapter 12 is that the hegemony of regarding the English or Western approach as being more valuable than Indigenous knowledge fails to realise the importance of culture and identity in developing and understanding knowledge. The key issue in this has been the break in education and home knowledge with many students leaving the village areas for schooling. Along with this has been the need for teachers and the community to realise the mathematics of cultures and the systematic ways of thinking that can be linked to school mathematics.

The Role of Groups and Relationships in Indigenous Mathematics

Throughout the chapters of this book, there is reference to a group or a complete group. In particular, it was noted by Saxe (2012) in discussing changes to *fu* in Oksapmin as a result of monetary systems, employment, and power of those with education, outside employment, or teacher status. The Manus type systems found mostly in the New Guinea Islands express numbers up to the group of 10. In particular, the Uisai of Bougainville note the completeness of a group at 9 and the beginning of a new one at 10. Several highlands groups such as the Mae dialect of Enga identify groups of four. Each cycle is given a specific name such as "dog", "pig" or for the culminating 60 "house on fire". Similarly the Mid-Wahgi Yu Wooi language of Jiwaka Province indicates specific gestural markers for groups of 4 and 8, while the Angel Heneng of the Mendi group of languages indicate specific number-naming of cycles (see Chapter 4). The representation of groups of numbers by the lexemes for one, two, three etc was common in a number of regions. Of interest too, was the use of different words for a group of 10 fish or 10 nuts etc as found in Motu and neighbouring languages. The decision to base the description of the languages on cycles is evidence of the importance of groups and thus groups of 5 and/or 20, and in some cases 2 and 10 have significance for the community influencing the numeric system. It should not be underestimated the significance of representations. In many groups, the fist for 5 or 4 was a significant gesture together with the joining of two fists. Interestingly the introduction of the Western monetary system of the pound or two dollar note was readily integrated into the digit tally systems and influenced the quantity for the group in Oksapmin and possibly other body-part systems. Finally we note the significance of groups as it appears in the 4-cycle systems in particular. In the case of Engan dialects, each group is named and the numbers for forming the group are related to the complete group. For example, 10 would be needing two to make the group of 12.

Mention should also be made of the subtractive nature of the so-called Manus systems in which various numbers from 7 to 9 were considered as short by 3, 2, or 1 to make the whole group. Establishing this link between home-language counting system and English pairs of numbers for 10 would be advantageous for students.

Expanding Understanding of Groups and Place Value

While place value of the base 10 system refers to the spatial ordered position of digits on paper, there are instances in the Indigenous languages of PNG where sides of the body are significant, finger positions are important in counting, or where people stand beside each other, literally or in words, and indicate a higher order as in counting in fives represented by hands or counting in twenties represented by people. In addition, the order of words with or without a morpheme for "joining" is also indicative not only of addition in most cases but also of multiplication as in pairs of 2s, 3s or 4s. The latter is evident in the Austronesian languages of Central Province. The use of two counters as occurs in the counting of large numbers of objects, one for the units and one for the groups of ten indicates another example of place value representation. This is recorded for the Hagen language as an example (Appendix B). Order of words was significant for informing number size in Kanum discussed in Chapter 7.

However, the use of quantifier classifiers (see Chapters 6 and 8) along with other classifiers of a group's characteristic is an important conceptual approach to place value that should be encouraged especially in these language groups but as a new approach for children in upper primary school in any country. Place value often causes many children difficulties in mathematics and understanding that the column is like a suffix in identifying the characteristic of the group (as indeed the words "hundreds", "thousands" etc actually are) will assist these children. This characteristic of identification is similar to the counting of "long, thin objects" or "four-footed animals" with a suffix for counting in the classifier systems. Approaching quantifier and qualifier classifications together may be an intriguing new way for some children to understand the concept of place value.

The fact that within New Guinea can be found counting systems that are based on groups other than 10 is significant as a contrast to the base 10 system. In learning about these different cyclic systems, students in school can appreciate that the base 10 system is indeed a human development rather than something to be learnt from "out there" and having no human or cultural source. It is also a subtle way of indicating that difference is acceptable as a broader purpose of education.

There are numerous incidences in the various languages where powers are represented in linguistic or physical terms. Larger numbers may be formed by powers of two multiplied by 10 or 100, as in some of the Oceanic languages as well as powers of 10. However, there are examples of powers of 20 and powers of 6. Body parts may be used to indicate powers or it is the context just as it were for the multiples especially in pairs (fairly ubiquitous in AN and NAN languages) or group counting such as indicated in Chapter 11 and in Western Province and other places with 3s and 6s.

Expanding Understanding of Parts and Wholes

There are various intentions in this heading. Often this is considered in terms of spatial shapes and arrangements. However, it is also used in the area of number to refer to fractions or of splitting up or decomposing a number and regrouping or recomposing a number. Pairs of numbers add to 10 so $7+8$ can be decomposed into $5+2+5+3=10+5=15$. It is also important in using place value for operations like addition where 10 as a unit can be formed or "traded" for 10 ones or in reverse for subtraction.

In the Indigenous systems of PNG, parts and wholes are frequently associated with cultural practices. For example, in dividing up a pig it is done according to relationships or involvement in the raising of funds or assistance with building a house etc. These may not be equal parts in terms of weight or volume or containers of pig fat (Owens, 2015). Thus the notion of a fraction as representing equal parts can be compared to unequal distributions. Similarly in sharing fish, taro, shell money or other countable items among families, these may be distributed according to need rather than equality making the meaning of equal sharing for the Western operation of division different to culturally embedded practices sharing with unequally numbered groups. Again the contrast can be a good pedagogical approach.

Nevertheless, it is in the languages themselves that there is strength in combining the teaching of number in the school language such as English with the language that often combines words very directly. So for learning combinations to 10, using the $5+n$ structure of the digit tally systems is advantageous. Other languages identify 10 plus a number which is somewhat easier to appreciate the connection than the English "twelve" or "eleven" for example. Combinations such as 70=3x20+10 in digit tally systems also strengthen the early two-digit arithmetic work; an interesting story for other children to think of their 20 digits as a whole group called "one group of 20" or "one person".

Expanding Representations of Number

It is not unusual for body parts to be used to represent numbers and to be used in the calculation process. Most children know how to use their fingers to do the 9 times table e.g., if calculating 9x7 then bend down the 7th finger from the right and the number of fingers to the left provides the tens place and those to the right the unit place. In ancient times Bede represented numbers on fingers to a large number. Owens' Vietnamese students used three parts for each finger so they could easily represent 15 reaching 20 with other parts of the palm.

However, the body-part tallies of PNG form a unique variety of systems along with some in the north east and south east of Australia (at least as known currently) in very different language groups. These body-part tally systems act as a virtual number line assisting students to recall or carry out calculations to numbers larger than 10. In Oksapmin, people moved directly to 10 position or they actually represented a second number using the symmetrical position on the other hand in order to maintain two number representations (Saxe, 2012). In nearly all interviews with people from many different language groups while counting they used their fingers as they said the numbers. Children in Matang's study (2005; Matang & Owens, 2014) used their toes to assist with arithmetic problems as well as their fingers. However, there are some examples of recording of negative numbers in terms of a virtual number line with, for example, pigs' husks hung on either side of centre point for two opposing clans' offerings or killed warriors.

It is with large numbers that people began to introduce other objects for large numbers so for example, words like *hip* (heap) and *bilum* have been used in highland areas, ropes, ropes with knots and bundled sticks for representing larger numbers (powers of 10) as was indicated by Muke (2000) in Yu Wooi (Mid-Wahgi) language. Yu Wooi also represented thousands extending the body-tally of multiples; 1 000 *simb daro* "left leg" or *hi ende sim angek poro begenj* "whole body parts of one person" and if 2 000 *hi tak* "two persons" etc. In the Bougainville area where base 10 systems occur, words for these large numbers like 1 000 or 10 000 may have been borrowed from neighbouring languages.

Base 10 blocks by Dienes and many other concrete materials represented 10s, 100s and 1 000s in Western schools. In fact multi-base blocks were introduced into PNG schools at one stage. Dienes and others could have considered that children would handle the different bases well but in fact any one child may not have a 2- or 5-base system in their language and no thought was given to associating these base systems with the child's language (education was in English). They were discarded when it was found that children could not easily use them for addition. However, sticks in the various languages studied by Lean could represent different numbers (e.g. Iqwaye). Furthermore, gestures were a common addition for number representation. For example, raising the arms for the complete group (Saxe, 2012), clenching fists and joining them together for eight as in Hagen or Mid-Waghi or for ten in many systems with a 5 cycle or wiping the two thumbs down the lips in Gawigl (of the Hagen language group) to signify 10 (Owens' field trip, 2001).

However, it was also found that groups of various sizes could be represented by the words for one, two, three etc. In most of these case, it was the pairs that were counted. In Yu Wooi, it was marking multiples of 20 (Muke, 2000). In Iqwaye, these numbers and a person represented powers of 20

(Mimica, 1988). In Oceanic languages, it is powers of 2 multiplied by 10 or powers of 10 that were identified as key markers of the system.

Cultural groups devised appropriate ways of developing their systems to take account of the mathematical activity of counting in cultural practices. These were generally isolated innovations but in some cases they were common in a regional group especially among Oceanic languages. This book brings attention to variations that enlighten western or school mathematical approaches. The abstract mathematics should be of interest to the student of mathematics. Variations from the base 10 system, for example, for considering large numbers, variations in valuing quantities in terms of research methodologies (qualitative with quantitative), and variations in the importance of number for specific purposes sits well with uses of mathematical systems that may be considered in pure mathematics. Nevertheless, in humanistic pedagogical approaches the contextual aspects of mathematics are valued (Pinxten, 2016) and that is indeed the situation in several of the cases expanded in this book such as the Iqwaye and Tolai mathematical thinking. However, as Pinxten recommends, this book foregrounds differences in counting systems that can assist students who have multiple heritages to appreciate and value the mathematics of their heritages and to assist the students to make connections and comparisons.

This book establishes the history and neighbouring variances of systems such as in the body-tally systems or the use of the hand for counting in context (e.g. yams). As Pinxten (2016) noted, counting systems are being transformed, reformulated and represented. These occur from establishing relationships with other groups rather than from invasion. It may result from a diffusion but not of the kind of spread discussed by Seidenberg (1960). The exception to this is the 10-cycle system of Proto Oceania that spread with the language. Diversity is more likely, for example, among the Trans New Guinea Phylum. However, there are innovations occurring especially in language Isolates or for specific purposes but in many cases for unknown reasons.

Key Issues about the History of Number Raised in This Book

The total number of languages spoken in New Guinea and Oceania is at least 1 400 and data have been acquired for three-quarters of these. Generally speaking, the picture that we obtain of the tallying and counting situation in this region is consistent with that found in the other major regions of the world: the Americas, Africa, and Asia in particular. The means by which human societies enumerate their world is not infinitely varied. The 2-cycle systems, the (5, 20) digit tally, and the 10-cycle system are found in New Guinea and Oceania as they are found elsewhere: so too are a number of variants of each type. This, however, is not to suggest that the process of enumeration in human societies is one of dull uniformity: the surveyed evidence of unique and unusual methods of tallying and counting ceremonial displays of wealth attests to the inventiveness of Indigenous societies in elaborating activities which might usually be regarded as mundane. Nevertheless, in the end, despite the richness apparent in the way counting and tallying is manifested, the most striking feature of this aspect of such societies is the shared similarities rather than the differences.

Primary Cycles

Lean suggested that the 2-cycle and the 10-cycle counting systems, together with the (5, 20) digit tally, share a primary status. In addition it seems likely that the complex body-part tally methods which appear to be unique to Australia and New Guinea may also be assigned primary status: there is no reason to assume that these are the more complex ancestral forms of the digit tally. The interaction of these primary means of enumeration produces the secondary, or hybrid, types. Thus the (5, 10) and

(5, 10, 20) systems are secondary derivatives resulting from the interaction of the 10-cycle numeral system with the (5, 20) digit tally. The (2, 5) or (2, 5, 20) systems, together with their variants, denoted by 2′ and 2″ in Chapter 3, are secondary derivatives of the 2-cycle numeral system and the (5, 20) digit tally. The users of the body-part tallies usually have, in addition, either a 2-cycle variant numeral system or, as we find in the Southern Highlands Province (PNG), 4-cycle systems. There are examples of hybrids of the 2-cycle systems and the body-part tallies which are such that the numerals 1, 2, 2 + 1, and 2 + 2 displace the first four names of the tally points. With the exception of this last case, the various systems that we have suggested belong to the hybrid class are found in many Indigenous societies in other parts of the world. The most economical interpretation for their existence is that the same mechanism applies for the generation of such systems as applies in New Guinea and Oceania, that is that the (5, 10), (2, 5), and other systems found, say, in the Americas and Africa, are also hybrid systems rather than primary systems in their own right.

In the works of some nineteenth century scholars writing on the counting systems of Indigenous societies, it is not uncommon to find disparaging or dismissive remarks which compare such systems unfavourably with the highly abstract number systems used in modern technological societies. This is particularly noticeable when reference is made to systems with a relatively small primary cycle. Conant (1896, p. 2), for example, said that "at first thought it seems quite inconceivable that any human being should be destitute of the power of counting beyond 2. But such is the case; and in a few instances languages have been found to be absolutely destitute of pure numeral words". Similar statements can be found in modern texts on the history of number: Struik's (reissued in 1987) *Concise History of Mathematics*, for example, speaks of number systems which essentially comprise the words "one, two, many". The evidence gathered for this study does not provide any basis of support for such statements. While it is certainly the case that the expression of grammatical number in the personal pronouns of many NAN and AN languages involves the use of singular, dual, and plural forms, that is a "1, 2, many" construction, Lean did not find a single instance of a language which has only the numerals 1 and 2 and which terminates precise enumeration at 2. As noted in Chapter 3, pure 2-cycle systems are found in New Guinea although they comprise only about ten percent of the 2-cycle variants. Pure 2-cycle systems, however, do not terminate at 2: higher numerals are formed as compounds of the basic numerals. It is also the case that 2-cycle systems are often associated with tallies: either the body-part tally or the digit tally. The possession of a relatively small primary cycle system does not imply that the users are unable to count to any extent or that they find counting difficult. As Fortune (1942) has noted with respect to the Mountain Arapesh people:

> To suppose that the paucity of the Papuan languages in root words for numerals makes counting difficult to the Papuan is quite incorrect. The Arapesh people count rather more quickly and better than the Melanesian Dobuans, who use a decimal system with many more root terms. (p. 59)

Cultural Contexts and Abstract Number

A further criticism of certain Indigenous counting systems was that they are tied to the concrete or qualitative aspects of the objects counted and do not possess the quality of abstraction characteristic of true number systems. This type of criticism has its origins in two different phenomena found in certain languages. The first is numeral classification, discussed in Chapter 8, in which a set of numeral roots is affixed by a classifier signifying the class to which the objects being enumerated belong. It is, however, normally the case that languages which employ such classifier constructions also possess a set of numeral morphemes which are used in the serial counting of any class of objects. The second phenomenon which some writers suggest indicates a lack of an abstract number system is the use of different names for multiples or standard collections of objects. Thus while one term may be used to denote 100 coconuts, a different term may be used to denote 100 dogs' teeth. The assignment of names to specific quantities of objects is most commonly found in the AN languages with well-developed counting systems having terms for large numbers. Barnes (1982) made the point that

> in branding the use of special objects or words for mensuration as primitive, writers such as Cassirer (1953) and Levy-Bruhl (1926) neglected to consider the respective requirements of oral versus written arithmetic. In the absence of writing, the employment of heaps or pairs of objects, as well as mnemonic aids such as standard multiples, serves purposes similar to the marking of figures with pencil or stylus (p. 20)

The concept of number and its practical applications play such an important and integral part of modern technological societies that it is often used as a means for judging the degree to which a given society, particularly an Indigenous society, has attained intellectual sophistication. In Chapter 8 we argued that Indigenous societies differ in the degree to which number is accorded a privileged position. Thus, while a particular society may possess the linguistic means for enumeration it does not follow that the members of that society value, or constantly engage in, the process of precise quantification. There are many societies in which individuals or groups engage in the accumulation of wealth items which are, in special ceremonial circumstances, displayed and distributed. While it is the case that certain wealth items which are accorded significant value, such as pigs, are counted precisely, other items such as collections of bananas, yams, taro, and the like, tend to be judged impressionistically. Ceremonial prestations involve displays of wealth which accord status and prestige to the donors. The recipient group must retain, in memory or tally, an account of goods distributed so that, at some subsequent time, a reciprocal prestation is made to the original donor group. Such accounting, however, varies from one society to another: some place importance on precise quantification of the goods involved while others do not. The brief counting ethnographies given in Chapter 8 provide examples of societies which differ according to the importance that they accord number and counting, ranging from the Ekagi, with a virtual obsession for counting, to the Loboda who have, according to Thune (1978), an essentially non-numerically oriented culture.

Ceremonial exchange within a given society is but one example of circumstances which may evoke the counting of goods. Another example is the trading of commodities. Here again, however, societies vary according to the degree to which marketing may involve counting. The bartering of goods is based on establishing equivalence between different types of commodities and requires a sophisticated sense of quantity and relative value: in practice, however, it does not usually evoke the counting of the individual objects involved. By contrast, those societies which have a monetarised economy in which traded items have a standard value in terms of the traditional currency, usually shells of some kind, do count in marketing and in ceremonial circumstances which involve the distribution of money: either individual shells or standard lengths of strung shell money may be counted (see Chapter 11 for an example). Such societies are most commonly found among the speakers of AN languages although there are a few instances of NAN-speaking groups having monetarised economies. Generally speaking, it is in these societies that we also find terms for large numbers of the order of a thousand or more.

Explaining Similarities and Differences and Critiquing Seidenberg's Theory

The similarities in the various types of counting systems that were found in widely dispersed languages located in the world's major continents gave rise, in the late nineteenth and early twentieth centuries, to speculation on why such similarities should exist. The independent inventionist view was that people everywhere invented their numeral systems: their similarities were due to common responses to the human condition. By contrast, there was the view that there was a time when language groups ancestral to those existing today did not possess the means of enumerating their world. Counting was invented under certain special circumstances in one of the ancient centres of civilisation from which it was diffused all over the world. This occurred not once but several times: after one primary system of counting was invented and diffused, further primary systems were developed and diffused as well. In some cases those systems which were diffused earlier were unaffected by subsequent diffusions and remained intact. In other cases the earlier systems were displaced by subsequent ones or were modified in a way which produced secondary or hybrid systems. This interpretation of how counting systems came to exist in human societies and which might be termed the "strong" diffusion hypothesis was first enunciated in detail by Seidenberg in 1960. Seidenberg's account of the prehistory of number has not been seriously challenged and still remains the prevailing view which finds its expression in recently published texts (e.g., Cooke, 2011), often limiting their sources to written texts or artifacts.

There were several components to the Seidenbergian hypothesis. First, there was his argument that counting had its origin in special ritual circumstances: this was elaborated in an article published in 1962. This aspect of his work has not been addressed at length here as we are not persuaded that the data collected for this study can be used to support or deny speculation on the origin of counting. Second, there was that aspect of Seidenberg's work which dealt with the type and nature of the counting systems and tallies which were invented and diffused, the essence of which is set out in his "genealogy" as given in Figure 9.2. Seidenberg's view was that the pure 2-cycle system was invented first. Subsequently the neo-2 or neo-2-10 system, of which there are two types, was invented; the diffusion of this system resulted either in the displacement of the original pure 2-cycle system or in its being left intact.

Seidenberg's explanation of how body parts came to be used for the act of enumeration is somewhat complex. He suggests that it has its origins in the practice of "parceling out various parts of the body to various gods" and was not initially associated with counting at all. Counting was essentially a linguistic act rather than a physical one. However, in certain societies there was, or there developed, a taboo on the verbal counting of particular things: it is often the case that the counting of people, for example, is proscribed. It was in circumstances such as these that verbal counting could be circumvented by the use of non-verbal, gestural tallying: the practice of parceling out the body to various gods and the act of counting came together in order to be able to enumerate objects under the verbal counting taboo. Two main types of tallying resulted from this: the complex body-part method and the simpler digit tally, the former being historically prior to the latter. After the diffusion of the (5, 20) digit tally, Seidenberg suggested that when the users of the tally came in contact with people possessing 10-cycle systems there developed such hybrids as the (5,10) system.

The third component of Seidenberg's work was that which places the invention of the primary counting systems in an historical context. His view of the invention and diffusion of counting and tallying has its origins in the work of Lord Raglan. Raglan (1939) placed the locus of the invention of various "civilised" practices in the ancient city-states of Sumeria which developed more than 5 000 years ago. Seidenberg suggested, following Raglan, that it is this time and place that we find the beginnings of counting and that in the succeeding millennia the various counting systems and tallies were invented and diffused throughout the world although the details of the diffusion process are not elaborated. In Chapters 8, 9 and 10, this study has addressed certain issues relating to the diffusionist view of counting, Seidenberg's in particular, and by an analysis of the current counting system situation in New Guinea and Oceania attempts to outline a prehistory of counting for this region. The conclusions reached provide some minimal measure of support for Seidenberg's views in certain respects; however, in other respects there are mainly significant departures from his views.

Reconstructing the History of Number

The reconstruction of the prehistory of counting in New Guinea and Oceania has been based on results deriving from work in the disciplines of archaeology and historical linguistics. Broadly speaking, the peopling of the region resulted from a series of migrations beginning with the Australoids about 50-60 000 years or more ago. These moved into the New Guinea region and southwards into Australia at a time prior to their geographical separation after the last ice age. At a much later date, perhaps 40-30 000 years ago, the first Papuan (NAN) language groups moved into New Guinea. We have suggested that both the 2-cycle system and the (5, 20) digit tally were present in the early history of the NAN languages and indeed were likely to have been introduced as part of the cultural baggage of these early immigrants. It may also be the case that the 2-cycle system entered Australia with the early Australoid migrations. The implication that the 2-cycle system has historical priority over other systems is in agreement with Seidenberg's view; however we have suggested that its origins lie much further back in time than posited by Seidenberg. The possession of a basic numeral stock and the means by which the basic numerals may be compounded to form larger numerals is seen in this study as an archaic feature of human language and not as a relatively recent invention. The presence of the 2-cycle system in South America, Africa, and parts of Asia may be interpreted as being the result of its being historically ancient and of its being carried into the major continents with the original migrations rather than being introduced at a much later time as a result of diffusion.

We have suggested that the (5, 20) digit tally was present at an early time in the NAN languages of New Guinea and indeed may have entered the region with the original NAN immigrants. The more complex body-part tally, however, may have been introduced to New Guinea by the migration of the speakers of languages ancestral to those which now constitute the Trans New Guinea Phylum. It is possible that this tally was also diffused southwards into the Australian continent. Seidenberg has surmised that of the two types of tally, the complex body-part method is the older. Whether this is the case is probably impossible to settle with certainty. What is suggested here, however, is that the digit tally was introduced into New Guinea and Australia prior to the introduction of the body-part tally. However, it is equally likely that they were developed as innovations in Proto NAN languages.

Subsequent to the introduction of these primary tallies and the 2-cycle system, several developments occurred in the New Guinea region. Many people possessing a 2-cycle system also made use of one of the two types of tally. Over time this resulted in the formation of hybrids so that, for example, the (2, 5) or (2, 5, 20) systems developed as well as a number of variants. The 4-cycle system developed from the digit tally due to the practice of certain groups which regarded the four fingers, but not the thumb, as constituting "one hand". It is possible that the rare 6-cycle system also developed from one of the tallies, perhaps by the practice of augmenting the five fingers with an additional tally point such as the thumb joint or the practice of bundling yams – 2 pairs of three. These developments which occurred in the numeration of the NAN languages prior to the arrival of the AN immigrants is given in the genealogy in Figure 10.2 in Chapter 10.

The next major development in the New Guinea region occurred about 7 000 years ago with the arrival of the AN language groups in the western part of West Papua. Migration by AN-speaking people to the east established the homeland of Proto Oceanic (POC) in, probably, the Willaumez Peninsula area of New Britain by about 5 500 BP. The reconstruction of POC by historical linguists indicates that the speakers of POC, in common with their ancestors, the speakers of Proto Austronesian (PAN), had a 10-cycle numeral system. It seems likely therefore that such a system was in use in the south-east Asian region, the homeland of PAN, by at least 7 000 BP when PAN was thought to have existed. Although it is not clear how the 10-cycle system originated, the fact that the PAN and POC words for both "five" and "hand" are identical suggests that it may have developed originally from the digit tally. In any case the arrival of the Austronesians introduced the 10-cycle numeral system to

New Guinea; subsequent migrations from the POC homeland eventually carried the system into the previously unoccupied areas of Island Melanesia, Polynesia, and Micronesia. The view adopted here is that the 10-cycle system accompanied the AN settlers as they peopled the Pacific: it was not introduced subsequently by diffusion from some external source.

Ross's reconstruction of the linguistic history of the AN languages in Papua New Guinea has various groups moving out of the POC homeland and on to the coastal and island regions of the PNG mainland. The situation arose in which the established NAN language groups, with their 2-cycle variant numeral systems and tallies, came into contact with the immigrant AN groups with their 10-cycle system. In this situation, diffusionists such as Crump (1990) took the view that the more efficient 10-cycle system would displace systems with a smaller primary cycle according to a sort of "survival of the fittest" principle. The evidence, however, indicates otherwise: on the mainland, at least, the 10-cycle system did not arrive and overwhelm the existing NAN systems. On the contrary, most AN groups accommodated in some way with the dominant NAN cultures, resulting in modifications to their original numeral system: loss of part of the numeral lexis and a reduction in the magnitude of the primary cycle. As a result of this, AN groups possessing 2-cycle variant systems, (5, 10) and (5, 10, 20) systems, and other variants in which the second pentad numerals are affected. This includes the type of system which Seidenberg has termed the neo-2-10 type but we do not view this system as an elaboration of the 2-cycle system. The summary picture of AN numeration is given in Figure 10.2 in Chapter 10. Given that the contact between AN and NAN groups has occurred over a period of 5 500 years or so, what is surprising is how little the 10-cycle system appears to have affected the counting systems of the NAN cultures of the New Guinea region.

With regard to enumeration, then, the nature of the diffusion process is rather more complex than that which the diffusionists seem to assume. The diffusion model implicit in Seidenberg's work seems to be a "relay baton" type with a counting system being passed from a donor group to a recipient group, the latter either not possessing a system or possessing one with a smaller primary cycle than that of the donor group. The outcomes when two such groups meet are either that the recipient group will adopt the new system or, if it already possesses one, its original system will be displaced or modified in some way. The view adopted in this study, which sees the primary systems and tallies as being carried with migrating language groups, regards a given group's counting system as an integral and relatively stable part of its culture. We have suggested that in the circumstance of linguistic speciation occurring over time, the cyclic structure of the counting system will normally remain invariant despite divergence in the numeral lexis between languages. This is one aspect of stability.

A further aspect of stability relates to the circumstances under which a counting system may be diffused from one group to another. In the event of two groups, which possess different types of counting system, interacting in some way, it does not inevitably follow that one group will adopt the counting system of the other, relinquishing, say, a lower primary cycle system for one with a larger primary cycle. Instead, we have discussed evidence which indicates that when an immigrant group settles in a region already occupied by an established cultural group, the former will often accommodate its own politico-economic institutions to those of the latter. For the immigrant group this may include the adaptation of its counting system to conform with that of the dominant group and, contrary to the expectation of the diffusionists, this may result in the loss of part of its numeral lexis and the reduction in the magnitude of the system's primary cycle as is the case, mentioned above, with certain AN groups. The conditions under which this may occur are not frequently met: two disparate cultural groups living in neighbouring regions may well retain their cultural identities even when engaging in a degree of trade. In these circumstances both groups are likely to retain intact their original counting systems. The main implication to be drawn from the various ways that different cultural groups interact is that diffusion is not an automatic process. This is particularly interesting in the cultures where oracy and bilingualism are valued. The contrast between the diversity of the Non-Austronesian languages and their counting systems and the widespread Oceanic languages with a high degree of similarity and created proto-Oceanic lexicon appears to be linked to the need for counting to large numbers for chieftain tribute. In more recent times, when money has been used for compensation and bride price in the New Guinea area, groups have developed new ways of reaching large numbers such as the Hagen use of knotted rope (Measurement Questionnaire, 2006) or the body parts to mark powers of 2 multiplied by 10 (Mimica, 1988).

Another aspect likely to affect integration, diffusion or innovation is the importance of language and the uniqueness of the counting system for the identity of a group. For example, the Duna, Hela Province have a unique asymmetrical body-part counting system linked to familial relationships (Sakopa, personal communication, 2015). However, the value placed on counting varies from place to place. Language openness, linguistic positioning, and importing languages act as part of identity (Dobrin, 2014; Dorian, 1994). A road for exchange of items is supported by friends or rather friends of friends along the road, all requiring high maintenance (Dobrin, 2014) and still observable today but along new roads and circumstances. By way of illustration, some of our 1980s students enjoyed speaking words backwards forming a multilingual friendship group. *Bilums* (continuous string bags) and other objects are symbols of belonging and relationships and designs are spread to build relationships; small gifts are regularly given to friends and family members. So just as a design, dance or object can be shared so could a system such as the counting system above and beyond the need for trade. Thus there may be some interchange of counting systems or words but each group maintained its identity and the sharing of friendships had stages along the way which may lead to maintenance of many varieties that were shared but not necessarily absorbed. Less tangible values play a stronger role in choice or spread of counting system than the more likely scenarios of the ecology such as the geography of mountains and means of movement, or the need for allies in warfare who may or may not have the same language.

Conclusion

The picture that has been reconstructed in this book of the counting and tally situation in New Guinea and Oceania supports the view that the prehistory of number covers a period of some tens of thousands of years, considerably longer than the 5 000 years suggested by Seidenberg. There was a long period of time when the dominant system in use was the 2-cycle one: the historical span of its use, far exceeding that of the 10-cycle system, attests to its success as a means of enumeration. This may have been due to the importance of pairs since there are a number of languages in which pairs are important, used in counting, or reflected in the system such as the Motu-types or the powers of two in the Oceanic languages. While there were periods of relative stasis in the counting system situation, there were also periods of flux and change when new systems were introduced or invented. In the era prior to the arrival of the European colonists there is evidence to suggest that various changes occurred to the counting system situation and that a degree of internal diffusion took place within the region. This is summarised in Figures 10.1 and 10.2. It is, however, in the colonial and post-colonial period that major and rapid changes have been induced in the traditional means of enumeration in many societies as a result of changes to their political and economic institutions and of the establishment of the pre-eminence of the introduced decimal system of enumeration of the colonial powers. It is in these circumstances that we see the operation of diffusion most decisively and its effect has been to set in train the largely irreversible process of the gradual decline and extinction of many of the region's counting systems which, until recently, have survived as a link to humankind's earliest intellectual history.

Rosa et al. (2016) emphasised that ethnomathematics is a significant part of mathematics education for political, social, mathematical, historical, conceptual, cognitive, epistemological and educational reasons. For all these reasons, this book on the history of number is significant. It shows that "mathematics is a social and cultural product…(with) contributions from both Western and non-Western civilizations" (Rosa et al., 2016, pp. 33-34). Nevertheless, the majority of the large number of research theses in ethnomathematics reported between 2005 and 2014 in Brazil and at three international conferences are in education (Rosa et al., 2016). Paraide's work in Chapters 11 and 12 particularly considers the pedagogical and epistemological aspects of pursuing ethnomathematics. She clarifies in her Tolai study how the social counting systems can initiate strong conceptual understanding in number and connect to school arithmetic and the base 10 system. Importantly, Tolai systems are valued knowledge that should be maintained. The political social-justice impetus for writing this history of number was for teachers around the world to take note of the importance of the rich fabric of oral histories of Indigenous communities. It is important for Indigenous students for their self-recognition and identity. It is important for all students to value Indigenous knowledge and to be engrossed in the history and diversity of counting systems upon which this book was developed. It expands students' mathematical knowledge. There are sophisticated and efficient systems besides the base 10 place value system. There are good reasons for their development in oral-based language communities and in maintaining interactions between communities. Lean's 22 year life journey collating and analysing the counting systems of Papua New Guinea and Oceania together with subsequent research makes this early and ancient intellectual history available to enrich mathematics education.

References

ATSIMA. (2015). *Creating connections & growing understandings: Aboriginal and Torres Strait Islander Mathematics Alliance conference 2014*. Adelaide, SA, Australia: ATSIMA.

Barnes, R. (1982). Number and number use in Kedang, Indonesia. *MAN, 17*(1), 1–22.

Bender, A., & Beller, S. (2006). Numeral classifiers and counting systems in Polynesian and Micronesian languages: Common roots and cultural adaptations. *Oceanic Linguistics, 45*(2), 380–403.

Bender, A., & Beller, S. (2012). Nature and culture of finger counting: Diversity and representational effects of an embodied cognitive tool. *Cognition, 124*(2), 156–182. doi:10.1016/j.cognition.2012.05.005.

Bender, A., & Beller, S. (2013). Mangarevan invention of binary steps for easier calculation. *Proceedings of the National Academy of Sciences of the United States of America*.

Bhabha, H. K. (1995). *The location of culture*. London, UK: Routledge.

Cassirer, E. (1953). *The philosophy of symbolic forms* (Vol. 1). New Haven, CT: Yale University Press.

Conant, L. (1896). *The number concept*. New York, NY: Macmillan.

Cooke, R. L. (2011). *The history of mathematics: A brief course* (2nd ed.). New York, NY: Wiley.

Crump, S. (1990). *The anthropology of numbers*. Cambridge, UK: Cambridge University Press.

Denham, T. (2011). Early agriculture and plant domestication in New Guinea and Island Southeast Asia. *Current Anthropology, 52*(S4), S379–S395. doi:10.1086/658682.

Dobrin, L. (2014). Language shift in an 'importing culture': The cultural logic of the Arapesh roads. In P. K. Austin & J. Sallabank (Eds.), *Endangered languages* (pp. 125–148). Fort McMurray, AB: Bouchier.

Donohue, M. (2008). Complexities with restricted numeral systems. *Linguistic Typology, 12*, 423–429. doi:10.1515/LITY.2008.043.

Dorian, N. C. (1994). Varieties of variation in a very small place: Social homogeneity, prestige norms, and linguistic variation. *Language, 70*(4), 631–696.

Dwyer, P., & Minnegal, M. (2016). Counting systems of the Strickland-Bosavi languages, Papua New Guinea. *Language and Linguistics in Melanesia, 34*(1), 1–36.

Evans, N. (2009). Two *pus* one makes thirteen: Senary numerals in the Morehead-Maro region. *Linguistic Typology, 13*, 321–335.

Fortune, R. (1942). Arapesh. *American Ethnological Society, 19*, 1–60.

Hallpike, C. (1979). *The foundations of primitive thought*. Oxford, UK: Clarendon Press.

Hammarström, H. (2009). Whence the Kanum base-6 numeral system? *Linguistic Typology, 13*(2), 305–319. doi:10.1515/LITY.2009.014.

Hammarström, H. (2010). Rarities in number. In J. Wohlgemuth & M. Cysouw (Eds.), *Rethinking universals: How rarities develop linguistic theory* (pp. 11–60). Berlin, Germany: De Gruyter Mouton.

Karp, A. (2014). The history of mathematics education: Developing a research methodology. In A. Karp & G. Schubring (Eds.), *Handbook on the history of mathematics education*. New York, NY: Springer.

Lean, G. (1992). *Counting systems of Papua New Guinea and Oceania* (Unpublished PhD thesis). PNG University of Technology, Lae, Papua New Guinea.

Leslie, A. M., Gelman, R., & Gallistel, C. R. (2008). The generative basis of natural number concepts. *Trends in Cognitive Sciences, 12*(6), 213–218. doi:10.1016/j.tics.2008.03.004.

Levy-Bruhl, L. (1926). *How natives think*. London, UK: George Allen & Unwin.

Matang, R. (2005). Formalising the role of Indigenous counting systems in teaching the formal English arithmetic strategies through local vernaculars: An example from Papua New Guinea. In P. Clarkson et al. (Eds.), *28th Conference of Mathematics Education Research Group of Australasia, Melbourne* (pp. 505–512). Adelaide: MERGA. https://www.merga.net.au/documents/RP562005.pdf.

Matang, R., & Owens, K. (2014). The role of indigenous traditional counting systems in children's development of numerical cognition: Results from a study in Papua New Guinea. *Mathematics Education Research Journal, 26*(3), 531–553. doi:10.1007/s13394-013-0115-2.

Mimica, J. (1988). *Intimations of infinity: The mythopoeia (cultural meanings) of the Iqwaye counting and number systems*. Oxford, UK: Berg Publishers.

Muke, C. (2000). *Ethnomathematics: Mid-Wahgi counting practices in Papua New Guinea* (Unpublished MEd thesis). University of Waikato, Waikato, New Zealand.

Owens, K. (2001). The work of Glendon Lean on the counting systems of Papua New Guinea and Oceania. *Mathematics Education Research Journal, 13*(1), 47–71.

Owens, K. (2015). *Visuospatial reasoning: An ecocultural perspective for space, geometry and measurement education*. New York, NY: Springer.

Paraide, P. (2010). *Integrating Indigenous and Western mathematical knowledge in PNG early schooling* (PhD). Deakin University, Geelong, Victoria, Australia.

Pawley, A., Attenborough, R., Golson, J., & Hide, R. (Eds.). (2005). *Papuan pasts*. Canberra, Australia: Pacific Linguistics, Australian National University.

Pinxten, R. (2016). *Multimathemacy: Anthropology and mathematics education*. Switzerland: Springer.

Raglan (Lord). (1939). *How came civilization?* London, UK: Methuen & Co.

Rosa, M., D'Ambrosio, U., Orey, D., Shirley, L., Alangui, W., Palhares, P., et al. (2016). Current and future perspectives of ethnomathematics as a program. *ICME-13 topical surveys*. Hamburg, Germany: SpringerOpen.

Ross, M. (1988). Proto Oceanic and the Austronesian languages of western Melanesia *Pacific Linguistics, C-98*.

Ross, M. (1989). Early Oceanic linguistic prehistory. A reassessment. *Journal of Pacific History, 24*(2), 135–149.

Salzmann, Z. (1950). A method for analyzing numerical systems. *Word, 6*, 78–83.

Salzmann, Z. (2006). *Language, culture, and society: An introduction to linguistic anthropology* (4th ed.). Boulder, CO: Westview Press.

Saxe, G. (2012). *Cultural development of mathematical ideas: Papua New Guinea studies*. New York, NY: Cambridge University Press.

Schoenfeld, A. (2007). Method. In F. Lester (Ed.), *Second handbook of research on mathematics teaching and learning* (pp. 69–107). Reston, VA: National Council of Teachers of Mathematics.

Seidenberg, A. (1960). The diffusion of counting practices. *University of California Publications in Mathematics, 3.*

Seidenberg, A. (1962). The ritual origin of counting. *Archives for History of Exact Sciences, 2*, 1–40.

SIL. (nd). Ethnologue: Languages of the world. Retrieved from http://www.ethnologue.com/

Smith, G. (1984). *Morobe counting systems: An investigation into the numerals of the Morobe Province, Papua New Guinea* (MPhil). Papua New Guinea University of Technology, Lae, Papua New Guinea.

Struik, D. (1987). *A concise history of mathematics*. New York, NY: Dover Publications.

Swadling, P. (1997). Changing shorelines and cultural orientations in the Sepik-Ramu, Papua New Guinea: Implications for Pacific prehistory. *World Archaeology, 29*(1), 1–14. doi:10.1080/00438243.1997.9980360.

Thornton, M., & Watson-Verran, H. (1996). *Living maths*. Abbotsford, Victoria, Australia: Yirrkala Community School and Boulder Valley Films.

Thune, C. (1978). Numbers and counting in Loboda: An example of a non-numerically oriented culture. *Papua New Guinea Journal of Education, 14*, 69–80.

Wassmann, J., & Dasen, P. (1994). Yupno number system and counting. *Journal of Cross-Cultural Psychology, 25*, 78–94.

Wurm, S., & Hattori, S. (Eds.). (1981). *Language atlas of the Pacific area. Part 1: New Guinea area, Oceania, Australia*. Canberra, Australia: Australian Academy of the Humanities.

Abbreviations

*	As a prefix, indicates a proto language word
/ʔ/	glottal stop
2′	compound for 3 = 2 + 1 or rarely 1 + 1 + 1 but numeral for 4
2″	distinct numeral for 3 but 4 is a compound, usually 2 + 2, rarely 5 - 1 or 2 + 1 + 1
AN	Austronesian languages
BC or BCE	Before Christ, that is before the Current Era taken as before the period of Christ
BP	Before the present
CE or AD	In the current era, that is after the year of the Lord (Domino/Dominum) Christ
CSQ, MQ	Counting System Questionnaire; Measurement Questionnaire
d.	dialect
IMP	Indigenous Mathematics Project
Manus type	Lean used this to refer to counting systems that used subtraction from 10 such as 7=10-3, 8=10-2, 9=10-1, often with the meaning e.g. for 7 as 3 needed to complete the group
MC	Micronesian
Motu type	Lean used this to refer to counting systems that used pairs such as 6=2x3, 7=2x3+1, 8=2x4, 9=2x4+1
NAN	Non-Austronesian (also called Papuan) languages
NCQ, CQN	Noun, classifier, quantifier; classifier, quantifier, noun
NQC, QCN	Noun, quantifier, classifier; quantifier, classifier, noun
NTM	New Tribes Mission, PNG
PAN	Proto Austronesian
PN	Polynesian
PNG	Papua New Guinea
POC	Proto Oceanic
QC, CQ	Order of quantifier-classifier; classifier-quantifier respectively
SHWNG	South Halmahera West New Guinea (AN Non-Oceanic language of the Central-Eastern Malayo-Polynesian, a subgroup of Proto-Malayo-Polynesian) after Tryon (2006)
SIL	Summer Institute of Linguistics
SOV	Order of words in a sentence: Subject Object Verb
SVO	Order of words in a sentence: Subject Verb Object
TNG	Trans New Guinea Phylum

Nomenclature

The Australian system of numbering is used. There is a space to mark the thousands so two thousand is written as 2 000; one million as 1 000 000. The exception is in reference to dates so the Census in 2000 is denoted without a space and likewise dates when books or articles were published.

In general, italics is used for a non-English language word with the exception of proper nouns such as the name of a language which is denoted with a capital in plain font although many of them have a non-English origin and are usually the language word for a place or the language. The English translation of a morpheme or word is put in inverted commas so there is no confusion in reading the sentence.

K. Owens et al., *History of Number*, History of Mathematics Education,
DOI 10.1007/978-3-319-45483-2

The translation "man" may be translated "person" in many cases.

Algebraic symbols are also put in italics e.g. $5+n$. + usually denotes addition but in linguistic contexts may represent an added morpheme e.g., group + n may represent group × n or group with a classifier suffix. A prefix plus the number morpheme may be written as $n+x$.

Appendix A
Counting System Questionnaire Used By Lean

Glen Lean

Purpose of the Questionnaire

The original aims of this study were, firstly, to elucidate the number and nature of the counting systems used in the traditional societies of Papua New Guinea (PNG) and, secondly, to investigate whether the accumulated data had any implications for understanding the history of number and for mathematics education.

PAPUA NEW GUINEA UNIVERSITY OF TECHNOLOGY

COUNTING SYSTEM RESEARCH PROJECT

QUESTIONNAIRE

SECTION 1 : LANGUAGE AND LOCATION

1. NAME OF YOUR VILLAGE :

2. CENSUS DIVISION :

3. SUB - PROVINCE : 4. PROVINCE :

5 a) NAME OF YOUR LANGUAGE :

5 b) ALTERNATIVE NAME (S) OF LANGUAGE :
(IF ANY)

SECTION 2 : TRANSLATION

ENGLISH	TRANSLATION (YOUR LANGUAGE)
FIRST	
SECOND	
THIRD	
MAN	
HIS HAND	
MY HAND	
A HALF	
A THIRD	
AND	
SOME	
A FEW	
MANY	

K. Owens et al., *History of Number*, History of Mathematics Education,
DOI 10.1007/978-3-319-45483-2

SECTION 3 : TRANSLATION OF NUMBERS

Please provide translations into your own language for each of the numerals given. If the translated words have any further meaning (such as a part of the body) please provide that also.

No	TRANSLATION	FURTHER MEANING
1		
2		
3		
4		
5		
6		
7		
8		
9		
10		
11		
12		
13		
14		
15		
16		
17		
18		
19		
20		
21		
22		
25		
30		
40		
50		
100		
200		
1000		
2000		

SECTION 4 : PHRASE TRANSLATION

1.	ONE BIG PIG :
2.	TWO BIG PIGS :
3.	FIVE BIG PIGS :
4.	THOSE TWO HOUSES :
5.	THOSE THREE HOUSES :
6.	THESE TWO MEN :
7.	THESE FIVE MEN :
8.	THESE TWO WOMEN :
9.	THESE FIVE WOMEN :
10.	THREE BANANAS :
11.	FOUR BANANAS :
12.	FIVE BANANAS :
13.	TEN BANANAS :

SECTION 5 : USE OF BODY PARTS FOR COUNTING

In many parts of PNG, when people count they keep tally by marking the numbers on parts of the body, i.e. on fingers, toes, and so on.

1. If this type of tallying is not practised in this language, please tick :
2. If this is practised in this language, on the diagram below, please indicate how the tally system works, i.e. indicate the number which corresponds to each body part used in the tally.

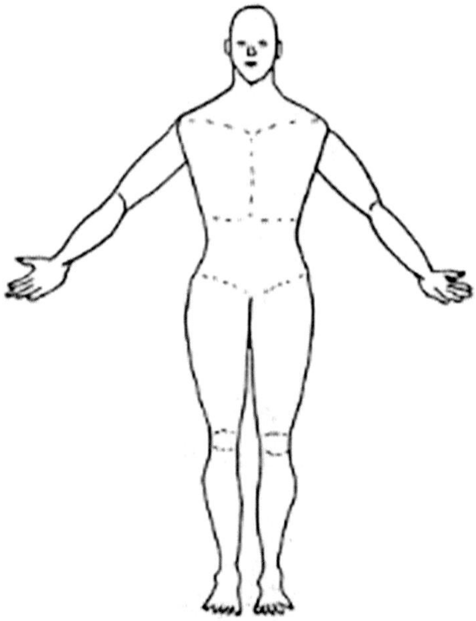

SECTION 6 : ALTERNATIVE WAYS OF COUNTING

In some PNG languages there is more than one set of words used for counting, i.e. there is one set of counting words used to count stones, for example, and a different set of words used to count coconuts, or pigs, or yams, etc. Sometimes a different set of words is used to count collections of things such as bundles of taro or coconuts.

1. In your language is there only one way of counting?

   (please tick): YES NO If YES, ignore part 2.
2. If there is more than one way of counting, please give examples below.

NAMES OF OBJECTS COUNTED

	1.	2.	3.
No.	TRANSLATION	TRANSLATION	TRANSLATION
1			
2			
3			
4			
5			
6			
7			
8			
9			
10			

Number words used for counting

You may use the space below for any further explanation of this :

THANK YOU FOR COMPLETING THE QUESTIONNAIRE

Appendix B
Details of Counting Systems Discussed in Chapter 8

Glen Lean and Kay Owens

Abstract Glen Lean collected over 1 200 counting systems from Papua New Guinea (PNG) and Oceania. His data were obtained from first contact records, later linguist data, and records such as government reports as well as from thousands of Counting System Questionnaires (CSQ) given to PNG University of Technology students and lecturers, school teachers, linguists and others. These he collated and described in terms of their frame words (basic words from which other counting words are made), operative patterns (being the way in which the words are connected), and cycles (such as 2, 5, 20) or other kinds of systems such as body-tally systems. Selected languages referred to in the chapters, especially for Chapters 8, have been included in Appendix B while Appendix C refers to languages mainly referenced in Chapter 10. Reference maps are found in Appendix E. The languages are included in the order in which they are mentioned in Chapters 8. Similar types of systems are mentioned together.

Keywords Details of counting system • Papua New Guinea and Oceania counting systems • Papuan, Austronesian, and Polynesian languages • non-base10 counting systems • cycles, frames and operative patterns of counting systems

Contents

K. Owens (✉)
Charles Sturt University, Dubbo, NSW, Australia
e-mail: kowens@csu.edu.au

G. Lean

K. Owens et al., *History of Number*, History of Mathematics Education,
DOI 10.1007/978-3-319-45483-2

Introduction

Appendices B and C provide a selection of the extensive details that Lean collated and analysed over 22 years in preparation for his thesis. He collected information on counting systems of 1 200 languages from Papua New Guinea, West Papua, and the rest of Oceania. This was indeed an extraordinary feat. Some of this material can be found in hard copy in a few libraries, mainly in PNG and previously on the web at http://www.uog.ac.pg. It formed Appendices A-D of his thesis. Since much of these data, now historical, are not readily available, counting system detail is included in these Appendices to give depth and support to the chapters of this book. It pays honour to the extraordinary work of Glendon Angove Lean.

Appendix B provides data on a number of languages that were referenced in particular in Chapters 8 and 9 to discuss how counting systems were diffused from one language to another in close proximity, especially within Papua New Guinea (PNG) or across large expanses of sea. However, there are a number of arguments that criticise the diffusionist theory of Seidenberg (1960) and provide for alternative perspectives on counting systems and their origins. There are significant alternative key forces for understanding language development and time frames within PNG and Oceania as discussed in Chapter 9. Lean used maps, both linguistic and geographic as well as local knowledge to bring data together from the many sources especially from the Counting System Questionnaires (CSQs).

Each language is introduced by reference to its location and demographics such as the number of people who speak the language or live in the language or tribal area. It is also introduced by a brief summary of details of the language before the discussion of the data sources and the data findings. Another table (sometimes more than one) provides the reader with the collated data on counting. It will be noticed that many languages have first contact data or linguistic data. Many of the early government records during colonial times contained language data including number data. These countries have not only thousands of languages but also hundreds of linguists such as from the Summer Institute of Linguistics (SIL) resident in many areas, often very remote.

Lean (1992) used the same approach to analysing each language, namely to note its frame words, operative patterns, and cycles to reflect some of the operative patterns (Salzmann, 1950). Frame words are the basic words from which other counting words are made. Operative patterns are the ways in which the words are connected. Cycles such as (2, 5, 20) indicate the basic cycle and superordinate cycles for a language. They indicate that subsequent numbers will use the cycle number in combination with other numbers. System types such as body-tally systems or digit-tally systems (associated generally with (2, 5, 20) cycles are noted. Further details are given in Chapter 2. Lean was then able to note similarities and differences between neighbours and to piece together information that would indicate influences and changes in the languages.

Note that words in languages other than English are italicised. Italicised language words are not placed in quote marks to avoid confusion with linguistic markers. The notes below Tables give the language name used by that author, often a place name or dialect which may now be recognised as a separate language.

Adzera

Adzera is spoken in the north-west region of the Morobe Province in the Upper Markham Valley (See Appendix E, Figure E9). Adzera, an Austronesian (AN) language, was classified by Holzknecht as a member of the Markham Family, Upper Markham Group (Holzknecht, 1989). Holzknecht's survey of the Markham Family languages is taken to be the current definitive statement of the linguistic situation of this part of the Morobe Province and it supersedes McElhanon's survey (1984), which dealt with the "Adzera Family". Adzera has six dialects and is spoken in about 70 villages. McElhanon (1984), p. 19) gave the total resident population of Adzera-speakers, based on statistics compiled in the 1970s, as over 15 000, thus making the language more widely used than any other language in the Morobe Province.

In 2011, the Umi/Adzera Ward had 36 600 people in 25 villages (National Statistical Office, 2014a). SIL Ethnologue (n.d.) suggests 31 000 speakers. A summary of the language features is given in Table B1.

Table B1
Summary of Adzera Language Features

Language type	Austronesian
Language family; subfamily	Markham Family, Upper Markham Group
Other names, dialects	Adzera: 1. Adzrac, 2. Amari, 3. Guruf & Ngariawan, 4. Ongac, 5. Tsumim, 6. Yarus
Villages	Kaiapit
Cycles	2, 5, 20
Operative patterns	digit tally from 10; 1 to 9 combine 1 and 2
Frame words	1, 2, 5?

In brief: Adzera, an AN language, has a non-AN system for 1 to 9 as it has a 2-cycle pattern with numbers like 6 being 2+2+2. "Five" does not contain a hand morpheme but 10 and subsequent numbers to 20 do. Also 15 to 20 contain a leg morpheme. After 10, digit tally with tallying as two hands and toes until 20 is reached as "my hands two and my feet two".

The number data obtained for Adzera derive from Holzknecht (1989), an SIL word list compiled at Kaiapit from three villages (undated), seven CSQ and Smith (1984). A selection of the available data is given in the Table B2. The Adzera system, which according to Holzknecht is the same for all dialects, possesses a basic numeral set (1, 2). All subsequent numerals from 3 to 9 are compounds of these so that 3, for example, has a 2+1 construction, 5 has a 2+2+1 construction, and so on. Although 5 does not contain a "hand" morpheme, 10, however, does and has the gloss "my hand half and my hand half completed". The number word for 15 has, in addition, a "foot" morpheme *faga*-and has the gloss "my hand half my hand half completed and my foot half". The number word for 20 has the gloss: "my hands two and my feet two".

Table B2
Adzera Systems

	A	B
1	*bisinta?, bitsinta, bits*	*bisinta*
2	*iru?run, iru?ruan, iru?*	*irurun*
3	*iru? da bits*	*iru da bits*
4	*iru? da iru?*	*iru da iru*
5	*iru? da iru? da bits*	*iru da iru da bits*
6	*iru? da iru? da iru?*	. . . .
:		
10	*dzi bangi marafain da dzi bangi marafain sib*	
15	*dzi bangi marafain da dzi bangi marafain da dzi faga-ng marafain*	
20	*bangi-ng? iru?run da faga-ng? iru?run*	

Note. ? = glottal stop. *Sources:* A. Holzknecht (1989). B. SIL Word List (undated). Village: Kaiapit.

The Adzera system thus has the character of a digit-tally one although it is unusual in that the basic numerals are compounded up to the numeral 9. The system possesses a 5- cycle but this does not come into operation until 10 is reached, the number word for 5 does not contain a "hand" morpheme. The Adzera system is also unusual in that its character is typical of a Non-Austronesian (NAN) or Papuan system and in this respect the systems of other members of Markham Family are similar. The phenomenon of AN counting systems being influenced by proximate NAN languages is indicated by the initial decimal nature of the AN systems being displaced with both 2- and 5- cycles.

Grand Valley Dani

Grand Valley Dani occupy upland valleys in the high mountains of West Papua. See Appendix E, Figure E5. Table B3 gives a summary of the language and counting system.

Table B3
Grand Valley Dani Summary

Dialects:	(6 ?).
Variant Names:	Ndani.
Classification:	NAN, Great Dani Family
Cycles	2 variant, 5,
Operative patterns	Various, 4=2+2; 1 to 9 mostly x+1 to 4; one dialect may be body-tally
Frame words	1, 2, 3, 5,

Table B4 gives the number data.There is diversity between the possible dialects. It seems that there is a (2′, 5) cycle system since 3 is a new word but 4 is mostly 2+2.

Table B4
Grand Valley Dani Systems

	A	B	C	D
1	*ambulit*	*ambui*	*bagi*	*opake-at*
2	*bile*	*mbere*	*bete*	*pere*
3	*kenaran*	*keneran*	*hynaken*	......
4	*embile embile*	*mbere-mbere*	*bete nen bete*	
5	*linggik*	*terak*	....	
6	*api-linggik*	*mbere-dak*		
7	*henok-bile*	-		
8	*henok-kenaran*	-		
9	*awu-linggik*	-		
10	*hanenggu*	*gut-terak*		

	E	F	G
1	*amboei*	*amboeét*	*ambuet*
2	*pere*	*bere*	*bere*
3	*pere amboei, hingiam*	*amboeet bere*	*ambuet bere*
4	*pere pere*	*(kenera ?)*	*kenera*
5	*isia (*= thumb*)*	*anom (*fist*)*	*enom*
6	*apokdal*	-	-
7	*sakbit*	-	-
10	......	*linin-ero (2* fists*)*	*linin-ero*
15		*linin-ero-enom*	....
20		*erovid-enom (*2 fists, 2 feet*)*	

Sources: A. Ndani: Galis (1960, p. 147); Holle Lists: Stokhof (1983, p. 232); B. unknown; C. Dani: van der Stap (1966, p. 155); D. Voorhoeve (1975, p. 105); E. Saeoweri-Hablifoeri: Le Roux (1950, p. 529); F. Swartvallei: Le Roux (1950, p. 529); G. Zentral Papua: Kluge (1938, p. 179), Central New-Guinea: Wirz (1924, p. 145).

However, the data from Saeoweri-Hablifoeri (Le Roux, 1950) do not give 6 and 7 as 5+1 and 5+2; furthermore the word for 5 is thumb which is usually associated with a body-part counting system. However, it might be that with the finger gestures bending down from the little finger as each word is counted that the thumb is reached and given the name. There are similar occurrences mentioned by Lean (1992) such as the Timbe and Selepet languages of Morobe. There is too little data to know if there is a body-tally system. System F suggests a digit-tally system is most likely although with current data there is not a new word for 20. It is known that other languages only indicate the word for 20 when they reach multiples of 20, such as 40 = 2 x "man".

Dobu

Dobu is an Austronesian language spoken on Dobu, Sanaroa and Tewara (or Uama) Islands, as well as in the southern part of Fergusson Island and parts of Normanby Island (See Appendix E, Figure E12). Lithgow (1976, p. 458) noted that: "Edugaura on Dobu Island is regarded as the 'real' Dobu language", and gave some indication of dialect variation between this and other locations of the language. Wurm and Hattori (1981, Map 9) had an estimate of 7 680 for the resident population of Dobu-speaking villages (about 1970) but this figure excludes the number speaking Dobu as a lingua franca. In 1998, SIL noted that there were 500 villages with 10 000 people; 6 000 being monolingual and 100 000 being second language users (1987 SIL). In 2011 census, Dobu Rural Ward had 22 981 people but other languages may be included. A summary of the Dobu language features and counting system is given in Table B5.

Table B5
Dobu Summary

Language type	Austronesian
Family, Subfamily	Milne Bay Family (Dobu C.S.)
Other names, Dialects	Dobu: Edugaura
Village	Loboda. CSQ: Asa, Naikwala, Yoo, Sawaedi, Sanaroa, Unai, Waluma(3), Gomina, Diu, Wedaweda, Kaiwau,Salakahadi, Bwa'ataya, Etana, (2 not given).
Cycles	5, 10, 20 or 5,10
Operative patterns	digit tally
Frame words	1 to 5, 10, 20 or 1-5, 10

In brief: Dobu is a digit tally system, uses ordinals, but is unlikely to use counting for comparing, for example, shell money but "size" is used.

Dobu was used in the D'Entrecasteaux Islands as a mission lingua franca by the Methodist Church since the end of the 19th century (Lawton, 1977) although it was apparently known fairly widely in the region prior to European contact in that it was used as a lingua franca in the Kula exchange cycle and is mentioned by Malinowski (1920, 1922) in this respect.

A vocabulary, including number words, of Dobu was published in the Annual Report on British New Guinea from 1891 to 1892. It was during this year that the Methodist Mission commenced work on Dobu Island; the head of the Mission, the Rev. W. Bromilow, subsequently published a vocabulary and grammar of the language in 1904. Further work was carried out by other missionaries, notably the Revs. J. Dixon, R.V. Grant, and J. K. Arnold, the last publishing in 1931 (Arnold, 1931). This grammar contains a short section on numbers and counting. Ray (1895) has the Dobu vocabulary, including number data, which appeared in the Cambridge Expedition Annual Report with additional data (Ray, 1907, p. 472) and again in Ray (1938a), all three Ray sources appear in Kluge (1941, p. 213).

During the Indigenous Mathematics Project (IMP), which began in PNG in 1976 (Lancy, 1983), Thune carried out anthropological field work at Loboda village on Normanby Island (in the Dobu region) and, in particular, studied the role of numbers and counting within Loboda society (Thune, 1978). An SIL team, the Lithgows, have also been involved in work on Dobu and data are available on the language from SIL, PNG. In addition to the data provided by these various sources, CSQs were completed by 18 informants from various locations in the Dobu region. The data for Dobu are presented in Tables B6 and B7.

Table B6
Counting Systems for Dobu

	A	B	C
1	*ebweuna(ebweu)*	*'ebweu*	*ebweu*
2	*eluwa(elua)*	*eluwa*	*erua*
3	*etoi*	*'etoi*	*eto*
4	*ata*	*'ata*	*ata*
5	*nima*	*nima*	*nima*
6	*nima ta ebweu(na)*	*nima ta 'ebweu*	*nima e bweo*
7	*nima ta eluwa*	.	*nima e rua*
8	*nima ta etoi*	*nima ta 'etoi*	*nima e to*
9	*nima ta ata*	*nima ta 'ata*	*nima ata*
10	*sanau(sanaui)*	*sanau*	*sanau*
11	*sanau ta ebweu(na)*	*sanau ta 'ebweu*	*sanau ta ebweu*
12	*sanau ta eluwa*	*sanau ta 'eluwa*	*sanau ta rua*
15	*sanau ta nima*	*sanau ta nima*	
20	*to'ebweu*	*'ebweu tomotai*	*rua sanau*
30	*to'ebweu ta sanau*		*eto sanau*
40	*tomoluwa*		*ata sanau*

Sources: A. 18 CSQ. B. SIL Data. C. Dobu: Bromilow (1893, p. 106), Ray (1895, pp. 35-38), Kluge (1941, p. 213).

Table B7
Further Counting Systems for Dobu

	A	B
1	*'ebweu*	*ebweuna*
2	*'erua*	*erua*
3	*'etoi*	*etoi*
4	*'ata*	*eata*
5	*nima*	*enima*
6	*nima'ebweu*	*enima-ebweuna*
7	*nima(ta)'erua*	*enima-erua*
8	*nima(ta)'etoi*	*enima-etoi*
9	*nima(ta)'ata*	*inima-eata*
10	*sanau*	*sanau*
11	*sanaui ta('aena)'ebweu*	*sanau ta ebweu*
12	*sanaui ta('aena)'erua*	*sanau ta aena erua*
15	*sanaui ta'aena ta mutu*	.
20	*to'ebweu*	*tai-to-ebweu or rua-sanau*

Sources: A. Dobu (Edugaura): Arnold (1928, pp. 37-38), Ray (1938a, p. 188), Kluge (1941, p. 213). B. Dobu: Kluge (1941, p. 213), Ray (1907, p. 472).

The earlier data in Table B6 both agree well with data in Table B7 which is Arnold's (1931) data collected at least 60 years before the data from CSQs which were also similar. Each of these systems has a (5, 10, 20) cyclic pattern. The number word for 5 is *nima* which is the same as the word for "hand": *nima* (with possessive suffixes). There is a distinct word for 10; *sanau*, which may be cognate with the typical Austronesian numeral *sangahul* (and its variants). The word for "man" is given by the CSQ informants as *tai* or *tomotai*; the SIL data, System B, Table B6, has *ebweu tomotai* "one man" for 20. The CSQ informants and Arnold give, however, *to ebweu*, presumably an abbreviated form of *tomotai ebweu* "man one". The number word for 40 is *tomoluwa* which appears to be an abbreviated form of *tomotai luwa* or "men two".

System C (Table B6), Bromilow's (1893) data, possesses a (5, 10) cyclic pattern and there is no indication of a 20-cycle as given by the other data. The numeral 20 is *rua sanau* "two tens", and this 10-cycle is repeated for the numerals 30 and 40.

Both Arnold (1931) and the CSQ informants provide information on ordinals. The CSQ informants report the ordinal first to be *nugana*, not derived from the corresponding cardinal, whilst Arnold gave '*ebweuna* "which is". All ordinals thereafter are derived from the corresponding cardinals so that we have the ordinals given in Table B8.

Table B8
Ordinals for Dobu

second	*'eluena or 'eruena*
third	*'etonina*
fourth	*'eyatana*
fifth	*'enimana*

Sources: Arnold (1931, p. 38).

Several CSQ informants report counting groups of fours. Coconuts are tied together in fours and a quartet is *puli* or *puli ebwen*. Two quartets is *puli eluwa*, and so on.

In his study of the people of Loboda village, on the north-eastern tip of Normanby Island, Thune (1978, pp. 69-80) described Loboda as being an example of a "non-numerically oriented culture". He noted that: "Although enumeration and counting are possible for the Loboda people, in fact they occur rarely and are, for the most part, unnecessary" (p. 69). Whilst, today, higher numbers may be used to discuss various things which occur in a European-introduced context, there are few circumstances, occurring traditionally, in which numbers above 4 or 5 would be necessary. In discussing various aspects of the traditional culture such as measurement of shell-money, measurement of time, exchange of objects between groups, and so on, each of which provides a situation in which numbers might theoretically be used, Thune observed that, in each case, the Loboda invoke alternative ways of describing their world which make use of relative rather than absolute scales or in which the qualitative aspects of objects are inextricably bound up with their quantification, and thus an abstract system of enumeration disassociated from the objects to be quantified is, on the whole, unnecessary and irrelevant. For example, in the ritual exchange of yams, a group giving yams should eventually receive an amount equivalent to what they gave. In this case, it is the overall size of the total pile of yams to be given which is significant rather than the number of individual yams in the pile. The size of a yam pile is recalled for purposes of repayment in terms the names of the people who received parts of it. Other categories of goods to be exchanged: pig, betel nut, stove goods, etc., are treated in the same way. This form of "name accounting" obviates the use of a precise enumeration of the items in a given category.

Arnold (1931) described the method of counting as follows:

> Beginning with the thumb of the right hand the digits are closed and then the left hand is similarly counted. Next the digits of the right foot beginning with the big toe are checked off with the first finger of the right hand and then the toes of the left foot are similarly treated which gives 20 or *to'ebreu*, literally "one man". (There may be slight variations in the order

of procedure). The Dobuan method of counting and their number names are steadily being superseded by English. The Dobuans count only to 40 for which their name is tomorua (two men). (p. 37)

Hagen, Medlpa

Medlpa is a language spoken around Mt Hagen, the Provincial capital of Western Province. It is an everyday language among the people. Another language/dialect is Gawigal spoken to the south. Considerable anthropological work was carried out by Strathern (Table B9) (Strathern, 1971, 1977).

Two examples of the Melpa (Medlpa) counting system are given; the first being unpublished data collected in 1978 during the IMP study, and the second being taken from Volume 1 of Vicedom and Tischner's (1948) Die Mbowamb, the data being collected prior to 1941 and therefore relatively unaffected by European influence.

Table B9
Hagen, Medlpa Summary

Papuan	Level Isolate, Central family, East New Guinea Highlands Stock
Other names, Dialects	**Hagen,** *Medlpa, Melpa*
Villages	Mt Hagen, Paglum, Baiyer River, Nebilyer
Cycles	4,8,10; 4,5,8
Frame Words	1, 2, 3, 8, 10 ; 1, 2, 3, 5, 8
Operative Patterns	5 to 7=4+n, 6=5+1or4+2, 7=5+2or4+3

On inspection of the data given in Table B12, System B there is evidence of the existence, in addition to the 8-cycle, of a vestigial 2-cycle, a 4-cycle, and a 5 cycle. It appears likely that the counting word for 3, *rakltika* or *raltika*, may well be a combination of the words for 2, *ralg* or *rakl*, and 1, given in Table B12 A as *tenta* or *tilipa* although in combination it may be abbreviated to *ti*. The counting word for 4, *timbokaka* or *tembokaka*, is distinct and does not have a 2+2 construction. We conjecture, then, that the first four counting words have the construction: 1, 2, 2+1, 4, i.e. a modified 2-cycle. The counting word for 5 in Table B12 is *pombi* or *pombi ngkutl* which may mean "thumb" (given by the IMP informant as *pamb* or *pomb*). This suggests that the four fingers of the left hand having been tallied, the tally does not proceed on to the next hand (as usual) but that the thumb may be tallied at this stage to give 5. The variants for counting 6 are *pombi ngkutl dende*, i.e. 5+1, or *ngutl rakl* which we take to have an implied 4+2 construction. No translation is given for the third variant *angelamb dende*. The two variants for 7, *pombi ngkutl rakl* and *kotlrakltika*, have respectively a 5+2 construction and an implied 4+3 construction, thus confirming the (apparently) optional use of either a 5-cycle or 4-cycle. Finally, the counting words 16 to 24 have the construction given in Table B10.

Table B10
Hagen, Medlpa, 8-Cycle

16	8x2 (the hands of two men)
17	8x2+1
18	8x2+2
19	8x2+3 (or 8x2+2+1)
20	8x2+4
21	8x2+5
22	8x2+5+1
23	8x2+5+2
24	8x3 (the hands of three men)

The data shown in Table B12 Melpa Source A display a number of interesting differences to those given from Source B. First, the cyclic structure of the system is rather more consistent than the system discussed above, the 5-cycle having disappeared leaving a regular (4, 8) cyclic pattern. However, superimposed on this is a 10-cycle. In the data discussed above, 10 could be tallied by adding the two thumbs on to the unit *engkaka* or *ki tende*, i.e. 8 or the "hands of one man", i.e. 8 fingers. The system, however, did not possess a 10-cycle. In the more recently collected data, the 10-cycle has been introduced by the mechanism of defining the unit "the hands of one man" to mean 10 fingers rather than 8. Thus 10 itself is *engaka pamb ralg pip*, i.e. "eight with two thumbs down", or *ki tenta* or *ki ti*, i.e. "hand(s) one" or "the hands of one man". The decades, then, are tallied using the "hands" unit so that 20 is *ki ralg* i.e. "hands two" or "the hands of two men", 30 is *ki raltika* i.e. "hands three" or "the hands of three men", and so on. It seems likely that the introduction of the 10-cycle may have occurred as a result of the influence of the European decimal system of counting.

The tally-system shown in Medlpa Source A thus has a (4, 8, 10) cyclic structure or possibly a (2', 4, 8, 10) structure if we take 3 to have a 2+1 construction. Tallying to 20, the counting words show the structure given in Table B11.

Table B11
Hagen, Medlpa Counting Method

1	1	11	(8+2)+1
2	2	12	(8+2)+2
3	3	13	(8+2)+3
4	4	14	(8+2)+4
5	4+1	15	(8+2)+4+1
6	4+2	16	(8+2)+4+2
7	4+3	17	(8+2)+4+3
8	8	18	(8+2)+8
9	8+1	19	(8+2)+8+1
10	8+2 or 10	20	10x2

Strathern (1977) had an account of the mathematics involved in the Melpa people's complicated system of ceremonial exchanges known as *moka* and he dealth with how objects are tallied during the actual ceremony as well as with the details of the ceremony itself. Strathern (1977) noted that:

> Eight and ten are standard numbers aimed at in exchanges. A man tries to give eight or ten items to a partner or set of partners. The wealth objects counted on the fingers in this way are various: pigs, money, shell valuables, cassowary birds, pork, and bamboo tubes of decorating oil, for example. When a large number of items is involved the action of counting off on the fingers can be repeated as necessary. When two men are together, one may keep a tally for the other by lending his fingers to the stask in hand. Thus the man recounting his items given reaches a set of ten and tells the other *kimbonim pendepa mukli*, you keep a record of the number of "hands" (i.e. the sets of eight or ten), and his friend does so by bending down his fingers as appropriate. Each of his bent fingers then represents a set of items, not just a single item. (p. 16)

Strathern also indicated that Melpa men kept a record of the number of shell-sets acquired and given away in *moka* by wearing tallies made out of slats of bamboo or cane, each slat representing a set of eight or ten shells exchanged. The bamboo tallies, worn on the chest as decoration, are not aides-memoire as such but are a public display of the extent to which the wearers have made *moka* and are thus an indication of their wealth or status. No permanent record is kept of the complex accounting, the debits and credits, that are part of *moka* but retention of these in long-term memory is enhanced by frequent discussion.

Table B12

Hagen (Medlpa) Counting Words

	A	B	C	C (meaning)
1	*tenta,tilipa*	*dende*	*tendta/tikpa*	one
2	*ralg*	*rakl*	*ragl*	two
3	*raltika*	*rakltika*	*raltika*	one more than two
4	*timbakaka*	*tembokaka* or *tembokaketle*	*timbikak*	four
5	*timbakaka pamb ti*	*pombi* or *pombingkutl*	*timbikak pumb ti pip*	four fingers and one thumb closing on top
6	*timbakaka pamb ralg*	*pombingkutl dende* or *ngkutl rakl* or *angelamb dende*	*timbikak pumb rgl pip/timbikak gugl ragl* or *pump gugl tendta*	four fingers and two thumb closing on top/two thumb closing on top of four fingers , ie an abstract to mean six
7	*timbakakagul raltika*	*pombingkutl rakl* or *kotrakltika*	*timbikak gugl ragltiki*	four fingers with three included
8	*engaka*	*engkaka* or *engkakaketle* or *ki dende*	*engag/ki tendta*	ki=hand=four fingers of both hands (excludes thumb)
9	*engaka pamb ti*	*pombi ti ngkutl*	*emgag pump to gugl*	one finger on top of four fingers of both hands
10	*engaka pamb ralg pip*, or *ki tenta,ki ti*	*pombinraklngkutl*	*engag pump ragl pip*	four fingers of both hands closed with two thumbs
11	*engaka pamb ralg pip pamb ol tenta*	*pombinrakltikangkutl*	*engag pumb ragl pip to pentipa*	two thumb closing on top of four fingers of both hands with one more included
12	*engaka pamb ralg pip pamb ol ralg*	*tembokakapoket*	*engag pumb ragl pip gugl ragl* or *engag wote gugl timbikak*	(8+2)+2 or 8+4
13	*engaka pamb ralg pip gul raltika*	*tembokakapoket pombin ti ngkutl*	*engag pumb ragl pip gugl ragltiki*	(8+2)+3
14	*engaka pamb ralg pip (gul) timbakaka pakit*	*tembokakapoket pombin rakl ngkutl*	*engag pumb ragl pip timbikak pukit*	(8+2)+4
15	*engaka pamb ralg pip timbakaka pakit pamb ti*	*tembokakapoket pombin rakltika ngkutl*	*engag pumb ragl pip timbikak pukit pumb ti* or *pumb ragltiki*	(8+2)+4+1 or 5x3
16	*engaka pamb ralg pip timbakaka pokit ralg gul*	*ki rakl*	*engag pumb ragl pip timbikak pukit gul ragl* or *ki ragl*	(8+2)+4+2 or 8x2 two hands
17	*engkaka pamb ralg pip timbakaka pakit raltika gul*	*ki rakl tende ngkutl*	*engag pumb ragl pip timbikak pukit gul ragltiki* or *ki ragl pump to gul*	(8+2)+4+3 or (8x2+1
18	*engaka pamb ralg pip wote engaka*	*ki rakl rakl ngkutl*	*engag pumb ragl pip wote engag* or *ki ragl gul ragl*	(8+2)+8 or (8x2)+2
19	*engaka pamb ralg pip engaka pamb ti*	*ki rakl rakltika ngkutl*	*engag pump ragl pip wote engag pump ti* or *ki ragl gul ragltiki*	(8+2)+8 or (8x2)+3
20	*ki ralg*	*ki rakl tembokaka*	*engag pumb ragl pip wote engag pump ragl pip* or *ki ragl timbikak pukit*	(8+2)+(8+2) or (8x2)+4 two hands with four fingers
21	*ki ralg wote tenta*	*ki rakl pombingkutl*	*engag pump ragl pip engag pumb ragl pip wote gul tenta* or *ki ragl timbikak pukit pumb to gul/i ragl timbikak pukit pumb to pip*	(8+2)+(8+2) +1 or (8x2)+4+1 two hands with four fingers with four fingers and one thumb included
22	*ki ralg gul ralg*	*ki rakl pombingkutl dende*	*engag pump ragl pip engag pumb ragl pip wote gul ragl* or *ki ragl wote timbikak pukit gul ragl*	(8+2)+(8+2)+2 or (8x2)+4+2

(continued)

Table B12
(continued)

	A	B	C	C (meaning)
23	*ki ralg gul raltika*	*ki rakl pombingkutl rakl*	*engag pump ragl pip engag pumb ragl pip wote gul ragltiki* or *ki ragl wote timbikak pukit gul ragltiki*	(8+2)+(8+2)+3 or (8x2)+4+3
24	*ki ralg timbakaka pakit*	*ki rakltika*	*engag pumb ragl pip engag pumb ragl pip timbikak pukit* or *ki ragltiki*	(8+2)+(8+2)+4 or 8x3 three hands with four fingers
25	*ki ralg timbakaka pamb ti*		*engag pumb ragl pip engag pumb ragl pip timbikak pukit wote gul tenta* or *ki ragltiki wote pumb to pip*	(8+2)+(8+2)+4+1 or 8x3+1
26	*ki ralg timbakaka gul ralg*		*engag pumb ragl pip engag pumb ragl pip timbikak wote gul ragl* or *ki ragltiki wote gul ragl*	(8+2)+(8+2)+4+2 or 8x3+2
27	*ki ralg timbakaka gul raltika*		*engag pumb ragl pip engag pumb ragl pip timbikak wote gul ragltiki* or *ki ragltiki wote gul ragltiki*	(8+2)+(8+2)+4+3 or 8x3+3
28	*ki ralg wote engaka*		*engag pumb ragl pip engag pumb ragl pip timbikak wote engag* or *ki raltiki wote gul timbikak*	(8+2)+(8+2)+8 or 8x3+4
29	*ki ralg engaka pamb ti*		*engag pumb ragl pip engag pumb ragl pip timbikak wote engag* or *ki ragltiki wote timbikak pukit pum to gul*	(8+2)+(8+2)+8+1 or 8x3+4+1
30	*ki raltika*		*engag pumb ragl pip engag pumb ragl pip wote engag gul ragl* or *ki ragltiki wote timbikak pukit gul ragl*	(8+2)+(8+2)+8+2 or 8x3+4+2 three hands with four fingers with four fingers and two fingers included
40	*ki timbakaka*		*engag pumb ragl pip engag pumb ragl pip engag pumb ragl pip engag pumb ragl pip* or *ki timbikak pumb tip pip*	(8+2)+(8+2)+(8+2)+(8+2) or 8x(4+1)
50	*ki timbakaka pamb ti*		same as 40 with additional *engag pumb ragl pip* or *ki timbikak pumb ti pip*	(8+2) repeated 5 times or [8(4+2)]+2
60	*ki timbakaka gul ralg*		same as 50 with additional *engag pumb ragl pip* or *ki timbikak pumb wote gul ragl*	(8+2) repeated 6 times or [8(4+3)]+3
70	*ki timbakaka gul raltika*		same as 60 with additional *engag pumb ragl pip* or *ki engag wote timbikak gul ragl*	(8+2) repeated 7 times or (8x8)+4+2
80	*ki engaka*		same as 70 with additional *engag pumb ragl pip* or *ki engag ragl pip*	(8+2) repeated 8 times or 8x(8+2) eight hands of four fingers and two thumbs included
90	*ki engaka pamb ti*		same as 80 with additional *engag pumb ragl pip*	(8+2) repeated 9 times
100	*kolg mong kiki ki tenta*		same as 90 with additional *engag pumb ragl pip* or *ki engag pumb ragl pip gul ragl wote timbikak pukit*	(8+2) repeated 10 times or 8(8+2+2)+4 eight hands of four fingers and two thumbs and two fingers and four fingers included
200	*kolg mong kiki kiki ki engaka*		*ki engag ragltiki pumb to pip*	8[(8x3)+1]

Sources: A. IMP Material (unpublished, 1978); B. Vicedom and Tischner (1948, Vol. 1, pp, 237-238); C. Mark (2003).

The original Melpa system of enumeration involved tallying on the fingers, but not initially the thumbs, of both hands. Tallying begins on the little finger of the left hand, which is bent down, and proceeds in order until the index finger is bent down then continues by bending down the little finger of the right hand continuing in order until the index finger is reached thus giving a total of 8 which is *engkaka* or *ki dende*, the latter translated literally meaning "hand(s) one", i.e. "the hands of one man". Thus 16 is *ki rakl*, "the hands of two men", and 24 is *ki rakltika*, "the hands of three men", this pattern indicating the operation of an 8-cycle.

Upon reaching the tally of 8, the two fists, thumbs protruding upwards, are brought together. Tallying may now proceed in one of two ways: either by proceeding through a further 8-cycle or by tallying on first one thumb (9) and then the other (10), the latter being *engaka pamb ralg pip* (as given in Table B12, Source A) which means "eight with two thumbs down".

Mountain Arapesh

Mountain Arapesh is spoken in villages situated along the north coast of the mainland and in a region extending inland to the mountain ranges to the south. Laycock (1973, p. 4) listed the villages within the language region, some of which lie in the West Sepik (Sandaun) Province. A more recent survey by Conrad (1978) suggests that several of these villages, including Matapau on the coast, are Southern Arapesh-speaking but given that the Arapesh Family consists of a long dialect chain, it is somewhat arbitrary to say where one language ends and the other begins. The total resident population of the 42 villages listed by Laycock, at the 1980 National Census, was 8 776. A summary of the language features is provided in Table B13 (National Statistical Office, 1982).

Table B13
Mountain Arapesh Summary

Language type	Papuan
Language family, subfamily	Arapesh Family, Kombio Stock, Torricelli Phylum
Other names, dialects	Mountain Arapesh: Kavu, Kawu, Dagur and Vatai, Bukiyip
Villages	Kauk
Cycles	2', 4, 10?
Operative patterns	3=2+1, 5=4+1
Frame words	1, 2, 4, 10?

In brief: The Mountain Arapesh has a 4 cycle system with possible 10 cycle. The variations are complex across the language group. The number words for 6 to 9, given by the CSQ informants, have a 5 + *n* construction.

Klaffl and Vormann (1905) contained a brief note and a word list of Kavu, i.e. Mountain Arapesh, which includes a small amount of number data. The Kavu data appear subsequently in Ray (1919a) under the same reference, and in Kluge (1938, p. 177) referenced Kawu. Friederici (1913, p. 41) had a set of number words of the language of Dagur and Vatai, on "the mainland of New Guinea, in the region of the Yuo and Muschu Islands"; these also appear to be examples of Mountain Arapesh. Friederici's data also appear in Ray (1919a) and in Kluge (1938, p. 178). A Mountain Arapesh word list is given in the New Guinea Report (1924).

The first comprehensive grammar of the language is given by Fortune (1942) and a more recent grammar by Gerstner (1963). The language is also referred to in surveys by Capell (1969), Glasgow and Loving (1964), and Conrad (1978). In addition to the data available from these various published sources, data are also given in four SIL word lists (compiled in 1975 at Kuak, Womsis, Waulogik, and Smain villages), an IMP questionnaire completed by SIL personnel, and 18 CSQs completed by informants from various parts of the Mountain Arapesh region. The data, selected from the sources above, are given in Tables B14, B15, B16, and B17.

Table B14
Mountain Arapesh Word Classes

Class Number	English Glossary	Examples: Singular	Plural
1	fireplace	*arugab*	*arugabys*
2	village	*wabor*	*waryb*
3	spear	*barawag*	*barawas*
	lobster	*arukweg*	*arukwegas*
4	species of hardwood	*waboku*	*wabemeb*
	cassowary	*unaruku*	*unarib*
	duck	*mabiteku*	*mabiteguhijer*
	fly	*amagoku*	*amagou*
5	spider	*daudam*	*daudeip*
6	species of bird	*kabaun*	*kabaub*
7	male child	*batauishin*	*batauishim*
	man	*araman*	*aramum*
8	long yam	*aboting*	*Abotish*
	stone axe	*bode* (-vowel)	*bodehas*
9	banana	*apap*	*apas*
	feast	*ilup*	*ilugwis*
10	bird	*aramir*	*aramiguh*
	shadow	*aboril*	*aboriguh*
11	brain	*banat*	*banatogu*
	broom	*garogit*	*garogitu*
12	vine	*maduh*	*madururuh*
	lizard	*natageuh*	*natagegwiruh*
	sugar cane	*arouh*	*araruh*
	ear	*atah*	*ateh*
	hand drum	*wiruh*	*wirih*

Note: Class 6 nouns do not imply masculine gender in meaning. *Sources:* Adapted from Fortune (1942, pp. 13-39).

Table B15
Mountain Arapesh Noun Classes

Class Number	1	2	3	4	Suffixes: Singular	Plural
1	*atub*	*biabys*	*biabis atub*	*nybatabys*	*b*	*bys*
2	*atubor*	*biarob*	*biarob atubor*	*nyabatarob*	*bor*	*rob*
3	*atug*	*biagas*	*biagas atug*	*nybatigas*	*g*	*gas*
4	*anoku*	*biou*	*biwato'*	*nybatiu*	*ku*	(various)
5	*atum*	*biep*	*biepatum*	*nybatip*	*m*	*eip, ip*
6	*atun*	*biub*	*bibatun*	*nybatib*	*n*	*b*
7	*anan*	*bium*	*bimatun*	*nybatim*	*n*	*m*
8	*engeng*	*biesh*	*bishating*	*nybatish*	*ng*	*sh*
9	*anap*	*bias*	*bisatup*	*nybatis*	*p*	*s*
10	*anar*	*biaguh*	*biaguhatur*	*nybagitur*	*r*	*guh*
11	*anat*	*biogu*	*bigatut*	*nybatigu*	*t*	*gu*
12	*atuh*	*biaruh*	*biaruhatuh*	*nybatuh*	*uh*	*ruh*
13	*atuh*	*bieh*	*bihatoh*	*nybatih*	*ah*	*eh*

Table B16
Mountain Arapesh Counting Words

	A	B	C
1	*atin*	*ate,atin*	*ating*
2	*piat*	*bie*	*biesh*
3	*pit'atin*	*bihate*	*biesh ating*
4	*nemat'it'*	*nebati*	*numbatish*
5	*anauvip*	*nebahate*	*anop wip*
6		*nebatvie*	
7		*nebat bihate*	
8		*nebat biebie*	
9		*nebate bihate bie*	
10	*vivis*	*anauhip*	

Sources. A. Kavu: Klaffl and Vormann (1905, p. 138), Ray (1919a, p. 330), Kluge (1938, p. 177), Kluge (1941, pp. 28-29) B. Dagur and Vatai: Friederici (1913, p. 41), Ray (1919a, p. 330), Kluge (1938, p. 178), Kluge (1941, p. 28ff.). C. SIL (1975). Village: Kauk.

Table B17
Mountain Arapesh Counting Words

	A	B	C
1	*ot-um*	*atin*	*at-,an-*
2	*bwi-yabal*	*bies*	*bi-*
3	*bwi-yabal ot-um*	*bies atin*	*bi- at-*
4	*nombat-ibal*	*nubatis*	*nybat-*
5	*nombat ot-um*	*anas wis*	
6	*onowip-ibal*	*anas wis atin*	
7	*onowip ot-um*	*anas wis bies*	
8	*onowip bwi-yabal*	*anas wis bies atin*	
9	*anap-ibal*	*anas wis nubatis*	
10	*anap ot-um*	*bies wis*	
11	*anap bwi-yabal*		
12	*wiwis-ibal*		
13	*wiwis ot-um*		
14	*wiwis bwi-yabal*		
15	*aiyag-ibal*		
16	*aiyag ot-um*		
17	*aiyag bwi-yabal*		
18	*elman-igu*		
19	*elman-igu ot-um*		
20	*elman-igu bwi-yabal*		

Sources: A. IMP/SIL Questionnaire (1978). B. 6 CSQs. Villages: Dagua(2), Urip(2), Wautogik(2) C. Numeral Roots : Fortune (1942, p. 58)

Fortune's (1942) study gives details of the noun-classification system employed in Mountain Arapesh and this is also referred to by Laycock (1975b, pp. 771-772). Fortune (1942) indicated that Mountain Arapesh possesses a system of noun-classification not dissimilar to that found in Southern Arapesh. Fortune distinguished 13 classes given in Table B14 (as opposed to 17 for Southern Arapesh) for which there is obligatory concordance for pronouns, adjectives, and numerals. In the case of the numerals there is a set of bound stems and to each of these is added a suffix characteristic of the class to which the objects being enumerated belong. Thus the data given in the Tables are each examples of a set of bound stems (numeral prefixes) suffixed with somewhat different noun suffixes, the variation in the latter arising, presumably, because the various elicitors of the data had their informants enumerate a particular class of objects, the objects and thus the class being somewhat different for each elicitor. For example, the data shown in Table B17 (System A), given by an SIL informant, are a set of number words employed when stones are counted.

Friederici's (1913) data (Table B16, System B) show a system with a (2, 4, 10) cyclic pattern, i.e. the first few number words have the construction : 1, 2, 2 + 1, 4, 4 + 1, 4 + 2, 4 + 2 + 1, 4 + 2 + 2, 4 + 2 + 2 + 1, 10. Friederici's number word for 10, *anauhip*, is somewhat at variance with the other data which have this (in variant forms) as the number word for 5, c.f. *anauvip* (Table B16, System A), *anop wip* (Table B16, System C), *anas wis* (Table B17, System B), *onowip-* (Table B17, System A). The translation provided for this by both the SIL and CSQ informants is "one hand".

The data provided by the CSQ informants indicate a system with a (2′, 5) cycle pattern. The first five number words of this system also accord with those of Systems A and C, Table B16, allowing for slight differences in orthography as well as (probably) variant noun-suffixes. The number words for 6 to 9, given by the CSQ informants, have a 5 + *n* construction: 6, for example, is *anas wis atin*, i.e. 5 + 1 (or "one hand one"). The existence of a 5-cycle in this system is clearly at variance with the cyclic structures of the systems as given by Friederici and by the SIL informant (System A, Table B17), the latter having a somewhat complex cyclic structure. It is possible that the 5-cycle has displaced the former 4-cycle (at least in the system used by younger informants) in order to accord more closely with the English and Tok Pisin decimal systems. The data provided by Fortune (1942), given below, support this view.

Fortune's (1942) book on Arapesh contains details of the noun classification system, referred to above, as well as a clear exposition of the several methods of enumeration employed by the Arapesh people. Here we will summarise some aspects of the noun classification system as Fortune found it in the 1930s. Laycock (1975b) indicated that there is evidence that the system may have changed, or is in the process of changing, in that some classes may have merged with others so that whereas Fortune delineated 13 noun classes, a current investigation may yield somewhat less.

Table B14, similar to Southern Arapesh, gives for each noun class one member of that class in both its singular and plural forms as well as the typical singular suffix and plural suffix characteristic of that class. As with Southern Arapesh, the noun classes do not appear to be a subdivision of the Arapesh universe by meaning, i.e. the criterion for class inclusion does not appear to be a semantic one (as we find in some of the Bougainville languages).

Fortune (1942, p. 58) gave a table of the first four Mountain Arapesh numerals used in counting objects belonging to the 13 noun classes (Table B15).

The four numeral roots (i.e. without classificatory suffixes) are: 1, *at-* or *an-*; 2, *bi-*; 3, *bi- at-*; and 4, *nybat-*. This basic numeral set thus has the form: 1, 2, 2 + 1, 4, and may be described as possessing a modified 2-cycle. Interestingly, the numeral 4 has the meaning "dog" (and thus is not, in the strict sense, a numeral, i.e. a word denoting a number only). However when *nybat-* is used for enumeration it functions somewhat differently than when it denotes "dog"; for example, the phrase "two fours" is not the same as the phrase "two dogs", the former being *biogu nybat* and the latter *biogu nybagu*. In general, *nybat* when used as a number retains its singular suffix in the plural whereas *nybat* in its meaning of "dog" behaves as a normal noun and takes the plural suffix *-gu* in the plural.

Enumeration beyond 4 is described by Fortune (1942):

> Counting from four to twenty-four proceeds on two roots for numerals and the word *nybat* treated as class 11 as follows: four one, four two, four three, *biogu* nybat (two fours or eight), eight one, eight two, eight three, *biogu atut* (= three fours or twelve) [n.b. the *nybat* is understood], twelve one, twelve two, twelve three, *bigi biogu* (two eights or sixteen), and so on to *biogu atuga biogu* or twenty and from there similarly to *anauwip* or twenty-four. In counting by threes in alternative system anauwip means two threes or six. (p. 59)

Thus the basic Mountain Arapesh system possesses a (2', 4) cyclic pattern or, rather, a (2', 4, 24) cyclic pattern if counting proceeds beyond 24. As indicated by Fortune, this system has *anauwip* for 24. In Table B16 System A has *anauvip* for 5, System B has *anauhip* for 10, and System C has *anop wip* for 5. In Table B17, System A has *onowip*- in the construction of 6, 7, and 8, and System B has *anas wis* for 5. This apparently confusing use of *anauwip* arises, in part, from its meaning of "one hand" and whether "one hand" is taken to mean 5 or 6, and also that, in addition to the basic system of counting, there is a system for counting groups of three objects.

In counting groups of three the procedure is described by Fortune (1942) as follows:

> three, three one, three two, *anauwip* or six. Then repeat to a second *anauwip* and call *wiwis* or twelve. This is supposed to be the count on the hands, five fingers and a thumb joint as well, to make six for each hand. Then *wiwis* means both hands, *anauwip* one hand. The same procedure is repeated for another twelve, theoretically the feet treated as the hands have been and the grand total is called a man, *araman*. (p. 59)

Fortune also noted that in using the basic 4-cycle system (i.e. counting in fours), finger and toe tallying does not occur. However when counting proceeds to several cycles of 24, each 24- cycle is recorded with a stick or peg before proceeding to the next.

Thus when *anauwip*, "one hand", is taken to mean the five fingers and the thumb-joint it functions as 6. The SIL data in Table B17 use *onowip* in this sense (the system given appears to be a combination of both the 4- cycle and 3- cycle systems described by Fortune). If, however, *anauwip* is taken to mean the five fingers of one hand (and not the additional thumb-joint) then it functions as 5 (*wiwis* or *vivis*, "both hands", is then 10). This indeed is what we appear to have in the data from Klaffl and Vormann (1905) given in Table B16. Also, the data provided by the CSQ informants (Table B17) and the SIL data collected in 1975 (Table B16, System C) both appear to be using *anauwip* in its 5 sense rather than its 6 sense.

Fortune (1942, p. 60) indicated that objects which are counted using the 4-cycle system are: "coconuts, small yams, taro, arm-rings, dogs' teeth, house poles, house posts, breadfruit, sago frond sheath, pots, plates, spears, arm-bands, bamboo lengths, sugar-cane lengths, eggs, birds, lizards, grubs, fish, [etc.]". The objects counted using the 3-cycle system are "betel nuts, thatch shingles, coconut, sago, and betel nut palms, big yams, packets of sago, wild game, sheets of sago bark, packages of vegetable greens, braids of tobacco, packets of tobacco [etc.]".

Although the number of numeral roots is somewhat restricted in Mountain Arapesh it is worth making the point that this does not necessarily mean that the process of enumeration is hindered or restricted in any way. Fortune (1942) noted that:

> Counting is done with great facility and ease with this conventionalisation of very few special roots. To suppose that the paucity of the Papuan languages in root words for numerals makes counting difficult to the Papuan is quite incorrect. The Arapesh people count rather more quickly and better than the Melanesian Dobuans, who use a decimal system with many more root terms. An unambiguous conventional method, with or without many root terms, is all that is necessary. (p. 59)

Ponam

Ponam is spoken only on Ponam Island which lies to the west of Andra Island and off the north coast of Manus Island. The population of Ponam Island at the 1980 National Census was 297 (National Statistical Office, 1982). The language is classified as a member of the Manus Family, North-West Islands Sub-Family (Wurm & Hattori, 1981, Map 14) (Table B18).

Table B18
Ponam Summary

Language type	Austronesian
Language family; subfamily	North -West Islands Sub - Family
Other names, dialects	Ponam
Villages	Ponam Island.
Cycles	10, 100, 1000
Operative pattern	7=10-3,etc
Frame words	1 to 6, 10, 100, 1000

In brief: Ponam has a regular Manus type of system but has numerals 7, 8, 9. The first three ordinals do not appear to derive from the corresponding ordinals but have the meaning "front", "middle", "behind". Generally the numerical classifiers have the structure "numerical root + class marker".

Z'Graggen (1975a, pp. 183-188) had Smythe's numerals taken at Ponam Island in the late 1940s. Ponam was one of the research sites for the Indigenous Mathematics Project (Lancy, 1978, 1983); part of the gathered Ponam data included information on the counting system in use and this was made available to Lean. Carrier's (1981) data were from Ponam Island and two CSQs were also obtained.

The Ponam numerals, Table B19 System A, form a regular Manus type of system and closely corresponds to the Andra system except that the Ponam */f/* has replaced the Andra */h/*, and for the numerals 7, 8, 9 the Ponam */h/* has replaced the Andra consonant cluster */ndr/*. Ordinals were reported

Table B19 *Ponam Counting Words*

	A	B	C
1	*si*	*si*	*si*
2	*luwof*	*luof*	*luof*
3	*talof*	*talof*	*talof*
4	*faf*	*fa:f*	*faf*
5	*limef*	*limef*	*limef*
6	*wonof*	*onof*	*wonof*
7	*ahatalof*	*ahadalof*	*ahatalof*
8	*ahaluwof*	*ahaluof*	*ahaluof*
9	*ahase*	*ahasi*	*ahase*
10	*sanguf*	*sanguf*	*sanguf*
11	*sanguf ne si*		*sanguf ne si*
12	*sanguf ne luwof*		
15	*sanguf ne limef*		
20	*lunguf*		*lunguf*
30	*tulunguf*		*tulunguf*
40	*fafunguf*		*fangfuf (fanguf)*
50	*limenguf*		*limenguf*
100	*sangat*	*sangat*	*sangat*
200	*longat*		*longat*
500	*limengat*		*limengat*
700			*ahatulungit*
1000		*sabau*	*sapau*
2000			*lovau*
3000			*tuluveu*

Sources: A. Data : 2 CSQs; IMP Field Notes (1978). Ponam Island. B. Ponam: Z'Graggen (Smythe), (1975a, 1975b, pp. 183-188). C. Carrier (1981, pp. 475-6).

by the IMP informant. The first three ordinals are respectively *marau, hakeu*, and *kein* which do not appear to derive from the corresponding ordinals. Carrier (1981, p. 466) indicated that these have the meaning "front", "middle", "behind".

Numeral classification exists in Ponam and some data on these, which derive from Carrier (1981, pp. 476-478), are given in Table B20. Generally the numerical classifiers have the structure "numerical root + class marker" as we find with Gele' and Andra-Hus. Unlike the latter, however, the Ponam system uses only the first four numeral roots and thereafter the basic counting system is employed.

Table B20
Ponam

	Halves,Parts	Bundles	Bags	Blades	Strings of Objects	Fish Hooks	Persons, Spirits	Holes, Cavities	Nets, Sails
1	*sabeh*	*sabis*	*safat*	*sahou*	*sahol*	*sakau*	*saman*	*samal*	*sapal*
2	*lofeh*	*lobis*	*lofat*	*lohou*	*lohol*	*lokau*	*njalui*	*lomal*	*lopal*
3	*tulufeh*	*tulubis*	*tulufet*	*tuluheu*	*tuluhel*	*tulukeo*	*tulumin*	*tutlumel*	*tulupel*
4	*fafeh*	*fabis*	*fafat*	*fahou*	*fahol*	*fakou*	*faman*	*famal*	*fapal*
5	*limef*	*limef*	*limef*	*limef*	*limef*	*limef*	*limef*	*limef*	*limef*

Sources: Carrier (1981, pp. 476-8)

Mengen

Mengen, or Maenge, is an AN language with three dialects: Mengen No.1 or Poeng, Mengen No.2 or Orford, and Bush Mengen or Longeinga. Mengen No.1 is spoken in the coastal region to the east and south of Pomio, East New Britain. Mengen No.2 is spoken in an area immediately to the south of the Sulka region in the vicinity of Cape Orford. Inland and north-west of Pomio the Bush Mengen dialect is spoken. Rath indicated that, at the time of writing, the total number of resident speakers of the Mengen No's 1 and 2 dialects was about 6,600 whilst about 1800 people spoke the Bush Mengen dialect. Mengen is classified as belonging, with Mamusi-Kakuna and Uvol, to the Mengen Family (Rath, 1980). Summary of information is presented in Table B21.

Table B21
Mengen Summary

Language type	Austronesian
Language family; subfamily	Mengen Family.
Other names; dialects	Mengen: Maenge.
Villages	Malo, Gaive, Malakur, Salel (Sampun)
Cycles	5, 10, 20
Operative patterns	6 to 9: 5 + n
Frame words	1 to 5, 10, 20

In Brief: Mengen is an Austronesian system with a digit tally system with (5, 10, 20) cycles and use of thumbs and fingers, hand, fists, toes, foot.

The first grammar of Mengen, including a short section on numbers, was published by Br. H. Müller in 1907. In the same year, Parkinson (1907) published including a comparative table of number words of Sulka, Mengen, and Tomoip. Both Müller's and Parkinson's data appear in Kluge (1941, pp. 194-195). In addition, the language has been mentioned in Chowning (1976), in Capell (1962, 1971), and in Laufer (1946–1949). Panoff (1969), an anthropologist who has done extensive work on the Maenge people, has an article on the Mengen counting system (Panoff, 1970). Rath (1980, p. 206) had the number words for 1 to 10 for the Mengen No.1 dialect. The data for Mengen, given in Tables B22 and B23, derive from these sources as well as from 13 CSQs and one IMP questionnaire.

Table B22
Mengen Counting Words

	A	B	C
1	*kena*	*tia ken*	*kena*
2	*lua*	*tia luo*	*'lua*
3	*mologi*	*tia moleg*	*mo'logi*
4	*tugulu*	*tia tugul*	*tu'gulu*
5	*lima*	*tane lim*	*'lima*
6	*lima va kena*	*tane* (or *kana) ken*	*lima ba kena*
7	*lima va lua*	*tane* (or *kana) luo*	*lima ba lua*
8	*lima va mologi*	*tane* (or *kana) moleg*	*lima ba mologi*
9	*lima va tugulu*	*tane* (or *kana) tugul*	*lima ba tugulu*
10	*tangulelu*	*kana lim* or *tangauna ta*	*tangu'lelu*
11	*tangulel va kena*		
15	*tangulel va lima*		
20	*tangulel lua (giaukaina)*	*tangaure ta luo* or *agagi tia ti mot*	
30	*tangulel mologi*	*tangaure ta moleg*	
40	*(giaukaina lua)*	*tangaure ta tagul*	
50	*(giaukaina lua va tangulelu)*	*tangauna iliman ta*	
60	*(giaukaina mologi)*	*tangauna kande ta ken*	

Sources: A. 11 CSQs, 1 IMP. Villages: Malo, Gaive, Malakur, Salel (Sampun). B. Mengen: Müller (1907, p. 88), Kluge (1941), pp. 194-195). C. Mengen: Rath (1980, p. 206).

Panoff's data (System A, Table B23) were acquired from informants speaking the Mengen No.1 dialect and who were aged mainly in their sixties; the system given, then, can be regarded as the "traditional" one for this dialect. The system has a (5, 10, 20) cyclic pattern with a basic numeral set (1 to 5, 10). The numerals from 6 to 9 have the operative pattern $5 + n$ where n takes the values 1 to 4 respectively. *giaukaena*, i.e. 20, is not a numeral but rather a number word which Panoff indicates can be analysed into two morphemes: *giau*, "human being", and *kaena*, "his/her foot" or "leg" or "foot and leg together". Panoff (1970) also notes that there were two methods of counting i.e. the counting of single units and the counting of groups of twenty. In the counting or, rather, tallying of single units:

> The Maenge begin counting on the right hand, which is held open. First, the index finger of the right hand touches the thumb of the same hand while one says *lua* (two). The middle and fourth fingers are then clustered together and bent downward while one again says *lua*. Finally, the little finger of the same hand is bent while one says *ne lima* (the ordinal for five). Once this has been done, one passes on to the left hand, using its fingers as tallies in the same way as those of the right hand and again calling the numerals *lua, lua, ne lima*. At this point one closes the fists, both hands being held up together, and says *tangulelu* (10). (p. 363)

When tallying proceeds beyond ten, the toes of the right foot are tallied next followed by the toes of the left foot until twenty is reached. The tallier places both fists on the toes and *giaukaena* is called.

The numerals 1, 3, and 4 are not typically AN in character although the numerals 2, 5, and 10 are, the last being cognate with *sangahul* except that the /*s*/ is replaced by /*t*/. The numeral 5, *lima* is not identical to the word for "hand", *kama*.

Table B23
Mengen Counting Words

	A	B
1	*kena*	*kena*
2	*lua*	*lua*
3	*mologi*	*moloi*
4	*tugulu*	*tugulu*
5	*lima*	*lima*
6	*lima va kena*	*lima ba kena*
7	*lima va lua*	*lima ba lua*
8	*lima va mologi*	*lima ba moloi*
9	*lima va tugulu*	*lima ba tugulu*
10	*tangulelu*	*tagunailu*
11	*tangulelu va kena*	*tagunailu ba kena*
15	*tangulelu va lima*	*tagunailu ba lua*
20	*giaukaena*	*giukainte*
30	*tiaukaena kena va tangulelu*	
40	*giaukaena lua*	
50	*giaukaina lua va tangulelu*	
100	*giaukaena lima*	

Sources: A. Mengen No.1: Panoff (1970, p. 359). B. Data: 2 CSQs. (Inland Pomio region).

At least half of the CSQ informants did not indicate the existence of a 20-cycle system. Instead they gave a system (see System A, Table B22) with a (5, 10) cyclic pattern and with the decades formed as multiples of 10.

Müller's data, Table B22, System B, are similar to those discussed above for the numerals 1 to 5. The numerals 6 to 9, however, do not explicitly have the numeral 5 appearing in them but rather have the construction x+1, x+2, etc., where x is *tane* or *kana*. The numeral 10 is given as either *kana lim*, i.e x+5, or as *tangauna ta*, or *tangau na ta*, these being cognate with the typical AN word for 10 *sangaun* with, as above, the /*s*/ replaced by /*t*/. The construction of the decades uses *tangau* so that 20 is *tangau re ta luo*, 30 is *tangau re ta moleg*, 50 is *tangau na iliman ta*, and so on. Müller also gave an alternative for 20, i.e *agagi tia ti mot* but does not provide a translation.

One CSQ informant from Sampun village in the Orford dialect region gives the numerals 1 to 9 as being almost identical to those of Table B22, System A. The numerals 10 and 20 are, however, given respectively as *tangaun* and *parun*. Two CSQ informants from the inland Pomio region gave the system shown in Table B23, System B. This may be an example of the Bush Mengen dialect.

Arosi

Arosi is found in the northwest of Makira Island, Makira Province, Solomon Islands. Honiara, the capital, is on an island to the north. Its neighbouring language is Fagani. It has two main dialects, Wango and Arosi, with others. Its population was recorded as 6 750 (SIL, 1999). The language is Austronesian, Oceanic, Southeast Solomonic and within the Malaita-San Cristobal group. It is still widely spoken with English. The literacy rate in the first language is around 60% of all ages with English slightly lower due to an active literacy program with a grammar and New Testament (Table B24).

Table B24
Arosi Summary

Language type	Austronesian, Oceanic
Language family; subfamily	South-East Solomonic, Malaita-San Cristobal Cluster
Other names; dialects	Arosi, Wango, other dialects
Villages	
Cycles	10, 100 etc
Operative patterns	base 10
Frame words	1 to 10, 100, 1000

Fox (1931) recorded the numerals and other vocabulary and grammar of this language. Capell (1971) incorporated into his discussion of the San Cristobel group in the 1960s and 70s. Table B25 provides the counting words for Wango with similarities from the recordings nearly a century apart. These can be compared to the Proto-Polynesian counting words developed by Pawley (1972) given in Table B26.

Table B25
Wango Counting Words

	A	B	C
1	*eta*	*e ta / e ta'ai / e ta'i*	*tai*
2	*rua*	*e rua*	*e rua*
3	*oru*	*e oru*	*e'oru*
4	*hai*	*e hai*	*e hai*
5	*rima*	*e rima*	*rima*
6	*ono*	*e ono*	*ono*
7	*biu*	*e biu*	*bi'u*
8	*waru*	*e waru*	*waru*
9	*siwa*	*e siwa*	*siwa*
10	*tangahuru*	*tangarhuru*	*tangahuru*
12	-	*e ta'i tangahuru mana rua*	
20	-	*e rua tangahuru*	*rua tangahuru*
100	*ta?i ?arangi*	*tangarau*	*tangarau*
1000.	*ta?i meru*	*meru*	

Sources: A. Tryon and Hackman (1983, pp. 125, 129, 133) B. Capell (1971, pp. 50-56) C. Wango: Codrington (1885, p. 511), Ray (1919b, p. 176).

Table B26
Proto-Polynesian Counting Words

	A.
1	**ha, *taha*
2	**rua*
3	**tolu*
4	**faa*
5	**lima*
6	**ono*
7	**fitu*
8	**walu*
9	**hiwa*
10	**hangafulu*

Sources: Pawley (1972, pp. 52-54).

Rennellese

Rennellese is a Polynesian language spoken on the most southerly island of the Solomon Islands. It is 200km by sea south-west of Arosi. Ray recorded data in 1917 and 1919 for the two dialects. Table B27 summaries the language classification data and Table B27 provides some of the number system indicating the use of large numbers (Table B28).

Table B27
Rennellese Summary

Dialects:	1) Rennell 2) Bellona
Variant Names:	Rennell: Mo-ngava; Belana: Mo-ngiki
Classification:	Polynesian

Table B28
Rennellese Counting Words

	A	B	C	D
1	*tahi*	*tahi, tasi*	*tasi*	*tasi*
2	*nggua*	*ngua, gua*	*ngua*	*gua [ngua]*
3	*tonggu*	*tongu, tolu*	*tongu*	*tol*
4	*ha*	*ha, va*	*ha/fa*	*va*
5	*a*	*ngima, lima*	*ngima*	*lima [*hand = *tonu]*
6	*ono*	*ono*	*ono*	*ono*
7	*hitu*	*hitu, vitu*	*fitu, whitu*	*vitu*
8	*banggu*	*bangu, walu*	*vangu*	*walu*
9	*iba*	*iwa, siwo*	*iva, iba*	*sivo*
10	*katoa/angahunggu*	*katoa*	*angahungu/katoa*	*katoa*
20	-	*katoa haka ngua*	*ngua ngahungu*	
30	-	------------	*tongu ngahungu*	
100	*nggau/tehua*		*noa*	
1000	*mano*		*afe*	

Sources: A. Tryon and Hackman (1983, pp. 126, 130, 134). B. Mo-ngava: Ray (1917, p. 174), Mo-ava: Ray (1919b, p. 69). C. Mo-ngiki: Ray (1917, p. 174); Mo-iki: Ray (1919b, p. 69); D. Rennell: Woodford (1907, p. 37).

Pukapuka

Pukapuka is a Polynesian language classified as a Samoic-Outlier spoken on an island north east of Samoa in Triangle Polynesia. Table B29 gives the language classification data and Table B30 the counting words that indicate the ways of counting of large numbers of objects and further discussed in Chapter 8.

Table B29
Pukapuka Summary

Classification:	Polynesian Group, Samoic-Outlier Section

Table B30
Pukapuka Counting Words

	A	B
1	*tayi*	*lua* : 4 coconuts
2	*lua*	*tolu* : 6 coconuts
3	*tolu*	*wa* : 8 coconuts
4	*wa*	*yepulupulu* : 10 coconuts
5	*lima*	*ono* : 12 coconuts
6	*ono*	*witu* : 14 coconuts
7	*witu*	*valu* : 16 coconuts
8	*valu*	*iva* : 18 coconuts
9	*iva*	*yaea* : 20 coconuts
10	*katoa, laungaulu*	*tolu* : 30 coconuts
11	*laungaulu ma tayi*	*wa* : 40 coconuts
20	*lau lua*	*lima* : 50 coconuts
30	*lau tolu*	*lau* : 100 coconuts
100	*lau*	....
200	*lua lau*	
300	*tolunga lau*	
400	*wanga lau*	
1000	*mano*	
2000	*lua mano*	

Sources: Beaglehole and Beaglehole (1938) A. p. 353. B. Counting coconuts by pairs

Tongareva

Tongareva is the largest and most remote island of the 15 Cook Islands lying 1 365 km north-north east of Rarotonga the capital. It is also known as Penrhyn Atoll named after Lady Penrhyn ship that sailed there in 1888. Besides having only 9 sq km of land and being less than 5m high, its population was severely depleted around 1864 when islanders were taken to Peru and not returned. Tongareva is a Polynesian language that also varies its counting according to items being counted. Summary data on language classification are given in Table B31 with some of the counting words given in Table B32.

The classifiers in Tongareva are mainly for important foods and other items had classifiers e.g. *tekau*: classifier for coconuts.

Table B31
Tongareva Summary

Variant Names:	Penrhyn
Classification:	Polynesian Group, Eastern Polynesian Section

Table B32
Tongareva Counting Words

	A
1	*tahi*
2	*rua*
3	*toru*
4	*haa*
5	*rima*

(continued)

Table B32
(continued)

	A
6	-
7	*hitu*
8	*varu*
9	*iva*
10	*ngahuru*
100	*rau*
2000	*mano*

Sources: A. Data from the Pollex file (The Comparative Polynesian Lexicon Project), printout dated 19 November 1991, courtesy of Bruce Biggs, Maori Studies Dept., University of Auckland.

Mota

Mota is an island of the Bank Islands in the north of Vanuatu. Mota became the church language throughout Vanuatu and as far north as Isabel in the Solomon Islands from the second half of the 19th century. The summary data are given in Table B33.

Table B33
Mota Summary

Dialects:	(2)
Classification:	AN, North-Central New Hebrides Group, East New Hebrides Sub-group

Table B34 presents the counting words from two sources, a century apart, showing remarkable agreement. These counting words were not used for counting game.

Table B34
Mota Counting Words

	A	B
1	*tuwale*	*tuwale*
2	*ni rua*	*ni-rua*
3	*ni tol*	*ni-tol*
4	*ni vat*	*ni-vat*
5	*tavelima*	*tave-lima*
6	*lavea tea*	*lavea-tea*
7	*lavea rua*	*lavea-tua*
8	*lavea tol*	*lavea-tol*
9	*lavea vat*	*lavea-vat*
10	*sangavul*	*sangavul*
20	*sangavul rua*	....
30	*sangavul tolu*	
40	*sangavul vat*	
100	*melnol*	
1000	*tar (tuwale)*	

Sources: A. Codrington (1885, pp. 235, 302-303). B. Tryon (1976, pp. 405-415).

Ere-Lele-Gele'-Kuruti

Wurm and Hattori (1981, Map 14) showed this language as having five dialects but it is noted that the dialects "may in fact constitute different languages". The dialects are 1) Ere, spoken on the south coast of Manus Island adjacent to the Nali region, 2) Lele, spoken on the north-east coast in the Lorengau area, 3) Gele', spoken in the central part of Manus Island, 4) Kuruti, spoken in the central and north-coast region of Manus Island, and 5) Koro, spoken on the north coast in a small area between the Kuruti and Lele regions. Laycock gave the following populations for the dialects at about 1970 as being: Ere (600), Lele (1 000), Gele' (1 000), Kuruti (1 800), and Koro (260). The language is classified as a member of the Manus Family, East Manus Sub-Family (Wurm & Hattori, 1981, Map 14). Table B35 gives the summary data for this group of languages/dialects.

Table B35
Ere-Lele-Gele'-Kuruti Summary

Language type	Austronesian
Language family; subfamily	East Manus Sub - Family
Other names	Kuruti: Kurti;
Villages	Liap, Kari, Lomoei, Dekimbat, Pundru.
Cycles	10, 100, 1000
Operative pattern	7=10-3,etc
Frame words	1 to 6, 10, 100, 1000

In brief. The counting systems of Ere, Gele', and Kuruti are all of the Manus type and probably other dialects too. A particle in construction of the thousands, Several Kuruti informants indicate a construction for tens of thousands basically using quantity classifiers. The numerals exhibit classification, with up to 36 classes in Buyang (Gele'). 43 groups also given for Gele'.

A large amount of information on the Gele' dialect can be found in Smythe's (n.d.) unpublished work Admiralty Island Linguistics, the manuscript of which is held at the SIL library in Ukarumpa. Volume 4 of this work deals with the grammar of Gele' whilst Volume 7 is an extensive vocabulary. Z'Graggen (1975a), which has Smythe's comparative word list of most of the Manus languages, also contains some Gele' data, referenced as Bujang. In addition to Smythe's data we also have available 23 CSQs completed by informants from all dialect regions, as well as four SIL word lists, each compiled in 1975, of the Ere, Gele', Lele, and Koro dialects respectively. The data on the basic counting systems of the five dialects are given in Table B36. Further data on the numeral classification found in the Gele' and Kuruti dialects are given in Tables B37, B38, B39, B40, B41, and B42.

Table B36
Ere-Lele-Gele'-Kuruti Counting Words

1	*sa- (se-, he-, ha-, s-, si-, ho-)*
2	*ru-*
3	*dul-*
4	*ha:-*
5	*li:m-*
6	*en-*
7	*endrdul- (ndro:dul-)*
8	*ndro:ru-*
9	*ndro:sa-*

The counting systems of Ere, Gele', and Kuruti are all of the Manus type. The Lele and Koro systems probably are as well but no information was provided on the numerals 6 to 9. The Gele' informants report a particle *-pou* for the construction of the thousands, e.g 1 000 is *e-pou*, 2 000 is *ru-pou*, 3 000 is *tul-pou*, and so on. Similarly, the Kuruti informants report the use of the particle *-pueu* for the

Table B37
Ere-Lele-Gele'-Kuruti (Gele') Classes

Class 1	Groups of 10. (Any kind of object)
Class 2	Groups of 100
Class 3	Groups of 1 000
Class 4	Groups of 10 000
Class 5	Groups of 100 000
Class 6	Groups of 1 000 000
Class 7	Days, parts of a day, e.g. nights.
Class 8	The 'basic' class, the most commonly used of all; used of all things not covered by other classes.
Class 9	Persons, animals.
Class 10	Houses, buildings.
Class 11	Long objects, e.g. trees, canoes.
Class 12	Cutting and slicing instruments, e.g. knife but not axe.
Class 13	Pools of water of any size, e.g. pond, lagoon, lake.
Class 14	Quarters of anything, e.g. a 1/4 - strip of rattan.
Class 15	Limbs, roots of trees, e.g. arms, legs.
Class 16	Actions involving the voice.
Class 17	The "other side" of anything, e.g. with strings of shell money it denotes half-strings.
Class 18	Made-up articles (used to distinguish them from the thing from which they were made).
Class 19	Halves of things which are jointed or folded down the middle. (?)
Class 20	Baskets, bags, containers.
Class 21	Complete sugar-cane plants, any complete plant.
Class 22	Flat, natural objects, e.g. leaf, ear.
Class 23	Small groups of natural objects, e.g. a hand of bananas.
Class 24	Any sort of path, thoroughfare, track.
Class 25	Spears.
Class 26	Districts (as opposed to villages). Also halves broken or cut transversely.
Class 27	Halves of natural things, cut off or broken longitudinally.
Class 28	Flat artifacts, e.g. paper, sheet, board.
Class 29	Taro.
Class 30	Holes of any description, inc. doorways, caves.
Class 31	Fire(s). Used for distinguishing between fire (*ji:h*) and firewood (*ji:h*).
Class 32	Villages.
Class 33	Rivers, creeks, streams, watercourses.
Class 34	Midribs of large leaves (coconut, sago). (?)
Class 35	Packets, bundles which are rolled or wrapped up (but not tied up).
Class 36	(Not given).
Class 37	Sets of garamuts (log-drums) comprising an 'orchestra' (usually seven).
Class 38	Groups of people comprising a family.
Class 39	Large groups of persons, animals, e.g. a shoal of fish.
Class 40	Heaps or piles of anything.
Class 41	Fathoms (i.e. the "standard" length between the fingertips of two outstretched arms).
Class 42	Cutting or chopping instruments, or the marks made by them.
Class 43	Small bits and pieces of anything.

construction of the thousands, the word for 1 000 being the exception. This is given as either *sede* or *ho-pou*; the higher thousands, however, show the structure: *ru-pueu* for 2 000, *li-pueu* for 5000. Several Kuruti informants also indicate a construction for tens of thousands as follows: 10 000, *po-sungoh*, and 30 000 is *po-tulungeh* or *po-tulngeh*.

All the Gele' and Kuruti informants report the existence of special number words for counting various classes of objects. Capell (1971) quoted Smythe who mentioned that the Manus Island languages show certain features which are atypical of the Melanesian languages generally. One of these features is that

> the numerals exhibit classification, with up to 36 classes in Buyang (Gele'). Such classification is known in the other languages but to a lesser extent: Ninigo, 16 classes, Aua and Wuvulu, as far as is known, exhibit two or three. (Capell, 1971, p. 247)

In fact, Smythe's volume on Gele' grammar contains data on 43 different classes but it is certain that there are many more.The numeral roots, which are suffixed by the classifier, take the basic forms given in Table B36.

For each of these 43 classes (sets) either the defining attribute of the class or the principal members of the class are given in Table B37.

In Table B38, there are 15 classes showing Gele' dialect counting words for 1 to 9.

Table B39 gives Gele' and Kuruti counting words from various sources.

Table B38
Numeral Classification in the Gele' Dialect

Number	Class 1	Class 2	Class 3	Class 4	Class 5
1	*sungwah*	*sangad*	*he:bwo*	*bwosungwah*	*bwosangad*
2	*rungeh*	*runged*	*ru:bwo*	*bwo:rngeh*	*bwo:rnged*
3	*dulngeh*	*dulnged*	*dulbwo*	*bwodulngeh*	*bwodulnged*
4	*ha:ngoh*	*ha:ngad*	*ha:bwo*	*bwoha:ngoh*	*bwoha:ngad*
5	*li:mngeh*	*li:mnged*	*li:bwo*	*bwoli:mneh*	*bwoli:mnged*
6	*angwah*	*anngad*	*enbwo*	*bwoanngwah*	*bwoanngad*
7	*ndro:dulngeh*	*ndro:dulnged*	*ndro:dulbwo*	*bwondrdulngeh*	*bwondrdulnged*
8	*ndro:rngeh*	*ndro:rnged*	*ndro:rbwo*	*bwondro:rngeh*	*bwondro:rnged*
9	*ensungwah*	*ansangad*	*ensebwo*	*bwoensungwah*	*bwoansangad*
Number	**Class 6**	**Class 7**	**Class 8**	**Class 9**	**Class 10**
1	*bwohe:bwo*	*sei*	*si:h*	*hamow*	*sim*
2	*bworu:bwo*	*ru:*	*rueh*	*rumow*	*rubwim*
3	*bwodulbwo*	*dul*	*daloh*	*dulmow*	*dulbwim*
4	*bwoha:bwo*	*ha:*	*ha:huh*	*ha:mow*	*ha:bwim*
5	*bwoli:bwo*	*li:m*	*li:mweh*	*li:mow*	*li:bwim*
6	*bwoenbwo*	*on*	*enoh*	*enmow*	*enbwim*
7	*bwondrdulbwo*	*endrdul*	*endrdaloh*	*ndro:dulmow*	*ndro:dulbwim*
8	*bwondro:rbwo*	*ndro:ru*	*ndro:rueh*	*ndro:rmuw*	*ndro:rbwim*
9	*bwoensebwo*	*ndrwa:sei*	*ndrwa:si:h*	*ensa:mow*	*ndrwa:sim*
Number	**Class 11**	**Class 12**	**Class 13**	**Class 14**	**Class 15**
1	*heij*	*habal*	*hebwil*	*hendreg*	*hagai*
2	*rueij*	*rubel*	*rubwil*	*rundreg*	*rugei*
3	*dulweij*	*dulbel*	*dulbwil*	*dulndreg*	*dulgei*
4	*ha:weij*	*ha:bal*	*ha:bwil*	*ha:ndreg*	*ha:gai*
5	*li:weij*	*li:bel*	*li:bwil*	*li:ndreg*	*li:gei*
6	*enweij*	*anbal*	*enbwil*	*enndreg*	*angai*
7	*ndro:dulweij*	*ndro:dulbel*	*ndro:dulbwil*	*ndro:dulndreg*	*ndro:dulgei*
8	*ndro:rweij*	*ndro:rbel*	*ndro:rbwil*	*ndro:rndreg*	*ndro:rgei*
9	*ndrwa:heij*	*ansabal*	*ensebwil*	*ensendreg*	*aneagai*

Sources: W.E. Smythe (undated): Admiralty Island Linguistics, Vol.4: Gele' Grammar.

Table B39
Gele' and Kuruti Counting Words

	A	B	C	D
1	*sih*	*si:h*	*sih*	*si*
2	*rueh*	*rueh*	*ruweh,rueh*	*ruwe*
3	*teloh*	*daloh*	*toloh*	*telo*
4	*hahu*	*ha:huh*	*hahu (hahu-u)*	*hahu*
5	*limueh*	*li:mweh*	*limueh*	*limue*
6	*anoh*	*enoh*	*onoh*	*enoh*
7	*ndroteloh*	*endrdaloh*	*ondrotoloh*	*n'dretelo*
8	*ndrorueh*	*ndro:rueh*	*ndroruweh*	*n'droruwe*
9	*ndrosih*	*ndrwa:si:h*	*ndrosih*	*n'druasei*
10	*sunguah*	*sungwah*	*sungoh*	*sugah*
11	*sunguah pe si*		*sungoh pe sih*	*sungah pe si*
12				*sungah pen ewe*
13				*sungah pe telo*
14				*sungah pe hahu*
15	*sunguah pe limueh*		*sungoh pe limueh*	*sungah pe limue*
16				*sungah pe enoh*
17				*sungah pe n'dretelo*
18				*sungah pe n'dronu we*
19				*sungah pe n'druosei*
20	*runguah*		*rungeh*	*rungeh*
30	*tulunguah*		*tulngeh*	*tulgeh*
40	*hangoh*		*hangoh*	*hangooh*
50				*limnegeh*
60				*angu*
70				*n'dretulgeh*
80				*n'drorungeh*
90				*angungah*
100	*sangat*	*sangat*	*sangat*	*sangat*
1000		*he:bwo*	*sede,ho-pou*	
2000			*ru-pueu*	

Sources: A. Gele' Data: 6 CSQs. Villages: Buyang, Tingou, Kawaliap. B. Buyang: Z'Graggen (Smythe), (1975a, pp. 183-188). C. Kuruti Data: 13 CSQs. Villages: Liap, Kari, Lomoei, Dekimbat, Pundru. D. Ere Kele (Gele'?) 1 MQ: Village Kawaliap/Tingou, from Lucy Muru.

Table B40
Gele' & Kuruti Counting Words

	Body parts of animals *pue cor*	In bundles *angan pue*	Baskets *anga pue*
1	*akai*	*epus*	*sa hat*
2	*ndrukei*	*n'drupus*	*ru het*
3	*tulkei*	*tulpus*	*tul het*
4	*hakai*	*hapus*	*ha hat*
5	*limkei*	*lipus*	*lim het*
6	*enkai*	*enus*	*en hat*
7	*n'drotukei*	*n'drotulpus*	*n'drot tuhet*
8	*n'drorukei*	*n'droherpus*	*n'droru het*
9	*angsakai*	*n'drohensepus*	*ansah hat*
10	*sungah*	*n'dro sungah*	*n'dre sngah*

(continued)

Table B40
(continued)

	Laplaps *Anan pu*	Days	Houses		Money (Coins)
1	*ampal*	*sei*	*sim*	10t	*em brul*
2	*rumpel*	*ruu*	*n'dru pim*	20t	*rum brul*
3	*tulpel*	*tul*	*tul pim*	30t	*tul brul*
4	*hapal*	*hai*	*ha pim*	40t	*ham brul*
5	*limpel*	*lim*	*lim pim*	50t	*lim brul*
6	*anpal*	*on*	*en pim*	60t	*en brul*
7	*n'drotul pel*	*n'drotolo*	*n'drotul pim*	70t	*n'frotul brul*
8	*n'dronu el*	*n'droruwe*	*n'droru pim*	80t	*n'dru brul*
9	*ansapal*	*ansungah*	*ansa pim*	90t	*ansem brul*
10	*sungah*	*sungah*	*sungah*	K1	*n'dre sngah*

Tables B41 and B42 are from a recently recorded set of numerals by Joseph Fisher (~2010) who recorded Kuruti mathematics for improving teaching of mathematics.

Table B41
Kuruti Numeral Classifiers

	Counting round objects/nuts/ fruits	Counting houses/huts	Counting living beings (humans/fish/ animals)	Counting trees/logs/sticks	Counting leaves/ears
1	*sih*	*sim/hopuing*	*homou*	*he-ei*	*hakap*
2	*ruweh*	*rupuing*	*rumu-u*	*rui-i*	*rikep*
3	*toloh*	*tulpuing*	*tulmu-u*	*tuli-i*	*tulkep*
4	*hahu-u*	*hapuing*	*hamou*	*haei*	*hakap*
5	*limueh*	*lipuing*	*limu-u*	*limi-i*	*likep*
6	*onoh*	*onpuing*	*onmou*	*ene-ei*	*ankap*
7	*ondro-toloh*	*ndro-tulpuing*	*ndro-tulmu-u*	*ndro-tuli-i*	*ndro-tulkep*
8	*ndro-ruweh*	*ndro-rupuing*	*ndro-rumu-u*	*ndro-rui-i*	*ndro-rikep*
9	*ndro-sih*	*onsupuing*	*onsomou*	*ense-ei*	*ansakap*
10	*sungoh*	*sungoh*	*sungoh*	*sungoh*	*sungoh*

	Counting days	Counting pieces of sliced-objects	Counting pools of water	Counting flat-like objects/scoop of food/clothing	Counting whole bunches of fruits or nuts
1	*sei*	*hombul*	*hopuil*	*hapal*	*hombung*
2	*ru-u*	*rumbul*	*rupuil*	*ripel*	*rumbung*
3	*tul*	*tulbul*	*tulpuil*	*tilpel*	*tulbung*
4	*hai*	*hambul*	*hapuil*	*ha-apal*	*hambung*
5	*lim*	*limbul*	*lipuil*	*lipel*	*limbung*
6	*onoh*	*onbul*	*onpuil*	*anpal*	*onbung*
7	*ondro-toloh*	*ndro-tulbul*	*ndro-tulpuil*	*ndro-tilpel*	*ndro-tulbung*
8	*ndro-ruweh*	*ndro-rumbul*	*ndro-rupuil*	*ndro-ripel*	*ndro-rumbung*
9	*ndro-sih*	*onsombul*	*onsopuil*	*ansapal*	*onsombung*
10	*sungoh*	*sungoh*	*sungoh*	*sungoh*	*sungoh*

(continued)

Table B41
(continued)

	Counting rivers/creeks	Counting sharp objects (spears/sticks/needles/ stitched roofing leaves)	Counting same species of trees or bamboos	Counting piles of objects/items	Counting bags
1	*handrang*	*homot*	*hapat*	*hondroh*	*sahat*
2	*rundreng*	*rumuet*	*ripet*	*rundreh*	*ruhet*
3	*tulndreng*	*tulmuet*	*tulpuet*	*tulndreh*	*tulhet*
4	*handrang*	*hamot*	*ha-apat*	*handroh*	*hahat*
5	*limndreng*	*limuet*	*lipuet*	*limndreh*	*limhet*
6	*andrang*	*onmot*	*anapt*	*ondroh*	*anahat*
7	*ndrotulndreng*	*ndro-tulmuet*	*ndro-tulpuet*	*ndro-tulndreh*	*ndro-tulhet*
8	*ndrorundreng*	*ndro-rumuet*	*ndro-ripet*	*ndro-rundreh*	*ndro-ruhet*
9	*ansandrang*	*onsomot*	*ansapat*	*onsondroh*	*ansahat*
10	*sungoh*	*sungoh*	*sungoh*	*sungoh*	*sungoh*

	Counting split Objects	Counting pieces of cut objects	Counting loosed pieces of objects	Counting body parts/bunches of fruits	Counting parcels
1	*hendrek*	*sehir*	*hene*	*hakai*	*hopus*
2	*rundrik*	*ruhir*	*runi*	*rukei*	*rupus*
3	*tulndrek*	*tulhir*	*tulni*	*tulkei*	*tulpus*
4	*handrek*	*hahir*	*handre*	*ha-akai*	*hapus*
5	*limndrik*	*limhir*	*limndri*	*limkei*	*lipus*
6	*endrek*	*onmor*	*endre*	*ankai*	*onpus*
7	*ndro-tulndrik*	*ndro-tulhir*	*ndro-tulni*	*ndro-tulkei*	*ndro-tulpus*
8	*ndro-rundrik*	*ndro-ruhir*	*ndro-runi*	*ndro-rukei*	*ndro-rupus*
9	*ensendrek*	*ansehir*	*ensendre*	*ansakai*	*onsopus*
10	*sungoh*	*sungoh*	*sungoh*	*sungoh*	*sungo*

	Counting firewood (loosed ones)	Counting taro (staple food)	Time/period (measured in terms of a pot of food to be cooked)	Months (from lunar cycles)	Other terms associated with numbers
1	*homor*	*hopou*	*kur sehir (kur = pot)*	*ndrou sih (ndrou=moon)*	
2	*rumuer*	*rupueu*	*kur ruhir*	*ndrou ruweh*	*hepe* =half
3	*tulmuer*	*tulpueu*	*kur tulihir*	*ndrou toloh*	*sehir* = half
4	*hamor*	*hapou*	*kur hahir*	*ndrou hahu-u*	*hombul*=half
5	*limuer*	*lipueu*	*kur limihir*	*ndrou limueh*	
6	*onmor*	*onpou*	*kur enhir*	*ndrou onoh*	
7	*ndro-tulmuer*	*ndro-tulpueu*	*kur ndro-tulihir*	*ndrou ndrotoloh*	
8	*ndro-rumuer*	*ndro-rupueu*	*kur ndro-ruhir*	*ndrou ndroruweh*	
9	*onsomor*	*onsopou*	*kur ndro-sehir*	*ndrou ndrosih*	
10	*sungoh*	*sungoh*	*kur puke sungoh*	*ndrou sungoh*	

Table B42
Kuruti Counting Words

	Kuruti		Kuruti		Kuruti
1	*sih*	30	*tulngeh*	7 000	*po-tulngeh*
2	*ruweh*	40	*hangoh*	8 000	*po-ndro-rungeh*
3	*toloh*	50	*limngeh*	9000	*po-onsopou*
4	*hahu-u*	60	*ongoh*	10 000	*po-sangat*
5	*limueh*	70	*ndro-tulngeh*	20 000	*po-runget*
6	*onoh*	80	*ndro-rungeh*	30 000	*po-tulnget*
7	*ondro-toloh*	90	*on-sungoh*	40 000	*po-hangat*
8	*ndro-ruweh*	100	*sede/sangat*	50 000	*po-limnget*
9	*ndro-sih*	200	*rupueu*	60 000	*po-anangat*
10	*sungoh*	300	*tulpueu*	70 00	*po-ndro-tulnget*
11	*sungoh-pe-sih*	400	*hapou*	80 000	*po-ndro-runget*
12	*sungoh-pe–ruweh*	500	*lipueu*	90 000	*po-ansangat*
13	*sungoh-pe-toloh*	600	*onpou*	100 000	*po-hopou*
14	*sungoh-pe-hahu-u*	700	*ndro-tulpueu*	200 000	*po-rupueu*
15	*sungoh-pe-limueh*	800	*ndro-rupueu*	300 000	*po-tulpueu*
16	*sungoh-pe-onoh*	900	*onsopou*	400 000	*po-hapou*
17	*sungoh-pe-ondro-toloh*	1 000	*po-sungoh*	500 000	*po-lipueu*
18	*sungoh-pe-ndro-ruweh*	2 000	*po-rungeh*	600 000	*po-onpou*
19	*sungoh-pe-ndro-sih*	3 000	*po-tulngeh*	700 000	*po-ndro-tulpueu*
20	*rungeh*	4 000	*po-hangoh*	800 000	*po-ndro-rupueu*
21	*rungeh pe sih*	5 000	*po-limngeh*	900 000	*po-onsopou*
22	*rungeh pe ruweh*	6 000	*po-ongoh*		

Source. Fisher (~2010)

Kilivila

Kilivila, or Kiriwina, is an Austronesian language spoken in the Trobriand Islands (see Appendix E, Figure E12 on maps). No dialects are given for the language by Wurm and Hattori (1981, Map 10). The total resident population of the Trobriand Islands (including the Lusancay Islands but not Kitava Island) was about 15 000 at the 1980 National Census (National Statistical Office, 1982). In the 2011 Census, Kiriwina Rural Ward had a population of 36 721 which may include a few islands with other languages. Brief details of the language Kilivila are given in Table B43.

Table B43
Kilivila Summary

Language type	Austronesian
Family, Subfamily	Milne Bay Family, Kilivila S.C.
Other names, Dialects	Kilivila: Kiriwina, Bwaioa, Bweyowa.
Village	Trobriands
Cycles	5, 10, 100
Operative patterns	6 to 9 = 5+n, 5 is not a hand morph
Frame words	1-5, 10, 100

In brief. Kilivila has a large number of classifiers for a variety of groups denoted by suffixes. There are at least 42 different suffixes used for referring to different things and these may be grouped into at least 8 groups. It also has (5, 10, 100) cycles with decades and hundreds being regular as a 10 base system.

A short vocabulary, including numerals, of Kiriwina appears in the Annual Report on British New Guinea from 1891 to 1892 (British New Guinea, 1893). The 1902 Annual Report also contains a grammar by Fellowes (1902). Subsequently, the delayed Annual Report for Papua 1914-1915 included a vocabulary of the Bwaioa Tribe compiled by Bellamy (1917, p. 178). Ray (1938a, p. 165) notes that a Government map locates Bwaioa in south-east Fergusson Island, but the location as given by Bellamy places the language in the Trobriand Islands.

The most important publications on the culture of the Trobriand Islands are those by Malinowski (Malinowski, 1917–1920, 1920, 1922). One of these deals with the numerals of Kilivila which are not numerals as such but number roots which are prefixed when counting certain classes of objects. Malinowski (1917–1920) noted that:

> In (Kilivila) the Demonstratives and Adjectives as well as the Numerals do not exist in a self-contained form, conveying an abstract meaning. There are no single words to express such conceptions as "this", "big", "long", "one", etc., in abstract. Thus, for example, there is no equivalent of the word "one", or of any other numeral. Whenever the number of any objects is indicated the nature of these objects must also be included in the word (p. 41).

The data, then, that we will present in the tables below will, on the whole, not be a set of pure numerals but will contain classificatory particles and numeral roots and will be a set of words used for counting a particular class of objects.

The data which appear in the various Annual Reports, referred to above, are reported in various publications by Ray (1895, 1907, 1938a, 1938b). Ray (1938a, p. 188) had Bellamy's Bwaioa data as well. Kluge (1941, p. 214) had the Kiriwina and Bwaioa numerals which appear in Ray's 1938 article. In addition to the information available from these various sources, 38 CSQs were received from informants from villages in the Trobriand Island Group. The data for Kilivila are presented in two Tables B44 and B45.

The Kilivila number roots, to which classificatory particles must be prefixed, are given in Column A of the Table B44, Malinowski's data, with CSQ data in parentheses, and a second Table B45 with Lawton's classifications, not all of which are numeral classifiers. Some additional CSQ data are given at the end of Table B44. The data are examples of words used for counting specific classes of objects. Both Malinowski and the CSQ informants provided information on the classificatory prefixes used in association with various classes. Malinowski provided a comprehensive summary of these, giving a total of 41 classificatory prefixes divided into 8 groups (Malinowski 1917–1920, p. 45).

Table B44
Classifiers in Kilivila (after Malinowski)

Group 1		
	A	
1	*tay (te)*	Human beings; males (men, boys)
2	*na*	Persons of female sex; animals (pigs)
3	*day (ke)*	Trees and plants; wooden things; long objects (canoes, sticks, poles)
4	*dway(kwe)*	Round, bulky objects; stones; abstract nouns (betel nut, houses, yams)
5	*ya*	Leaves; fibres; objects made of leaf or fibre; flat and thin objects (coconuts, spherical containers, clothes, string)
6	*sisi*	Boughs (branches)
7	*li*	Forked branches; forked sticks
8	*kavi*	Stone blades
9	*kwoya (mweya)*	Human and animal extremities (legs, arms); fingers of a hand
10	*luva*	Wooden dishes
11	*kwoyla (kwela)*	Clay pots (cups, containers)
12	*kada*	Roads
13	*kaduyo*	Rivers, creeks, sea passages

Table B44 (continued)

14	*vilo*		Villages
Group 2			
15	*kila*		Clusters ("hands") of bananas
16	*sa*		Bunches of betel nut
17	*bukwa*		Bunches of coconut
Group 3			
18	*pila*		Parts of a whole; divisions; directions; (books)
19	*vili*		Parts twisted off
20	*bubwa*		Parts cut off by transversal cutting
21	*utu*		Parts cut off; small particles
22	*si*		Small bits
	(sisili)		(Slices of meat, bread)
	(kabila)		(Parts of meat cut off from animals)
Group 4			
23	*kabulo*		Protuberances; ends of an object
24	*nutu*		Corners of a garden
25	*niku*		Compartments of a canoe
26	*kabisi*		Compartments of a yam house
27	*nina*		Parts of a song, of a magical formula
28	*mayla*		Parts of a song; of a magical formula
29	*kubila*		Large land-plots-ownership divisions
30	*siwa*		Sea portions-ownership divisions with reference to fishing rights
31	*kala*		Days
32	*siva*		Times (no. of occurrences)
Group 5			
33	*kapwa*		Bundles-wrapped up (packages)
34	*oyla*		Batches of fish
35	*um'mwa*		Bundles of taro
36	*kudu*		Bundles of lashing creeper (dried reeds)
37	*yuray*		Bundles of four coconuts, four eggs, four water-bottles
	(kupwa)		(Fish counted in twos)
	(kayo)		(Crabs counted in twos)
Group 6			
38	*kasa*		Rows
39	*gili*		Rows of spondylus shell disks on a belt
40	*gula*		Heaps
Group 7			
41			Numerals without a prefix are used to count baskets of yams
Group 8			
42	*uwa*		Lengths, the span of two extended arms, from tip to tip (fathoms)
			Additional CSQ Data
	ta	Baskets	
	tam	Vines	
	yata	Watermelons, pumpkins	
	bwa	Short or thick solids	
	kaula	Groups of 20	

Source. A. These data are from Malinowski (1917–1920) with CSQ data given in parentheses.

Although Malinowski stated that the list in Table B44 can be considered as a "complete enumeration and not as an exemplification only" (p. 44), the data provided by the CSQ informants indicate that there are a number of further particles which are in common use. Some of these are listed below (although four have been included in the tables above when it has been apparent that they may be fitted into Malinowski's grouping scheme). However, subsequent work by the Lawtons and the Trobriand translators have provided many more. Since the numbering systems are quite different, these will be given separately. Often they coincide with the CSQ data. One initial comment by the Lawtons were the recognition of early classes about animate objects, namely the separation of animate and inanimate but then of the human and nonhuman together with male and female. Other categories are of parts or other, namely activity, partition and arrangement. Details are given in Table B45.

Table B45
Kilivila Classifiers (after Lawton)

	Basic Property Specifiers	
Cl 1	*to*$_1$-	Human
Cl 2	*na*$_1$-	Nonhuman
Cl 3	*kai-*	Rigid/long
Cl 4	*ya-*	Flexible/thin
Cl 5	*kwai-*	Thing
	Subclassifiers	
Cl 6	*to*$_2$-	Male human
Cl 7	*na*$_2$-	Female human
Cl 8	*gudi-*	Immature human
Cl 9	*kwela-*	Pot-like
Cl 10	*kova-*	Fire
Cl 11	*kailikova-*	Fireplace
Cl 12	*tam-*	Sprouting
Cl 13	*sobulo-*	Growing
Cl 14	*sega-*	Branching
Cl 15	*tuto-*	Time
Cl 16	*siva-*	Number of times
Cl 17	*lilou-*	Journey
Cl 18	*yam*$_1$-	Day
Cl 19	*kala-*	Passage of day
Cl 20	*bugi-*	Passage of night
Cl 21	*biga-*	Word
Cl 22	*kaiga-*	voice
Cl 23	*ligila-*	Group action
Cl 24	*mweli*$_1$-	Practices
Cl 25	*miga-*	Appearances
Cl 26	*wouyo-*	Newness
Cl 27	*kumlo-*	Oven
Cl 28	*nigo-*	Nest
Cl 29	*kavi-*	Tool
	(c) Pieces cont.	
Cl 84	*givi-*	Serve of fish
Cl 85	*kununu-*	Serve of greens
Cl 86	*yivi*	Serve of food pieces
Cl 87	*gini-*	Mouthful of food
Cl 88	*kapu-*	Mouthful of drink
	(d) Multiple reference	
Cl 89	*kabulo-*	Section/half
Cl 90	*katupo-*	Section/quarter
Cl 91	*pila-*	Part/piece
	By arrangement	
	(a) Inherent arrangement	
Cl 92	*tubo-*	Generation
Cl 93	*kumila-*	Clan
Cl 94	*dila-*	Family line
Cl 95	*kila-*	Hand of bananas
Cl 96	*buko*$_2$	Fruit cluster
Cl 97	*biko-*	Coconut bunch
Cl 98	*sa-*	Nut bunch
	(b) Non-inherent arrangement	
	(i) Distributional	
Cl 99	*budo-*	Group, crowd
Cl 100	*deli-*	Group moving
Cl 101	*gulo-*	Group, heap
Cl 102	*gugulo-*	Gathering
Cl 103	*yuwo-*	Group

(continued)

Table B45
(continued)

Cl 30	*pwa-*	Excrement	Cl 104	*tupila-*	Fleet
Cl 31	*igi-*	Wind	Cl 105	*duli-*	Bundle, cluster
Cl 32	*vilo*	Place	Cl 106	*seluva-*	Bundle being tied
	Residue		Cl 107	*luva-*	Tied bundle
Cl 33	*iga-*	Name	Cl 108	*ta-, Ø-*	Basket, basketful
Cl 34	*kuno-*	Rain	C109	*kapo-*	Parcel
	Group II classifiers—modification		C110	*kapuli-*	Group of parcels
	By activity		C111	*luba-*	Bundle of rolls
Cl 35	*bubulo-*	Made	C112	*mweli*$_2$-	Bundle of leaves
Cl 36	*buko*$_1$-	Buried	C113	*dodiga-*	Load
Cl 37	*bulu-*	Half-submerged	C114	*kaiyuvai-*	Layer
Cl 38	*beku-*	Floating submerged	C115	*pupai-*	Layer of filth
Cl 39	*gabu-*	Burning	C116	*keivala-*	Batch drying
Cl 40	*no-*	Blow	C117	*mmo-*	Conical bundle
Cl 41	*nutu-*	Kneaded	C118	*sipu-*	Tangle
Cl 42	*ponina-*	Punctured	C119	*wela-*	Fish (quantity)
Cl 43	*pwasa-*	Rotten	C120	*kudu-*	Band of fibres
	By partition		C121	*suyo-*	Things strung through holes
	(a) Topographical		C122	*kappupu-*	Grove
Cl 44	*udila-*	Land tract	C123	*lukuva-*	Growing bundle
Cl 45	*kubila-*	Land plot	C126	*poulo-*	Grove, group
Cl 46	*kalivisi-*	Large garden division	C125	*umila-*	Grove (one species)
Cl 47	*gubo-*	Garden division		(ii) Configurational	
Cl 48	*vala-*	Small garden division	C126	*tavi-*	Loose coil
Cl 49	*lupo-*	Smaller garden division	C127	*kupa-*	Loose coil
Cl 50	*kadida-*	Very small garden division	C128	*teni*	Tight coil
Cl 51	*pulu-*	Garden mound	C129	*katukuni-*	Reel
Cl 52	*kalipo-*	Site	C130	*bili-*	Roll
Cl 53	*kailiku-*	Section	C131	*tabili-*	Roll
Cl 54	*kada-*	Track	C132	*gili-*	Row
Cl 55	*seuyo-*	Lagoon	C133	*kasa-*	Line
Cl 56	*soulo-*	Fishing spot		(iii) Quantitative	
	(b) Parts within wholes		C134	*uva-*	Span measure
Cl 57	*lada-*	Small fishing spot	C135	*yuma-*	Length
Cl 58	*sisi-*	Bough	C136	*puli-*	Bunch (2–6)
Cl 59	*lila-*	Small bough	C137	*katuluwo-*	Large group
Cl 60	*lilivi-*	Forked stick	C138	*uwo-*	Two-bundle
Cl 61	*liku-*	Canoe division	C139	*kalo-*	Two-bundle (crustacean)
Cl 62	*lipu-*	Tier	C140	*kupo-*	Two-string
Cl 63	*buliga-*	Storey	C141	*yulai-*	Four-bundle
Cl 64	*kabisi-*	Section	C142	*kasila-*	Ten-group (wealth)
Cl 65	*livisi-*	Shelf	C143	*buluwo-*	Ten-group (animals)
Cl 66	*tabudo-*	*Room*	C144	*kaulo-*	Ten-group (animals)

(continued)

Table B45
(continued)

Cl 67	*kaduyo*	Entrance	C145	*ika-*	Ten-group (strings of fish)
Cl 68	*moya-*	Limb	C146	*kaluwo-*	Ten days
Cl 69	*kwaya-*	Severed limb	C147	*kwailuwo-*	Tens of things
Cl 70	*yam_2*	Hand			
Cl 71	*nina-*	Idea			
Cl 72	*mavila-*	Verse			
	(c) Pieces				
Cl 73	*bubo-*	Cut across			
Cl 74	*vili-*	Untwisted			
Cl 75	*lapou-*	A third			
Cl 76	*gum-*	Small piece			
Cl 77	*gibu-*	Sufficient			
Cl 78	*kuwo-*	Crumb			
Cl 79	*utu-*	Scrap			
Cl 80	*kabila-*	Large cut of meat			
Cl 81	*kipu-*	Cut of meat			
Cl 82	*sisili-*	Cut of meat			
Cl 83	*kaya-*	Half piece of food			

Source: Lawton (1991, pp. 265-269).

Malinowski observed that there is only a single case of enumeration in which abstract numerals can be employed, i.e without the use of a classificatory prefix, and this is the counting of baskets of yams (Group 7 above). He said:

> the counting of basketfuls of yams in Kiriwina is counting par excellence. The whole social life of the native is bound up with systems of mutual payments, in which yam payments stand first and foremost. (p. 53)

Malinowski also observed that counting above a thousand is easily done by the Trobriand Islanders and "on occasions of great annual harvest gifts to a chief, the figures of baskets given come well into five figures" (p. 54).

With regard to the number roots themselves, those for 3, 4, and 5 are typically Austronesian (2 as well if we take *yuwa* to be cognate with *lua*). The number root for 5 is *-lima* which is not the same as the word for "hand", "my hand" being *yamagu*, "his hand" being *yamala*. The number words for 6 to 9 have the construction 5+*n* where *n* takes the values 1 to 4 respectively. The cyclic pattern of the systgrowingem is (5, 10, 100). The construction of the decades 10 to 90 is done by suffixing *-luwa* with the respective cardinals 1 to 9 so that *-luwatala* is 10, *-luwayu* is 20, *-luwatolu* is 30, and so on. In this respect, *-luwa-* seems to be used in the same way as a classificatory prefix is, the class for which it is used being, perhaps, groups of ten. The number word for 100, *lakatutala* does not appear, in the data presented in the Tables B46 and B47, to have the obligatory classificatory prefix. Malinowski indicated that this is because, in practice, high numbers are never used with anything except yam baskets.

The data collected in 1890 and 1915, System C (Table B46) and System A (Table B47) agree well, on the whole, with the data provided by the CSQ informants (Table 47, System B). A slight anomaly exists in both systems in that on reaching the numeral 10 the classificatory prefix changes from *kwai-* to *ka-* or *ko-*. There is, incidentally, little doubt that Bellamy's Bwaioa data (Table B47, System A) are those of Kilivila; about half the CSQ informants refer to their language as Bweyowa. The government map (see Ray, 1938a, 1938b, p. 154) that places Bwaioa in Fergusson Island seems clearly to be in error.

Table B46
Kilivila Counting Words

	A	B	C
1	*tala*	*pilatala*	*kwai-tala*
2	*yuwa*	*pilayyu*	*kwai-yu*
3	*tolu*	*pilatolu*	*kwai-tono*
4	*vasi*	*pilavasi*	*kwai-vasi*
5	*lima*	*pilalima*	*kwai-nima*
6	*lima tala*	*pilalima pilatala*	
7	*lima yuwa*	*pilalima pilayyu*	
8	*lima tolu*	*pilalima pilatolu*	
9	*lima vasi*	*pilalima pilavasi*	
10	*luvatala*	*piluvatala*	*kolo-wa-tala*
20	*luvayya*	*piluvayyu*	*kolo-wa-iyu*
50	*luvalima*	*piluvalima*	*kolo-wa-nima*
60	*luvalima luvatolu*	*piluvalima piluvatolu*	

Sources: A. Numeral Roots: Malinowski (1917–1920, p. 63). B. Numerals used for counting portions of a subdivided whole: Malinowski (1917–1920, p. 63). C. Kiriwina: British New Guinea (1893, Appendix U), Ray (1895, pp. 35-38).

Table B47
Kilivila Counting Words

	A	B	C
1	*kwaitala*	*-tala*	*kwetala*
2	*kwaiiu*	*-yu(wa)*	*kweyu*
3	*kwaitolu*	*-tolu*	*kwetolu*
4	*kwaivasi*	*-vasi*	*kwevasi*
5	*kwailima*	*-lima*	*kwelima*
6	*kwailima kwaitala*	*-lima –tala*	*kwelima kwetala*
7	*kwailima kwaiiu*	*-lima –yuwa*	*kwelima kweyu*
8	*kwailima kwaitolu*	*-lima –tolu*	*kwelima kwetolu*
9	*kwailima kwaivasi*	*-lima –vasi*	*kwelima kwevasi*
10	*kaluatala*	*-luwatala*	*kweluwatala*
11	*kaluatala kwaitala*	*-luwatala –tala*	*kweluwatala kwetala*
12	*kaluatala kwaiiu*	*-luwatala yuwa*	*kweluwatala kweyu*
15	*kaluatala kwailima*	*-luwatala –lima*	*kweluwatala kwelima*
20	*kaluau*	*-luwayu*	*kweluwayu*
30	*kaluatolu*	*-luwatolu*	*kweluwatolu*
50	*kalualima*	*-luwalima*	*kweluwalima*
100	*lakatutala*	*lakatutala*	*lakatutala*

*Sources:*A. Bwaioa: Bellamy (1917, p. 178), Ray (1938a, p. 188), Kluge (1941, p. 214). B. Number Roots: 38 CSQs. C. Counting round objects, e.g stones: CSQ s.

Ralph and Margaret Lawton with numerous Kilivila speakers including Ben and Lepani have been studying the language and translating the Bible for many years. They shared some details on the link between measurement and number as given in Table B48 (from their personal conversation with me, 2009). These data for arm measures were also confirmed by a lecturer at Madang Teachers College. Lawton mentioned that measuring at arm height could be different to measuring at ground level.

Table B48
Counting and Measuring in Kilivila

	Horizontal	Vertical
1	*mweyatala sikwekula* "end of fingers"	*kwaibakwaila* "mid foot"
2	*okwailopoula* "end of cup of palm"	*okweibunela* "ankle"
3	*okabotakuwa* "wrist"	*wakwapa* "mid lower leg"
4	*inutywetywa* "mid forearm"	*okweitutu* "knee"
5	*okatupoi* "elbow"	*okaboyaula ylm*
6	*oyumakwasi* "mid upperarm"	*okuvilu* "hips"
7	*umatala* "arm pit"	*oposola* "waist"
8	*isividoga* or *i'kaidigu*	*ovisiyala* "breast"
9	*tomwaidona*	*ovitakola* "arm pit"
10	*ikouma imoi oyumakwasi* "mid upper arm"	*okeiyola* "neck"
11	*ikouma imoi okatupoi* and also *ikouma imou imoi hpai*	*otaigala* "ear"
12	*ikouma imoi omituwetuwa*	*okuluwotala* "top of head"
13	*ikouma imou okabotakuwa*	
14	*ikouma imou okweilopoula yamila*	
15	*ikoma imou kweyatala sirwekula*	
16	*uvatala*	

Source: Record made while talking with Ralph and Margaret Lawton, Ben and Lepani. Any errors would be due to Owens' recording.

The details here refer to horizontal at arm height, the height of platforms around the yam displays, and the vertical heights. Measures are mainly used for measuring the yams although they are used for measuring houses. It should be noted that there are not plurals of the smaller units as they were identifying different numbers represented by lengths. There is a body link between numbers. A "heap" was also used for either 400 or 800 (e.g. yams). Yams are carried in baskets and piled with small ones in the centre and the large ones kept for display on the outside. The longer variety do not last long and so are eaten or displayed, often painted for the chief in red or white. The long-lasting ones were stored and displayed on platforms often covered. As each basket is added, a palm leaf on a frond is torn off, the tenth one being left, etc. Thus they were representing large numbers of yams in baskets.

Lawton pointed out that not only were the numbering in a sense related to body tally or ordered measurement but the exchange of yams for fish was more about relationships than about amount putting the value of counting into perspective. People did not count ages or months as the child's age was described rather than counted as given in Table B49.

Table B49
Descriptions for children's ages in Kilivila

ikenu	*he is lying down*
ikanukova	*he rolls over*
isili	*he sits up*
to ikabi kaiyala	*he holds and poises a spear (4 year old)*

Source: Record made while talking with Ralph and Margaret Lawton, Ben and Lepani.

Bradshaw (1994), in reviewing Lawton's (1991) publication noted that the Trobriands participated in the *kula* trade and share many Massim cultural and linguistic characteristics, they maintained much of their own linguistic structures such as SOV and lacks postpositions. He supports Chowning's (1989) view that the Trobriand Islands may be a later migration to the area from the Solomon Islands or East New Britain although it would mean that some of the verb compositions and patterns may be borrowed from neighbouring languages. This would contradict Ross' (1992) argument that it existed long enough to vary. The order of subject, verb and object is not as variable as Senft (1986) suggested.

Muyuw

Muyuw, or Murua, is an Austronesian language belonging to the Kilivila group which is spoken on Woodlark Island and environs, on the Marshall Bennett Islands, and as far west as Kitava Island which lies about 30 kilometres to the east of Kiriwina Island in the Trobriand Group (see Map 106, Insets 1 and 2). The population of the Muyuw-speaking region at the 1980 National Census was a little over 5000 (National Statistical Office, 1982). The Murua Rural Ward had 10 191 in the 2011 census (National Statistical Office, 2014b) but that includes other language groups such as Budibudi (289). Summary language information is given in Table B50.

Table B50
Muyuw Summary

Language type	Austronesian
Family, Subfamily	Milne Bay Family, Kilivila S.C.
Other names, Dialects	Muyuw: Muyuw, Wamwan, Nawyem, Longwaw, Kitava. Variant : Murua, Marshall Bennet I., Woodlark I.
Village	CSQs: Woodlark Is, SIL Gwasop (Guasopa)
Cycles	5, 10, 100
Operative patterns	Classifiers
Frame words	1 to 5, 10, 100

In brief. Muyuw has classifiers mentioned based on tree, house, and animals/females. Classifiers are not generally used for 1. Also Muyuw uses numeral prefixes for decands and hundreds.

Lithgow (1976, p. 452) said of the language:

> It consists of a dialect chain from east to west which becomes progressively more similar to the Kilivila language. The dialect names are Muyuw (eastern Woodlark), Wamwan (central Woodlark), Nawyem (western Woodlark), Longwaw (Gawa, Kwaewata, Iwa, Yanaba) and Kitava (Kitava Island).

The earliest published vocabulary of Muyuw (referenced Murua) appears in the Annual Report on British New Guinea from 1889 to 1890 (British New Guinea, 1890b, Appendix X) and this includes several numerals. This vocabulary reappears in Ray (1895), Ray (1938a), and the numerals are included in both of these as well as in Ray (1907, p. 470); all are referenced as Murua. Ray (1938a) also had a vocabulary, including number data, referenced Marshall Is., i.e Marshall Bennett Island(s) which lie in the Muyuw region. This vocabulary was compiled by Symons (1919). The number data from Ray's (1938a) article are collected in Kluge (1941, p. 214). An SIL team, the Lithgows, worked on the language for about 10 years and were based at Woodlark Island. They have published several articles on the language and, in 1974, published a Muyuw Dictionary. The SIL library at Ukarumpa has word lists of the various Muyuw dialects and these include some number data. Lithgow (1976) also provided information on the language, in particular on noun classifiers. Four CSQs were completed by Muyuw informants. The data for the language are shown in Tables B51 and B52.

Table B51
Muyuw Noun Classifiers & Numeral Suffixes

	kay-(tree)	*buna*-(house)	*na*-(animals,females)
1	*kay-tan*	*buna-tan*	*na-tan*
2	*ka-y*	*bune-y*	*na-y*
3	*kay-ton*	*buna-ton*	*na-ton*
4	*kay-vas*	*buna-vas*	*na-vas*
5	*kay-nim*	*buna-nim*	*na-nim*

Lithgow (1976, p. 465) indicated that Muyuw, like other members of the Kilivila Group, features a large number of noun classifiers which combine with numeral suffixes. He provided the examples given in Table B51. Counting words are provided in Table B52.

Table B52
Muyuw Counting Words

	A	B	C	D	E
1	*katanok,kweitan*	*katanok*	*koitan*	*kweitara*	*kotanoga*
2	*kwei-y(u)*	*akwei*	*kweyu*	*kweiyu*	*kaia*
3	*kwei-ton*	*kweiton*	*kweiton*	*kweitoru*	*koitolu*
4	*kwei-vas*	*kweivas*	*kweivas*	*kweivasi*	*kwaivasi*
5	*kwei-nim*	*kweinim*	*kweinim*	*kweirima*	*kwairimi*
6	*kwei-nim katanok*		"as for 1"	*kweirima kweifara*	*kwairima koita*
7	*kwei-nim kwei-y*		"as for 2"	*kweirima kweiyu*	*kwairemi kwaita*
8	*kwei-nim kwei-ton*		"as for 3"	*kweirima kweitoru*	*kwaeremi kwaitola*
9	*kwei-nim kwei-vas*		"as for 4"	*kweirima kweivasi*	*kwairemi kwaivasi*
10	*sinawa-tan*	*sinawatana* or *kasaeatan*	*sinawatan* or *kasuratan*	*kweruwatara*	*koruwata*
11	*sinawa-tan katanok*				
12	*sinawa-tan kwei-y*				
15	*sinawa-tan kwei-nim*			*kweiruwatara kweirima*	
20	*sinawa-y*		*sinawaiyu*	*koruwayu*	*koruwayu*
30	*sinawa-ton*		*sinawaton*	*koruwatoru*	
50	*sinawa-nim*				
100	*lakatu-tan*			*lakatutara*	
200	*lakatu-y*				

Sources: A. 3 CSQs (villages on Woodlark Is.). B. SIL Word List (undated), Gwasop. C. Murua: British New Guinea (1890b, Appendix X7). D. 1 CSQ (Gawa Is., Marshall Bennett Group)., E. Marshall Group: Symons (1919, p. 87, Appendix E), Ray (1938a, p. 188), Kluge (1941, p. 214).

Two of the classifiers, *kay-* and *na-*, are the same as for Kilivila and appear to function in the same way. Many of the number words shown in Table B52 have the classifier *kwei-* prefixed to the numeral root, the class for which *kwei-* is used having the rather catch-all defining attribute of "general things, objects". The classifier is prefixed to all number roots from 2 to 9, *katanok* the numeral 1 being an exception, and in Lithgow's data the number root for 1 is not prefixed, nor are the remaining decades up to 90. The hundreds, *lakatu-tan*, *lakatu-y* have the same "hundred" prefix as in Kilivila, *lakatu-* being a quantitative classifier for hundreds.

The number roots for 3, 4, 5, respectively *-ton, -vas, -nim*, are fairly typically Austronesian. The cyclic pattern for, say, System A is (5, 10, 100).

Budibud

Budibud, a further member of the Kilivila Group, is spoken only in the Laughlan or Nada Islands lying at the extreme eastern part of the Milne Bay Province. At the 1980 National Census, the population of the Laughlin Islands was 209 (National Statistical Office, 1982). In the 2011 census, 289 were listed for Budibudi village alone. Table B53 summarises information on the language.

Table B53
Budibud Summary

Language type	Austronesian
Family, Subfamily	Milne Bay Family, Kilivila S.C.
Other names, Dialects	Budibud: Nada, Laughlin I.
Village	Laughlan or Nada Islands
Cycles	5, 10
Operative patterns	10-10 or 10s two, 10s three, teens different
Frame words	1 to 5, 10

The Annual Report (British New Guinea, 1892, p. 139) contains some number data. These appear in Ray (1895, pp. 35-38), in Ray (1907, p. 470), and in Ray (1938a, 1938b, p. 189). The data from the last publication appear in Kluge (1941, p. 215). A further vocabulary of Budibud, containing data on 1 to 5 and 10 to 20, and the tens 30 to 90, was published by Symons (1917) (see Table B54). There are some similar patterns to Muyuw. There may be several errors or misprints in System A of Table B54, e.g. the number 17 *kionima* should read *koinima*, and the number 15, *koineen* should read *koinim* or *koinima*. System A seems to have a somewhat irregular (5, 10) cyclic pattern. Although 10 is given as *asiratana* (both Systems), this is not used in the formation of the number words for 11 to 19 where we have *sinawatan* instead. The number words for 20 and 30 (System A), unlike those of System B, appear to have a formation respectively 10-10 and 10-10-10 although this interpretation is uncertain. The number word for 50, *sinawanima*, does not follow this pattern and is almost identical to the corresponding Muyuw numeral *sinawanima*, i.e 10-5. System B, whilst virtually identical to System A for the numbers 1 to 5, and 10, differs in the construction of the decades so that 10 is *asira-tana*, 20 is *asira-waiu*, 30 is *asira-tola*, and 50 is *asira-lima*, i.e we have a 10 - *n* construction or "ten one", "tens two", "tens three", "tens five" respectively.

Table B54
Budibud Counting Words

	A	B
1	*atonak*	*atanok*
2	*akwaiu*	*akwaiu*
3	*akwaiutola*	*akwaitola*
4	*akwailas*	*akwailas*
5	*akwailima*	*akwailima*
10	*asiratana*	*asiratana*
11	*sinawatan-kanok*	
12	*sinawatan-koi*	
13	*sinawatan-koiton*	
14	*sinawatan-koivas*	
15	*sinawatan-koineen*	
16	*sinawatan-koinima-tanok*	
17	*sinawatan-siawai-kionima-koi*	
18	*sinawatan-koinima-koiton*	
20	*sinawatan-sinawai*	*asirawaiu*
30	*sinawatan-sinawai-sinawatoi*	*asiratola*
50	*sinawanima*	*asiralima*

Sources: A. Laughlin I.: Symons (1917, Appendix 4q, p. 179). B. Nada (Laughlin I.): British New Guinea (1892, Appendix GG4, p. 139), Ray (1895, pp. 35-38), Ray (1907, p. 470), Ray (1938a, p. 189), Kluge (1941, p. 215).

The numerals 2 to 5 all possess the prefix *akwai-* and, although we have no information on this, it is possible that the prefix is a classifier, perhaps for a rather general class as we have in Muyuw. Subsequently, in the construction of the number words for 12 to 15 the numeral roots are prefixed by *koi-* rather than *akwai-*.

There are further irregularities in System A in Table B54 in that the numeral roots for 3, 4, and 5 are, respectively *-tola, -las*, and *-lima*, whilst subsequently, in combination (as with 13, 14, and 15) we have *-ton, -vas*, and *-nim(a)*. The word for "hand" or "arm" is, according to Symons, *niman* (presumably "hand-his").

Nasioi

Nasioi is a Papuan (NAN) language spoken in the south-eastern portion of Bougainville Island with two dialects (Allen & Hurd, 1963, pp. 51-52; Wurm, 1975), Nasioi proper and Simeku, each with its own sub-dialects. Nasioi is spoken by 14,100 (at about 1970) (Wurm, 1982, p. 238) and has the greatest number of speakers in the North Solomons Province. Nasioi and Nagovisi comprise the Nasioi Family of languages and together with the Buin Family, comprise the East Bougainville Stock which had a total of 35,200 speakers at about 1970 (Wurm, 1982, p. 238). Table B55 provides some summary information on the language.

Table B55
Nasioi Summary

Classification	Papuan, East Bougainville Stock, Nasioi Family
Other names, Dialects and sub-dialects	1. Nasioi proper: Koromira, Lantanau, Oune, Orami, Pakia-Sieronji 2. Simeku: Mainoki, Korpei
Villages	Kurai, Moroni, Sipuru, Bairima, Noruapa, Pankama, Marai, Koveng, Karuru, Biremeko, Guava, Kuria
Cycles	5, 10,100,1000
Frame Words	1 to 5, 10, (50), 100, 1000
Operative Patterns	6 to 9 = 5 + *n*

In brief. A complex classification system is used in East Bougainville Stock. The classifiers also indicate classifications in the culture. Besides the hand + *n* pattern for 6 to 9, it also has powers of 10. *Kokore* is word for chicken or domestic fowl as well as 1000 and used in several South Bougainville languages but not consistently for the same power.

In the early literature on Nasioi, the Nasioi proper sub-dialect and the Koromira sub-dialect were treated as two distinct languages. Rausch (1912) thus has short grammars and vocabularies of both Nasioi and Koromira; in both cases number data are given. McAdam (1926) also has vocabularies, including number data, of Nasioi and Koianu which, judging from the villages at which the data were collected, are respectively the Nasioi proper and Koromira sub-dialects. Both sets of McAdam's data appear in Kluge (1941, p. 28g). Frizzi (1914) had data on Nasioi, and the language is briefly discussed by Oliver (1949). Hurd and Hurd (1966) of the Summer Institute of Linguistics produced a Nasioi Language Course. Data from CSQs are presented in Table B56 giving Nasioi proper subdialect of Nasioi proper dialect.

Wurm (1975, pp. 792-793) indicated that one of the main characteristics of East Bougainville Stock is a complex classification of nouns into upwards of forty classes and that different sets of numerals are present for different classes of nouns. All CSQ informants gave some information on various numeral sets. Rausch (1912) also gave data on the numerals for different noun classes for both Nasioi proper and Koromira.

System A in Table B56 appears to be a commonly used counting system associated with a general noun class. It has a (5,10) cyclic pattern with the numerals 6 to 9 formed from compounds of the numeral 5, *panoko*, and the numerals 1 to 4. The numeral 6 is *panoko keta narung taa* which is translated as "five-from one more". There is a distinct word for 10 *kivora* and this is employed to construct the decades: 20 is *kenanka kivora*e "two tens", 30 is *benaumo kivora* "three tens", and so on. There is a distinct word for "hundred" *daku* and this is used in a consistent way to construct the hundreds up to 1000: 200 is *kenanka daku*, etc. There is also a distinct word for "thousand" *kokore*, and this is used to construct the subsequent thousands. System B in Table B56 given as a variant by two CSQ informants differs in words for 1 to 4 and used for counting birds and fish. Rausch has comprehensive data on the counting of various noun classes for both the Nasioi proper sub-dialect and the Koromira sub-dialect. System C in Table B56 is a selection of the data for Nasioi proper which appeared in Rausch (1912, pp. 109-111).

Table B56
Counting Words of Nasioi Proper Sub-Dialect of Nasioi Proper Dialect

	A	B	C
1	*narung*	*nau*	*narung*
2	*kenanka*	*keura*	*kenanka*
3	*benaumo*	*bekuri*	*benaumo*
4	*karenaumo*	*karekuri*	*karenaumo*
5	*panoko*	*panoko*	*panoko*
6	*panoko keta narung taa*	*panoko keta nau*	*panoko keta narunta*
7	*panoko keta kenanka taa*	*panoko keta keura*	*panoko keta kenanka*
8	*panoko keta benaumo taa*	*panoko keta bekuri*	*panoko keta benaumota*
9	*panoko keta karenaumo taa*	*panoko keta karekuri*	*panoko keta karenaumota*
10	*kivora*	*kivora (kibora)*	*kivora (narung kivora)*
11	*kivora eta narung taa*		*kivora eta narunta*
15	*kevora eta panoko*		*kivora eta panoka*
20	*kenanka kivora*		*kenanka kivora*
50	*panoko kivora*		*panoko kivora (narunga ape)*
100	*narung daku*		*kenanka ape (narung daku)*
1000	*narung kokore*		*kokorei*

Sources: A. 12 CSQs Villages: Kurai, Moroni, Sipuru, Bairima, Noruapa, Pankama, Marai. B. 5CSQs Villages: Koveng, Karuru, Biremeko, Guava, Kuria. C. Rausch (1912, p. 108).

The data provided by the CSQ informants agree with Rausch's data, although the former dealt with a less comprehensive set of classes and were mainly restricted to the counting words used for coconuts (class F), pigs (class B), coins (class C, in which case perhaps the defining attribute is "small objects"), shirts and blankets (class J), and bags (class I). Numeral classification is not commonly found in PNG languages. In the North Solomons Province it is mainly found in the NAN languages of the East Bougainville Stock but it is found in AN languages of the Manus Province and the Milne Bay Province.

The essential difference between data lies in the first four numerals. In other respects the two systems are much the same, having identical words for 5, *panoko*, the same (5, 10) cyclic pattern, and the same words for 10, having /*b*/ for /*v*/ in *kibora*.

There is a common structure discernible in each of the numeral sets associated with a particular noun class shown above. This structure takes the form shown in Table B57.

Table B57
Suffix Patterns Common in East Bougainville Stock and AN Languages of Manus and Milne Bay Provinces

1 *na*-(suffix)	6 to 9: compounds of 5 and 1 to 4.
2 *ke*-(suffix)	10 *na*-(suffix)
3 *be*-(suffix)	20 *ke*-(suffix)
4 *kare*-(suffix)	30 *be*-(suffix)
5 *panoko*	40 *kare*-(suffix)

Table B58 provides the way in which these suffix structures are evident in Nasioi for a number of the classification sets.

Table B58
Counting Words in Nasioi

Set	A. People	B. Large Four-Footed Animals e.g. pigs, dogs	C. Birds and Small Mammals	D. Fish	E. Fruit, Taro etc	F. Hollow Objects e.g. Bamboo
1	*narung* (used for males) *nani* (used for females)	*navoro*	*nau*	*nau*	*nau*	*naro*
2	*kenankara*	*kevoroka*	*keura*	*keura*	*keuka*	*keroka*
3	*benaura*	*bevoroi*	*bekuri*	*bekuri*	*bekupi*	*beropi*
4	*karenaura*	*karevoroi*	*karekuri*	*karekuri*	*karekupi*	*kareropi*
5	*panoko*	*tapiu*				
10	*nanai*	*navon*	*nava*	*nanang*	*narampu*	*navaku*
20	*kenaika*	*kevonta*	*kevara*	*kenanka*	*kerampura*	*kevakuku*
30	*benaipi*	*bevontu*	*bevari*	*benampi*	*berampuri*	*bevakupi*
40		*karevontu*	*karevari*	*karenampi*	*karerampurii*	*karevakupi*
50	*nimaunu*					

	G. Tools, Implements, e.g. stone axes	H. Rope-like objects e.g. ropes, bowstrings	I. Bag-like objects e.g. fish nets, sacks	J. Leaf-shaped or feather-like objects	K. Years and months	L. Days
1	*narp*	*navin*	*nara*	*nane*	*navera*	*naming*
2	*kerika*	*kevinta*	*keraka*	*keneka*	*keveraka*	*kemunta*
3	*beripi*	*bevintu*	*berapi*	*benepi*	*beverapi*	*bemuntu*
4	*kareripi*	*karevintu*	*karerapi*	*karenepi*	*kareverapi*	*karemutu*

M. Counting words of various types of wooden objects. The decades are the same as for class F above for counting hollow objects such as bamboo and sugarcane.

	Wood, Trees	Posts, Tree Trunks	Flat Objects (planks)	Fruit Trees
1	*nave*	*namo*	*navento*	*navari*
2	*kevera*	*kemoka*	*keventoka*	*kevarira*
3	*beveru*	*bemopi*	*beventopi*	*bevariru*
4	*kareveru*	*karemopi*	*kareventopi*	*karevariru*

Note. Nima is commonly found in Austronesian languages to mean 5.

It can be seen that for each of the numeral sets only five distinct words are used to construct the set. The suffixes shown are not invariant throughout the set but nevertheless do show a pattern which tends to be consistent within a given set. Some examples are given in Table B59.

Table B59
Suffixes for various sets in Nasioi

	Set B	Set C	Set D	Set E	Set F
1	*-voro*	*-u*	*-u*	*-u*	*-ro*
2	*-voroko*	*-ura*	*-ura*	*-uka*	*-roko*
3	*-voroi*	*-kuri*	*-kuri*	*-kupi*	*-ropi*
4	*-voroi*	*-kuri*	*- kuri*	*- kupi*	*-ropi*
10	*-von*	*-va*	*-nang*	*-rampu*	*-vaku*
20	*-vonta*	*-vara*	*-nanka*	*-rampura*	*-vakuku*
30	*-vontu*	*-vari*	*-nampi*	*-rampuri*	*-vakupi*
40	*-vontu*	*-vari*	*-nampi*	*-rampuri*	*-vakupi*

We might expect a single counting word belonging to a given class of nouns to contain information regarding both the numeral and the class of nouns. Regarding the numeral component, there are, as indicated above, only five variables and these are used for counting both units and decades. The suffix, then, ought to contain information about the class of nouns together with an indicator that enables one to distinguish whether a unit or a decade is being referred to. The structure of each suffix for the units may thus take the form of a morph which indicates the class in question and which is only used in counting units, together with a further suffix which coordinates with the numeral component. The structure of each suffix for the decades will similarly take the form of a morph which indicates the class in question and which is only used in counting tens, together with a further suffix which must coordinate with the numeral component. Thus for the units, the pattern is:

1 *na* + M1 (classifier morph used for units)
2 *ke* + M1 + S1 (two-suffix)
3 *be* + M1 + S2 (3/4-suffix)
4 *kare* + M1 + S2 (3/4-suffix)

and for the tens:

10 *na* + M2 (classifier morph used for decades)
20 *ke* + M2 + S3 (two-suffix)
30 *be* + M2 + S4 (3/4-suffix)
40 *kare* + M2 + S4 (3/4-suffix) .

Rausch (1912, p. 111) listed a number of objects each of which forms a class in itself, i.e. it is the only member of the class. Two examples will be given here to illustrate the structure of the counting words as suggested above. Parcels or packages, called *miku*, form a single noun class; the first four counting words are:

1 *na-miku*
2 *ke-miku-ra*
3 *be-miku-ri*
4 *kare-miku-ri*

In this case, the classifier morph is exactly the same as the noun. Houses, called *pava*, also form a single noun class and the counting words are:

1 *na-va*
2 *ke-va-ra*
3 *be-va-ri*
4 *kare-va-ri*

In this case the classifier morph *-va-* is not the same as the noun *pava* but is presumably derived from it. Unfortunately Rausch does not provide further information on the decades.

The ordinals used in Nasioi are derived from the corresponding cardinals and they vary according to the noun class in question. The ordinals for the class "Houses" given above, are:

1st: *tutunava*
2nd: *kevaranava*
3rd: *bevarinava*
4th: *karevarinava*
5th: *panokonava* etc

Further data for the Korpei sub-dialect and Koromira sub-dialect of Nasioi proper are also available from CSQs and given in Table B60. System B in Table B60 does not have the usual word for 10 *kivora*, and indeed the construction of the decades is unusual with 10 *narambu*, 20 *kerambura*, and 30 *beramburi*. However, according to Rausch's data on Nasioi proper, these words are used when counting fruit, taro, yams, etc. In counting this noun class, Rausch gave 10 as *narampu*, 20 *kerampura*, 30 *berampuri*. McAdam's numerals 1 to 4, however, are not those given by Rausch for counting fruit. If we take Rausch's data to be reliable, it is possible that McAdam, who does not indicate the existence of noun classes, has elicited part of the basic counting system for Koromira together with some words which are used for counting fruit.

Table B60
Counting words of Korpei and Koromira subdialect of Nasioi proper.

	Korpei		Koromira A	Koromira B
1	*nareri*	1	*nari*	*nara*
2	*kerikara*	2	*kiannka*	*keinta*
3	*been*	3	*bero*	*baro*
4	*karen*	4	*kairo*	*kairo*
5	*panoko*	5	*panoko*	*panoko*
6	*panoko ita nareri*	6	*panoko ita nari*	*panoko eta nari*
7	*panoko ita kenikara*	7	*panoko ita kianuka*	*panoko eta keinta*
8	*panoko ita been*	8	*panoko ita bero*	*panoko eta baro*
9	*panoko ita karen*	9	*panoko ita kairo*	*panoko ita kairo*
10	*kibora*	10	*narambu*	*kijora*
11	*kibora ita nareri*	11	*narambu ita nari*	.
15	*kibora ita panoko*	15	*narambu ita panoko*	*panokoe keinta kivora*
20	*nan narea*	16	*panoko kiramba ita nari*	*panokoe keinta kivore eta nara*
100	*narea raku*	20	*kerambura*	*keinta kivora*
		30	*beramburi*	....

Source: A. 2 CSQs. Villages: Borumai, Sirobai. B. Koianu: McAdam (1926, Appendix B, p. 87), Kluge (1941, p. 28g). C. Koromira: Rausch (1912, p. 964).

Rausch's System C (Table B60) shows an unusual construction for the numerals 15 and 16. For example, 15 *panokoe keinta kivora* in which *panokoe* derives from 5, *panoko*, and *keinta kivora* is "two tens". It thus appears that 15 is expressed, not as 10+5 as it is in the Nasioi proper system, but as 20-5. For 16 we have *panoko keinta kivora eta nara* which appears, on this supposition, to be (20-5)+1. McAdam's data for 16 show *panoko kiramba ita nari* where *kiramba* may be derived from *kerambura*, the word for 20. In that case, McAdam's 16 also appears to be constructed as (20-5)+1. Subtraction constructions will be found in the neighbouring Buin Family languages given below.

Siwai Proper

Siwai, or Motuna as it is commonly known, is spoken in the south-western part of Bougainville Island. Wurm (1982, p. 238) gave the language as having two dialects: Siwai proper and Baitsi. Allen and Hurd (1963, pp. 54-55) listed 68 villages in which Siwai proper is spoken and at the 1980 National Census, 62 of these had a total population of 7 585. They also listed (p. 56) 5 villages in which the Baitsi dialect is spoken and these, at the 1980 Census, had a population of 938 giving a total population for the language, at that time, of something in excess of 8 500 speakers. (National Statistical Office, 1982). Table B61 summarises the information on this language.

Table B61
Siwai Language Summary

Papuan	Buin Family, East Bougeanville Stock
Other names, Dialects	Baitsi; Dialects: Siwai Proper and Baitsi
Villages	Kunu, Naronai, Maisua, Kaparo, Usokoli, Kumuki, Kakatokori, Haisi, Koropo, Purikon
Cycles	5, 10
Frame Words	1 to 5, 10
Operative Patterns	6 to 9= *n* to 10

In brief. Siwai has a word for hand used for 5 in all noun classifier operative patterns.

Siwai, together with Buin, comprise the Buin Family, part of the East Bougainville Stock of Papuan languages. McAdam, the District Officer based at Kieta, obtained in 1924-25 a vocabulary, including number data, of Siwai proper from Father Grisward, who had extensively studied Buin. McAdam also collected a vocabulary of the Baitsi dialect during the same period (McAdam, 1926, pp. 81-2, 91-93). Both sets of McAdam's number data appear in Kluge (1941, p. 289). The American anthropologist D.L.Oliver studied the Siwai people during the Peabody Museum Expedition to Bougainville in 1938. The results of this appear in his book on the Siwai (Oliver, 1967, pp. 62-63, 100-101), and in a series of reports (Oliver, 1949). The former contains some brief information on the noun classes and counting in Siwai. Some discussion of Siwai also appears in Allen and Hurd (1963). Data were received from 18 CSQ informants and these, together with McAdam's data, appear in Table B62.

Table B62
Siwai Proper Counting Words

	A	B	C
1	*no'o*	*nawa*	*nau*
2	*ki'iko*	*kiwano*	*kina*
3	*pekang*	*pewangu*	*bekanga*
4	*korikang*	*aoriwangu*	*karikangu*
5	*agumuka*	*angumuka*	*panoko*
6	*no'oki narang*	*mawa-kinarang*	*naiangenora*
7	*kii ke narang*	*kewane narang*	*keiakeinora*
8	*pekang i narang*	*piwangu ki narang*	*bekangoinora*
9	*korikang I narang*	*koriwangu ki narang*	*karekangoinora*
10	*narang*	*narang*	*narum*
11	*no'o ki kiranno*	*narang nawa lugowa*	....
15	*agumuka ki kiranno*	*narang angumaku lugowa*	
20	*kiranno*	*kiranno*	
30	*perangu*	*perangu*	
40	*korirangu*	....	

Sources: A. 18 CSQs. Kunu, Naronai, Maisua, Kaparo, Usokoli, Kumuki, Kakatokori, Koropo, Purikon. B. Siwai: McAdam (1926, p. 82), Kluge (1941, p. 28g). C. Baitsi: McAdam (1926, p. 93), Kluge (1941, p. 28g).

It is not certain whether System A is a basic counting system for counting a general class of objects, or whether it is used for a specific, well-defined noun class. It is, however, the counting system given by most of the CSQ informants. It possesses a (5,10) cyclic pattern: there are distinct words for the numerals 1 to 5 and 10. The number words for 6 to 9 are constructed in the same way as are those for Nagovisi. The number word for 6 is *no'oki narang* "one towards ten"; that for 7 is *kii ke narang* "two towards ten" and so on. The number words for 11 to 19 are constructed in a similar way: 11 is *no'oki kiranno*, "one towards twenty"; 15 is *agumuka ki kiranno* "five towards twenty". The number word for 16 is *no'oki narang i kiranno* "one towards ten towards twenty", and so on.

The decades are constructed in a fashion similar to that used for the units. There are distinct words for 10 to 50, and 100. The numeral 30 is *perangu*, the numeral 40 is *korirangu,* 50 is *pukii*. The numeral 100 is *pore*. The construction of the decades 60 to 90 is such that 60 is *narang i pore* "ten towards one hundred", 70 is *kiranne pore* "twenty towards one hundred", and so on. There are distinct words for 1000, *kukuraku noweri muh tugah*, where *kukuraku* is a domestic fowl. The word for 1000 in Buin also means "domestic fowl". No informant indicates why this is so. Oliver (1967) noted:

> Potentially, then, it is possible for a Siuai to talk precisely in terms of, say, the 22nd house, 178 coconut palms, or 1000 leaves. In actual usage, however, these natives seldom number most items past ten, the general qualifier "many" (*ponna*) is specific enough for most uses, except for such entities as pigs, sheets of thatch, shell money, and other items which are commonly used in large numbers. (p. 101).

Oliver also listed forty classes of nouns and the most common objects found within each class (1967, p. 63). He did not, however, give the details of how the noun classes are counted. The data on this, given below, derive from the CSQ informants.

A. Counting shell money: 1. *no'ori* 2. *kirogo* 3. *perii'h* 4. *korinakui*
B. Counting trees, bamboo, sugarcane: 1. *nomun* 2. *kimuno* 3. *pemunuuh* 4. *korimunuuh*

In counting each class, the word for 5, *agumuka*, is usually used unchanged for all classes. *Agumuka*, or *angumuka*, as some informants gave it, derives from the word for "hand", which is *angu*. Oliver (1967, p. 100) gave *angumuka* as meaning "hand-like".

McAdam's number words for Siwai proper (System B, Table B62) are different to those given in System A. Since different suffixes are used for the numerals 1 to 4, it is fairly certain that this system is a set of number words used for counting a noun class different to the class counted in System A, but McAdam does not provide details. There is an error in the numeral four which should be *koriwangu*, not *aoriwangu* as given. The construction of the number words for 11 to 19 does not follow that which is apparent in System A. The latter gives these as counting "towards twenty" whereas System B shows counting apparently proceeding "from ten".

McAdam's number words for the Baitsi dialect (System C) resemble those of Nasioi rather than those of either Siwai proper or Nagovisi, both of which border Baitsi.

Buin Proper

Buin is spoken in the southernmost portion of Bougainville Island. Wurm (1982, p. 238) showed the language as having two dialects: Buin proper (or Telei) and Uitai (or Uisai). Allen and Hurd (1963, pp. 56-7) indicated that the former is spoken in 94 villages, and the latter dialect in 11 villages. The population statistics for the two dialects, which are given in Wurm (1982, p. 238), derive from the 1970 National Census and are 8300 for Buin proper and 1200 for the Uitai dialect. Buin and Siwai together comprise the Buin Family which is part of the East Bougainville Stock of Papuan (NAN) languages. Table B63 summarises information on the language.

Table B63
Buin Proper (Uisai) Summary

Papuan	Buin Family, East Bougeanville Stock
Other names, Dialects	Telei, Terei, Rugara); 2. Uitai (Uisai). Variant: Wisai
Villages	Tarapa, Okoiragu, Bogisago, Ugubakogu
Cycles	10, 100
Frame Words	1 to 6, 9
Operative Patterns	7=10-3, 8=10-2

In the German colonial period the people of the region around Buin were extensively studied. This material, on the ethnography, art, music, and language of the Buin people, may be found in Schmidt (1909), Wheeler (1910–1911), Parkinson (1907), Frizzi (1914), and Thurnwald (1909), to select but a few. Material dating from the Australian colonial period may be found in Chinnery (1931), Blackwood (1931–1932), and Oliver (1949). Grisward (1910) provided a grammar of Buin, or Telei, which includes the numerals of the language. Parkinson (1907, p. 479) also has Buin numerals, referenced as Kronprinzengebirge, which are also collected in Kluge (1941, p. 254). Laycock has worked extensively on the language and has prepared a grammar and dictionary and has Buin numerals in an article on PNG counting systems Laycock (1975a, p. 226). SIL data are extensive. CSQs were received from 25 informants.

System A in Table B64, given by the CSQ informants, has several unusual features. There are distinct words for the numerals 1 to 6. The construction of 7 and 8 are, respectively, *paigami tuo* and *kiitako tuo*, i.e "three less than ten" and "two less than ten", so that a subtractive process operates for these numerals as is the case for the Manus type (as Lean, 1992) called these subtractive-patterned) counting systems.

Table B64
Counting Words

	A	B	C
1	*nonumoi (noikei)*	*nonumoi*	*monumoi*
2	*keitako*	*kitako*	*kikako*
3	*paigami*	*baigami*	*paigami*
4	*korigami*	*korigami*	*koregami*
5	*upugami*	*ubugami*	*uvugami*
6	*tugigami*	*tugigami*	*tugigami*
7	*paigami tuo*	*baigami tuo*	*paigami tuo*
8	*keitako tuo*	*kitako tuo*	*kitako tuo*
9	*kampuro*	*kamburo*	*kamburo*
10	*kiburo(kipuro)*	*kiburo*	*kuvuro*
11	*noikei lugo*	*kiburo nonumoi lugobumoi*	....
15	*upugami lugo*	*kiburo ubugami lugobumoi*	
20	*kikoko*	*kikoko*	
30	*paimaku*	*baimaku*	
40	*korimaku*	*korimaku*	
50	*upumaku*	*ubumaku*	
100	*pore*	*pore*	
1000	*kukurei noikei*	*kukurei*	

Sources: A. 25 CSQs. Villages: Tarapa, Okoiragu, Bogisago, Ugubakogu, Ibinai, Laitaro, Mamarommino, Morula, Nabaku, Tsimbo, Malabita B. Telei : Grisward (1910, p. 88). C. Kronprinzengebirge: Parkinson (1907, p. 479), Kluge (1941, p. 254).

The numeral 9, *kampuro*, does not continue this pattern but, as Laycock (1975a, 1975b, p. 226) points out, is a separate word meaning something like "completed". This was confirmed by ethnomathematician Kaleva in personal communications in 2000 (Table B64). Laycock continued: "Ten is then the first number in a number-set for counting tens, as hundred is the first in a set for counting hundreds" (p. 226). The word *kampuro* completes the decades number-set, meaning 90, and also completes the hundreds number-set, meaning 900 (Table B65).

Table B65
Decade and Hundred Sets in Buin (Uisai)

	Decades Set		Hundreds Set
10	*kipuro*	100	*pore*
20	*kikoko*	200	*kiporigo*
30	*paimaku*	300	*paiporegi*
40	*korimaku*	400	*koriporegi*
50	*upumaku*	500	*upuporegi*
60	*tugimaku*	600	*tugiporegi*
70	*paimaku tuo*	700	*paiporegi tuo*
80	*kikoko tuo*	800	*kiporigo tuo*
90	*kampuro (lopore)*	900	*kampuro*

The construction of the numerals 11 to 19 follow a consistent pattern. The numeral 11 is *noikei lugo* (or *noikei rugo*) where *lugo* (or *rugo*) means "on top of", i.e 11 is "one on top of" with "ten" being understood. The numeral 15 is *upugami lugo* "five on top of", and so on. A similar pattern is seen for the numerals 21 to 29 (and 31 to 39, etc). The numeral 21 is *kikoko girai noikei lugo* "twenty and one on top of"; the numeral 25 is *kikoko girai upugami lugo* "twenty and five on top of", and so on.

There are distinct words for the numerals 1000 and 10 000, the former being *kukurei* and the latter *tarina* or *taarina*. The word for "thousand" *kukurei*, has the meaning *domestic fowl* as it does as well in Siwai and Nasioi.

System B in Table B65, Grisward's Buin numerals, collected over 100 years ago, are in most respects identical to those in System A (Grisward consistently uses /i/ for System A's /i/). The construction of the numerals 11 to 19 is different to that shown in System A. The numeral 11 is *kiburo nonumoi lugobumoi* "ten one-on top of" so that the "ten" is explicit rather than understood. The System A numerals appear to be an abbreviation of this. Grisward (1910, pp. 89-90) also gave information on the number-sets used in counting various noun classes as given in Table B66.

Table B66
Counting Different Groups

	Men	Months	Animals	Trees
1	*nonumoru*	*nonorobira*	*noikei*	*naukeo*
2	*keorakino*	*kebirako*	*keitako*	*keutako*
3	*baireia*	*baibira*	*baiem*	*bairui*
4	*korineia*	*korebira*	*koiem*	*keirui*
5	*ubureia*	*ububira*	*ubum*	*uburui*
6	*tugineia*	*tugibira*	*tuginoi*	*tugirui*
7	*baireia tuo*	*baibira tuo*	*baiem tuo roi*	*bairui tuo*
8	*keorakino tuo*	*kebirako tuo*	*deitako tuo roi*	*kentako tuo*
9	*kamburo*	*kamburo*	*kamburo roi*	*kamburo*
10	*kiburo*	*kimburo*	*lagaturoi*	*kimburo*

The general structure of a counting-set for counting the units of a particular noun class (although there are exceptions to this) appears to be:

1 *no*-M1
2 *ke*-M2
3 *bai*-M3
4 *kori*-(or *ko*-)M3
5 *ubu*-M3
6 *tugi*-M3
7 *bai*-M3 *tuo*
8 *ke*-M2 *tuo*
9 *kamburo*

where M1, M2, M3 are (usually) three different classifier morphs which are used only with a particular noun class and which, presumably, give an indication of the class in question. In counting tens for a given class a different counting-set is used, i.e. the same structure is used but a different set of classifier morphs apply.

Yele

Yele, or Yeletnye, is a Non-Austronesian or Papuan language spoken only on Rossel Island, the eastern most island of the Milne Bay Province (see Appendix E, Figure E12). Wurm and Hattori (1981, Map 10) showed the language as having two dialects, East and West. Henderson (1975, p. 818) noted that the language contrasts markedly with Austronesian languages of the same area in phonology, grammar and vocabulary; it shares only 6% of its vocabulary with its nearest neighbour, Sud-Est. Henderson said: "Only a few people who have settled on the island can use Yeletnye with any degree of fluency and the language has developed the reputation of being impossible to learn" (Henderson, 1975, p. 818). The total citizen population of Rossel Island at the 1980 National Census was 1679 (National Statistical Office, 1982). Table B67 summaries the language information.

Table B67
Yele Summary

Language type	Papuan
Family, Subfamily	Family Level Isolate, Yele-Solomons Stock
Other names, Dialects	Yele 1. Eastern, 2.Western.
	Variants: Yela, Yeletnye, Rossel, Wamiu, Kwai,Olanga
Village	Rossel Island
Cycles	4-cycle inside 10 for 1000s
Frame words	1 to 10,

In brief. Yele shows evidence of borrowing of words or morphemes from neighbouring AN suffixes. Particularly difficult sounds in the language makes it difficult to reproduce consistently. There is evidence of a 4-cycle system especially between 5000 and 9000.

The first publication of Yele vocabulary was in the Annual Report on British New Guinea from 1889 to 1890 (British New Guinea, 1890c, Appendix X10). This was subsequently reproduced in Ray (1895) referenced as Roua. A further vocabulary of the language was published in the Annual Report for 1893 to 1894 (British New Guinea, 1895). Ray used both of the Annual Report vocabularies to give a description of the language in the Cambridge Expedition to Torres Straits (Ray, 1907, p. 383) but subsequently noted (Ray, 1938b, p. 377) that the vocabularies, and hence his description, were "imperfect". The Assistant Government Anthropologist for Papua in the 1920s, Armstrong (Armstrong, 1923, 1924, 1928), contributed significantly to the study of the anthropology of the Rossel Islanders. Additional Yele vocabularies, originally collected by Strong, were published in Ray (1938b) referenced as Wamui, Kwai, Olanga and S.W.Rossel. These vocabularies included number data which appear in Ray (1938a, p. 190), under the same references, together with data from the two Annual Report vocabularies for Yela 1 (1890 and 1895); Armstrong's (1923) vocabularies, Rossel E. and W.

All the data on Yele numbers mentioned above were collected prior to 1922. No further information appears to have been published on the language until an SIL team, the Hendersons, who spent some five months on field research at Rossel Island in 1971, published an account of Yeletnye in 1975. Additional data are also held in the SIL library at Ukarumpa. Three informants from Rossel Island also completed CSQs. The data for the language are given in Tables B68 and B69.

Table B68
Counting Words for Yele

	A	B	C
1	*ngme*	*mo*	*ngme,mwin*
2	*mio*	*mio*	*miyo*
3	*pyiile*	*pyili*	*pyile*
4	*paadi (baadi)*	*pari*	*paadi,peard*
5	*limi*	*limi*	*limi,lim*
6	*weni*	*sini*	*wini,wen*
7	*pyidu*	*pidi*	*pyudu,piud*
8	*waali*	*wale*	*waali,weal*
9	*tyu*	*tya*	*tiu*
10	*y:a*	*ya*	*ya*
11	*y:a me ngme*	*mo ngwa*	*ya ma ngme*
12	*y:a me mio*	*ma mio*	*ya ma miyo*
15	*y:a me limi*		*ya ma limi*
20	*myo-y:a (y:a ma y:a)*	*ma ya,yulemio*	*miyo ya*
30	*pyole-y:a*	*yulemio ya,yipyili*	*pyolo ya*
40	*poro-y:a*	*yipyili ma ya,yalapari*	*podo ya*
50	*limo-y:a*	*yalapari ma ya,yili*	*limo ya*
60	*wone-y:a*	*(yili ya)*	*ywono ya*
100	*yone-y:a*	*yono ya*	*ma yono ya*
200	*myoyone-y:a*	*mio yono ya*	*miyo yono ya*
1000	*yoneyone-y:a*	*yono yono ya*	*ma yonoyono ya*
2000		*mio yono yono ya*	*miyo yonoyono ya*

Sources: A. Yeletnye: Henderson (1975, pp. 824-5) and SIL data. B. Rossel: Armstrong (1928, p. 77). C. 3 CSQs.

The considerable variation between the various systems shown in the Tables may possibly be partly attributed to the peculiarities of Yele phonology and the difficulty of recording the language with a suitable orthography. Ray (1938a, p. 168) quoted Sir William Macgregor's description of Yele as "the most intractable in the possession, and using sounds that cannot be expressed by any combination of letters in the English language". Armstrong (1928, pp. 218-9) similarly has a (pejorative) description of Yele by the Acting Administrator Murray: "The language is extraordinarily unmusical in sound; it is full of nasals and gutturals, and cannot be better described than as resembling the snarling of a dog interspersed with hiccoughs".

With reference to System B, Table B68, Armstrong (1928) said:

> In the above system the terms for 60, 70, 80, and 90 are rather a blemish on a notation which is otherwise systematically decimal. In actual counting the fifties are taken separately; that is, having reached 50 the word *yili* is said with emphasis, and counting commences again from 1, not 51. Having reached 50 again, the two fifties are put together to make 100 *yono ya*. The term *yili* is an end-term, which can be used for other numbers that may be subsequently added. It is the word ordinarily used for many, and it occurs again as the first of the set of terms for the thousands. Possibly the tens from 20 to 100 could also be expressed by the terms *mio ya, pyalo ya*, etc., since 100 is expressed by *yono ya* and *yono* belongs to the series of terms *mio, pyalo, ...yono*. (p. 78)

Indeed, as Armstrong suggests, the tens do seem to be expressed in the form *n-ya* where *n* is a cardinal from 2 to 10, or strictly speaking, a word derived from the cardinal so that, for example, 50 is *limo ya* rather than *limi ya*. Systems A and C (Table B68) and System C (Table B69) show examples of this type of decade construction. System A (Table B69) also shows this type of construction for the numbers 50 to 100 but a different type of construction for the numbers 20 to 40.

Table B69
Yele Counting Words

	A	B	C
1	*me*	*ngmeni*	*mme,mweni*
2	*miu*	*miwa*	*miuwa*
3	*pieli*	*piele*	*pieli*
5	*padi*	*pai*	*pali*
5	*limi*	*limi*	*limi*
6	*uweni*	*weni*	*uweni*
7	*pidi*	*pidi*	*piri*
8	*uwali*	*weli*	*uwali*
9	*tiu*	*tiwa*	*tiu*
10	*iya*	*iya*	*iya*
20	*iya-yulumiu*	*maiya*	*miuwa-iya*
30	*iye-pieli*	.....	*pioila-iya*
40	*yala-padi*		*pala-iya*
50	*lima-iya*		*lima-iya*
60	*uwona-iya*		*owona-iya*
100	*yana-iya*		.....

Sources: A. S.W.Rossel: Ray (1938b, p. 378), Ray (1938a, p. 190), Kluge (1941, p. 281). B. Yela(2): British New Guinea (1895, Appendix GG2), Ray (1938a, p. 190), Kluge (1941, p. 281). C. Kwai: Ray (1938b, p. 378), Ray (1938a, p. 190), Kluge (1941, p. 281).

The number words for 1 to 10 appear to be decidedly Austronesian, rather than Papuan, in character. We note, for example, the numeral 5 *limi* which is cognate with the typical AN *lima*. The decimal nature of the counting system (i.e the (10) cyclic pattern) is also unusual, although not unknown, in the Papuan languages. It is possible, then, that the Rossel Islanders have borrowed their numerals from some Austronesian source. The two nearest Austronesian languages are Nimowa and Sud-Est both of which employ numeral classifiers which is not a feature of Yele. A comparison of the Yele numbers with the number roots of the neighbouring languages is given in Table B70.

Table B70
Number Roots for Nimowa (d. North-West Tagula), Sud-Est and Yele

	Nimowa	Sud-Est	Yele
1	*-taga*	*-dava*	*ngme*
2	*-iwo*	*-iwo*	*mio*
3	*-to*	*-to*	*pyile*
4	*-fadi*	*vare*	*paadi(baadi)*
5	*-lima*	*-lima*	*limi*
6	*-woni*	*-wona*	*weni*
7	*-fidi*	*-piri*	*pyidu*
8	*-wa*	*-wo*	*waali*
9	*-siwo*	*-siwo*	*tyu*
10	*(sia)-wate*	*-yowolo*	*y:a*

In Table B70 the number roots for Nimowa have been taken from Ray's (1938a, 1938b) data, the North-West Tagula (i.e. Western Point) data, since several of those appear to have some resemblance to the Yele data. The Sud-Est numbers derive from Lithgow's (1976b) data. Whilst several of the Yele numerals appear to be cognate with number roots of both Sud-Est and Nimowa, it would appear that the latter language has a slightly closer correspondence than the former (compare, for example, the numerals 4 and 7).

Certainly not all the Yele number words appear to be Austronesian in character. Armstrong gave examples of counting thousands, none of which appear to be cognate with AN numerals (see Table B71).

Table B71
Yele Words for Thousands

1 000	*yili*	6 000	*mwadwong*
2 000	*dwong*	7 000	*mwateme*
3 000	*teme*	8 000	*mwadab*
4 000	*dab*	9 000	*mwadi*
5 000	*mwayili*	10 000	*mwadi mwadab*

It will be observed that a 4-cycle operates within this sequence in that the number words for 5 000 to 9 000 each contain, respectively, the number words for 1 000 to 4 000 preceded by the prefix *mwa-*. Armstrong (1928) also noted that:

> A curious feature of this last series of terms, the combination of the terms for 8 000 and 9 000 to express 10 000, is explained in the legend, which attributes the invention of counting to Wonajo, who wished to count the *nko* (shell money) that he had made. Having counted up to 9 000 he grew weary, and, unable to think of a fresh word for 10 000 adopted the novel, if unmathematical, device of using in justaposition the words for the last two thousands. (p. 78)

Armstrong observed that counting to such high numbers is associated with the counting of shell money (*nko*) but is not otherwise required. Two types of shell money (*nko* and *ndap*) are used on Rossel Island and these are associated with important ceremonial activities. Unlike the Tolai of East New Britain who tend to measure lengths of their shell money (very small shells threaded onto cane) when dealing with large quantities, the Rossel Islanders use relatively large shells which are counted singly or in groups of ten. For ceremonial purposes the Tolai may count hundreds of fathoms of shell-money; the Rossel Islanders count the individual shells and thus counting may proceed into the thousands. Within the island as a whole the total amount of shell-money of either type was constant (at the time Armstrong was writing), no new shell money being created or existing money being destroyed or traded elsewhere. Within the society there was a practice of borrowing and lending shell money and associated with this a complex set of rules governing repayment of loans. Given also that the *nko* and *ndap* are also graded into a multiplicity of types which have differing associated values, it is clear from Armstrong's account of the monetary system and ceremonial that the Rossel Islanders require a reasonably high degree of numeracy for day-to-day traditional purposes and that, in this respect, they have a culture quite different from the "non-numerate" culture of the Loboda as given in the account by Thune (1978).

Kuot

Kuot or Panaras is spoken on both the east and west coasts of the mainland, immediately to the south of the Nalik region. Of the 21 languages of the New Ireland Province, Kuot is unique in being the only Papuan (NAN) language. Table B72 summaries the language information and Table B73 provides the counting system.

Table B72
Kuot Summary

Language type	NAN Family-Level Isolate, New Britain Stock, East Papuan Phylum
Other names; dialects	Dialects: 1) Kul, 2) Naiyama, 3) Letatau. Variant; Panaras, Kul
Villages	Panaras
Cycles	10
Operative patterns	30=3x10, 50=5 x 10, 18=10+5+3
Frame words	1 to 10

In brief. Kuot has (5, 10) cycles with decades and higher teens eg. 18=10+5+3 with similar pattern for other decades

Table B73
Kuot Counting Words

	A	B
1	*an*	*an, are, iovo*
2	*haras*	*aras*
3	*naien*	*ayen, iyin, nayen*
4	*nagala*	*a ale, na ale*
5	*muanam*	*muanam, noane*
6	*gunamun*	*gunamun, gunamuru*
7	*gamura*	*gamura*
8	*gamin*	*gamin*
9	*gamiala*	*gamiala*
10	*mana buruan*	*ga-maneburuan*
11	*mana buruan ga na murit*	*manumburuan yovo*
12	*mana buruan ga na rain*	
15	*mana buruan ga muanam*	
20	*na rain mana burua laven*	
30	*naien mana burua lap*	*naiyen manemburuan lap*
50	*muanam mana burua lap*	
100	*mana buruan ma mana buruan lap*	

Sources: A. Lean's Field Notes (1983). Village: Panaras. B. Kul: Deutsche Marine-Expedition, Kluge (1941, p. 190).

Wurm (1982) classified Kuot as a Family-level isolate belonging to the New Britain Stock of the East Papuan Phylum. Wurm (1982) noted that:

> One of the difficulties in establishing the phylum and the various groups included in it is the very extensive lexical, and in part also structural and typological, Austronesian influence upon quite a few of the languages. High percentages of the basic vocabularies of such languages consist of Austronesian loans which seriously impede comparisons involving the original Papuan lexical element in these languages. (p. 231)

The possibility of borrowing lexical items from Austronesian neighbours should be kept in mind in dealing with the Kuot data given below. 10-cycle systems common to AN languages are not typical features of the NAN languages found on the New Guinea mainland. Laycock gave the language population of Kuot as 904 (Wurm & Hattori, 1981, Map 14). Capell (1962) regarded Kuot as having three dialects but did not give their locations.

Die deutsche marine-expedition 1907–1909 (1908) took examples of Kuot numerals and these appear in Kluge (1941, p. 190) under the reference Kul, the name of one of Capell's three dialects. Data for Kuot also derive from the author's field notes taken in 1983 at Panaras village on the west coast of New Ireland. The data are presented in Table B74.

Table B74
Kuot Counting Words

	A	B
1	*an*	*an, are, iovo*
2	*haras*	*aras*
3	*naien*	*ayen, iyin, nayen*
4	*nagala*	*a ale, na ale*
5	*muanam*	*muanam, noane*
6	*gunamun*	*gunamun, gunamuru*
7	*gamura*	*gamura*
8	*gamin*	*gamin*
9	*gamiala*	*gamiala*
10	*mana buruan*	*ga-maneburuan*
11	*mana buruan ga na murit*	*manumburuan yovo*
12	*mana buruan ga na rain*	
15	*mana buruan ga muanam*	
20	*na rain mana burua laven*	
30	*naien mana burua lap*	*naiyen manemburuan lap*
50	*muanam mana burua lap*	
100	*mana buruan ma mana buruan lap*	

Sources: A. Lean's Field Notes (1983). Village: Panaras. B. Kul: Deutsche Marine-Expedition, Kluge (1941, p. 190).

System A appears to be a 10-cycle system although it has a number of irregularities. On the one hand it has distinct words for the numerals 1 to 10. A case can be made that the Kuot system has some features of a 5-cycle system in that the words for 8 and 9, respectively *gami-n* and *gami-ala* derive from the words for 3 and 4, *nai-en* and *nag-ala*. The word for 10, *mana burua(n)*, is used to construct the numerals 20, 30, 40, 100, etc. The words for 1 and 2, respectively *an* and *haras* are not used, however, in the construction of the numerals 11, 12, 21, 22, 31, 32, etc. For these, *murit* is used for 1 and *rain* is used for 2. The numeral 21, for example, is: *na rain mana buruan laven ga na murit* "two tens and one". Beyond the numeral 10 there is an irregularity to decimal structure (if indeed it is) for all numerals constructed with 6, 7, 8, or 9. The words for the numerals 16 to 19 are, for example:

16 *mana buruan ga ivoa kinoman*,
17 *mana buruan ga liova kineven*,
18 *mana buruan ga maiova kinep naien*,
19 *mana buruan ga maiova kinep nagala*.

Similar constructions are given for the numerals 26 to 29, 36 to 39, and so on. The construction for 18 contains the word *naien,* 3, and the construction for 19 contains *nagala*, 4; both these constructions suggest a (5, 10) cyclic pattern, i.e 18=10 + 5 +3, and 19 = 10 + 5 + 4. If this is the case, the words *kinoman* and *kineven* are being used to represent the numerals 1 and 2 rather than the expected *murit* and *rain*, and this is a further apparent irregularity in the system.

As mentioned earlier, the counting systems of NAN languages do not usually have a 10-cycle structure. This structure is, however, common among AN languages. It seems possible, then, that Kuot has been influenced by its Austronesian neighbours. A somewhat parallel case exists in the Louisiade Archipelago of the Milne Bay Province where Yele (or Yeletnye), the language of Rossel Island and which is NAN, appears to have been influenced by its Austronesian neighbours. The Yele counting system is a 10-cycle system and it also clearly shows lexical borrowing: the word for 5 is *limi* for example. Kuot, however, does not appear to have borrowed words for its numerals but its cyclic structure may well have been influenced by neighbouring languages with systems possessing (10) or (5, 10) cyclic patterns.

The Kul numerals, collected about 75 years ago, show good agreement with the recently collected data. The Kuot word for *hand* is *kilan-* (with possessive suffixes) which bears no apparent relation to the word for 5 *muanam*. There is a word for "first" *ilouke*, best translated as "foremost", and a word for "second" *ilouluan*, but no simple constructions for further ordinals. There is a word for "fraction or part *upau*.

Appendix C
Details on Languages for Chapter 10

Glen Lean and Kay Owens

Abstract Glen Lean collected over 1 200 counting systems from Papua New Guinea (PNG) and Oceania. His data were obtained from first contact records, later linguist data, and records such as government reports as well as from thousands of Counting System Questionnaires (CSQ) given to PNG University of Technology students and lecturers, school teachers, linguists and others. These he collated and described in terms of their frame words (basic words from which other counting words are made), operative patterns (being the way in which the words are connected), and cycles (such as 2, 5, 20) or other kinds of systems such as body-tally systems. Like Appendix B, Appendix C provides a selection of the extensive details that Lean collated and analysed over 22 years in preparation for his 1992 thesis. This was indeed an extraordinary feat. PNG material can be found in hard copy in a few libraries, mainly in PNG and was available on the web at http://www.uog.ac.pg. The data and analysis formed Appendices A-D of his thesis. Since much of these data, now historical, are not readily available, counting system detail is included in this Appendix to supplement the text, mainly for Chapter 10 although some are also referred to in Chapters 8 or 9. The data help to support the arguments of the chapters. Reference maps are found in Appendix E. Similar types of systems are mentioned together. The appendix pays honour to the extraordinary work of Glendon Angove Lean.

Keywords Counting systems of PNG and Oceania • Austronesian Oceanic counting systems • Non-Austronesian (Papuan) counting systems • non-base 10 counting systems • cycles, frames, and operative patterns of counting systems

Contents

K. Owens (✉)
Charles Sturt University, Dubbo, NSW, Australia
e-mail: kowens@csu.edu.au

G. Lean

K. Owens et al., *History of Number*, History of Mathematics Education,
DOI 10.1007/978-3-319-45483-2

Introduction to the Non-Austronesian (NAN) Trans New Guinea (TNG) Systems

This appendix is set out mainly referring to the arguments of Chapter 10. The Non Austronesian (NAN or Papuan) Trans New Guinea Phylum is the biggest phylum in New Guinea with many Families. It stretches from West Papua to most of the provinces of Papua New Guinea (PNG). The chapter begins with languages with 10 cycles. Other systems are also considered and these unusual patterns indicate modifications have occurred. The discussion in Chapter 10 revolves around how these variations may have occurred.

NAN Examples from West Papua of 10-Cycle Digit Tally Systems

Baham and Iha

Baham, Iha, and Semimi, all of which belong to the Mairasi-Tanah Merah Stock, are located in western West Papua in coastal regions which are also inhabited by non-Oceanic AN neighbours. Besides 6 to 10 being 5+*n,* 10 is a distinct numeral (not 5+5). However, they also have a 20-cycle which is evident from the words for 30 and 40 being 20x1+10x1 and 20x2 respectively suggesting a modification from a digit-tally system. Summaries of the language features for Baham, and Iha are given in Table C1. Semimi is similar.

Table C1
Baham and Iha Summary

	Baham	Iha
Language type	Papuan (Non-Austronesian)	Papuan (NAN)
Language family, subfamily	West Bomberai Family	West Bomberai Family
Other names, dialects	Patimuni	Kapaur, Kapauer
Cycles	5, 10, 20	5, 10, 20
Operative patterns	6–9 is 5 + *n* where *n* = 1–4, 20 = 20 × 1, 30 = ten × 3, 40 = 20 × 2	6–9 is 5 + *n* where *n* = 1–4, 20 = 20 × 1, 30 = ten × 3, 40 = 20 × 2
Frame words	1, 2, 3, 4, 5, 10, 20	1, 2, 3, 4, 5, 10, 20

The counting words for Baham and Iha are given in Tables C2 and C3. Similar operative pattern and cycles occur in both. There are distinct numerals to 5, then numbers from 6 to 9 are 5+*n* where *n* = 1 to 4. There is a distinct numeral for 10 as well as for 20, and 40 is 20 x 2, 30 is 10 x 3 indicating that decade numbers are used.

Table C2
Baham Counting Words

	A	B	C
1	*kujogono*	*kujegono*	*ogwono*
2	*kureik, kuwareik*	*kwareik*	*wirik*
3	*kuwendik*	*kwundik*	*aundir*
4	*kurianggera*	*kurianggra*	.....
5	*kutumbu*	*kutumbu*	
6	*kutumbu-kujogono*	*kutumbu kujogono*	
7	*kutumbu-kuwareik*	*kutumbu koreik*	
8	*kutumbu-kuwendik*	*kutumbu koiwendik*	
9	*kutumbu-kurianggera*	*kutumbu kurianggora*	
10	*kukora, kuwara*	*kukora*	
11	*kuwara ni ogono*	....	
20	*kendi-ogono*		
30	*kendi-ogono kuwara*		
40	*kendi n wareik*		

Sources: A. Patimuni, Baham: Galis (1960, p. 139). B. Patimuni: Cowan (1953, p. 33). C. Baham: Anceaux (1958, p. 120).

Table C3
Iha Counting Words

	A	B
1	*herewo, yewo, nangko, kwo*	*herewo*
2	*hererik, yulit, nangri, (he)rik*	*hererik*
3	*hereteri, yuteri, nangteri*	*hereteri, -terie*
4	*herenggara, yugara, nanggara*	*herenggara, herengara*
5	*(here) tumbu, nangtumbu*	*heretembu*
6	*tumbu-herewo*	*heretembu herewo*
7	*tumbu-hereri*	*heretembu hererik*
8	*tumbu-hereteri*	*heretembu hereteri*
9	*tumbu-herenggara*	*heretembu hereng(g)ara*
10	*herekpara, pra*	*herekpra, pra'a*
11	*herekpara-herewo*	-
20	*tundi-djowo*	*to mdijowo*
30	*tundi-djowo para*	*tomdidjowo pra'a*
40	*tundi-djari (k)*	*tomdidjoeriek*
50	....	*tomdidjoeriek pra'a*

Sources: A. Iha: Galis (1960, p. 138); Voorhoeve (1975, p. 101). B. Kapaur: Cowan (1953, p. 33), d'Armandville (1903, p. 3); Kapauer: Ray (1912b, pp. 340-341).

NAN Examples from Madang Province PNG of 10-Cycle Digit Tally Systems

Pay, Tani, and Ulingan, all members of the Madang-Adelbert Range Sub-Phylum, are located on the coast north of Madang near Malala Harbour in a region in which the AN language Medebur (Laycock, 1976, p. 416; SIL word list and 1 CSQ) is situated.

Pay

Summary information on the language and counting system for Pay is given in Table C4

Table C4
Pay Summary

Language type	Papuan
Language family, subfamily	Kaukombaran Family, Pihom Stock, Madang-Adelbert Range Sub-Phylum, Trans New Guinea Phylum
Other names, dialects	Pay: Hatzfeldthafen, Tombenam,Dagoi, Alam.
Villages	Rurunat
Cycles	5, 10?, 20
Operative patterns	digital tally
Frame words	1 to 5,10,20

In brief. Pay has (5, 10, 20) cycle system forming a digit tally system.

Data for Pay are available from published sources and from an SIL word list compiled at Rurunat village in 1975, and from one CSQ completed by an informant also from Rurunat.

There is reasonable agreement as to the first five number words as given by Schmidt, Zöller, the SIL word list, and the CSQ informant. Werner's data are somewhat at variance with the rest having *ngaia* for 1 (as opposed to *udala, undala, nda, ndap*), *ngner* for 2 rather than *nger, arob* for 3 rather than *ngarop*. The data provided by Schmidt, Werner, and the CSQ informant agree in that the construction of the number words for 6 to 9 is 5 + *n* where *n* takes the values 1 to 4. The number word for 10 is given variously as *ombener, a-umbene*, or *anabe nger* "hand(s) two". The number word for 20 is given as *moande* (Schmidt, Werner) and this has the meaning "man" (the SIL list has *muade* for "man", as has the CSQ informant). The Pay system thus appears to be a digit-tally one with a (5, 10?, 20) cyclic pattern. The data are presented in Table C5.

The CSQ informant indicates that certain objects are counted in fours, e.g. flying fish, whilst others are counted in pairs, e.g. coconuts.

Table C5
Pay Counting Words

	A	B	C	D	E
1	*ngaia*	*udala*	*udala*	*nda*	*undala, ndap*
2	*ngner*	*nger*	*nger*	*nger*	*nger*
3	*arob*	*ngarop*	*ngarop*	*ngarop*	*ngarop*
4	*ngnarambam*	*ngarabam*	*ngarabam*	*ngarambam*	*ngaramban*
5	*kur*	*ikur*	*ikur*	*ikur*	*ikur*
6	*ikunara-unda*	*ikur ubelene udala*			*ikur mara unda*
7	*ikunara-ngner*	*ikur ubelene nger*			*ikur maran nger*
8	*ikunarang-arob*	*ikur ubelene ngarop*			*ikur maran ngarop*
9	*ikunararam-bam*	*ikur ubelene ngarabam*			*ikur maran ngaramban*
10	*a-umbene*	*anabe nger*		*ombener*	*ombener*
11	*a-umbene ara unda*				
20	*moande*				*moande*

Sources: A. Hatzfeldthafen: Zöller (1890, p. 127) B. Tombenam: Schmidt (1900, pp. 360-361), Ray (1919a, p. 330), Kluge (1938, p. 174) C. SIL Word List (1975). Village: Rurunat. D. Dagoi: Werner (1909, p. 111), Ray (1919a, p. 330), Kluge (1941, p. 18) E. 1 CSQ Village: Rurunat.

Tani

Tani is spoken in 23 villages (one of these is part Saki-speaking and another part Ulingan-speaking) (Z'Graggen 1975b, p. 26) which are situated in a region lying largely to the south of Hatzfeldthaven. Wurm and Hattori (1981, Map 7) showed the language classified (with Pay and Saki) as a member of the Kaukombaran Family, Pihom Stock. The total resident population of Tani-speakers (allowing for the two-language villages) was 2 295 at the 1980 National Census (National Statistical Office, 1982). The 2011 census gives 846 residents for the unit of Simbine (National Statistical Office, 2014a). Table C6 provides summary data for the language of Tani.

Table C6
Tani Summary

Language type	Papuan
Language family, subfamily	Kaukombaran Family, Pihom Stock, Madang-Adelbert Range Sub-Phylum, Trans New Guine
Other names, dialects	Tani: Miami, Moando : Kaukombaran Family.Miani, Maiani
Villages	Alisuab, Rarin, Simbine
Cycles	2', 5
Operative patterns	digital tally? 3=2+1, 4=2+2
Frame words	1,2,5

Some ethnographic and linguistic data of the Moando language region, i.e. Tani, may be found in Tranel (1952); this includes some number data. We have, in addition, an SIL word list compiled at Alisuab village in 1975 which contains some number words, as well as two CSQs. The data are presented in Table C7.

Table C7
Tani Counting Words

	A	B	C
1	*utua*	*udia*	*udia*
2	*ngar*	*ner*	*ner*
3	*arop*	*arop*	*arop*
4	*arabam*	*arebani*	*arebam*
5	*una pakak*	*igur*	*igur*
6	*dawira utua*		*igur tavira udia*
7	*dawira ngar*		*igur tavira ner*
8	*dawira arop*		*igur tavira arop*
9	*dawira arabam*		*igur tavira arebam*
10	*wabota*	*igur da igur*	*uwapo nener pakak*
11	*tuburomoro utua*		
12	*tuburomoro ngar*		
15	*tuburomoro una pakak*		
18	*tuburomoro dawira arop*		
20	*moato utua*		

Sources: A. Tranel (1952, p. 469). B. SIL Word List (1975). Village: Alisuab. C. 2 CSQs. Villages: Rarin, Simbine.

The Tani system is a digit-tally one with a (5, 20) cyclic pattern or a (5, 10, 20) cyclic pattern (as given by Tranel). The basic numeral set appears to be (1, 2, 3, 4, 5), the word for 5, *igur*, not being the same as the word for "hand", *-uwapo* or *-uapo*. The number words for 6 to 9 have a 5 + *n* construction where *n* takes the values 1 to 4 respectively: the CSQ informant, for example, has *igur tavira udia* where *tavira* means "other" or "other hand", whilst Tranel (1952) had *dawira utua*, the 5 being understood. Tranel had *una pakat*, i.e. "thumb (completed, cut off)", for 5 as opposed to *igur* as given by the other informants. Tallying proceeds to 10, 5 + 5 (SIL data) or "hand(s) two (completed, cut off)", (CSQ informants), and then to 20, *moato utua*, i.e. "man one" (the fingers and toes of one man).

Ulingan

Spoken in a region lying south-east of Hatzfeldthaven, Ulingan belongs to the Kumilan Family, Pihom Stock, (Wurm & Hattori, 1981, Map 7). There are 12 villages which are Ulingan-speaking and the village of Rarin, in addition, is part-Ulingan and part-Tani-speaking (Z'Graggen, 1975b, p. 26). At the 1980 National Census, the total number of resident speakers of the language was 1,736 (National Statistical Office, 1982). The 2011 Census gives 1090 residents for Ulingan. Table C8 summaries some information on the language.

Table C8
Ulingan Summary

Language type	Papuan
Language family, subfamily	Kumilan Family, Pihom Stock, Madang-Adelbert Range Sub-Phylum, Trans New Guinea Phylum.
Other names, dialects	Ulingan: Mawake, Mauwake
Villages	Moro, Susure, Amiten
Cycles	2' ?, 5, 20
Operative patterns	digit tally, 4=2+2?
Frame words	1 to 5, 20

In brief. 5, 10 cycle and possibly digit tally but no data beyond 10. Some data imply a modified 2 cycle

The data available for Ulingan derive from an SIL word list compiled at Moro village in 1980, as well as from 2 CSQs. The data are presented in Table C9. There appear to be three, or possibly four, basic numerals: 1 is *kusow (kusov, kuisow)*, 2 is *erup*, 3 is *arow (arov)*, and 4 is *erepam*, either a distinct numeral or, possibly, a reduplication of 2, implying that the system has a modified 2- cycle. The number word for 5 contains a "hand" morpheme, *wapena (vapena, wapen)*; the SIL list also has this appearing in the number word for 10. The CSQ informants indicate that the system is a digit-tally one although no data are available beyond 10.

Table C9
Ulingan Counting Words

	A	B
1	*kuisow*	*kusov,kusow*
2	*erup*	*erup*
3	*arow*	*arov,arow*
4	*erepam*	*erepam*
5	*wapen(a) inawiya*	*wapen inowa*
6		*okaipe kami kuisow*
7		*okaipe kami erup*
8		*okaipe kami arow*
9		*okaipe kami erepam*
10	*wapen(a) okwaiwi-pa-okwaiwi-pa inek*	*imeka kusow*

Sources: A. SIL Word List (1980). Village: Moro. B. Data : 2 CSQs. Villages: Susuri, Aminten.

Medebur

Medebur is the Austronesian language neighbouring and apparently influencing the above three languages – Pay, Tani and Ulingan. With Sepa and Manam, Medebur forms the Manam Sub-Family of the Siassi Sub-Family (Z'Graggen, 1975b, p. 26) (Wurm & Hattori, 1981, Map 7). The language is spoken in two villages (Z'Graggen, 1975b, p. 40) situated on the north-east coast of the mainland about 25 kilometres south-east of Hatzfeldthaven. At the 1980 National Census, the two Medebur-speaking villages, Medebur and Toto, had a total population of 336 (National Statistical Office, 1982) and in 2014, the recorded population of the two units was 619. Table C10 summarises the language information.

Table C10
Medebur Summary

Language type	Austronesian
Language family, subfamily	Manam Sub-Family, Siassi Family.
Other names, dialects	Medebur
Villages	Medebur, Toto
Cycles	5, 10?, 20?
Operative patterns	digit tally ?
Frame words	1 to 5

Laycock (1976, p. 416) had a set of Medebur numerals in his comparative word list of the AN languages of the Sepik region. In addition, number data are available from an SIL word list compiled at Medebur village in 1974, as well as from one CSQ completed by an informant also from Medebur. The data are given in Table C11.

Table C11
Medebur Counting Words

	A	B	C
1	*takana*	*takana*	*takana*
2	*aru*	*aru*	*aru*
3	*tol*	*tol*	*tol*
4	*bilitak*	*piltak*	*piltak,filtak*
5	*lim*	*lim*	*lim*
10		*lim ilak*	*limlak*

Sources: A. Laycock (1976, p. 416). B. SIL Word List (1974). Village: Medebur. C. 1 CSQ. Village: Medebur.

On the whole, the Medebur numerals are characteristically AN with the exception of 4, *piltak* (*filtak, bilitak*). The numeral 2 is *aru* compared with Manam's *rua* and Sepa's *lua*, these being typically AN in character. The numeral 5 is lim which is distinct from the word for "hand", *nima-* (with possessive suffixes). The system appears to possess a 5- cycle: the CSQ informant indicates that the numbers 6 to 10 have a 5 + *n* construction where *n* takes the values 1 to 5 respectively but there is insufficient data to confirm digit tally system

NAN Example from Morobe Province PNG of 10-Cycle and Digit Tally System

This example comes from the Umboi-Siassi Island area of Morobe with links also to West New Britain and the mainland coast. Kovai is a NAN Isolate with neighbouring AN languages. The languages have influences from NAN to AN and vice-versa.

Kovai

Kovai is a Non-Austronesian (NAN) Isolate of the Finisterre-Huon Stock spoken in 13 villages situated in the western half of Umboi Island, the main island of the Siassi Group, which lies between the eastern coast of the Morobe Province and the western extreme of New Britain. At the 1980

National Census the total population of the Kovai-speaking villages was 2 522 (National Statistical Office, 1982). Their neighbours speak Mangap, Mutu, Barim which are all AN. Table C12 summarises the language information.

Table C12
Kovai Summary

Language type	Papuan
Language family; subfamily	Isolate, Finisterre-Huon Stock, Trans New Guinea Phylum
Other names, dialects	Kovai
Villages	Umboi Island; Arot, Aiyau, Mararamu
Cycles	5, 10, 20
Operative pattern	Digit tally
Frame words	1 to 5, 10, 15, 20

In brief. Whilst the system is a digit-tally one (informants for one of the authors, Lean, tallied the digits on the right hand first, the left hand next, then the right foot and finally the left foot) there appear to be distinct words for both 10 and 15; the word for 20 does not translate as "man". 5 is "my one hand".

One of the authors, Lean, has field notes on the Kovai counting system collected at Arot village in 1973. Other unpublished data derive from an SIL word list compiled at Mararamu village in 1975, and six CSQs completed by informants from four villages. G. Smith (1984, p. 188) also had a set of Kovai number data. A selection of data is given in the Table C13.

Table C13
Kovai Counting Words

	A	B
1	*mungon*	*munugon*
2	*lolon*	*lolon*
3	*alwon*	*albon, aulbon*
4	*ilon*	*ilon*
5	*milin mungon*	*milin munugon*
6	*milin mungon manan mungon*	
7	*milin mungon manan lolon*	
8	*milin mungon manan alwon*	
9	*milin mungon manan ilon*	
10	*melalin*	*melalin*
11	*melalin manan mungon*	
12	*melalin manan lolon*	
13	*melalin manan alwon*	
14	*melalin manan ilon*	
15	*bab*	*bab*
16	*bab manan mungon*	
17	*bab manan lolon*	
18	*bab manan alwon*	
19	*bab manan ilon*	
20	*riring mungon*	*riring munugon*
30	*riring mungon longon melalin*	
40	*riring lolon*	*riring lolon*

Sources: A. Lean's Field Notes (1973). Arot. B. 3 CSQs. Villages: Aiyau (2), Mararamu.

In several respects the Kovai system is an unusual one as far as NAN systems go. The frame pattern is (1, 2, 3, 4, 5, 10, 15, 20) and the cyclic pattern is (5, 10, 20). Whilst the system is a digit-tally one (one of the authors, Lean's, informants tallied the digits on the right hand first, the left hand next, then the right foot and finally the left foot) there appear to be distinct words for both 10 and 15; the word for 20 does not translate as "man".

The "hand" morpheme in Kovai is *mel-*: his/her hand is *melon*, my hand is *milin*, their hands: *melowon*, our hands: *melowin*. The number word for 5 then, *milin mungon (milin munugon)*, is "my hand one", i.e. the fingers of one hand. The number word for 10, *melalin*, appears to contain a "hand" morpheme and indeed is given by the CSQ informants to mean "two hands"; this may be a reduplicated form of *mel-*with a "-my" possessive suffix.

The existence of a distinct number word for 15, *bab*, is unusual. Whether or not this has a body-part referent is uncertain.

The possible influence of the Kovai-speakers' AN neighbours can be seen in the use of *riring* for 20. At the western end of West New Britain, the Kilenge and Maleu speakers use *ringring* for 1 000. The Mutu-speakers on the islands off the south of Umboi using *dingding* or *ndingnding* to mean 100 (according to some informants) or 200 (according to others). It is possible that the Kovai have borrowed the word from groups with which they trade. Why, however, the borrowed word should be assigned to 20 rather than, say, a large number of the order it has in the AN languages, is not clear.

As is also common with the AN groups in the region, the Kovai-speakers count coconuts in groups of four. One quartet of coconuts is *gaun mungon (gaun munugon)*, two quartets is *gaun lolon*, and so on.

Mangap

Mangap is an AN language spoken in seven villages situated on the eastern end of Umboi Island, and in one village on Sakar Island, both islands belonging to the Siassi Group (McElhanon, 1984, p. 18). The current linguistic classification of Mangap is that it is a member of the Island Group, Vitiazan Sub-Family, Siassi Family (McElhanon, 1984, p. 13). At the 1980 National Census, the total population of the Mangap-speaking villages was 2 256 (National Statistical Office, 1982). Table C14 summarises information on the language.

Table C14
Mangap Summary

Language type	Austronesian
Language family; subfamily	Island Group, Vitiazan Sub-Family, Siassi Family
Other names, dialects	Mangap. (Kaimanga, Iangla (Rook I) 1. Kaimanga, 2. Mangap.
Villages	Rook I, Yangla
Cycles	5, 20
Operative pattern	Digit tally
Frame words	1 to 5, 20

In brief. AN with characteristics of digit tally system. While 5 is a numeral word, 20 seems to derive from "man".

Dempwolff (1905) had a word list of Mangap, referenced Kaimanga, but this has no number data. The Australian Government Anthropologist Chinnery (1928, pp. 99-100) compiled word lists of the AN languages spoken on Umboi (Rook) Island and each of these include some number data (see also Barim and Mutu below). G. Smith (1984, p. 155) also had number data for Mangap. Further data derive from two SIL words lists, one compiled at Kampalap village in 1968, the latter word list being subsequently published in Hooley (1971). In addition, one CSQ was completed by an informant from Yangla village A selection of the data are given in Table C15.

Table C15
Mangap Counting Words

	A	B
1	*ta*	*ta*
2	*ru*	*ru*
3	*tel*	*tel*
4	*pang*	*pang*
5	*lama ta*	*lamata*
6	*lama ta ta*	*lamata mi ta*
7	*lama ta ru*	*lamata mi ru*
8	*lama ta tel*	*lamata mi tel*
9	*lama ta pang*	*lamata mi pang*
10	*lamuru*	*lamuru*
11	*lamuru ta*	*lamuru mi ta*
12	*lamuru ru*	*lamuru mi ru*
13	*lamuru tel*	*lamuru mi tel*
14	*lamuru pang*	*lamuru mi pang*
15	*lamoru mata*	*lamorumata*
16	*lamoru mata ta*	*lamorumata mi ta*
17	*lamoru mata ru*	*lamorumata mi ru*
20	*tomota*	*tomota*
30	.	*tomota lamuru*
40	*tomto ru*	*tomto ru*

Sources: A. Iangla, Rook I.: Chinnery (1928, pp. 99-100). B. Yangla: 1 CSQ.

The data given by Chinnery, collected some 90 years ago, are almost identical to those given by the CSQ informant (Smith's data, too, closely resemble both of these). These indicate that the Mangap system possesses a (5, 20) cyclic pattern. The basic numeral set is (1, 2, 3, 4, 5): whilst it appears to be the case that digit-tallying occurs, nevertheless the word for 5, *lamata*, is not identical to the word for "hand"; *namana* is "hand-my", *nomong* is "hand-his"; thus 5 appears to be a numeral with no body-part referent. 5 itself appears to have the construction *lama* + *ta*, i.e. 5-morpheme + one or 5 x 1. 10 is *lama* + *ru*, or 5 x 2, but 15 is *lamorumata* which does not have, say, a 5 x 3 construction, a 5 x (2 + 1) construction being suggested instead. 20 is *tomota*: the word for "man" in Mangap is *tomoto* and 20 may be seen as a contraction of *tomoto ta*, i.e. man one. 40 is *tomto ru*, i.e. a contraction of *tomoto ru*, men two. 20 thus is a number word rather than a numeral.

The CSQ informant indicates that fish are counted in pairs, one pair being *lulun ta*, two pairs *lulun ru*, and so on. Coconuts are counted in fours, one quartet being *me ta*, two quartets *me ru*, and so on. Prawns and other shellfish are counted in fives, one quintet being *zeran ta*, two quintets being *zeran ru*, and so on.

The SIL data show two variants for the numeral 4. This is *pang* which, with slight variants, is found in a number of West New Britain AN languages. The SIL data have both *pal* (1968) and *fahang* (1975). Five appears to be a numeral with no body-part referent. 5 itself is 5-morpheme + one or 5 x 1. 10 is *lama* + *ru*, or 5 x 2, but 15 is *lamorumata* which is 5 x (2 + 1). 20 is *tomota*, the word for "man"; 40 is "two men" (Table C15).

Mutu

Whilst Kovai and Mangap are languages largely confined to the main Siassi island of Umboi, Mutu is largely spoken on the small islands off the southern end of Umboi: Mandok, Aramot, Malai, Tuam, and Mutu-Malau. On Umboi it is spoken in Salaboran village and there are some Mutu speakers in the village of Yaga which is otherwise Mangap-speaking (McElhanon, 1984, p. 18). Mutu is an AN language which McElhanon has classified (with Gitua on the mainland) as a member of the Bariai Sub-Family, Siassi Family, whose other members are found in West New Britain (see Table C16). At the 1980 National Census the total population of the Mutu-speaking villages was 1704 (National Statistical Office, 1982). In 2007, SIL recorded 35 000 speakers. Mandok Unit had a population of 821, Tuam Unit 401, Malai Unit 482 and Gitua Unit 1176 at the 2011 Census (National Statistical Office, 2014a).

Table C16
Mutu Summary

Language type	Austronesian
Language family; subfamily	Bariai Sub-Family, Siassi Family
Other names, dialects	Mutu, (Tuom, Rook Island)
Villages	Islands off south of Umboi- Tuam Island
Cycles	5, 10, 20
Operative pattern	6 to 9 = 5 + n
Frame words	1 to 5, 10, 20

In brief. 5 and 10 are numeral words. 6 to 9 are 5 + *n*. 100 is common with neighbours with whom they trade in West New Britain. 10 is similar to many New Britain, New Ireland and Manus words for 10. 20 "man" morpheme is used in 40 etc.

An early example of Mutu number data can be found in Schmidt (1900, pp. 359-360) who cited Zöller and Schellong as the sources of Mutu word lists. Schmidt referenced the language as Rook Island. Dempwolff (1905) had two Mutu word lists, one referenced Mantok, the other referenced Tuom: only the latter word list contains number data. Ray (1919a, p. 327) contained both the Schmidt and Dempwolff data under the references given. Similarly, Kluge (1941, pp. 201-202) had the Tuom number data and the Rook Island data. Chinnery's (1928) Siassi word lists include data on Mutu referenced Tuam, Malei, and Aramot, Siassi Group. Other, unpublished, data available derive from two SIL word lists, one compiled at Mandok in 1968 and one compiled on Tuam Island (undated), and from Lincoln's (1976) Rai Coast Word Lists, as well as three CSQs, completed by informants from Tuam. Smith (1984, p. 162) also had a set of Mutu number data.

The data shown in Table C17 show remarkably close agreement. These indicate that the Mutu system has a (5, 10, 20) cyclic pattern. The basic numeral set is (1, 2, 3, 4, 5, 10): 5 is a numeral and is not the same as the word for "hand" *nima-*. The numeral 10 *sangavul (sangul)* is cognate with the numeral 10 and its variants found in other parts of New Britain, New Ireland, and Manus. The number word for 20 *tamot es* contains a "man" morpheme *tamot* and has the gloss "man one", 40 is *tamot ru*, i.e. "men two".

There is some confusion among informants regarding the meaning of *dingding (ndingnding)*. The CSQ informants uniformly give this as 100. Smith (1984) gave it as 400 but indicated situations where it may be constructed to have other meanings. Some Mandok informants in Lae have indicated that *dingding* is used in counting pairs or in twos and means "100 pairs", i.e. 200. We have noted above that a word *ringring* is used by the Kilenge and Maleu of West New Britain, who trade with Mutu-speakers, the sense usually meaning 1 000 although some informants give the meaning as 400 (20 x 20). The Kovai have *riring* as 20.

Table C17
Mutu Counting Words

	A	B	C	D	E
1	*es*	*ts (es?)*	*es*	*ez, es*	*es*
2	*ru*	*ru*	*ru*	*ru*	*ru*
3	*tol*	*tol*	*tol*	*tol*	*tol*
4	*pang*	*pang*	*pang*	*pang*	*pang*
5	*lim*	*lim*	*liim*	*lim*	*lim*
6	*lim be es*	*lim be es*	*lim be es*	*lim be (or ve) es*	*lim be es*
7	*lim be ru*	*lim be ru*	*lim be ru*	*lim be ru*	*lim be ru*
8	*lim be tol*	*lim be tol*	*lim be tol*	*lim be tol*	-
9	*lim be pang*	*lim be pang*	*lim be pang*	*lim be pang*	-
10	*sangul,sangavul*	*sangawul*	*sangul*	*sangavul*	*sanggul*
11		*sangawul we es*		*sangavul ve es*	*sanggul be es*
12		*sangawul we ru*		*sangavul ve ru*	*sanggul be ru*
13		*sangawul we tol*			
14		*sangawul we pang*			
15		*sangawul we lim*		*sangavul ve lim*	*sanggul be lim*
16		*sangawul lim be es*		*sangavul ve lim be es*	*sanggul be lim be es*
17		*sangawul lim be ru*			
20	*tamote*	*tamot es*		*tamot es*	*tamot es*
30		*tamot es sangawul*		*tamot es sangavul*	*tamote be sanggul*
40				*tamot ru*	*tamotru*
100				*dingding es*	*tamot lim*
400					*ndingnding*

*Sources:*A. Dempwolff (1905, p. 217), Ray (1919a, p. 327), Kluge (1941, p. 201) B. Tuam, Malei and Aramot : Chinnery (1928, pp. 99-100). C. Lincoln (1976). D. 3 CSQs. Tuam I., E. G. Smith (1984, p. 162).

Barim

Barim is spoken in three villages: Barim, situated on the west coast of Umboi Island, Saupanam, on the south-west of Umboi, and Aronai-Mutu, a small island off the south-west coast (McElhanon, 1984, p. 17). Barim and Mangap are classified as belonging to the Island Group, Vitiazan Sub-Family, Siassi Family. See Table C18 for a language summary.

Table C18
Barim Summary

Language type	Austronesian
Language family; subfamily	Island Group, Vitiazan Sub-Family, Siassi Family
Other names, dialects	
Villages	3 Villages, Barim, Saupanam on Umboi and Aronai-[Mutu on island
Cycles	5, 10, 20
Operative pattern	6 to 9 = 5 + *n*
Frame words	1 to 5, 10, 20

In brief. The word for 5, *lim*, is not the word for hand, *bagin*. 10 is a numeral whilst 20 is not, *tamota* "man one". 5 to 9 are 5+*n*, 10 to 19 are 10+*n*. There are multiples of 20.

Table C19 provides data on the counting words for Barim.

Table C19
Barim Counting Words

	A	B
1	*ta*	*ta*
2	*ru*	*ru*
3	*torl*	*tol*
4	*pay*	*pang*
5	*lim*	*lim*
6	*lim gbe ta*	*lim be ta*
7	*lim gbe ru*	*lim be ru*
8	*lim gbe torl*	
9	*lim gbe pay*	
10	*sayaul*	*sangul*
11	*sayaul we ta*	*sangul be ta*
12	*sayaul we ru*	*sangul be ru*
13	*sayaul we torl*	*sangul be tol*
14	*sayaul we pay*	*sangul be pang*
15	*sayaul we lim*	*sangul be lim*
16	*sayaul we lim we ta*	
17	*sayaul we lim we ru*	
20	*tamot ta*	*tamota*
30	*tamot ta we sayaul*	*tamota sangul*
40	*tamot ru*	*tamotru*

Sources: A. Chinnery (1928, pp. 99-100) B. Smith (1984, p. 152).

At the 1980 National Census, the total population of the Barim-speaking villages was 480 (National Statistical Office, 1982). In 2011 Census, the Barim Unit had a population of 861 and Aronai Unit of 481 (National Statistical Office, 2014a), probably including some Mutu area.

Chinnery (1928) had a word list of Barim which includes number data. Hooley (1971) also had a word list of Barim collected in 1968, however this contains only the numerals 1 to 4. Smith (1984, p. 152) had number data for Barim collected in the early 1980s.

With slight differences in orthography, Smith's data and Chinnery's data show close agreement. Both indicate a system with a (5, 10, 20) cyclic pattern and a basic numeral set (1, 2, 3, 4, 5, 10): the word for 5, *lim*, is not identical to the word for "hand", *bagin*. 10 is also a numeral whilst 20 is not: *tamota* has the gloss "man one".

NAN Examples from Central Province, PNG with 10-Cycle System and Neighbouring AN Counting Systems

Kwale and Magi, both belonging to the Central and South New Guinea Stock and which are situated in the Central Province (PNG). Kwale is located in a region adjacent to that of the AN Sinagoro while Magi is spoken along the south coast which is also inhabited by the speakers of the AN Magori, Ouma, and Yoba languages. The proximity of AN and NAN languages is evident in the map of Central Province, Figure E1.

Magi

Magi is a Papuan (NAN) language which is spoken along the south coast and several islands off the coast of the Central Province, the region extending from just east of Cape Rodney through to Gadaisu village, on Orangerie Bay, in the Milne Bay Province. Wurm and Hattori (1981, Map 9) listed nine dialects for Magi and these have been studied by Thomson (1975a, 1975b). The total resident population of Magi-speakers was, presumably in the mid-1970s, about 5 300 (Thomson, 1975a, p. 55). The Magi-speaking villages listed by Thomson had, however, a total population of 4 418 at the 1980 National Census (National Statistical Office, 1982). Summary data for Magi language and counting system are given in Table C20.

Table C20
Magi Summary

Language type	Papuan
Language family; subfamily	Mailuan Family
Other names, dialects	Magi: 1.Domara, 2.Darava, 3.Asiaoro, 4.Derebai, 5.Island, 6.Geagea,7.Borebo, 8.Ilai, 9.Baibara. Variant : Mairu, Mailu
Villages	Mailu Is., Domara, Pediri
Cycles	5, 10
Operative pattern	6 to 9 is 5 + *n* where n is 1 to 4
Frame words	1 to 5, 10

In brief. The Magi system seems to have undergone change under Austronesian influence to 10, 100 cycle. 10 is numeral from Sinagoro. Trading influence causing decimalisation is likely and counting in fours occurs.

Two Magi vocabularies were published in the Annual Report on British New Guinea from 1890 to 1891, one referenced as Domara, the other as Mairu (British New Guinea, 1892). Both of these appear under the same references in Ray (1895). In the Annual Report for Papua, 1910-11, Strong (1911) had a Magi vocabulary, with number data, referenced as Mailu while Saville (1912) also published some numerals in his Mailu grammar. Domara from the Annual Report and Mailu from Saville are included in Ray's (1938a, 1938b) work together with a further vocabulary referenced Dedele. Kluge (1941, p. 22) had the number data from these three vocabularies, taken from Ray. Two articles by Thomson (1975a, 1975b) on the Magi language and its dialects contain some number data (for the Island dialect). In addition to the published data, we have some data from an SIL word list compiled at Boru village (Damara dialect) in 1975, and three informants from three different dialect areas completed CSQs. The data are shown in Table C21.

The Magi counting system shows features which are quite unusual in NAN counting systems. There is, for example, no sign of a 2-cycle and we have distinct words for the first five numerals. The numeral 5 itself, *ima* or *ima omu*, is Austronesian; the word for hand is *ima* as well and it seems likely that it is this word which has been borrowed and, since the "hand" morpheme is employed in counting, the original Magi numeral has been displaced. It is not immediately obvious that any other borrowing of numerals has taken place (except perhaps in the numeral 100, discussed below).

It is apparent, then, that the Magi counting system has undergone some Austronesian influence. The influence, however, is not so much discerned in the number words themselves as in the cyclic infrastructure of the counting system. The Magi system has a (5, 10) cyclic pattern rather than, say, a (2, 5, 20) pattern that is common in NAN systems. The number words for 6 to 9 have a 5+*n* construction where *n* takes the values 1 to 4. There is a distinct "ten" morpheme, *nanau*, and this is used to construct the higher decades so that the numerals 10 to 90 have the construction *nanau*+*n* where *n* takes the values 1 to 9. It is the existence of this 10-cycle and absence of 2- and 20-cycles which suggests an AN influence.

Table C21
Magi Counting Words

	A	B	C	D	E
1	*?omu*	*omu*	*ombua*	*omu(omuna)*	*om*
2	*?ava*	*ava*	*awa*	*ava*	*ava*
3	*aiseri*	*aiseri*	*aisheri*	*aiseri*	*aiseri*
4	*tourai*	*tourai*	*taurai*	*tourai*	*tourai*
5	*ima ?omu*	*ima*	*ima*	*ima omu*	*ima om*
6	*ima lilia ?omu*	*ima lilia omu*	*liliomo*	*ima lili omu*	*ima lilia om*
7	*ima lilia ?ava*	*ima lilia ava*	*liliawa*	*ima lili ava*	*ima lilia ava*
8	*ima lilia aiseri*	*ima lilia aiseri*	*liliaisheri*	*ima lili aiseri*	*ima lilia aiseri*
9	*ima lilia tourai*	*ima lilia tourai*	*liliataurai*	*ima lili tourai*	*ima lilia tourai*
10	*nanau ?omu*	*nanau omu*	*nanaom*	*nanau omu*	*nanau om*
11	*nanau ?omu ?omu*	*nanau omu omu*	*liliomombua*	*nanau omu omu*	*nanau lilia om*
12	*nanau ?omu ?ava*	*nanau omu ava*	*awa*	*nanau omu ava*	*nanau lilia ava*
13					*nanau lilia aiseri*
14					*namau lilia tourai*
15	*nanau ?omu ?-ima ?omu*	*Nanau omu ima*		*Nanau omu ima*	*nanau ima om*
16					*nanau ima lilia om*
20	*nanau ?ava*	*nanau ava*	*nana awa*	*nanau ava*	*nanau ava*
30		*nanau aiseri*	*nana aisheri*	*nanau aiseri*	*nanau aiseri*
40		*nanau tourai*	*nana tauai*	*nanau tourai*	*nanau tourai*
50		*nanau ima*	*nana ima*	*nanau ima*	*nanau ima*
60		*nanau ima-lilia ava*	*nana lilioma*		*nanau ima lilia ava*
100		*nanau gabana-omu*	*nana gabana*	*(wuwuru omu)*	
200		*nanau gabana-ava*			

Sources: A. Island d.: Thomson (1975b, pp. 73-4; 1975a, p. 645). B.Mailu: Saville (1912, p. 429), Ray (1938a, p. 184), Kluge (1941, p. 22). C. Domara: British New Guinea (1892, p. 115), Ray (1895, pp. 35-8; 1938a, 1938b, p. 184), Kluge (1941, p. 22). D. 3 CSQs. Villages: Mailu Is., Domara, Pediri. E. Onagi (~2008).

The data shown in Systems B and C in Table C21 indicate that there is not (or, rather, was not) a distinct number word for 100. System B has *nanau gabana omu* and System C has *nana gabana* whereas we might expect, say, *nanau nanau (omu)*, i.e "ten tens". *gabana* is the "ten" morpheme for Sinagoro (Sinaugoro). It seems possible, then, that the *gabana*, appearing in the Magi numeral 100 and which clearly has a "ten" meaning, has been borrowed from AN Sinagoro. None of the CSQ informants gave *nanau gabana (omu)* for 100; one, however, gave *wuwuru omu*, suggesting the existence of a distinct "hundred" morpheme. No corroborative evidence, however, exists to determine this.

A change in cyclic infrastructure has been observed in other NAN counting systems. It is not necessarily the case that an encounter of NAN and AN language groups will result in the NAN counting system undergoing a change in cyclic infrastructure under the AN influence. Indeed, we appear to have evidence to the precise contrary. Some NAN systems, however, do undergo change but, presumably, only if there is some necessity to do so, e.g for purposes of trading with (predominantly) AN groups.

In the case of Magi, the need for a more efficient or sophisticated counting system may have arisen as a result of the Mailu trading expeditions (the Ilo voyages) to various AN groups along the southern coast and eastwards to the Suau-speaking people and other groups in the Milne Bay Province. Counting by fours is reported by the CSQ informants. Coconuts, for example, are grouped in fours, one quartet being *gau omu*, two quartets, *gau ava*, and so on.

Onagi noted that recently *omu* had become *om* such as in *garu om* "good night". He also noted that people counted up to very large numbers such as 10 000. Table C22 gives some of these larger numbers. "Ten" is *nanau;* "hundred" is *nanau gabana nanau* so *n* hundreds is expressed as "100 + *n*"; a thousand is "100 + 10" so that *n* thousands is "100 + 10 + *n*". The multiplicative power of the language in base 10 is evident while retaining numbers 6 to 9 as 5 + *n* where *n* is 1 to 4. Thus the language is best described as having (5, 10, 100) cycle system.

Table C22
Large numbers in Magi

	Magi
400	*nanau gabana nanau tourai*
500	*nanau gabana nanau ima om*
600	*nanau gabana nanau ima lilia om*
700	*nanau gabana nanau ima lilia ava*
800	*nanau gabana nanau ima lilia aiseri*
900	*nanau gabana nanau ima lilia tourai*
1 000	*nanau gabana nanau nanau*
2 000	*nanau gabana nanau nanau ava*
3 000	*nanau gabana nanau nanau aiseri*
4 000	*nanau gabana nanau nanau tourai*
5 000	*nanau gabana nanau nanau ima om*
6 000	*nanau gabana nanau nanau ima ava*
7 000	*nanau gabana nanau nanau ima aiseri*
8 000	*nanau gabana nanau nanau ima tourai*
9 000	*nanau gabana nanau nanau ima lilia tourai*
10 000	*nanau gabana nanau nanau nanau*

Source: Onagi (~2008).

Table C23 indicates that counting food and other items requires a different perspective or use of terms for specific groups (also found in Motu). Some common words like *otoai* "a few things" are used including for food items but the specific groupings have specific words.

Table C23
Counting Food Items and Other Items in Magi

Food item	Units in Magi	Meaning in English
orebe "fish"	*laveru om turi om*	few (not many) a string of fish
tebele "taro"	*gaima om doa*	four taros
ama "coconuts"	*gau om* *gogore om*	four coconuts 20 or more coconuts
ueni "betel nuts"	*butu om* *lala om*	one portion of the bunch full bunch of betel
odei "sago"	*pai om* *koma om/ata om*	one roll of sago similar to a roll but smaller
lavata "banana"	*ubu om* *wuwuru om* *maau*	one hand of bananas a bunch of bananas half a hand of bananas
oba "shell money"	*maina om* *toba om*	string of 2 shells a string of 30 shells
eu "firewood"	*deidei on/moa om* *veaveva*	one bundle of fire wood few loose sticks of firewood
bora<u>a</u> "pig"	*bega om*	a group of four big pigs

Source: Onagi (~2008). *Note:* <u>a</u> indicates stress on that vowel.

Kwale

Kwale, the other member of the Kwalean Family, is spoken in a region lying to the east of the Humene area and bordered by the Kemp Welch River, the western border of Sinagoro. The language has two dialects, Kwale proper, spoken in the north, and Garia, spoken in the south (Wurm & Hattori, 1981, Map 9). Dutton (1970, p. 906) provided a list of Kwale-speaking villages and these, at the 1980 National Census, had a total population of 956 (National Statistical Office, 1982). Table C24 summarises information on the language.

Table C24
Kwale Summary

Language type	Papuan
Language family; subfamily	Kwalean Family
Other names, dialects	Kwale : 1.Kwale, 2.Garia. Variant Names : Garihe
Villages	Geresi(Garia d.)
Cycles	5, 10
Operative pattern	6 to 9 are 5 + n where n=1 to 4
Frame words	1 to 5, 10

In brief. Kwale has a (5, 10) cycle where no 2 or 20 cycle is present. The 5 cycle is still evident. It is expected to have been influenced by the neighbouring Austronesian language.

Ray (1929) had two Kwale vocabularies, one referenced as Kwale, the other as Garia. The number data from this article appear in Kluge (1941, p. 28). Dutton (1970) had some data on Kwale in his study of the Rigo district languages. A comparative vocabulary of these contains some Kwale number data. In addition to the published data, two informants from Geresi village (Garia dialect) completed CSQs. The Kwale data are presented in Table C25.

Table C25
Kwale Counting Words

	A	B	C	D
1	*teba*	*'tiba*	*tepo*	*tiba*
2	*aheu*	*a'heu*	*aheu*	*aheu*
3	*ugidu*	*u'gidu*	*igitu*	*ugitu*
4	*gazaga*	*ga'zaga*	*gajaga*	*gazjaga*
5	*fu'o*	*'fu?o*	*fua*	*fu'o*
6	*fu'o to teba*	*fu?otu 'tiba*	*fua tepo*	*tiba-fale*
7	*fu'o to aheu*	*fu?otu 'aheu*	*fua aheu*	*tiba-fale-aheu*
8	*fu'o to ugidu*			*tiba-fale-ugitu*
9	*fu'o to gazaga*			*tiba-fale-gazjaga*
10	*gabanana*		*gabana*	*gabanana*
11	*gabanana to teba*			
15	*gabanana to fu'o*			
20	*gabanana aheu*		*oda-gabanana*	*gabana-aheu*

Sources: A. 2 CSQs. Village: Geresi (Garia d.) B. Dutton (1970, pp. 947-8). C. Garia: Ray (1929, p. 81), Kluge (1941, p. 28). D. Kwale: Ray (1929, p. 82), Kluge (1941, p. 28).

Systems A and D, with some differences in orthography, are in agreement regarding the data for 1 to 5. Interestingly, there is no sign of a 2-cycle and there are five distinct words for the first five numerals. There is also a distinct numeral 10, *gabanana* and the numeral 20 is *gabanana aheu*, i.e "tens two". The numerals 6 to 9 have the construction 5+*n* where *n* takes the values 1 to 4 so that the numeral 6, for example, is *ru'o to teba*, i.e five+conjunction+one. System D has a different version for the numerals 6 to 9 but this, too, appears to have an $x+n$ construction. The system thus has a (5, 10) cyclic pattern. The word for hand, *evore* is not the same as the numeral 5, *fu'o*.

We have noted elsewhere that the absence of a 2-cycle and, more importantly, the existence of a 10-cycle and absence of a 20-cycle, is somewhat unusual in Papuan (NAN) counting systems. It has been suggested previously that the decimalisation of an NAN system may occur under the influence of proximate AN languages. In the case of Kwale it is possible that there has been some change to its counting system brought about by the influence of the neighbouring Sinagoro, an AN language, the possibility suggested by the borrowing of the Sinagoro numeral 10, *gabanana*, into the Kwale system.

Sinagoro

The Sinagoro region lies to the east of the Humene and Kwale areas and the Motu-speaking village of Kapakapa. The language comprises a chain of some 17 dialects or communalects and in the 1970s was spoken by more than 12 000 people (Pawley, 1976, p. 303). The 2000 census suggested 18 000 people in this area (SIL, n.d.). Table C26 summarises information on the language.

Table C26
Sinagoro Summary

Language type	Austronesian
Language family; subfamily	Motu-Sinagoro Section, Central Family
Other names, dialects	Sinagoro: 1.Ikolu, 2.Balawaia, 3.Saroa, .Babagarupu, 5.Kwabida, 6.Taboro, 7.Kwaibo, 8.Alepa, 9.Omene, 10.Tubulamo, 11.Ikega, 12.Boku, 13.Buaga, 14.Wiga, 15.Vora, 16.Kubuli, 17.Uruone. Variant : Rubi(?), Tarova, Sinaugolo
Villages	
Cycles	5, 10, 100, 1000
Operative pattern	6 to 9 is 5 + n where n = 1 to 4
Frame words	1 to 5, 10, 100, 1000

In brief. Sinagoro is a Motu type with 6 to 9 being 2x3, 2x3+1, 2x4, 2x4+1 but also having 6 to 9 as 5 + *n* where *n* is 1 to 4. There are data on ordinals using possible prefixes and suffixes. Teens and decades are regular and built on 10.

A vocabulary, including numerals, of Sinagoro appeared in the Annual Report on British New Guinea from 1890 to 1891 referenced as Sinaugolo. The numerals from this reappear subsequently in Ray (1895, 1907, 1929) and in Kluge (1941, p. 208). Ray (1895) also had a vocabulary (and numerals) referenced Tarova or Saroa and Ray (1907, 1929) had a further vocabulary (and numerals) referenced Rubi. Both of these appear to be examples of Sinagoro. The Rubi numerals appear in Kluge (1941, p. 208) and the Tarova numerals as well (p. 218). Seligmann (1912-13) had a grammar of Sinaugolo but this does not contain numeral data. Dutton's 1970 article on the languages of the Rigo area contains data on Sinagoro and the language is included in a comparative vocabulary of the Rigo languages which includes a set of numerals. Kolia (1975) published a grammar sketch of one of the Sinagoro dialects, Balawaia, and this, too, has some numeral data. In addition to these published data, two SIL word lists were compiled at Kemabolo village (Balawaia dialect) and at Gareba village (Tubulamo dialect) in 1975 and these have the numerals 1 to 5, and 10. Eighteen informants from various dialect areas within the Sinagoro region completed CSQs.

A sample of Sinagoro data is given in Table C27 for the Balawaia dialect (Sources A and B) and Table C28 (Sources C and D) for the other data.

Table C27 *Sinagoro Counting Words – Balawaia Dialect*

	A	B
1	*ta,tebona*	*sebona*
2	*lualua*	*lualua*
3	*toitoi*	*toitoi*
4	*vativati*	*vasivasi*
5	*imaima*	*imaima*
6	*taulatoitoi*	*taulatoi*
7	*taulatoitoi tebona*	*taulatoi sebona*
8	*taulavativa*	*titaulavasi*
9	*taulavativati tebona*	*taulavasi sebona*
10	*gabanana*	*galana*
11	*gabanana tebona*	*galana sebona*
20	*gabana lua*	*gala lualua*
30	*gabana toitoi*	*gala toitoi*
40		*gala vasivasi*
50		*gala imaima*
100	*tinauna*	*sinauna*
200		*sinau lualua*
1000	*dagarana*	*dagarana*

Sources: A. Balawaia d.: Kolia (1975, p. 147). B. 4 CSQs. (Balawaia d.).

Table C28
Sinagoro Counting Words

	C	D
1	*sebona*	*sebona*
2	*lualua, lwalwa, lwa*	*ruarua*
3	*toitoi, toi*	*kweikwei*
4	*vasivasi, vasi*	*vasivasi*
5	*imaima, imata, ima*	*imaima*
6	*imaima sebona, taula toitoi*	*imaima sebona*
7	*imaima lualua, taula toitoi sebona*	*imaima ruarua*
8	*imaima toitoi, taula vasi vasi taula*	*imaima kweikwei*
9	*imaima vasivasi, vasi vasi sebona*	*vasivasi*
10	*gabanana, gabana, galena*	*gananana*
11	*gabanana sebonagalana sebona*	*gabanana sebone*
12	*galana lualua*	
13	*galana toitoi*	
14	*galana vasi vasi*	
15	*galana ima ima*	
16	*galana ima sebona*	
20	*gabana lualua, gabana lwalwa, gala lualua*	*gabana ruarua*
30	*gabana toitoi, gabana toitoi, gala toitoi*	*gabana kweikwei*
40	*gabana vasivasi, gabana vasivasi, gala vasi vasi*	*gabana vasivasi*
50	*gabana imaima, gala ima ima*	*gabana imaima*
100	*sinaona, sinauna*	*sinauna*
200	*sinau lualua*	*sinau ruarua*
1000	*dagalana*	*dagarana*

Sources: C. Sinaugolo: British New Guinea (1892, Appendix GG6), Ray (1895, pp. 35-38, 1907, p. 470, 1929, p. 79), Kluge (1941, p. 208); 2MQ 2008 Village: Matairuka (Tubulamo language); D. 4 CSQs. (Taboro d.) Village: Tauruba (Balawaia language).

For the first five numerals, Systems A to D are, on the whole, in general agreement with the exceptions of interchanges of /l/ and /r/, /s/ and /t/, and in that the CSQ informants from the Taboro dialect region gave *kweikwei* for the numeral 3 rather than *toitoi* as given in Systems A, B, and C. The most notable feature of these (typically Austronesian) numerals is the repetition which occurs in the numerals 2 to 5 as in *ruarua, toitoi, vasivasi*, and *imaima*, a feature not found in the AN languages in the west of the Province.

The major difference between the various systems occurs with the numerals 6 to 9. Systems A and B in Table C27 both examples of the Balawaia dialect, are of the Motu type (although without a distinct numeral 7) so that the numeral 6 has the construction 2x3, the numeral 8 has the construction 2x4, and the numerals 7 and 9 have the constructions 2x3+1 and 2x4+1 respectively. Systems C and D in Table C28 have these numerals with the construction 5+*n* where *n* takes the values 1 to 4 respectively.

From the CSQ data it would appear that the counting systems of Sinagoro fall into one of these two types. Whether or not those dialects of Sinagoro which have the Motu type of system have been influenced by Motu proper and have thus adopted a system which they did not originally possess, is not known.

The numeral 10 is *gababana* or *galana* (System A). The higher decades, however, have the construction *gabana*+*n* where *n* takes the values 2 to 9 respectively. System B has *gala*+*n*. There is a distinct "hundred" morpheme, *sinau*. The numeral 100 is *sinauna* (or *tinauna*) and thereafter the centuries have the construction *sinau*+*n* where *n* takes the values 2 to 9 respectively. There is a distinct numeral 1 000 *dagarana* (or *dagalana*).

The CSQ informants and Kolia (1975, p. 147) both give data on ordinals. Kolia has *gonena* for first and thereafter the ordinals derive from the corresponding cardinals as in second, *valuana*; third, *vatoina*; fourth, *vavatina*; fifth, *vaimana*. The CSQ informants gave *guinena* for first, *vagaluana* for second, *vagatoina* for third, and so on, in each case using the prefix *vaga*- and the suffix -*na*. Kolia's ordinals closely resemble those of the Tolai language of East New Britain.

Magori

Magori is situated in the far east of Central Province on the coast. Magori is an AN language and with Ouma, Yoba, and Bina forms the Magori Section of the Central Family, Magori being the only one of these for which Lean had some data. Dutton (1973, p. 48) listed two Magori-speaking villages, Deba and Magori, which had a total population, at the 1980 National Census, of 142 (National Statistical Office, 1982). Wurm recorded 100 speakers in 2000. Table C29 provides a summary on the language.

Table C29
Magori Summary

Language type	Austronesian
Language family; subfamily	Magori Section, Central Family
Other names, dialects	Magori
Villages	
Cycles	5, 10
Operative pattern	6 to 9 is 5 + *n* where n is 1 to 4
Frame words	1 to 5, 10

In brief. Magori is an endangered language. The word for hand is used for 5.

A Magori vocabulary was published in the Annual Report for Papua 1917-18 (Bastard, 1919). The number data from this subsequently appeared in Ray (1938a, p. 184) grouped with the Papuan languages rather than the Melanesian (AN) languages. These data also appear in Kluge (1941, p. 23). The only other data available for Magori comprise several numerals taken from an SIL word list compiled in 1973. The data are given in Table C30.

Table C30
Magori Counting Words

	A	B	C
1	*daiton*	*deson*	*da,deson*
2	*dere*	*de*	*de*
3	*yampo*	*iampo*	*eampo*
4	*degede*	*de-gi-de*	*de-ge-de*
5	*nani yamunaet*	*nanin-emene*	*nanin-eamu*
6	*nani yamu daiton*	*nanin-emene-e-deson*	*nanin-eamu-tenua-deson*
7	*nani yamu dere*	*nanin-emene-e-de*	*nanin-eamu- tenua-de*
8	*nani yamu yampo*	*nanin-emene-tenua-*	*nanin-eamu- tenua- eampo*
9	*nani yamu degede*	*de-ge-de*	*nanin-eamu-tenua-de-ge-de*
10	*aonagaet (ao boen)*	*aunait*	*au-naed, au-boin*
11	*aonagaet pusinawan daiton*		
12	*aonagaet pusinawan dere*		
15	*aonagaet pusin yamunaet*		
16	*aonagaet pusin yamu daiton*		
20	*apane daiton*		*apane-de,aunaed-de*

Sources: A. Magori: Bastard (1919, p. 91), Ray (1938a, p. 184), Kluge (1941, p. 23). B. SIL Word List (1973).

The Magori number words certainly do not appear to be typically AN and it is not surprising that Ray has them grouped with his Papuan number data although with the note that the numeral 5, *ima*, is Melanesian (i.e AN). On the other hand the cyclic structure is not typically Papuan. We have distinct words for the first five numerals of which only the numerals 4 and 5, *wati* and *ima* could be taken to be AN; there is no trace of a 2-cycle. There is also a distinct numeral 10, *nanau*, and the numeral 20, *nanau bu-au*, i.e *tens two*, indicate the operation of a 10-cycle. We note that *nanau* is the "ten" morpheme in Magi. The construction of the numerals 6 to 9 indicate the operation of a 5-cycle; the construction is *lili+n* where *n* takes the values 1 to 4 respectively, and *lili* (or *lilia*) also is used to construct the corresponding numerals of Magi as in *ima+lilia+n*.

The question is: to what extent have Magori and the other members of the Magori Section, being the most proximate AN languages to Magi, influenced the Magi counting system, and vice-versa. Thomson (1975a, p. 56) suggested that "there was a good deal of influence exerted on Magi from the Magori group and the AN III languages to the east". In Lean's, 1992 discussion of Magi, several features of the Magi counting system which are not typically Papuan and which may have occurred as a result of AN influence, i.e the borrowing of the numeral 5 (or the word "hand"), *ima*; the existence of a distinct numeral 10, *nanau*, conceivably borrowed from Magori (or one of the Magori Section languages); the absence of a 2-cycle and the operation of a 10-cycle. On the other hand, Magori has at least one numeral, 3, *aiseli*, which may have been borrowed from Magi (or one of the Mailuan Family languages). There is, however, no reason to believe that any other numerals have been borrowed even though they do not appear to be typically AN. (Magori is a threatened language according to SIL Onagi, in a personal communication, told me that the Magori had requested that he make a dictionary for their language now that he had done one for his own Magi language. They were keen not to lose their language and identity though few in number and feeling somewhat dominated by their neighbour.). Why Magori-speakers should borrow one single NAN numeral is a mystery and, although on the face of it, this seems to be the case, it would be less of a mystery if *aiseli* was in fact a Magori numeral to start with and which has been borrowed by Magi. There is no evidence to support this view however.

(2, 5, 10) Cycle Patterns

Koita, Koiari, and Domu, which have counting systems that are 2-cycle variants with secondary 5-cycles but with tertiary 10-cycles. These are also members of the Central and South New Guinea Stock and are located in the Central Province. The first two are situated outside the Port Moresby region which is dominated by the AN Motu speakers while Domu is situated on the south coast near Cape Rodney and is adjacent to the region inhabited by the AN Keapara speakers (see the map in Figure E1).

Koita

Koita is a Papuan (NAN) language and a member of the Koiaric Sub-Family, the other members being Mountain Koiari and Koiari. Koita is spoken in a region which extends inland from the south coast and which runs from Redscar Bay in the west, where its Western dialect is spoken, to east of Port Moresby around which its Eastern dialect is spoken. There are enclaves of Motu-speaking villages within the Koita region which extend along the south coast and around Port Moresby. Wurm and Hattori (1981, Map 9) give the population of Koita speakers as 2 260 (possibly based on the 1966 Census figures). SIL (n.d.) give the number as 2 700 in 2000. A summary on the language is given in Table C31.

Table C31
Koita Summary

Language type	Papuan
Language family; subfamily	Koiaric Sub-Family, Koiarian Family
Other names, dialects	Koita
Villages	Papa, Kilakila, Kido.
Cycles	2, 5, 10, 100, 1000
Operative pattern	8=10-2?. 9=10-1?
Frame words	1, 2, 5, 6, 7, 10, 100, 1 000

In brief. Koita has (2, 5, 10, 100) cycles which are unusual in Papuan languages suggesting influence from Motu except for words for 100 and 1 000. It retains a hand morpheme for 5.

A comparative vocabulary of Koiari and Koita, referenced respectively as Koiari Goto and Koita Ga, was published in the Annual Report on British New Guinea from 1889 to 1890 (British New Guinea, 1890a, 1890b). The vocabularies of both languages, including number data, subsequently appear in Ray (1895) and a grammar in Ray (1907) based on notes taken by the Rev. Chalmers in 1880. Seligman's (1910) book has a discussion of the Koita people and Ray (1929, p. 81) has a set of Koita number data which derive from data given in the Cambridge Expedition and in the Annual Report.

Dutton (1969) had a discussion of the Koita people and their language with further work (Dutton, 1975) on the grammar including numbers. An SIL word list of Koita was compiled at Papa village in 1974 and this contains data for 1 to 5, and 10. In addition, three CSQs were received from Koita informants, two from the Eastern dialect area and one from the Western dialect area. A selection of Koita numerals is given in Table C32.

The data, on the whole, show remarkably good agreement indicating that little change seems to have occurred in the numerals over the last century. The Koita counting system has some interesting features somewhat untypical of a NAN language. It is apparent that both 2- and 5-cycles operate

Table C32
Koita Counting Words

	A	B	C	D
1	*kobuaiku*	*koubugabe, kobabe,be*	*kobabe*	*igagu, kobua be*
2	*abu*	*abu*	*abu*	*abu*
3	*abigaga*	*abugaga*	*abiga*	*abi gaga*
4	*abuabu*	*abuabu*	*ababu*	*ababu*
5	*adakasiva*	*ada kasiva*	*adakasiva*	*adakasiva*
6	*agorokiva*	*agorokiva*	*agorokiva*	
7	*yatirigava*	*yatirigava*	*iatirigava*	
8	*abuguveita*	*abuguveite*	*abu gu veite*	
9	*igaguveiti*	*ikaguveite*	*igau veite*	
10	*utube*	*utu be*	*utube*	*utube*
11	*utube hakira-koboga*		*utube hakira-kobabe*	
12	*utube hakira- abu*		*utube hakira-abu*	
15	*vakira-adakisiva*	*utu ada-kasiva*	*utube hakira-adakasiva*	
20	*uta abu*	*utu abu*	*utu abu*	
30	*uta abigaga*		*utu abiga*	
40	*uta abaabu*		*utu ababu*	
50	*uta adakasiva*		*utu adakasiva*	
100	*tinahube*		*tinahu be*	
1000	*dahabe*		*daha be*	

Sources: A. Koita Ga: British New Guinea (1890a, p. 131), Ray (1895, pp. 35-38, 1929, p. 81), Kluge (1941, p. 27). B. Koita: Dutton (1975, p. 353). C. CSQs. Villages: Papa, Kilakila, Kido D. SIL (1974). Village: Papa

within the system, which is commonly found in NAN languages. There are distinct words for the numerals 1 and 2; the numerals 3 and 4 have the construction 2+1 and 2+2 respectively. The number word for 5 is *ada kasiva* where *ada* is the word for hand. For 10 we have a distinct word *utu be (utube)*. The "ten" morpheme *utu* is then used to construct the other decades so that 20 is *utu abu* "tens two", and 30 is *utu abiga (uta abigaga)* "tens three", and so on. Thus a 10-cycle operates within the system which is somewhat unusual for a NAN language.

The numerals 6 to 9 are quite irregular and there is no apparent 5+*n* construction for these. There appears to be two distinct words for 6 and 7, *agorokiva* and *iatirigava* (*yatirigava*). The number words for 8 and 9 appear to contain, respectively, the numerals 2 and 1: System A has *abu-gu-veita* and *iga-gu-veiti*; System B has *abu-gu-veite* and *ika-gu-veite*; System C has *abu-gu-veite* and *igau veite*. This suggests that these numerals are formed by a subtractive process, two less, one less, also an unusual feature in a NAN system.

There are also distinct words for 100 and 1 000, respectively *tinahu be* and *daha be*. These, however, appear to have been borrowed from Motu, the corresponding numerals of which are *sinahu* and *daha*.

Koita, then, has the unusual cyclic pattern (2, 5, 10, 100, 1 000). It may be a reasonable hypothesis to suggest that the Koita counting system has been largely influenced by its pre-eminent neighbour Motu. If this is so, the influence has not been such that the original counting system has been displaced or that wholesale borrowing of numerals has occurred: this is only apparent with the numerals 100 and 1 000. The influence may be discerned in the introduction of a 10-cycle which is not normally a feature of a NAN counting system but which is present in Motu (although the "ten" morpheme for that language is *ahui*, the numeral 10 itself being *gwauta*). If this is the case then we have a counter-example to the cyclic regression, noted in Milne Bay which affects AN languages (possibly) under the influence of proximate NAN languages. The decimalisation of NAN counting systems has, however, been apparent in Yele (Milne Bay Province) and in Kuot (New Ireland).

Koiari

Koiari, or Grass Koiari to distinguish it from the language spoken by the mountain people, is a further member of the Koiaric Sub-Family. It possesses two dialects, Western and Eastern, and Wurm and Hattori (1981, Map 9) provide estimates of the speakers of these: respectively 1133 and 643 (based on the 1966 Census statistics). The language is spoken on the Sogeri plateau and southwards to the east of Port Moresby and to the southern coast. Around 2000, the population was 1700 (SIL, n.d.). A summary is given in Table C33.

Table C33
Koiari Summary

Language type	Papuan
Language family; subfamily	Koiaric Sub-Family, Koiarian Family
Other names, dialects	Koiari: 1.West, 2.East (N.E,S.E). Variant : Koiali, Koiari oto,Sogeri, Iarumi,Eikiri, Maiari, Favele.
Villages	D. Savetana.
Cycles	2, 5
Operative pattern	
Frame words	1, 2, 5

As indicated above in the section on Koita, a comparative vocabulary of Koita and Koiari was published in the Annual Report on British New Guinea from 1889 to 1890 (British New Guinea, 1890a, 1890b). The Koiari (Koiari Goto) vocabulary, including numerals, subsequently reappeared in Ray (1895). A short grammar of Koiari is given in Ray (1907) and this, too, has some number data. Ray's article of 1929 has data on Koiari referenced as Koiari, Sogeri, Larumi, Eikiri, Maiari, and Favele, and there is some number data on each of these. Apart from the data appearing in these published sources, there is also some number data given in an unpublished SIL word list compiled at Savetoma village in 1973. A selection of Koiari data is given in Table C34.

Table C34
Koiari Counting Words

	A	B	C	D
1	*igau*	*igau*	*igau,igane*	*igau*
2	*abuti*	*abuti*	*abute,(abuia)*	*abuti*
3	*abuti-igau*	*buti-gati*	*abuti-a-igau (abui-igani)*	*abuti la'iuta*
4	*abuti-abuti*	*abuti-ta-abuti*	*(obobova)*	*abuti-abuti*
5	*abuti-abuti-igau*	*adafakibi*	*(adavaimore)*	*ada fakibe*
6	*abuti-abuti-abuti*	*fakibi-igau*		
7	*abuti-abuti-abuti-igau*	*adahakibi-abuti*		
8		*adahaki-fakibe abuti ta abuti*		
10	*obua*		*adavaimore-adavaimore*	*ada fakibe fakibe*
11	*obua igau*			
12	*obua abuti*			
20	*obuaobua*			

Sources: A. Ioiari Goto: British New Guinea (1890a, p. 131), Ray (1895, pp. 35-38, 1929, p. 81), Kluge (1941, p. 27).B. Larumi: Ray (1929, p. 81), Kluge (1941, p. 27). C. Sogeri: Ray (1929, p. 81), Kluge (1941, p. 27). D. SIL (1973). Village: Savetana.

System A has a 2-cycle operating for the construction of the numbers 1 to 9 and there exists a distinct word for 10, *obua*. The other systems also have a 2-cycle but in each case a 5-cycle comes into operation, the number words for 5 being related to the word for "hand" *ada*. Systems C and D both have number words for 10 which appear to have 5+5 (or hand-hand) construction.

The number word for 8 in System B, *adahaki-fakibe abuti ta abuti*, has a 5+2+2 construction and thus appears to be an error.

Domu

Domu is a Papuan language and a member of the Mailuan Family which is spoken in the Cape Rodney area immediately to the east of the Keapara region (see map in Appendix E, Figure E1). No dialects are given for Domu in Wurm and Hattori (1981 Map 9). Dutton (1973, p. 21) listed three villages (or census units) which are Domu-speaking and these, at the 1980 National Census, had a total population of 479 (National Statistical Office, 1982). In 2000, there were 950 (SIL, n.d.). A summary is given in Table C35.

Table C35
Domu Summary

Language type	Papuan
Language family; subfamily	Mailuan Family
Other names, dialects	Domu : Merani, Dom
Villages	D SIL Dom. CSQ Tutubu
Cycles	2', 5, 10
Operative pattern	6 to 9 is 5+n where n = 1 to 4
Frame words	1, 2, 3, 5

In brief. Domu has a modified 2 cycle with distinct numeral for 3 but 4=2+2, 5 cycle with 6 to 9 being 5 + *n* where *n* is 1 to 4. 10 cycle is suggest by "one ten" for 10 which is unusual for a Papuan system. 5 contains a morpheme for "arm", not "hand".

Two Domu vocabularies, including number data, are given in the Annual Reports for Papua, one in the 1917-18 Report referenced as 'Merani' and the other in the 1918-19 Report referenced as Dom (Bastard, 1919, 1920). Both of these are included in Ray (1938a) and the number data from the vocabularies are found in Kluge (1941, p. 21). The only other Domu data available are taken from an SIL word list compiled at Dom village in 1975, and from one CSQ. The data are presented in Table C36. The three systems are in agreement regarding the first five numerals. There are distinct words for the first three numerals but the numeral 4 has a 2+2 construction and thus we have a modified 2-cycle operating. The number word for 5 contains a morpheme for "arm" (not "hand") *eni*.

Systems B and C indicate that the number words for 6 to 9 have an $x+n$ construction where n takes the values 1 to 4 respectively. For System A, this pattern extends to 10 and we thus also have a 5-cycle operating within the system. System B has *pua mi* for 10 (System C does also) and *pua hauna* for 20 (System A has *puhauna*) so that *pua* appears to be a "ten" morpheme which is combined with *mi,* derived from the numeral 1, *miau,* to form "ten one", and which is combined with the numeral 2, *hauna,* to form the numeral 20, i.e "ten two". We thus also have a 10-cycle operating which is an unusual feature for a Papuan language.

Table C36
Domu Counting Words

	A	B	C
1	*mi-au*	*mi-au*	*miau*
2	*hauna*	*hauna*	*hauna*
3	*aiseni*	*aisen*	*aiseni*
4	*hauna hauna*	*hauna hauna*	*hauna hauna*
5	*eni-apu*	*eni apu*	*eniapu*
6	*epiminu-apu-fogai*	*pogai mi*	
7	*fogai hauna*	*pogai hauna*	
8	*fogai aise*	*pogai aisen*	
9	*fogai hauna hau*	*pogai hauna hauna*	
10	*fogai eni-apu*	*pu ami (pua mi)*	
20	*puhauna*	*pu ha hauna (pua hauna)*	

Sources: A. Merani: Bastard (1919, p. 88), Ray (1938a, p. 183), Kluge (1941, p. 21). B. Dom: Bastard (1920, p. 112), Ray (1938a, p. 183), Kluge (1941, p. 21). C. SIL (1975). Village: Dom. 1 CSQ Village: Tutubu.

Motu

Motu, probably the best documented language of the Central Province, is an Austronesian language spoken in about 14 villages scattered sparsely along one hundred kilometres of the south coast of the mainland from Galley Reach eastwards to the Sinagoro region (see Map in Appendix E, Figure E1). Pawley (1976, p. 303) indicated that the lingua franca is spoken by close to 200 000 people whilst 'pure' Motu is spoken by about 14 000 (as at the mid-1970s). Although Wurm and Hattori (1981, Map 9) do not distinguish dialects for Motu, work by Taylor in the 1970s suggests that at least two dialects exist, Eastern and Western, and possibly a further two. In 2008, Taylor suggested 39 000 speakers and Holmes in 1989 that there were 120 000 using Hiri Motu as a second language (SIL, n.d.). A summary of information on the language is given in Table C37.

Table C37
Motu Summary

Language type	Austronesian
Language family; subfamily	Motu-Sinagoro Section, Eastern(Motu) Sub-Family, Central Family
Other names, dialects	Motu: 1.Eastern, 2.Western
Villages	Hanuabada, Porebada,etc
Cycles	10, 100, 1000 etc
Operative pattern	6 to 9 =2x3,2x3+1, 2x4, 2x4+1
Frame words	1 to 5, 10, 100, 1 000, 10 000

In brief. Language used as a lingua franca in Papua. While retaining a 5 cycle system it has the form of 6=2x3, 7=2x3+1, 8=2x4, 9=2x4+1. It has distinct numerals for 10, 100, 1 000, 10 000, 100 000. Prefixes are used for counting different classes, e.g. men, women, long things, coconuts, animals. Ordinals are obtained by prefixing cardinal numbers for second etc.

Lawes' (1885) grammar and vocabulary is the earliest major work on Motu. The grammar notes contain a section on numerals and his later work includes some comparative vocabulary for several languages (Lawes, 1890). A modified version of this appeared in subsequent editions of Lawes' Motu Grammar and in Ray (1895) and in Ray (1907), the latter giving some grammatical features and having a section on numerals. Ray (1929) also contained Motu data including a set of numerals, and these subsequently appear in Kluge (1941, p. 208). A further grammar and dictionary of Motu was published by Lister-Turner and Clark (1931) with subsequent editions revised by Sir Percy Chatterton. In addition to the considerable published data available, 75 informants from the various Motu villages completed CSQs. The data are presented in Table C38.

Lean defined the type of counting systems with pairs as the Motu type. Referring to Table C38, the first five numerals are recognisably Austronesian in character. The numerals 6 and 8 are formed by the doubling of the numerals 3 and 4 respectively. Motu has *tauratoi* for the numeral 6 which has a 2x3 construction, and *taurahani* for the numeral 8 which thus has a 2x4 construction. Motu, though, unlike Roro, has a distinct numeral 7, *hitu*. Roro has the numerals 7 and 9 having the constructions 2x3+1 and 2x4+1 respectively. Motu has the same feature for the numeral 9, *taurahani ta*.

Table C38
Motu Counting Words

	A	B
1	*tamona*	*ta*
2	*rua*	*rua*
3	*toi*	*toi*
4	*hani*	*hani*
5	*ima*	*ima*
6	*tauratoi*	*tauratoi*
7	*hitu*	*hitu*
8	*taurahani*	*taurahani*
9	*taurahani ta*	*taurahani ta*
10	*qauta*	*gwauta*
11	*qauta ta*	*gwauta ta*
20	*rua ahui*	*ruahui*
30	*toi ahui*	*toi ahui*
50	*ima ahui*	*imahui*
60	*tauratoi ahui*	*tauratoi ahui*
70	*hitu ahui*	*hitu ahui*
80	*taurahani ahui*	*taurahani ahui*
100	*sinahu*	*sinahu ta*
200	*sinahu rua*	*sinahu rua*
1 000	*daha*	*daha ta*
10 000	*gerebu*	
100 000	*domaga*	

Sources: A. Motu: Lawes (1895, p. 8), Lawes (1890, pp. 158-167), Ray (1895, pp. 35–38), Ray (1907, p. 470; 1929, p. 79), Kluge (1941, p. 208). B. 75 CSQs. Villages: Hanuabada, Porebada, etc

Motu has a distinct numeral 10, *gwauta*, which is then used to construct the numerals 11 to 19. The remaining decades 20 to 90 do not, however, contain *gwauta* in their construction but have a separate "ten" morpheme *ahui* so that their construction is *n*+*ahui* where *n* takes the values 2 to 9 respectively. However, Owens in 2014 during fieldwork found in Tubusereia that the word for a group of a specific kind of object was used for 10 rather than a generic numeral.

There is a distinct "hundred" morpheme, *sinahu*, and the centuries 100 to 900 have the construction *sinahu*+*n* where *n* takes the values 1 to 9 respectively. There is also a distinct "thousand" morpheme *daha* and the thousands from 1 000 to 9 000 have the construction *daha*+*n* where *n* takes the values 1 to 9 respectively. Lawes' data also give distinct morphemes for both "ten thousand" *gerebu*, and "one hundred thousand" *domaga*.

The CSQ informants indicate the existence of ordinals and give "first" *gini guna* or *gini gunana*. The subsequent ordinals derive from the corresponding cardinals by prefixing the cardinal with *ia*- and suffixing it with -*na*. The ordinal "second" is thus *iaruana*, "third" *iatoina*, and so on.

Lawes (1885) also gave examples of the modification of numerals when certain objects are counted. He noted, for example:

> Persons take a prefix in low numbers, as, *ra, tau rarua*, 2 men; *ta, hahine tatoi*, 3 women; *ha, kekeni hahani; la, memero laima*, 5 boys. It does not go beyond 5, but is used with those units with tens, as, *aposetolo qauta rarua*, the twelve apostles.
>
> Things of length, such as spears, poles, etc., are counted differently. The numerals have *au* prefixed, as, *auta, aurua*, and so on up to nine, which is *autaurahaniauta*, and ten, *atalata*. After ten they are (expressed as in Table C39.)
>
> Of fish, pigs, and wallaby, the ordinary numerals are used up to ten, which is *bala ta*; 20, *bala rua*; beyond 20, the common numerals are used, as *toi ahui, hani ahui,* up to 100, which is the same as ordinary, *sinahu*. Coconuts are counted by *varo* (strings), as, *varo ta*, 10; *varo rua*, 20, etc. (p. 9)

Table C39
Motu Counting Prefixes

20	*rabu rua*
30	*rabu toi*
40	*rabu hani*
50	*rabu ima*
60	*rabu tauratoi*
70	*rabu hitu*
80	*rabu taurahani*
90	*rabu taurahani ta*
100	*sinahu*

Keapara

Immediately to the south of the Sinagoro region lies the Keapara area. Keapara is an Austronesian language and comprises some nine dialects or communalects spoken along a group of largely coastal villages which extend from the Hood Bay area, where the Hula dialect is spoken, eastwards to Cape Rodney where the Lalaura dialect is spoken (see Map in Appendix E, Figure E1). Pawley (1976, p. 302) indicated that Keapara is spoken by more than 16 000 people but this estimate appears to be based on the 1966 Census statistics (see Dutton, 1970, p. 901). Dutton (pp. 900-01) listed the Keapara-speaking villages (or Census units) and the population of these at the 1980 National Census was 14 605 (National Statistical Office, 1982). Tryon (2006) gave the figure of 19 400 in three villages. There was a number around 14 7000 just in the villages of the dialect names in the 2011 census. In total in the Rigo Coast and Aroma Wards there were 25 966 and 28 639 people respectively (National Statistical Office, 2014b), possibly including other languages. Language information is summarised in Table C40.

Table C40
Keapara Summary

Dialects :	1.Hula, 2.Babaga, 3.Kalo, 4.Keapara, 5.Aloma, 6.Maopa, 7,Wanigela, 8.Kapari, 9.Lalaura.
Variant Names:	Aroma, Galoma, Kerepunu, Bula'a, Keakalo, Karo, Loyalupu.
Classification:	AN, Motu-Sinagoro Section, Central Family.
Cycles	10, 100 1 000 etc
Operative patterns	Motu type: 6=2x3, 7 = 2x3+1 etc,
Frame words	1 to 5, 10,

In the Annual Report on British New Guinea from 1889 to 1890, Lawes' (1890) comparative vocabulary of New Guinea Dialects includes two examples of Keapara referenced as Kerepunu and Aroma. A revised version of this vocabulary appears in Lawes' Motu Grammar (3rd Ed.), published in 1896, with the two Keapara vocabularies referenced Keapara and Galoma. Further Keapara vocabularies, including numerals, appear in several other Annual Reports. Guise (1892) had one referenced Bula'a; another one, referenced Keakalo, is in the Annual Report for 1892-93 (F. Lawes, 1894). Ray (1895) had these various vocabularies (and numerals), i.e Bula'a, Keapara, Kerepunu, Aroma, and a further one, Hula. Ray (1907, p. 470) had several sets of Keapara numerals, deriving from the above sources, referenced Hula, Keapara, Galoma and Keakalo. These reappear in Ray's, 1929 article and also in Kluge (1941, p. 208). A Keapara vocabulary appears in the Annual Report for Papua, 1918-19, referenced as Karo (Bastard, 1920) and this reappears in Ray (1938a) under the same reference. Kluge (1941, p. 210) also had the Karo numerals.

Dutton's (1970) article on the languages of the Rigo district contains a comparative vocabulary of these and includes Keapara. In addition to the various published sources, numeral data are available in three SIL word lists of, respectively, the Kalo dialect (1974), the Aloma dialect (1975), and the Babaga dialect (1976). Fifty-five CSQ informants from the various Keapara dialect areas also provided data on numerals. A selection of data is given in Tables C41 and C42.

Eight dialect groups were represented by the CSQ informants (the exception being the Lalaura dialect). Of the eight, five were selected for inclusion in the data tables and these give a reasonable representation; the excluded counting systems are not markedly different from those shown.

It is apparent that there are important differences, in some respects, between the counting systems of the five Keapara dialects shown in Table C41 (Systems A to D) and in Table C42 (System A). With regard to the first five numerals, the only one which is the same for all dialects is the numeral 5,

Table C41
Keapara Counting Words

	A	B	C	D
1	*kopuna*	*kwauna*	*opuna*	*apuna*
2	*lualua*	*roula*	*lualua*	*lualua*
3	*koikoi*	*toitoi*	*oioi*	*oioi*
4	*vaivai*	*vativati*	*vaivai*	*vaivai*
5	*imaima*	*imaima*	*imaima*	*imaima*
6	*kaulakoi*	*tuluatoi*	*aula oi*	*uala oioi*
7	*mapere kaulavai*	*tuluatoi kwauna*		*aula oioi apuna*
8	*kaulavai*	*tolavativati*	.	*aula vaivai*
9	*mapere ka galana*	*tolavativati kwauna*	.	*aula vaivai apuna*
10	*galana*	*galana*	*galana*	*kapanana*
11	*galena kopuna*	*galena kwauna*	*galena opuna*	*kapanana apuna*
12	*galena lualua*	*galena roula*	*galena lualua*	*kapanana lualua*
15	*galena imaima*	*galena imaima*	*galena imaima*	*kapanana imaima*
20	*gala lualua*	*gala roula*	*galu lualua*	*kapana lualua*
30	*gala koikoi*	*gala toitoi*	*gala oioi*	*kapana oioi*
40	*gala vaivai*	*gala vativati*	*gala vaivai*	*kapana vaivai*
100	*inavuna*	*(tinauna)*	*inavuna*	*inavuna*
200	*inavu lualua*	.....	*ina vu lualua*	*inavu lualua*
1000	*rana*	*rana*	*ragana*	

Sources: A. Hula d.: 17 CSQs. Village: Hula (17). B. Babaga d.: 5 CSQs. Villages: Babaka (4), Makerupu (1). C. Keapara d.: 4 CSQs. Villages: Karawa (3), Keapara (1). D. Kapari d.: 5 CSQs. Village: Kapari (5).

Table C42
Keapara Counting Words

	A	B	C
1	*kwapuna*	*ka,koapuna*	*abuna*
2	*ruala*	*lualua*	*lualua*
3	*toitoi*	*koikoi*	*oioi*
4	*vativati*	*vaivai*	*baibai*
5	*imaima*	*imaima*	*imaima*
6	*taula toitoi*	*kaula koi*	*aula oi*
7	*taulatoitoi kwapuna*	*mapere aula vaivai*	*aula oi wabuna*
8	*taulavativati*	*kaula vaivai*	*aula vaivai*
9	*taulavativati kwapuna*	*mapere ka gahalana*	*aula wai wabuna*
10	*gagalana*	*gahalana*	*kapanana*
11	*gagalana kwapuna.....*	.....	
12	*gagalana ruala*		
15	*gagalana imaima*		
20	*gagala ruala*		
30	*gagala toitoi*		
40	*gagala vativati*		
100	*tinavuna*		
200	*tinavu ruala*		
1000	*ragana*		

Sources: A. Kalo d.: 7 CSQs. Village: Kalo (7). B. Bula'a: Guise (1892), Ray (1895, pp. 35-38), Kluge (1941, p. 218). C. Galoma and Keakalo: Ray (1929, p. 79), Kluge (1941, p. 208).

imaima. The word for "hand", incidentally, is *gima* (for most dialects) or *ima*. For most dialects the numeral 2 is *lualua* but Babaga has *roula* and Karo has *ruala*. The numeral 3 is *oioi* (Keapara and Kapari), *toitoi* (Babaga and Kalo) or *koikoi* (Hula). The numeral 4 is generally *vaivai* but Babaga and Kalo both have *vativati*.

The numeral 10 is generally given as *galana* or *gala-na*, the "ten" morpheme being *gala* which is used to construct the higher decades. The numeral 10 in the Kapari dialect, however, is *kapanana*, similar to the corresponding Sinagoro numeral and the Kalo dialect has *gagalana*. In each case the *-na* suffix indicates "one" ten.

There is a distinct numeral 100, generally *inavuna* or *inavu-na* where *inavu* is the "hundred" morpheme. The numeral 200 is *inavu inalua*. The Kalo dialect has *tinavuna* for the numeral 100 and one CSQ informant from the Babaga dialect area gave *tinauna*, similar to the corresponding Sinagoro numeral. There is also a distinct numeral 1 000 which is *rana* in the Hula and Keapara dialects and *ragana* in the Kapari and Kalo dialects.

The most striking differences between the various counting systems, however, occur in the construction of the numerals 6 to 9. The majority of the Keapara dialects have counting systems of the Motu type. Babaga, Kapari, and Kalo, for example, all have numerals 6 to 9 with the constructions 2x3, 2x3+1, 2x4, and 2x4+1 respectively; unlike pure Motu there is no distinct numeral 7. The Hula dialect, however, is an unusual variant of the Motu type; whilst both the numerals 6 and 8 are formed by the doubling of the numerals 3 and 4 respectively, the numerals 7 and 9 are both formed by a subtractive process, the numeral 7 having the (implied) construction 2x4-1 and 9 the construction 10-1. We have not previously encountered this type of construction for the numeral 7; the use of subtraction from 10 to form the numeral 9 is, however, a feature of the Manus type of counting system where 7, 8, and 9 have, respectively, the constructions 10-3, 10-2, 10-1.

The Hula dialect informants give consistent data regarding ordinals. "First" is given as *kune* which is not derived from the corresponding cardinal. The remaining ordinals do, however, derive from their corresponding cardinals by prefixing these with *va-* and suffixing them with *-na* so that "second", for example, is *va-lualua-na* and "third" is *vakoikoina*, and so on.

The CSQ informants also indicate that special words are used for counting certain objects in bunches or groups of ten. For example, the Hula dialect informants indicate that in counting ten coconuts tied together there is a "ten" morpheme *walo-* used to construct the counting words as in *walo-na*, "one group of ten" *walo-lua* (or *walo-lualua*), "two groups of ten" *walo-koi* (or *walo-koikoi*), three groups of ten, and so on. For sago the "ten" morpheme is *gewa-*, for bananas it is *gi-* or *gii-*, and for betel nut it is *gule-*.

Sko and Sangke

Sko and Sangke are languages spoken in West Papua North Coast close to the PNG border. See the map of this area in Appendix E, Figure E4. Language information is given in Table C43.

Table C43
Sko and Sangke Summary

	Sko	*Sangke*
Variant Names:	Tumawo, Tumavo, Seka, Sakou.	
Classification:	NAN, Sko Family, Sko Phylum- Level Stock.	NAN, Sko Family, Sko Phylum- Level Stock.

Table C44 provides the counting words for Sko (two differing sets) and Sangke.

Table C44
Sko and Sangke Counting Words

	Sko A	Sko B	Sangke
1	*hi*	*ahi*	*ofa*
2	*hintung*	*hitjun*	*hime, njimu*
3	*hentong*	*hetun*	*njena*
4	……	*nabu*	*noug*
5		*naplan*	*ui*
6		*naplahi*	*notji-o*
7		-	*notji-nje*
8		-	*notji-njena*
9		-	*notji-noug*
10		*amplahari*	*ui-tjawi*
20		*amplanaplan*	*no-wa-kna*

Sources: A. Sko: Galis (1960, p. 145) B. Seka: Ray (1912a, 1912b, pp. 340-341); Kluge (1938, p. 179). *Sangke*: Galis (1960, p. 145).

Fas, Baibai, Biaka, and Kwomtari

The Fas-speaking region stretches from the coast, not far from the Sera-speaking villages, both inland and to the west in the West Sepik (Sandaun) Province. Laycock (1973, p. 43) listed 19 villages in which the language is spoken and at the 1980 National Census these had a total population of 1 596 residents (National Statistical Office, 1982). The 2000 census had 2 500 (SIL, n.d.).

Baibai is spoken in three villages (Laycock, 1973, p. 43.) located in a region to the south of the Fas area and to the west of the Kwomtari region. The three villages listed by Laycock had a total population of 286 at the 1980 National Census (National Statistical Office, 1982). The 2000 census recorded 340 (SIL, n.d.).

Laycock (1973, p. 43) listed 11 Kwomtari-speaking villages which are located in a region immediately to the east of the Baibai region. At the 1980 National Census these villages had a total population of 920 (National Statistical Office, 1982). SIL (n.d.) gave six villages has having 600 based on some 1988 SIL data.

Three Biaka-speaking villages are listed by Laycock (1973, p. 43) and these lie in a region to the south of the Kwomtari area, adjacent to the Anggor-speaking region. At the 1980 National Census these three villages had a total population of 482 (National Statistical Office, 1982). See Appendix E, Figure E13 for the position of these languages.

Wurm and Hattori (1981, Map 6) have Fas and Kwomtari classified as a member of the Kwomtari Family, Kwomtari Stock and Baibai and Biaka as being a member of the Kwomtari Stock, Baibai Family. Table C45 summarises the language information for these four languages.

Table C45

Fas, Baibai, Biaka, and Kwomtari Summary

Language	Fas	Baibai	Biaka	Kwomtari
Language type	Papuan	Papuan	Papuan	Papuan
Language family; subfamily		Baibai Family, Kwomtari Stock.	Baibai Family, Kwomtari Stock.	Baibai Family, Kwomtari Stock.
Villages	Bembi.			
Cycles	2, 5	?	5	?
Operative patterns	10=5+5	Body parts?	10=5+5	Body-parts
Frame words	1,2,5	1 to 5	1 to 5	1 to 5

The only data available for Fas, Baibai, and Biaka derive from SIL word lists (undated). These are given in Table C46. With the limited amount of data available the Fas system may nevertheless be seen to possess a (2, 5) cyclic pattern (or a modified 2-cycle: the Fas number word for 3 is uncertain). There are distinct words for 1 and 2 and 4 has a 2 + 2 construction. The number word for 5 is *kaipupamo* which contains a "hand" morpheme *kai*; 10 has the construction 5 + 5. It is thus possible that the Fas system is a digit tally one, superimposed over pair-counting.

Table C46

Fas, Baibai, Biaka, and Kwomtari Counting Words

	Fas	Baibai	Biaka	Kwomtari, A	Kwomtari, B
1	*pasakandai*	*giesa*	*maime?*	*mamo?*	*mitene*
2	*tienabam*	*naebaeia*	*arame?*	*ara?*	*aser*
3		*nambakiesa*	*magope?*	*kawulira*	*amdir*
4	*ienabam tienabam*	*nambakaros*	*amitai?*	*kawuwari?*	*niso*
5	*kaipumamo*	*anayuwa (?)*	*yaritu?*	*ipu?*	*akmo*
10	*kaipupamo kaipupamo*		*yaritu? yaritu?*		

Sources: Fas: SIL Word List (undated). Baibai. SIL Word List (undated). Biaka: SIL Word List (undated). Kwomtari A.SIL Word List (undated). Village: Mango? (Wesono?). Kwomtari B.: 1 CSQ. Guriaso Community School.
Note: /?/ - glottal stop.

There is no discernible cyclic pattern in the few number words given for Baibai. The word for 5 is *anayuwa* and this is not related to the word for "hand" *`degi*. It is possible, therefore, that this is not a digit tally system but may possibly be a body-part tally one, the first few words given being the names of fingers.

Kwomtari data derive from an SIL word list (undated). In addition we have a further set of number words which were obtained from a CSQ completed by an informant from Guriaso Community School which is situated in the eastern part of the Kwomtari region. Loving and Bass (1964, p. 2) noted that significant dialectical variations occur within Kwomtari and this may partly account for the difference between the SIL data and the CSQ data, the former deriving from the western part of the Kwomtari region.

Laycock (1973, p. 42) noted of the Kwomtari Phylum languages in general that: "Number systems tend to be binary or body-parts, with a mixed system in Kwomtari (decimal counting on two hands?)". The CSQ informant indeed does indicate that a decimal system operates (i.e. a 10-cycle system with no sub-ordinate cycles) but at the same time says that the system is a body-part tally one that utilises tally parts on the body as well as the fingers of both hands. From the limited amount of data given in the SIL word list it is not possible to discern a definite cyclic structure. We may note, however, that the number words for 3 and 4 both contain the prefix *kawu-* and also that the number word for 5, *ipu?*, is not identical to the word for "hand" *ame?*. This latter point suggests (but, of course, does not show) that the system is not a digit-tally one. The difference between Systems A and B is sufficiently great to suggest that it may not be due to dialectical differences alone.

For Biaka, there are five distinct number words for the first five cardinals with no indication of a 2-cycle. The number word for 5, *yaritu?*, is not identical to the word for "hand" *ame?*. The number word for 10 is *yaritu? yaritu?*, i.e. 5 + 5. While this apparent 5-cyclic structure suggests that the system may be a digit-tally one, it is also possible that we have an example of a body-part system which has been modified or truncated as a result of the influence of the English or Tok Pisin decimal systems. Further data, however, are needed to confirm or deny this.

Rocky Peak, Iteri, Bo, Nimo

Rocky Peak is spoken in a small region of the Rocky Peak mountains situated in the extreme west of the East Sepik Province. Conrad and Dye (1975, p. 9) indicated that the Rocky Peak-speaking villages comprise: "Iwau, Agrame and Uwau and at least two other villages". Agrame and Uwau had a total population of 121 at the 1980 National Census (National Statistical Office, 1982). See Appendix E, Figure E13 for the position of these languages. Iteri is spoken in a region adjacent to the Rocky Peak area in the extreme west of the East Sepik Province. Laycock (1973, p. 45) had a brief mention of a language Yinibu which is spoken in about six villages, one or more of which may be Iteri-speaking. Conrad and Dye (1975, p. 9) listed five Bo-speaking villages (none of which are listed as Census Units for the 1980 Census) which are situated in West Range in the extreme west of the East Sepik Province. Conrad and Dye (1975, p. 9) indicated that Nimo, also a member of the Arai Phylum-Level Family, possesses at least two dialects spoken in a total of about seven villages situated in the Left May River area in the extreme west of the East Sepik Province. SIL (2003) have 6 900 speakers for Rocky Peak – Iyo and Iteri as 480 while NTM in 1998 has Bo – Right May river as 85 and Nimo – Left May as 350 (SIL, n.d.). Table C47 summarises the language information for these four languages.

Table C47
Rocky Peak, Iteri, Bo and Nimo Summary

Language	Rocky Peak	Iteri	Bo	Nimo
Language type	Papuan	Papuan	Papuan	Papuan
Language family, subfamily	Arai (or Left-May) Family	Arai (or Left May) Family	Arai (or Left May) Family	Arai (or Left May) Family
Other names, dialects	Laro, Iyo, Yinibu	Yinibu	Po	1. Nimo-Wasnai d., 2. Nakwi d.
Villages	Iwau			
Cycles	?	?	?	?
Operative patterns	?	?	?	?
Frame words	1 to 5, ?	1 to 5?, ?	1 to 5?, ?	1 to 5?, ?

Note. No body morphemes were given to indicate whether it might be a digit tally or body-part tally system

Number data for Rocky Peak derive from two SIL word lists: one (unpublished) compiled at Iwau village (undated), and the other published in Conrad and Dye (1975). The data are given in Table C48. Without further data it is not possible to determine precisely the nature of the Rocky Peak system. The word for "hand" *nai* does not appear in either the number words for 5 or 10 as we might expect for a digit-tally system. The word for "thumb" *namulu* is also not related to the number word for 5; it is sometimes the case with body-part tally systems that the two are identical. There is insufficient data for Iteri, Bo and Nimo to decide the system.

Table C48
Rocky Peak, Iteri, Bo and Nimo Counting Words

	Rocky Peak: A	Rocky Peak: B	Iteri	Bo: A	Bo: B	Nimo
1	*soso*	*suso*	*sus(ae)s(ae)*	*soso*	*soso*	*siaesa*
2	*tiso*	*tiso*	*lis(ae)?*	*tisa*	*tisa*	*ti:*
3	*douso*	*touso*	*taus(ae)?*	*tousa*	*tousa*	*to:to*
4	*neneiso*	*nineso*	*ninais(ae)*	*aisa*	*aisa*	*eyi*
5	*noso*			*nosa*		
10	*nanis*			*nanisa*		

Sources: Rocky Peak A. SIL Word List (undated). Village: Iwau. B. Conrad and Dye (1975, pp. 21-24). Iteri, Bo: B and Nimo from Conrad and Dye (1975, pp. 21-24). Words in Bo: A for 5 and 10 were given in an SIL word list (1972).

Mianmin

The language is spoken in a sparsely populated region in the southern part of the West Sepik (Sandaun) Province. An estimate of the resident Mianmin-speaking population, based on the 1970 Census, is about 2 000. SIL (n.d.) give 1 400 based on the 2000 census. Table C49 summarises Mianmin's language information.

Mianmin is classified as a member of the Ok Family (Mountain Ok Sub-Family) and is given as having at least two dialects: Wagarabai (or North Mianmin) and South Mianmin (Wurm & Hattori, 1981, Map 6). Laycock (1973, p. 50) treated these as separate languages called, respectively, Wagarabai and Mianmin. A study of the Ok Family of Languages, including Mianmin, was carried out by Healey in the early 1960s, the details of which appear in his doctoral dissertation (Healey, 1964). The data obtained on Mianmin counting derive from grammar notes on the language by SIL workers, J. Smith and Weston (1975) who noted that:

> counting begins with the left thumb, followed by the fingers of the left hand; then up the left-hand side of the body and down the right. In coming down the right side, people become

Table C49
Mianmin Summary

Language type	Papuan
Language family; subfamily	Ok Family, Mountain Ok Sub-Family.
Other names; dialects	Mianmin: 1. North Mianmin, 2. South Mianmin.Wagarabai.
Villages	
Cycles	2, 27
Operative patterns	Body parts
Frame words	1, 2, 7 to 14

In brief. Mianmin has a body-part tally system with 27 tally points and a 2 cycle system. The words for the first 6 numbers in the 2-cycle system are the first words of the body-part tally system in one data source.

vague in their counting, and the word for "many" *homon* is used in preference. In fact, *homon* is commonly used for anything over five. (p. 50)

In the Mianmin body-part tally system (see Table C50), the first six number words belong to a 2-cycle numeral system and do not denote the names of the body-parts. These are, nevertheless, tallied on the fingers beginning with the thumb of the left hand and (presumably) proceeding along the index, middle, ring, and little fingers, 6 being tallied on the left wrist. Thereafter, the body parts are used for tallying.

Table C50
Mianmin Body-Part Tally System

	Part 1	2-cycle system
1	*ele-yem*	this-alone
2	*asu*	two
3	*asu-matna*	two-one more
4	*asu-ke asu-ke*	two-and two-and
5	*asu-ke asu-ke mak-e*	two-and two-and other
6	*asu-ke asu-ke asu-ke*	two-and two-and two-and
	Part 2	Body-Part Tally
7	*ban-lim*	forearm-on
8	*hetlefab*	inner elbow
9	*tumin*	shoulder joint
10	*nakal* or *kwing-lim*	shoulder, shoulder-on
11	*tam-lim*	side of face-on
12	*klon-lim*	ear-on
13	*kin-lim*	eye-on
14	*munung-lim*	nose-on
15	*kin-milim*	eye-other side
16	*klon-milim*	ear-other side
17	*tam-milim*	side of face-other side
18	*nakal-milim*	shoulder-other side
19	*tum-milim*	shoulder-joint-other side
20	*hetlefab-milim*	inner elbow-other side
21	*ban-milim*	forearm-other side
22	*gong-milim*	wrist-other side
23	*kweit-awok-milim*	hand-thumb-other side

Sources: A. Smith and Weston (1975, pp. 50-52).

Although Smith and Weston give no data beyond the tally word for 23, it seems likely that the original Mianmin system possessed a 27-cycle and utilized 13 body-parts on the left-hand side of the body, the nose, which is the mid-point of the cycle (lying on the body's vertical axis of symmetry), and then a further 13 body-parts on the right-hand side of the body, the symmetrical counterparts of the first 13. See Figure C1.

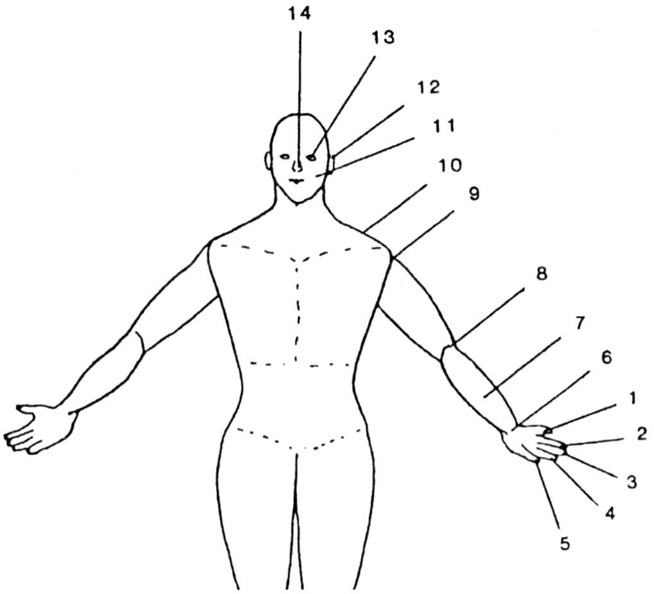

Figure C1. Mianmin body-part tally system to 14.
Source: Lean (1992).

Similar examples of body-part tally systems in which the first 4, 5 or 6 number words belong to a 2-cycle numeral system, thereafter the tally words denoting parts of the body occur also in Western Province. It has been suggested that this type of system is a hybrid one, that certain cultures may possess, initially, both a simple 2-cycle numeral system and a body-part tally system, and that subsequently the system becomes such that the first few numerals displace the first few body-part words (those for fingers) thus producing the hybrid.

Musian and Amto

Musian is classified, with Amto, as belonging to the Amto-Musian Phylum-Level Family; it is spoken in a village near the Abau region (Wurm & Hattori, 1981, Map 6; Laycock, 1973, p. 53) (see Appendix E, Figure E13). At the 1980 National Census, the Musian village, Seiawi, had a resident population of 111 (National Statistical Office, 1982) and by 2011, a population of 312 (National Statistical Office, 2014a). Amto is spoken in three villages (Laycock, 1973, p. 53) situated in the Rocky Peaks Census Division and near the southern part of the Abau region. Of the three villages listed by Laycock, however, only Amto village appears as a Census Unit and this had a resident population of 205 at the 1980 National Census (National Statistical Office, 1982) and by 2011, there were 413 (National Statistics Office, 2014). New Tribes Mission (NTM) gave 300 in 2006. Table C51 provides summary information for Musian and Amto.

Table C51
Musian and Amto Summary

Language	Musian	Amto
Language type	Papuan	Papuan
Language family; subfamily	Amto-Musian Phylum-Level Family.	Amto-Musian Phylum-Level Family.
Other names; dialects	Musian: Musa.	Amto
Villages	Seiawi	
Cycles	?	2, 5
Operative patterns		Digit tally?
Frame words	1 to 5, 10	1, 2, 3, 5

The only number data obtained for Musian derive from an undated SIL word list collected at Seiawi. These are given in Table C52. The word list gives *ka* for "hand" and it may be that this morpheme appears in the number word for 5 (and also for 4); it does not, however, appear in the number word for 10, *?ogwalo*. With the data given, no cyclic pattern is apparent. While it is possible that the system is a digit-tally one it seems more likely that the words given form part of a body-part tally system but further evidence is required.

An SIL word list was compiled at Amto village in 1972 and this contains the number words for 1 to 5, and 10 which are given in Table C52. While it is possible that a morpheme for 5 also appears in the number word for 10, thus implying the operation of a 5-cycle, this is not certain and further data are required.

Table C52
Musian and Amto Counting Words

	Musian	*Amto*
1	*samo?*	*ohu*
2	*himolo*	*kiyah*
3	*luwelo*	*kri:ya*
4	*katukwialo*	*kiyapei*
5	*katilo*	*pu:m'ka*
10	*?ogwalo*	*wapumka*

Sources: Musian. SIL Word List (undated). Village: Seiawi. Amto: SIL Word List (1972). Village: Amto. *Note:* (/?/ = glottal stop).

West Papua Examples

Meax

This language is found on the northeast coast, west of Manokwari town in Papua Barat Province of West Papua. There were 14 800 speakers from SIL notes in 2000. Some language information is summarised in Table C53. This and other languages mentioned can be found on the map in Appendix E, Figure E2.

Table C53
Meax Summary

Variant Names:	Mansibaber, Mejach, Mejah, Meyach, Arfak, Mansibaber, Meah
Classification:	NAN, Meax Family, East Bird's Head Phylum-Level Stock, Sentani
Cycles	5, 10?, 20
Operative patterns	6 to 9 = 5+*n*
Frame words	1, 2, 3, 4, 5, 10, 20

The counting words are given in Table C54, derived from three sources.

Table C54
Meax Counting Words

	A	B
1	*ergens, arges*	*ervêns*
2	*ergek, argek*	*ervêk*
3	*ergomu, argomu*	*or vômu*
4	*tokoru*	*tochkôru*
5	*tjindja*	*tsjintsja*
6	*tjinda-ergens*	*tsjintsjervens*
7	*tjinda-ergek*	*tsjintsjervek*
8	*tjinda-ergomu*	*tsjintsjervomu*
9	*tjinda-tokoru*	*tsjintsjerdachkoru*
10	*setk(a), sjetk(a)*	*sêtk*
20	*snok*	*snok*
21	....	*snokervens*
22		*snokervek*

Sources: A. Mejach: Galis (1960, p. 141). B. Mansibaber: Cowan (1953, p. 10); Wirz (1923, p. 208).

Mantion

Mantion adjoins Meak to the east stretching south east from the Anggi lakes to Momi town on the northwest of Cenderawasih Bay (see Appendix E, Figure E2). The Ethnologue shows an Isolate Hatam nestled between them on the coast. Lean's 1992 showed Mantion as a Family-Level Isolate and the Ethnologue as an East Bird's Head Phylum-Level Sentani or Mantion. There are four dialects in 50 villages with a population of 12 000 in 1987 (SIL). Language information is summarised in Table C55 and the counting system words in Table C56.

Table C55
Mantion Summary

Dialects:	1) Manikion , 2) Mantion
Variant Names:	Manikion
Classification:	NAN, Family-Level Isolate
Cycles	5, 20
Operative patterns	6 to 9 = 5+*n*
Frame words	1, 2, 3?, 4, 5, 10, 20

Table C56
Mantion Counting Words

	A	B	C
1	*hom*	*hom*	*hom*
2	*huai*	*hwai*	*huai, hoia*
3	*homui*	*homoi*	*homoi*
4	*hoku, hoüku*	*houku*	*hoku*
5	*sergem*	*serikem*	*serkem, sirkem*
6	*sergem-hom*	*serikem hom*	*serkem homi, sengam*
7	*sergem-huai*	*serikem hwai*	*serkem huai*
8	*sergem-homui*	*serikem homoi*	*serkem homoi*
9	*sergem-hoku*	*serikem houku*	*serkem hoku*
10	*sisa*	*sisa*	*sesa*
20	*situ-hom*	....	....
30	*situ-hom-sisa*		
40	*situ-huai*		

Sources: A. Manikion: Galis (1960, p. 140) B. Mantion: Cowan (1953, p. 13). C. Manikion: Cowan (1953, p. 13), Ray (1912a, pp. 340-341), Kluge (1938, p. 180).

Yava

Yava is spoken on Yapen Island in West Papua at eight north coast, 2 interior and 18 south coast villages across the centre of the island (see Appendix E, Figure E2). Its population in 2011 (SIL) was 10 000 and widely spoken in the markets, school and services. Language information is summarised in Table C57.

Table C57 *Yava Summary*

Dialects:	15
Variant Names:	Mantembu, Yapanani, Mora, Turu, Yawa, Yawa Umat Dialects: East, West, South, North, Central (Mora)
Classification:	NAN, Stock-Level Isolate
Cycles	5, 10, 20
Operative patterns	6 to 9 = $x+n$
Frame words	1, 2, 3?, 4, 5, 10, 20

Table C58 provides the counting words indicating that the digit tally has a (5, 10, 20) cycle system.

Table C58
Yava Counting Words

	A	B
1	*intabe, intabo, utabo*	*entabo*
2	*biru, djiru, jiru*	*djiru*
3	*biru (?), mandei*	*mandei*
4	*mambi*	*mambi*
5	*na, nai, radani*	*radani* or *nai*
6	*kawintabe, kaüdjentabo*	*kaudjentabo*
7	*kaüdjiru*	*kaudjiru*
8	*kaümainde, kaümandei*	*kaumandei*
9	*kaümambi*	*kaumambi*
10	*sabuï*	*sabui*
11	*sabuï-e anentabo*	......
20	*batan-intabe, tenam(bé)*	

Sources: A. Turu: Galis (1960, p. 142), Voorhoeve (1975, p. 120) B. Cowan (1953, p. 7).

Bauzi

Bauzi is spoken in an extensive inland area of Yapen Island between the Mamberamo and Rouffaer rivers and further afield including Vakiadi, Noiadi, Danau Bira, Solom, Kustera, Neao and Itaba villages (see Appendix E, Figure E4). It has three dialects and is widely spoken by other language groups. The population was 1 500 in 1991 (SIL). Table C59 summarises some language information and Table C60 provides the counting words for Bauzi.

Table C59
Bauzi Summary

Variant Names:	Bauri, Baudji. Dialects: Gesda Dae, Neao, and Aumenefa
Classification:	NAN, East Geelvink Bay Stock-Level Family.
Cycles	2?, 5, 20
Operative patterns	6 to 9 = 5+*n*. Digit tally
Frame words	1, 2, ?, 5, 10, 20

Table C60
Bauzi Counting Words

	A	B
1	*vamtia*	*vametea*
2	*beasu*	*behasu*
3	......	?
4		?
5	*auohole*	"palm finished"
6	*au mei viva*	"palm another digit"
7	*au mei behasu viva*	"palm another two digit"
8		-
9		-
10	*au ahim fole*	"palm pair finished"
11	*naba bu vametea viva*	"foot *(bu)* one digit"
15	*naba meida ahebu fole*	"foot one all finished"
20	*naba ahim fole*	"foot pair finished"
21	*dat meida anekeha- vametea viva*	"man another hand- one digit"

Sources: A.Voorhoeve (1975(c), p. 122). B. Briley (1977, p. 30), Briley (1977, p. 30).

Tarunggare

Tarunggare (Tunggare) is a threatened language with only 500 speakers in 1993 (Doriot, SIL ethnologue). It is located north east of Nabire town around Ruwiami Point on south Cendrawasih Bay, in north central part of West Papua (see Appendix E, Figure E2). It has 70% lexical similarities to neighbouring languages. Table C61 summarises some language information.

Table C61
Tarunggare Summary

Classification:	NAN, E. Geelvink Bay Stock-Level Family

Table C62 gives the counting words with Stokhof providing two lists with variations for four and five.

Table C62
Tarunggare Counting Words

	A	B	C
1	*dua, dua?a*	*doewa ä*	*doe'wa'ha*
2	*amaite*	*a maitih*	*amai'te*
3	*duate*	*nateahai*	*na'te'a*
4	....	*liau*	*die'aj, die'aauu*
5		*doeatai*	*nie'a'hoe orre te ha*
6		*doewaa*	....
7		*moraate*	
8		*nehamoe*	
9		*ateaha*	
10		*doeatai*	

Sources: A. Tarunggare: Galis (1960, p. 143), Voorhoeve (1975(c), p. 120). B. Tarunggareh: Holle Lists: Stokhof (1983, p. 36). C. Tarunggareh Area: Holle Lists: Stokhof (1983, p. 41).

Warenbori and Pauwi

Warenbori lies on the north coast of Yapen Island, West Papua, west of Mamberamo river to Manini point. In 1998, SIL records noted 600 speakers. Table C63 summarises information on the languages. Pauwi (Yoke) is spoken in an area with a population of 200, northeast of Rombebai Lake, east of Mamberamo River on Yapen Island, West Papua, and is a threatened language. Lean suggests Ray and Kluge's data for Pauwi is in fact Warenbori (see Appendix E, Figure E3).

Table C63
Warenbori and Pauwi Summary

	Warenbori	Pauwi
Classification:	NAN,Phylum Level Isolate	NAN, Phylum - Level Isolate; Lower Mamberamo
Variant Names:	Warembori	Yoke, Bitovondo, Yauki
Villages	Poiwai, Waembori, Taamakuri, Bonoi	Mantarbori

Language features include SVO similar to Austronesian languages. Warenbori counting words are given in Table C64 from several sources. 7 is $x + 2$ and 8 is $x + 3$. Its neighbouring languages have a clear $x + n$ for 6 to 9 and there are no further data to give cycle patterns other than 5. There are little available data for Pauwi (Table C64) but interestingly in one case the word for 2 was *ru* reflecting the common AN Oceanic words for 2. Note that Lean has placed Ray and Kluge's data as Warenbori.

Table C64
Warenbori and Pauwi Counting Words

	Warenbori: A	Warenbori: B	Pauwi: A	Pauwi: B
1	*pasari, (iseno)*	*pa-sari*	*oschenu*	*osxenu*
2	*pari, (kainduo)*	*pa-ri*	*kaiamba*	*ru*
3	*parosi, (iwonto)*	*pa-rosi*	*bimessi*	
4	*parasi*	*pa-rasi*	*bimengsi*	
5	*parinisi*	*pa-rinisi*	*boangsi*	
6	*ponensi*	*ponensi*		
7	*pengmong-gari*	.		
8	*pengmong-garosi*	.		
9	*petiserai*	.		
10	*putaonsi*	*putaonsi*		

Sources: Warenbori A. Warembori: Galis (1960, p. 143), Voorhoeve (1975(c), p. 122). Warenbori B. Pauwi: Ray (1912a, 1912b, pp. 340-41), Kluge (1938, p. 180). Pauwi A. Pauwi: Galis (1960, p. 143), Anon (1913, p. 258). Pauwi B. Voorhoeve (1975(c), p. 122).

Yuri

Wurm and Hattori (1981, Map 6) have Yuri classified as a Phylum-Level Isolate. It is spoken in a region adjacent to the border of West Papua immediately to the south of the Dera and Anggor regions (see Appendix E, Figure E13). Laycock (1973, p. 52) listed seven Yuri-speaking villages; one of these, Sugumoru, is not listed as a Census village in the 1980 National Population Census data, and another, Auya (Auia), is listed as Auia 1 and Auia 2. The total population of the Yuri-speaking villages which were censussed in 1980 was 668 (National Statistical Office, 1982). Auia 1 had a population of 273 and Kambriap 624 in 2011 (National Statistical Office, 2014a). Table C65 summaries the language information.

Table C65
Yuri Summary

Language type	Papuan
Language family; subfamily	Phylum-Level Isolate.
Other names; dialects	Yuri
Villages	
Cycles	23
Operative patterns	Body parts
Frame words	1 to 12

Laycock (1973, p. 52) noted that Yuri, a tonal language, possesses a 23-cycle body-part tally system but gives no further details. The number words obtained for this system derive from an SIL word list, dated 1974/75, which was compiled at Auya 2 village and one CSQ completed by an informant from Kambriap village. These are given in Table C66.

Table C66
Yuri Counting Words

	A	B
1	*anggarang(k)*	*angkarang (*left little finger)
2	*anangk*	*anang* (ring finger)
3	*yina'nggobwe(i)*	*yining* (middle finger)
4	*iyvor*	*iyor* (index finger)
5	*a'ndik*	*andi* (thumb)
6		*togoam* (wrist)
7		*fok-kini* (lower arm)
8		*yukup* (elbow)
9		*sara-ak* (upper arm)
10	*fa're sambram*	*fare (*shoulder)
11		*mum* (left breast)
12		*indir* (sternum)
13		*anger mum* (right breast)
14		*anger fare* (right shoulder)

Sources: A. SIL Word List (1974/75). Village: Auya No.2. B. 1 CSQ. Village: Kambriap.

The number words given in System A are transcribed from phonetic to Roman orthography. The first five number words are used when tallying on the fingers. The number word for 5 *a'ndik*, is not identical to the word for "hand" *ye(i)*, which is as we would expect for a body-part system. System B provides the details of the body-part tally system and confirms Laycock's comment that it possesses a 23-cycle. The system employs only body-parts on the upper part of the body. Beginning with the little finger of the left hand, tallying proceeds in order along the fingers to the thumb and then to the wrist, lower arm, elbow, upper arm, shoulder and left breast. The midpoint of the tally is the sternum at a tally of 12. Tallying then proceeds using the symmetrical counterparts of the first eleven tally-points starting with the right breast and ending, at a tally of 23, with the right little finger.

Nagatman and Busa – Phylum-Level Isolates

Nagatman and Busa are Phylum-Level Isolates (Wurm & Hattori, 1981, Map 6). Nagatman is spoken in a region lying immediately to the south of the Kwomtari region. Laycock (1973, p. 52) listed eight Nagatman-speaking villages and at the 1980 National Census these had a total population of 580; the 2011 Census Nagatiman village alone had 394 people. Busa is spoken in four villages, (Laycock, 1973, p. 52) which are situated in a region lying to the south of the Nagatman area. Only three of the Busa-speaking villages are given on the 1980 Census Maps (Birimei is not shown) and these had a total population of 198 at the time of the Census (National Statistical Office, 1982). Some information on the language is summarised in Table C67.

The information available on the Nagatman counting system is contradictory. Laycock (1973, p. 53) said that the "number system is quinary" while a note given in an SIL word list compiled at Dila village in 1972 indicates that the system is a somewhat unusual body-part one. The number words taken from the SIL list are given in Table C68.

Table C67
Nagatman Summary

Language	Nagatman	Busa
Language type	Papuan	Papuan
Language family; subfamily	Phylum-Level Isolate.	Phylum-Level Isolate.
Other names; dialects		
Villages	Dila	Rawei
Cycles	? 74?	?
Operative patterns	Body parts	?
Frame words	1 to 5	1 to 5

In brief. For Nagatman, there are few counting words available to confirm the SIL note that the counting proceeds up the arm and then down the side of the body to 36 and presumably up the other side and down the arm to reach 74. This is a very unusual body-part tally system and not apparently found elsewhere.

Table C68
Nagatman and Busa Counting Words

	Nagatman	Busa
1	zuwa?	*otutu*
2	tele?	*tinana?*
3	?auna?	*wurana*
4	nilina?	*aiti*
5	tubali?	*yuve' nati*
10	wurza?	

Sources: Nagatman. SIL Word List (1972). Village: Dila. Busa. SIL Word List (1980). Village: Rawei. *Note:* (/?/ = glottal stop).

The note on the SIL word list says that tallying begins on the left side of the body, proceeding up the left arm (presumably after the fingers have been tallied) and then down the left side. The toes are tallied and upon reaching the big toe the tally stands at 36. The side of the left foot is then tallied (although the number of tally-points is not indicated) and the symmetrical points on the right side of the body are tallied (presumably starting at the side of the right foot and proceeding in reverse order until tallying ends on the fingers of the right hand). The data given in Table C68 are not sufficient to either confirm or deny the SIL note. We may observe, however, that the tally-system described is very unusual indeed firstly because of the magnitude of the cycle (at least 74), and secondly because body-parts on the lower part of the body are used as tally-points.

Laycock (1973, p. 52) said of the Busa counting system that it "appears to be fully decimal (which is rare in New Guinea among NAN languages); however, it is possible that the ten numbers obtained are simply the first ten of a body-parts system". The only data obtained here derive from an SIL word list compiled at Rawei village in 1980. These are given in Table C68.

There are five distinct words for the first five cardinals with no evidence of a 2-cycle. The number word for 5 *yuve' nati*, is not the same as the word for "hand" *noh*. This suggests that the system is not a digit tally one and it is possible, as Laycock conjectures, that the words are part of a body-tally system.

Anggor

Anggor, a NAN language classified as a member of the Senagi Family, Senagi Stock (Wurm & Hattori, 1981, Map 6) is spoken in a region lying immediately to the south of the Amanab area (see Appendix E, Figure E13). Laycock (1973, p. 48) listed 17 Anggor-speaking villages; at the 1980 National Census 16 of these had a total population of 1 399 and so the total resident population for the language is probably about 1 500 (National Statistical Office, 1982). By 1990 census SIL (n.d.) note 11 villages with 1 270. In 2011, Bibriari alone had 792 (National Statistical Office, 2014a). Table C69 summarises some of the language information.

Table C69
Anggor Summary

Language type	Papuan
Language family; subfamily	Senagi Family, Senagi Stock
Other names; dialects	Angor; Anggor: Senagi, Bibriari, Watapor.
Villages	Senagi, Bibriari, Watapor.
Cycles	2, 23
Operative patterns	Body parts
Frame words	1 to 12, 20 to 30

An SIL team, R. and S. Litteral, have been working on Anggor since 1965 and have published a number of articles on the language (Patrick, 1981). An IMP/SIL counting systems questionnaire was completed in 1978 (for the Indigenous Mathematics Project); some number data are also available in an SIL word list compiled at Wamu village in 1976. In addition, one CSQ was received from the headmaster of Angor Community School situated near Bibriari Village. The data are presented in Table C70.

Table C70 *Anggor Counting Words*

	A	B
1	*mam* (little finger)	*mam* (little finger)
2	*yimbu (yumb)* (ring finger)	*yumbo* (ring finger)
3	*ngim* (middle finger)	*ngem* (middle finger)
4	*yimbu yimbu* (index finger)	*yumbo yumbo* (index finger)
5	*hondihiyafundimb* (thumb)	*hondefemb* (thumb)
6	*hondirasi* (wrist)	*lesi* (wrist)
7	*waranduhu* (elbow)	*hamend* (ulna)
8	*warambuhu* (shoulder)	*walanduh* (elbow)
9	*titi* (breast)	*wadagidal* (upper arm)
10	*(ngopoa) titi* (other breast*) (amberamindomb)*	*walambuh* (shoulder)
11	*warambuhu* (other shoulder)	*tet* (breast)
12	*waranduhu* (other elbow)	*mesend* (sternum)
13	*hondirasi* (other wrist)	*tet* (other breast)
19		*hond* (thumb)
20		*batend* (index finger)
21		*momongesel* (middle finger)
22		*aked kemenend* (ring finger)
23		*aked* (little finger)

Sources: A. SIL Word List (1976) for 1 to 5. IMP/SIL Questionnaire (1978). B. 1 CSQ. Angor Community School.

Both systems A and B in this Table indicate that the Anggor system is a body-part tally one; the two systems shown, however, differ regarding the number of points tallied and therefore in the overall cyclic pattern of the system. System A, the IMP/SIL data, has a half-cycle of 9; there is no unique midpoint lying on the body's axis of symmetry and the remaining 9 tally-points are the symmetrical counterparts of the first 9. The numbers 14 to 18 are tallied on the fingers of the hand (as for system B) and the number words are the names of the fingers (also as for System B). The first four number words for both Systems A and B are not, however, the names of the fingers but are four numerals. There are three distinct numerals for 1, 2, and 3; 4, however, has a 2 + 2 construction. The numerals thus possess a modified 2-cycle.

System B has a full cycle of 23 and there is a unique midpoint (the sternum) at the 12th tally-point. The remaining 11 tally-points are the symmetrical counterparts of the first 11.

Other examples of body-tally systems occurring with 2-cycle systems of numerals are found in the Western Province. There are also examples of systems which are such that the first four or five numerals have been grafted (so to speak) onto the body-part system, the numerals displacing the names of the fingers. The names of the fingers are used, however, at the end of the tally-cycle when tallying proceeds on the fingers of the other hand. Both Systems A and B, as given in Table C70, are further examples of this latter type.

Morobe Province Examples

Labu

Labu is classified by McElhanon (1984, p. 13) as a member of the Siassi Family, Vitiazan Sub-Family, Huon Gulf Group, although Holzknecht (1988) indicated that this classification may need to be revised. Along the coast to the east and further south, Bukauac, another member of this language Sub-Family, is spoken. The language is spoken in three villages situated on the western coast of the Huon Gulf and these had, at the 1980 National Census, a total resident population of 1329 (National Statistical Office, 1982). SIL (n.d.) gives the population as 1 600 in the three villages. According to the 2011 Census, these villages had 2 475 residents, possibly some being squatters from other regions or married partners and relatives. Labubutu is across the Markham River from the capital of Morobe Province, Lae (see Appendix E, Figure E9). The language that was at Lae itself is extinct. In the 1970s and 1980s, Labu was spoken extensively in the village and village technologies such as lime making, canoe making, sailing, and taro gardening were well maintained. Language information is summarised in Table C71.

Table C71
Labu Summary

Language type	Austronesian
Language family; subfamily	Siassi Family, Vitiazan Sub-Family, Huon Gulf Group
Other names, dialects	
Villages	Labubutu, Labutali, Labumiti
Cycles	5, 10, 20
Operative patterns	digit tally
Frame words	1 to 5, 10, 20

In brief. The digit tally system has (5, 10, 20) cycles. 6 to 9 are 5+n where n is 1 to 4, 10 is a distinct numeral and numbers like 30 are 20 + 10.

The number data for Labu derive from Holzknecht (1988), Hooley (1971), two CSQs, and Smith (1984, p. 147). A selection of the available data are given in Table C72.

Table C72
Labu Counting Words

	A	B	C
1	*togwato*	*togwato*	*tugwatu*
2	*salu*	*salu*	*salu*
3	*sidi*	*sidzi*	*sidi*
4	*soha*	*soha*	*suha*
5	*maipi*	*maipi*	*maipi*
6	*maipi anendi togwato* or *haipi anendi togwato* or *maipa tomolo*		*maipi anendi tugwatu*
7	*maipi anendi salu* or *haipi anendi salu* or *maipa salu*		*maipi anendi salu*
10	*nomusu*		*numusu*
11	*nomusu togwato*		*numusu anendi tugwatu*
20	*asamoni*		*asamuni*
30	*asamoni nomuso*		*asamuni namusu*
40	*asamo salu*		*asamu salu*
60	*asamo sidi*		*asamu sidi*

Sources: A. Holzknecht (1988). B. Hooley (1971); SIL Word List (1968), Village: Labubutu. C. Smith (1984, p. 147).

Labu has a number system with at least four basic numerals: it is uncertain whether 5 *maipi* is a numeral or a number word with a body-part referent; it does not, however, mean "hand" which is *nama*. Similarly, the status of 10, *nomusu*, as a numeral is uncertain. Holzknecht indicates that 20, *asamoni*, contains *samo* meaning "all", "whole", and may be interpreted to mean "whole man". Yalu, a neighbouring language across the Markham River has *artsamu* for 20 and Holzknecht notes that this appears to be a borrowing from Yabim *nga? samu*, "a whole man".

The cardinals 6 to 9 appear to be expressible in several alternative ways, as given in System A. The basic construction, however, has the form 5 + *n* where n takes the values 1 to 4 respectively. The number word for 10 appears to be distinct and not a combination such as 5 x 2 or 5 + 5 and thus the system possesses a 10- cycle in addition to a 5- cycle. There is, in addition, a superordinate 20-cycle and 30, for example, has the construction 20 + 10, and 40 is 20 x 2.

Mumeng

Mumeng, a further member of the Buang Family, has five dialects and is spoken in 25 villages which had, according to statistics compiled in the 1970s, a total population of 9 186 (McElhanon, 1984, p. 14, pp. 20-21). The Mumeng-speaking villages are situated to the south-west of Lae. In the 2011 Census, Mumeng Rural, excluding Piu and Kapin, had a total of 22 942 but this possibly includes other languages. The recorded populations for Patep 1 875, Yanta 1 148, Zenag 2 400, Dambi 542 and Mumeng Station 487, and Kumalu 1 206 which alone total 7 655. See map in Appendix E, Figure E9. Table C73 gives some language information.

Table C73
Mumeng Summary

Language type	Austronesian
Language family; subfamily	Buang Family
Other names, dialects	Mumeng: 1. Yanta, 2. Zenag, 3. Kumalu, 4. Patep, 5. Dambi.
Villages	Dambi, Latep (Zenag d).; Tawangala (Yanta d)
Cycles	2', 5, 20
Operative patterns	digit tally
Frame words	1, 2, 3, 5, 20

In brief. Mumeng has a modified 2-cycle with continuing (5, 20) cycles of a digit tally system. There are slight variations between dialects so 5 in one dialect is "half" implying "half of the two hands". The hand morpheme is used for 5 to 10 with 5 being "hands half", 15 is "hands two and feet half".

Number data are available for several of the Mumeng dialects. We have, for example, an SIL word list compiled at Dambi village (Dambi dialect); an IMP/SIL questionnaire, compiled in 1978, and an SIL word list compiled in 1975, both for the Patep dialect; an SIL word list compiled in 1968 at Latep village (Zenag dialect); and an SIL word list compiled in 1968 at Tawangala village (Yanta dialect). Smith (1984, p. 182) and Hooley (1971) also have some Mumeng data. Some of the available data are given in Table C74.

Table C74
Mumeng Counting Words

	A	B	C
1	*ti*	*ti*	*ti*
2	*yuu*	*yu*	*yo*
3	*yon*	*yon*	*yan*
4	*yuu di yuu*	*maentina*	*ondeo*
5	*vige vilu*	*valu*	*banggive'lu*
6	*vige vilu di se? ti*		*banggi'yo*
7	*vige vilu di se? yuu*		
8	*vige vilu di se? yon*		
9	*vige vilu di se? yuu di yuu*		
10	*vige yuu*		
15	*vige yuu di vixa vilu*		
20	*kehe ti*		

Sources: A. IMP/SIL Questionnaire (1978). Patep d. B. SIL Word List (1965). Dambi d. C. SIL Word List (1968). Yanta d. *Note:* /?/ = glottal stop.

There is some variation between dialects as to the number of basic numerals. The Patep dialect has a basic numeral set (1, 2, 3) and 4 has the construction 2 + 2. The Dambi and Yanta dialects, though, have basic numeral sets (1,2,3,4). In all dialects 5 is a number word and this usually contains a "hand" morpheme: *vige,venge (*Patep), *banggi- (*Yanta). In the Patep dialect, 5 is *vige vilu,*i.e. "hand(s) half" where *vilu* means half; the Yanta dialect *banggi-ve'lu* "hand(s) half", in this case *ve'lu* being *"half"*. In the Dambi dialect we have simply *valu* "half", for 5, the "hand(s)" being understood.

In the Patep dialect, the number words (phrases) for 6 to 9 have the construction 5 + *n* where *n* takes the values 1 to 4 respectively. 6, for example, is *vige vilu di se? ti* "hand(s) half and cross-over-to one". Tallying on the fingers continues to 10, *vige yuu* "hand(s) two" (the Yanta dialect has *banggi' yo* "hand(s) two"). Tallying then proceeds on the toes. The number word for 15 is *vige yuu di vixa vilu* "hand(s) two and feet half" where *vixa* is a "foot" ("feet") morpheme, the *x* = *ch* as in the Scots *loch*. The 20 "man"- cycle is completed with *kehe ti (*Patep dialect) where *kehe* is "base", "source", "foundation". Several other dialects (not given) have "man one" instead. The Mumeng system, thus is a digit-tally one with a (5, 20) cyclic pattern, or, in the case of the *Patep* dialect, a (2', 5, 20) cyclic pattern.

Mapos Buang

Mapos, is spoken in 19 villages (McElhanon, 1984, p. 20) which are situated in a region lying to the east of Mumeng and to the south-west of Lae. McElhanon's estimate of the number of speakers of the language, based on statistics compiled in the 1970s, is 6 666. From the 2011 Census, Mapos 1 and Mapos 2 had a population of 983, Mambump 401, and the whole Buang Rural (which possibly incorporates other Buang Family languages) had 9 200. Mapos Buang has two dialects: Mapos proper and Mambump, the former accounting for about two-thirds of the population. The language is classified as a member of the Buang Family (AN). See map in Appendix E, Figure E9. Further information on the language is summarised in Table C75.

Table C75
Mapos Buang Summary

Language type	Austronesian
Language family; subfamily	Buang Family
Other names, dialects	Mapos Buang : 1. Mapos, 2. Mambump
Villages	
Cycles	2', 5, 20
Operative patterns	digit tally
Frame words	1, 2, 3, 5, 20

The number data for Mapos Buang derive from several sources: an IMP/SIL questionnaire completed in 1978, four CSQs, and Smith (1984, p. 181). Articles by Girard (1968–1969) and Hooley (1978) also contain some number data. A selection of data from these sources is given in Table C76.

Table C76
Mapos Buang

	A	B	C
1	*ti*	*ti*	*ti*
2	*luu*	*lu*	*lu*
3	*loo*	*lal*	*lal*
4	*luu be luu*	*lumbalu*	*lembulu*
5	*nemadvahi*	*orund vandu*	*orond walu*
6	*nemadvahi videk ti*	*orund vandu mbti*	*orond walu vendik ti*
7	*nemadvahi videk luu*		*orond walu vendik lu*
10	*nemad luho*	*orund luo*	*orond luro*
15	*nemadluho nemadvahi*	*orundluo mb varang dwadu*	*orond luro varen walu*
20	*meho dahis ti*	*miran dadus ti*	*dadus ti*

Sources: A. IMP/SIL questionnaire (1978); Hooley (1978). B. Smith (1984, p. 181). C. Girard (1968–1969, p. 161).

The system shown is a digit-tally one with a basic numeral set (1, 2, 3): the numeral 4 has a 2 + 2 construction. The system thus has a modified (or quasi-) 2- cycle. The number words for 5 and 10 each contain a "hand" morpheme: in System A this is *nemad*. The gloss for 5 is "our hand(s) half" and that for 10 is "our hands, both of them" or "hands two".

The data provided by both Smith and Girard indicate that, after 10, tallying proceeds on the toes. System B has for 15: "hands two and foot half", as does System C. System A, however, has, somewhat unusually, "hands two hand half". The tally cycle is complete at 20: System A has "person complete" and System B has "person complete one".

Manga Buang

Manga Buang, or Manga, a further member of the Buang Family, is spoken in eight villages (McElhanon, 1984, p. 20) which are situated to the east and south-east of Mumeng (see Appendix E, Figure E9. McElhanon's estimate of the Manga-speaking population, based on statistics compiled in the 1970s, is 2 688. In 2011, Mangga village itself had 604 people (National Statistical Office, 2014a). Table C77 summarises some information on the language.

Table C77
Manga Buang Summary

Language type	Austronesian
Language family; subfamily	Buang Family
Other names, dialects	Manga Buang
Villages	Manga, Kwasang, Bayematu
Cycles	2', 5, 20
Operative patterns	digit tally
Frame words	1, 2, 3, 5, 20

In brief. Manga Buang is a (2', 5, 20) digit tally system but five is "hands half" or "thumb" without the apparent use of a leg morpheme in 15, but 20 is "person complete".

The number data for Manga Buang derive from an IMP/SIL questionnaire compiled in 1978, from an article by Girard (1968–1969), and from four CSQs. These are given in Table C78. The Manga Buang system is a digit-tally one with a basic numeral set (1, 2, 3): the numeral 4 has the construction 2 + 2. The number word for 5 has the gloss "hand(s) half" (Systems A and C) or "thumb" (Systems A and B). The number word for 10 (for all systems) has the gloss "hands two".

Although the system is a digit-tally one, the number word (phrase) for 15 does not, as is usual, contain a "leg" or "foot" morpheme: Systems A and C have, for example, "hands two and hands half". The number word for 20, however, has the gloss "complete one", i.e. one complete person, which thus implies that the 20 digits of one person have been tallied. The system thus appears to have a (2', 5, 20) cyclic pattern.

Table C78
Manga Buang Counting Words

	A	B	C
1	*ti*	*ti*	*ti*
2	*yuuh*	*iur*	*yu, yuug*
3	*yaal*	*lar*	*yar, yal*
4	*yu mbe yuu*	*iumbulu*	*yu mbe yu*
5	*namaa vaalu* or *butoov*	*butov*	*nama valu*
6	*namaa vaalu be vindak ti*	*vindak ti*	*nama valu vindak ti*
7	*namaa vaalu be vindak yuuh*	*vinda iur*	*nama valu vindak ti*
10	*namaa yuuh*	*nema iur*	*nama yu*
15	*namaa yuuh ambe laam namaa vaalu*	*kambi valu butov*	*nama yu ambe lam nama valu*
20	*doo ti*	*do ti*	*do ti*

Sources: A. IMP/SIL Questionnaire (1978). B. Girard (1968–1969, p. 161). C. 4 CSQs. Villages: Manga (2), Kwasang, Bayematu

Vehes and Piu

Vehes and Piu are members of the Buang Family. Vehes is spoken in a single village, Buissi, situated on the coast of the Huon Gulf, about 25 kilometres due south of Lae (McElhanon, 1984, p. 21), south of Mapos Buang area (see Appendix E, Figure E9). At the 1980 National Census, the resident population of Buissi was 81 (National Statistical Office, 1982). McElhanon also indicated that Piu is spoken in a single village, Piu, located to the west of Mumeng in a region which is otherwise South Watut-speaking. At the 1980 National Census, Piu had a population of 135. By 2011, there were 320 residents in Piu (National Statistical Office, 2014a, 2014b). Language information for Vehes and Piu are summarised in Table C79.

Table C79
Vehes and Piu Summary

Language	Vehes	Piu
Language type	Austronesian	Austronesian
Language family; subfamily	Buang Family	Buang Family
Other names, dialects	Vehes : Buasi, Buissi	Piu
Villages		Piu
Cycles	2', 5, 20 ?	2', 5, 20
Operative patterns	digit tally	digit tally?, 6 to 9 is "and 1", "and 2" implying hand and 1 etc
Frame words	1, 2, 3, 4, 20?	1, 2, 3, 5, 20?

In brief. Vehes, an Austronesian language, has a digit tally system with modified 2 cycle and 5 and an implied 20 cycle but insufficient data to confirm this. A morpheme for "hand" is used for 5 and 10. Piu data imply a digit tally for construction of 6 to 9 as hand (implied) and *n* and hand morpheme is implied in 10.

The number data for Vehes derive from Hooley (1971, pp. 116-7) which were collected at Buissi in 1968. These are given in Table C80. The Vehes system appears to be a digit-tally one with a basic numeral set (1, 2, 3). The numeral 4 has a 2 + 2 construction and thus the system possesses a modified (or quasi) 2-cycle. The number words for both 5 and 10 each contain a hand morpheme nama-; 5 has the gloss "hand(s) half" and 10 has the gloss "hands two". The system thus also possesses a 5-cycle. Insufficient data are available to determine whether, as is likely, the system also possesses a superordinate 20- cycle as well.

The number data available for Piu derive from an SIL word list compiled in 1965, from Hooley (1971), and from Smith (1984, p. 183). These are given in Table C80. The Piu system has at least three basic numerals: 1, *tika*, 2, *lu*, and 3, *yan*. It is possible that 4 has a 2 + 2 construction: this is given variously as *lutalu, ndalu*, and *luntalu*. Hooley (1971) gave the word for "hand" as *vanggi-* and this appears explicitly in the number word for 10, *vanggi lu*, i.e. "hand(s) two". It is uncertain whether 5, "*vate*" contains a "hand" morpheme or not; Smith, however, translates this as "hand one". The number words for 6 and 7 are, respectively, *serkti* and *serkalu* which Smith translates as "and one", "and two" which thus have implied 5 + 1 and 5 + 2 constructions respectively. The number word for 20 (System B) is *yuka ti*: the word for "man" is given by Hooley as *mug* and thus 20 is clearly not "man one" although the *ti* implies the existence of a 20- cycle. The system thus appears to have some features of a digit-tally system with a (2', 5, 20) cyclic pattern. However, there is no evidence of a "foot" or "leg" morpheme appearing in the number words for 11 to 19.

Table C80
Vehes Counting Words

	Vehes	Piu. A	Piu. B	Piu. C
1	*timu*	*tika*	*tika*	*tika*
2	*u.y*	*lu*	*lu*	*lu*
3	*yar*	*yan*	*yan*	*yan*
4	*uyakuy*	*lutalu*	*ndalu*	*luntalu*
5	*nama' varu*	*vate*	*vate*	*vate*
6			*serkti*	
7			*serkalu*	
10	*namauyin*	*tangilu*	*serkvate*	*vanggilu*
11			*vanggilu da serkti*	
15			*vanggilu da vate*	
20			*yuka ti*	

Sources: Vehes: Hooley (1971, pp. 116-7). Piu: A. SIL Word List (1965). Village: Piu. B. Smith (1984, p. 183). C. Hooley (1971, pp. 116-7).

New Ireland Province, PNG

Lihir

Lihir is spoken on Lihir, Mali, Masahet and Mahur Islands in the Lihir group situated off the east coast of the mainland of New Ireland. The 1980 Census statistics give the total (citizen) population of the Lihir group as 5 503 (National Statistical Office, 1982). By the 2011 Census, Lihir had become a mining town. A summary of information on Lihir language is given in Table C81.

Table C81
Lihir Summary

Language type	Austronesian
Family, Subfamily	Patpatar-Tolai Sub-Group, New Ireland-Tolai Group
Other names, Dialects	Lihir : Lir, Sung
Village	Lataul, Lakamelen, Makapa, Ton (Lihir Island) Bulamue (Masahet Island).
Cycles	5, 20
Operative patterns	
Frame words	

In brief. Lihir is unusual in being (5, 20) cycle. Another system involves counting in groups of 4s as one group of 4 yams. Half and ordinals are used.

Lihir (Lir) has been studied by Neuhaus (1954). Peekel (1909) gave a number of Lihir numerals which are collected in Kluge (1941, p. 191). In addition, the data for Lihir, shown in Table C82, are derived from questionnaires, from the Lean's field notes collected from Lihir informants at Namatanai in 1974, and from an SIL Word List compiled on Masahet Island in 1985.

The Lihir system (A) is unique among the New Ireland counting systems in that it has a (5, 20) cyclic pattern with no evidence of a 10-cycle. Peekel (1909) reported a 5-cycle system used on Tabar Island, however no contemporary informants report its existence. One informant (one of the authors, Lean's, field notes, 1974) reported that an alternative for "twenty" (4x5) was "one complete man" and that this was used for constructing higher numbers. Neuhaus' data confirm this. The Peekel (1909) numerals 2 to 7 appear identical to the more recently collected data.

There is a word for "fraction" or "part": *a puk*. The first three ordinals are *kinemuo, kinedon*, and *kinemuil*, which do not appear to be directly obtained from the respective cardinals.

Table C82
Lihir Counting Words

	A	B	C
1	*a-uo (a kuo), ko'u*	*a min*	*a ko min, a min, a wuo*
2	*(a) laklio, laklu*	*a lakalio*	*a liak lio, a lo*
3	*(a) laktol, laktul*	*a lakatol*	*a laktol, a buet*
4	*(a) burut, ambrut*	*a burut*	*a burut*
5	*(a) liem , alyem*	*a liem*	*a liem*
6	*(a) liem ka uo*	*a liem ka min*	*a liem k'a min, a tanges*
7	*(a) liem ka laklio*	*a liem ka laklio*	*a liem k'a liaklio*
8	*(a) liem ka laktol*		*a liem k'a laktol*
9	*(a) liem ka burut*		*a liem k'a burut*
10	*a lo liem*	*lohem*	*a lo liem*
11	*a lo alem ka uo*		*a lo liem k'a min*
15	*a laktol liem*		*a laktol liem*
20	*a burut liem, a chiktun*		*a lo liem k'a lo kiak,*
21			*a ziktun*
40			*a lo ziktun*
100			*a limen ziktun*

Sources: A. 5 CSQs; Lean's Field Notes (1974); SIL (1985) : B. Lihir*:* Peekel (1909, p. 92), Kluge (1941, p. 191). C. Neuhaus (1954).

The Lihir word for "four" is *a burut*. In Barok, Patpatar and Sursurunga, for example, *burut* is the word used for counting objects in fours. In the Lihir group there also exists the practice of counting in fours, in this case the word *burut* being used for a collection of four coconuts, for example. In counting yams, the word *pis* is used instead: *a pis* (1x4 yams), a *laklio pis* (2x4 yams), and so on.

Tomoip

Tomoip is spoken in an area lying to the south-west of the Sulka region in New Ireland. It is an Austronesian language, although its grammar has certain NAN features, and is classified (Wurm & Hattori, 1981, Map 13) as a Family-level isolate. Table C83 gives some language information.

Table C83
Tomoip Summary

Language type	Austronesian
Language family; subfamily	Family-Level Isolate.
Other names; dialects	Tomoip: Tumuip, Tomoive
Villages	
Cycles	2, 5
Operative patterns	3 = 2' + 1' : 4 = 2' + 2'
Frame words	1,2, 5, 20

In brief. This island language has a (2, 5, 20) cycle system.

Some Tomoip number data are found in Parkinson (1907, p. 780) and these were subsequently included in Kluge (1941, p. 197). The language is mentioned in Ray (1927) and in (Capell, 1962), in both cases being classified as Non-Austronesian. A word list of Tomoip also appears in Laufer (1946–1949). Some data are also given in Lindrud (1980, p. 179). The data are shown in Table C84. The few number words given by Parkinson are certainly not typically AN except perhaps that for 5, *ko liem*. The structure of the first four numerals appears to be 1, 2, 2'+1', 2'+2', where 1' is related to 1 and 2' is related to 2 (Table C84).

Table C84
Tomoip Counting Words

	A
1	*denan*
2	*ro huru*
3	*horum detu*
4	*horumu horum*
5	*ko liem*
10	*liem*
20	*tamdil*

Sources: A. Tumuip: Parkinson (1907, p. 780), Kluge (1941, p. 197).

Faga-Uvea, New Caledonia

Faga-Uvea is a Polynesian Outlier in the Loyalty Islands, New Caledonia. A summary of the language and system is given in Table C85.

Table C85
Faga-Uvea Summary

Variant Names	West Uvea
Classification	Polynesian
Other names; dialects	Uvea
Villages	
Cycles	5, 10, 20
Operative patterns	6 to 10 = $x+n$, n = 1 to 5
Frame words	1, 2. 3. 4. 5

The counting words for Faga-Uvea are given in Table C86 based on data collected around 1920.

Table C86
Faga-Uvea Counting Words

	A	B
1	*tahi*	*tahi*
2	*lua*	*lua*
3	*tolu*	*tolu*
4	*fa*	*fa*
5	*lima*	*lima*
6	*tahiatupu*	*tahi-o-hna-tupu*
7	*luaonatupu*	*lua-o-hna-tupu*
8	*toluonatupu*	*tolu-o-hna-tupu*
9	*faonatupu*	*fa-o-hna tupu*
10	*limaonatupu*	*lua-lima*
11	*tahi a koje*	-
12	*lua a na koje*	*lua-lima-ma-lua-fo-gi-o-hnatupu*
16	*tahi a hano*	------------
17	*lua a hano*	
40	*lua te henua*	

Sources: A. Leverd (1922, p. 99). B. Uvea: Ray (1919a, 1919b, 1919c, p. 69), Ray (1919b, p. 69).

Enga

Enga is the main language of Enga Province. Table C87 summarises some of the complex systems.

Enga had in excess of 170 000 resident speakers (estimated from the 1980 National Census Statistics, National Statistical Office, 1982) but 230 000 in the 2000 census. Enga has eleven dialects (Wurm & Hattori, 1981, Map 11), the Mai dialect being the most numerous, followed by Layapo and Kandepe. Kyaka Enga is spoken in a region extending from the eastern part of the province into the Western Highlands Province around Baiyer River and is the main dialect spoken outside the Enga Province.

Table C87
Enga Summary

Papuan	Trans New Guinea Phylum East New Guinea Highlands Stock, West Central Family, Enga Sub-Family
Other names, Dialects	Enga Kopona, Layapo, Sau, Kaina, Mai,Yandapo,Kandepe,Malamuni,Iniai,Tayato, Kyaka
Villages	Maramuni dialect, north of Province
Cycles	4, 60; 10; 12?
Frame Words	1 to 8 (4-cycle units to 60); 1 to 7, 10; 1 to 12
Operative Patterns	cycle unit $x+n$, with $n=1$ to 3; 8=10-2?,9=10-1?

The dialects may not have different systems but are presented here to separate the sources. The earliest published example of Enga counting data appears in an article by Kirschbaum (1938) referenced as Maramun which is taken to be the Maramuni dialect of Enga spoken in the extreme north of the province almost to the border of the East Sepik Province. In 1951, Crotty published his First Dictionary of the Tchaga Language, Tchaga being his nomenclature for Enga. The anthropologists P. Wirz (1924), G.A.M. Bus (1951) and M.J. Meggitt (1957/1958, 1958) have published accounts of the Enga people and their culture but not necessarily with data on counting. The linguist A. Lang published her definitive Enga Dictionary in 1973. These are given in Table C88.

Table C88
Engan Counting Words

	A	B	C
1	*mendai*	*mend(ai)*	*mondei*
2	*lapo,lapota, lapoma*	*rab(um)*	*rap*
3	*tema,tepo, tepoma*	*teb*	*top*
4	*kitomende, kitumende*	*kirumend*	*kidumonto*
5	*yangi pundu, yau,yaunge*	*juungk, juingki*	*you*
6	*tokange*	*t(r)ogangk*	*kunju*
7	*kalange,sekaita*	*karangk*	*map*
8	*tukulapo, mangelao waketau*	*tugurab*	*poyup*
9	*tukutepo*	*tugurib*	*kunju(?)*
10	*akalita mendai, mange pundu*	*tugureponje-mendai*	*yenap*
11	-	-	*tugulap*
12	-	-	*kalap*

Sources: A. Lang (1973). B. Crotty (1951). C. Kirschbaum (1938, p. 278), Kluge (1938, p. 173), (Kluge, 1941, p. 18(O))

In addition to these published sources, CSQ data were obtained and a field trip by Lean in which he recorded a 60-cycle system in the Mai dialect. Kopamu (2005) gave some other mathematical terms and noted the issues of Engan dialects having different expressions which is confusing for establishing an Engan mathematical register. The details below illustrate some of these known differences.

The 60-cycle system is particularly interesting, the cycle consisting mainly of a sequence of 4-cycles. Only the first few counting words are numerals. The sequence of 4-cycles begins at 9 and each cycle typically has the construction: "cycle unit+1", "cycle unit+2", "cycle unit+3", "cycle unit+end", where + means an added word lime an adjectival suffix or quantitative classifier and not mathematical addition. There are 13 cycle units which are not numerals as such but may be words or phrases with some non-numerical meaning. For example, in counting the 4-cycle (29, 30, 31, 32) the cycle unit is "dog":

29 *yanapun me(n)dai* "dog"+1
30 *yanapun lapo* "dog"+2
31 *yanapun tepo* "dog"+3
32 *yanapun gato (etete)* "dog"+end

Details of the 60-cycle system are given in Tables C89 and C90. Table C89 are translations for the groups of four for the complete 60-cycle and Table C90 gives the full list of number words as collected by Lean.

Table C89
Names for Groups of Four in Engan

	Enga	English Meaning
1	*me(n)dai*	one (numeral)
2	*lapo*	two (numeral)
3	*tepo*	three (numeral)
4	*kitume(n)de*	four (numeral)
5	*yungi,yugi*	time
6	*tokage*	here is the flat object, e.g. bridge
7	*kalage*	seven (numeral)
8	*tukulapo*	two arrows; *tuku* arrow or core *tukutepo*
9-12	*tukutepon (ya)+n*	three arrows
13-16	*mapu mapun(ya)+n*	kaukau,i.e. sweet potato
17-20	*yupu yupun(ya)+n*	ground
21-24	*watakapu watakapun(ya)+n*	wild
25-28	*paipu paipun(ya)+n*	come and go
29-32	*yanapu yanapun(ya)+n*	dog
33-36	*kamapu kamapun(ya)+n*	open ground
37-40	*kujupu kujupun(ya)+n*	I cut
41-44	*kalipu kalipun(ya)+n*	I retract my foreskin; in Tok Pisin: *mi skinim kok*
45-48	*lapalu lapalun(ya)+n*	being said
49-52	*menaini menainin(ya)+n*	pig
53-56	*akipu akipun(ya)+n*	what can I say? or what more can I do?
57-60	*kaeapalu kaeapalun(ya)+n*	I stop

Source: Lean's field trip. *Note.* Several CSQ informants also provide data on the 60-cycle and these largely agree with those given above with some variations. Others have commented that 60 is "house on fire".

The alternative to the 60-cycle is a 10-cycle system as, for example, given by Lang and shown in Table C88. This, in fact, appears to be a modification of the 60-cycle system: the counting words for 1 to 8 are essentially the same (Lang gives dialect variations as well) as those given in Table C89 and C90 from the Mai informants. In counting 9 and 10 in the 60-cycle the cycle unit *tukutepo* is used so that these are respectively *tukutepo(n)+1*, *tukutepo(n)+2*. In the modified 10-cycle version, 9 is simply the cycle unit *tukutepo*, and 10 is *akalita* or *akalita mendai*, both of which contain the morpheme for "man" *akali*.

Table C90
Engan (4. 60) Cycle System

	Enga		
1	*me(n)dai*	31	*yanapun(ya) tepo*
2	*lapo*	32	*yanapun(ya) gato*
3	*tepo*	33	*kamapon(ya) me(n)dai*
4	*kitome(n)de*	34	*kamapon(ya) lapo*
5	*yungi (yugi)*	35	*kamapon(ya) tepo*
6	*tokage*	36	*kamapon(ya) gato*
7	*kalage*	37	*kujupun(ya) me(n)dai*
8	*tukulapo*	38	*kujupun(ya) lapo*
9	*tukutepon(ya) me(n)dai*	39	*kujupun(ya) tepo*
10	*tukutepon(ya) lapo*	40	*kujupun(ya) gato*
11	*tukutepon(ya) tepo*	41	*kalipun(ya) me(n)dai*
12	*tukutepon(ya) gato*	42	*kalipun(ya) lapo*
13	*mapun(ya) me(n)dai*	43	*kalipun(ya) tepo*
14	*mapun(ya) lapo*	44	*kalipun(ya) gato*
15	*mapun(ya) tepo*	45	*labalun(ya) me(n)dai*
16	*mapun(ya) gato*	46	*labalun(ya) lapo*
17	*yupun(ya) me(n)dai*	47	*labalun(ya) tepo*
18	*yupun(ya) lapo*	48	*labalun(ya) gato*
19	*yupun(ya) tepo*	49	*menainin(ya) me(n)dai*
20	*yupun(ya) gato*	50	*menainin(ya) lapo*
21	*watakapun(ya) me(n)dai*	51	*menainin(ya) tepo*
22	*watakapun(ya) lapo*	52	*menainin(ya) gato*
23	*watakapun(ya) tepo*	53	*akipun(ya) me(n)dai*
24	*watakupun(ya) gato*	54	*akipun(ya) lapo*
25	*paipun(ya) me(n)dai*	55	*akipun(ya) tepo*
26	paipun(ya) lapo	56	*akipun(ya) gato*
27	*paipun(ya) tepo*	57	*kaeapalun(ya) me(n)dai*
28	*paipun(ya) gato*	58	*kaeapalun(ya) lapo*
29	*yanapun(ya) me(n)dai*	59	*kaeapalun(ya) tepo*
30	*yanapun(ya) lapo*	60	*kaeapalun(ya) gato*

Source: Lean's data from Mai dialect informants. Many CSQs supported some of these words/patterns.

The CSQ informants fall into two groups: those who report a 60-cycle counting system and those who report a 10-cycle system. It is possible that the latter is a relatively recent modification of the former, the modification occurring as a result of the influence of the decimal English or Tok Pisin systems. Several CSQ informants also provide data on the 60-cycle and these largely agree with those given by Lang (earlier dialect data) with some variations. For example, each cycle unit is suffixed but there was some variation in the suffix given: *-n*, *-nga*, or *-na*. One informant indicated that the suffix *–nya* means "you take"; another indicated that the suffix *-na* means "you eat". In the former case each 4-cycle has the construction "cycle unit+you take+*n*". In the 60-cycle the cycle unit *tukutepo* is used so that these are respectively *tukutepo(n)*+1, *tukutepo(n)*+2. In the modified 10-cycle version, 9 is simply the cycle unit *tukutepo*, and 10 is *akalita* or *akalita mendai*, both of which contain the morpheme for "man", *akali*.

Kirschbaum's data show some resemblance to other Enga data as far as counting from 1 to 5 is concerned. Thereafter, however, the resemblance ceases except for 11, *tugulap*, which is similar to the Enga *tukulapo,* i.e. 8. One feature of Kirschbaum's data, not included in Table C88, is that each of the number words is correlated with a body-part; this does not accord with the other Enga data for which we have no report of body-part tallying. The numbers and associated body-parts are given in Table C91. Ipili is a body-part tallying system in the Engan Province but only the word for 5 is similar so there is an old connection, the words varying considerably.

Table C91
Enga Dialect Body-Tally System

	Enga	Body Part
1	*mondei*	little finger of right hand
2	*rap*	fourth finger of right hand
3	*top*	third finger of right hand
4	*kidumonto*	second finger of right hand
5	*you*	thumb of right hand
6	*kunju*	forearm
7	*map*	elbow
8	*poyup*	shoulder
9	*kunju(?)*	neck
10	*yenap*	ear
11	*tugulap*	eye
12	*kalap*	nose

Source: Kirschbaum (1938, p. 278).

The 10-cycle system of Kyaka Enga (Table C92) does not, on the other hand, appear to be a modified version of a 60-cycle system. Indeed this appears to be a finger-count system, the counting word for 5 being *kingipaki* where *kingi* is a "hand" morpheme. The counting words for 8 and 9 contain, respectively, the numerals 2 and 1, *lama* and *mendaki*, suggesting that these are formed by a subtractive process, i.e. 10-2, 10-1. The counting word for 10 is *akalisa* which contains a "man" morpheme, *akali*, suggesting that "one man's hands" is the 10-cycle unit. Subsequent decades are formed using *akalisa* so that 20 is *akalisa lama*, 30 is *akalisa tema* (Table C92).

Table C92
Lean's CSQ Data on Engan 10-Cycle Systems

	A	B
1	*mendaki*	*mendai*
2	*lama*	*lapo, lama*
3	*tema, trema, rema*	*tepo, tema*
4	*kisima*	*kitumende, kituma*
5	*kingipaki*	*kondape, kondape mende*
6	*pakinyamange*	*yangi mange*
7	*yanda ipingi*	*sakaita*
8	*mange lama karo*	*tukulapo*
9	*mange mendaki*	*mange mendai wakitao*
10	*akalisa*	*akalita*
11	*akalisa ipisu mendaki*	*akalita kisa mendai*
12	*akalisa ipisu kingipaki*	*akalita kisa kondape*
20	*akalisa lama*	*akalita lapo*
30	*akalisa tema*	*akalita tepo*

Source: A. 9 CSQs from Kyaka Enga, Baiyer River area, B. 9CSQs from Layapo, Wapenamanda area.

One other CSQ informant agrees with the cycle units as given in the Mai data but differs in associating the units with particular 4-cycles, i.e. in the table above, the cycle unit for 37 to 40 is *kujupu*, that for 41 to 44 is *kalipu*, that for 45 to 48 is *lapalu*, and that for 49 to 52 is *menaini*. The CSQ informant gives the cycle units in a different order so that for the same sequence of 4-cycles we have, respectively, the units: *menaini, kujupu, kalipu*, and *lapalu*. The Layapo dialect data as given in Table C92 also comprise a 10-cycle system. The first four numerals are identical to those of the 60-cycle system; 5, 6, 7 are, however, quite different; 8 is *tukulapo* as found in the 60-cycle system; 9 is similar to the Kyaka Enga numeral and contains the numeral 1 which, as suggested above, may indicate that 9 has a 10-1 construction. The counting word for 10 is *akalita* which contains the morpheme for "man", *akali*. The higher decades are multiples of this so that 20 is *akalita lapo* or "two tens", 30 is *akalita tepo* or "three tens". As suggested in the case of Kyaka Enga, the Layapo system may also be a finger-count system with the "hands of one man" being 10. The word for 5 contains a hand morphene. A man morphene is used for ten, suggesting one man's hands is used indicating a finger-count system.

In Table C93, Kopamu's (2005) data indicate that a 6-cycle system may be evident.

Table C93
Engan Counting System based on 6-Cycle and Mathematical Terms

Symbol	English	Enga Tokples	Literal meaning
1	One	*mendai*	
2	Two	*lapo*	
3	Three	*tapo*	
4	Four	*kiromende*	
5	Five	*yungi*	
6	Six	*tokange*	
7	Seven	*kalange*	
8	Eight	*tukulapo*	Six and two
9	Nine	*tukutepo*	Six and three
10	Ten	*kingira*	A complete group of fingers
>	Greater than	*nyo ana pyandenge*	
5>2	Five is greater than two	*yungi ima lapo ana-pyandenge*	
<	Less than	*nyo kisa pyasingi*	
2<5	Two is less than five	*lapo ome yung kisa-pyasingi*	
=	Equal	*ongo*	You have, exactly
	That one	*minelene ongo*	That one exactly
	Balance	*kapakapa*	In equal quantity
≠	Not equal	*ongo da*	Equal not
	Congruent shapes	*opale-erere*	Having equal features
	Belongs to	*ongonya*	
	Does not belong	*ongonya da*	Belonging not
+	Plus (add)	*kisade*	On top put
	Addition	*kis-de*	
–	Subtraction	*neparenges*	Take away
		tepo saranya yungi neparenges ongo lapo da-jarae	Two act-on-by five take-away we-have two not-enough
–2	Negative 2	*lapo da-jarae*	Two not-enough-of
0	Zero	*mende-nasingi*	Does not exist or have anything
	No solution	*mende-da*	None of these
×	Times (multiply)	*aka*	
½	Half	*parayaingi*	
¼	One quarter	*kia mendai*	
1/8	One eighth	*lamba mendai*	
n/m	*n* over *m* (*n* divided by *m*) or fraction	*m kunjapaenya n*	
%	Percent	*kunjapaenya mendai*	Fraction of something
10%	Ten percent	*kingara kunjapaeanya mendai*	
()	Brackets	*urupa sranya*	
	Empty set	*andanda or imbu*	

Source. Kopamu (2005).

Appendix D
Details of Body-Part Tally Systems

Kay Owens

Abstract Appendix D refers to material from Glen Lean's thesis and the work on the East Strickland Family and Bosavi Family by Peter Dwyer and Monica Minnegal. It provides a Table generated from a spreadsheet on the languages as well as details of three of the languages.

Keywords Body-part tally systems • Sepik languages • Western Province Papua New Guinea languages • Southern Highlands Papua New Guinea languages

Glen Lean provided the most extensive work on the body-part tally systems in his 1992 thesis (Lean, 1992) and this was further discussed by Owens (2001) with additional materials provided by Saxe over the years (Saxe, 1981, 2012). Recently Dwyer and Minnegal (2016) have been collating data on languages in Western Province and Southern Highlands Province. They made use of materials they had collected around 1980 and later as well as the work, in particular, by Daniel Shaw (2009) and Britten and Sören Årsjö (Årsjö & Årsjö, 2005), all of whom have spent considerable time in the area. To provide an example of the data, Tables D1 presents data on three neighbouring languages from the south west of Papua New Guinea. Along with a large number of systems from languages across Papua New Guinea, details of their classification are given. It should be noted that the classification of these languages is on-going. Lean generally followed Wurm and Hattori's classifications. The SIL Ethnologue gave a number of these languages as Trans New Guinea Phylum whereas Lean had recorded that they were Sepik-Ramu Phylum, Sepik Sub-Phylum, Sepik Hill Stock. Ross also attempted to classify and relate the languages. In Lean's appendices, he frequently gave just the Family indicating that where they fit in terms of phyla was either less relevant or more problematic than noting immediate Family members. Dwyer and Minnegal also included a table giving the percentage of language similarities between the languages based on a standard comparison.

Nomad (Kubo), Ebolo, and Bedamuni

Bedamuni and Nomad are found in the Western Province. Nomad is near the Strickland River while Etolo reside mostly in the Southern Highlands close to the borders of the Western Province and Hela Provinces (near Huli speakers). Dwyer and Minnegal (2016) said:

> Bosavi family includes Edolo, Bedamuni, Onobasulu, Kaluli, Sonia, Eibela and Kasua and the East Strickland family includes at least Konai, Febi, Kubo, Samo, Gobasi (Gebusi) and Odoodee with some authors including Honibo and Oybae as separate languages to the south-west of Nomad. Population sizes of these language groups vary from less than 500 (e.g. Febi and Eibela) to reach 4,000 (Kaluli) and around 10,000 (Bedamuni). With the exception of Eibela, population sizes are larger for Bosavi language groups than for East Strickland language groups. Kelly (1993) brings together a great deal of information on the ethnography of the Strickland-Bosavi language groups. (p. 3)

K. Owens et al., *History of Number*, History of Mathematics Education,
DOI 10.1007/978-3-319-45483-2

Table D1
Etolo, Bedamuni, and Kubo (Nomad dialect) Body-Part Tally Systems

	Etolo		Bedamuni	Kubo	
1	*age*	right hand, little finger	*age*	*dano*	right little finger
2	*agedu*	right hand, ring finger	*agedu*	*biau*	right ring finger
3	*asota*	right hand, middle finger	*asoda*	*komadia*	right middle finger
4	*piadu*	right hand, index finger	*biadu*	*dasodia*	right index finger
5	*pi*	right hand, thumb	*bi*	*haudia*	right thumb
6	*gafe*	right hand, palm	*gafe*	*waiendia*	right wrist
7	*gafotan*	right arm, wrist (ventral)	*lobobasolo* [cf. Etolo *gafetanu*]	*djuadia*	right lower arm
8	*goton*	right lower arm (ventral, half way along length)	*godo*	*dumadia*	right elbow
9	*segen*	right arm, inner elbow	*sesege*	*dobedia*	right upper arm
10	*nabun*	right bicep	*nabu*	*oidia*	right shoulder
11	*gita*	right shoulder (laterad)	*gida*	*gabaidia*	right side of neck
12	*giwi*	at junction of right clavicle with neck	*fago* [cf. Etolo *giwi*]	*dudia*	right ear
13	*garon*	neck, right side	*galo*	*dihodia*	right eye
14	*gehe*	right ear, lower lobe	*ge*	*midia*	nose
15	*pa*	right temple	*ba:*	*diho iodia*	left eye
16	*si*	right eye	*si*	*du iodia*	left ear
17	*mimogo*	tip of nose	*mi*	*gabai iodia*	left side of neck
18			*lei si*	*oi iodia*	left shoulder
19			...	*dobe iodia*	left upper arm
20				*duma iodia*	left elbow
21				*djua iodia*	left lower arm
22				*waien iodia*	left wrist
23			*lei gida*	*hau iodia*	left thumb
24			*nabu*	*daso iodia*	left index finger
25			...	*kamai*	left middle finger
26				*sasamai*	left ring finger
27				*sasafai*	left little finger
33			*age*		

Source. Dwyer and Minnegal (2016)

Summary of Most Body-Part Tally Systems

Diagrams from Lean's (1992) thesis appendix are given in Figure D1. The following pages present Table D2 which is the summary of languages with body-part tally systems. Some of these are identified as having 2-cycle systems in addition to the body-part tally system. This table was developed by Owens and Imong from Lean's (1992) data but some Western Province language classifications have been updated. Significantly there is a large diversity suggesting influences from relationships with neighbours and innovations. The number of systems is also noteworthy. It would also seem that body-part tallies were in early proto-languages as suggested in Chapter 10 and were not a later innovation.

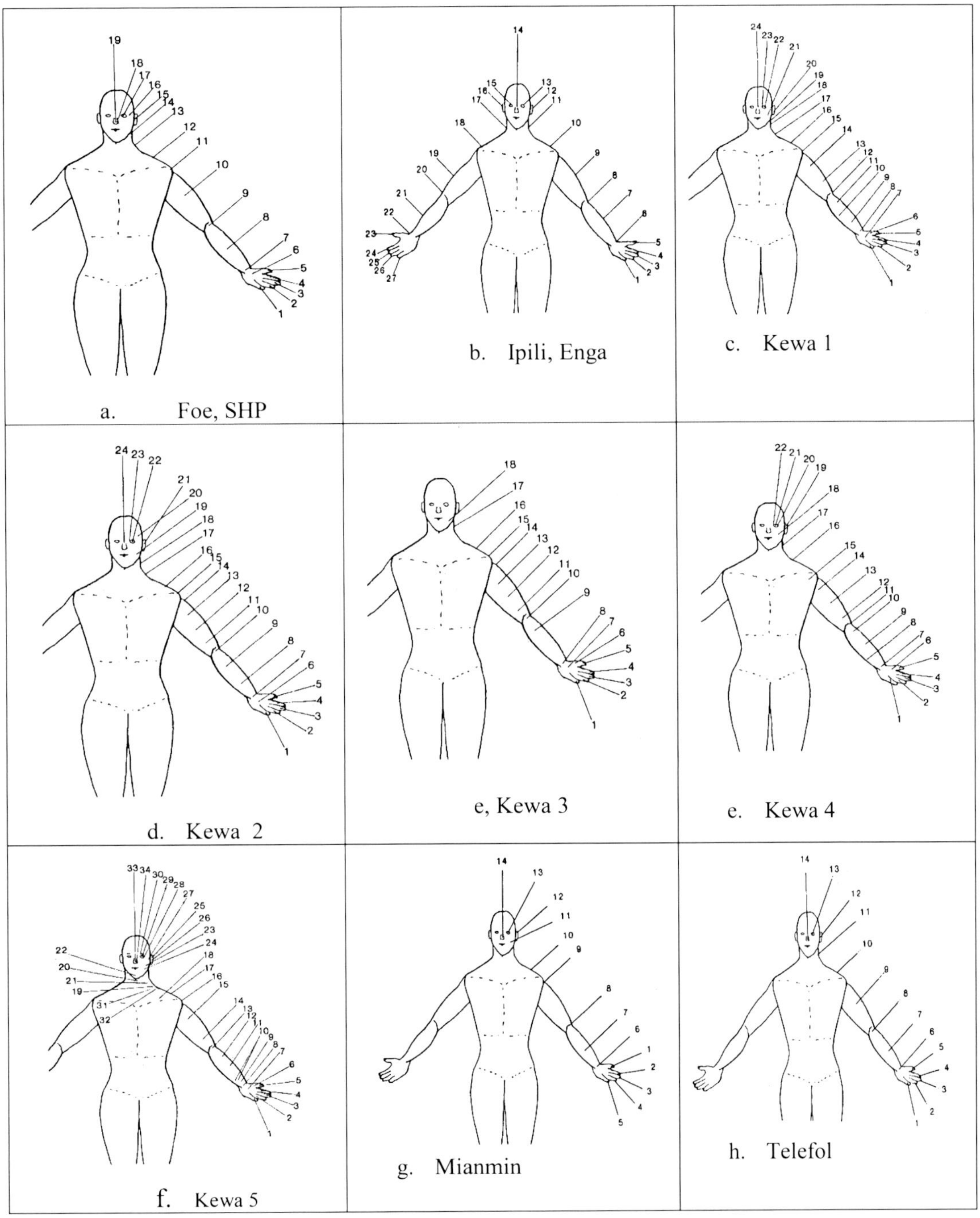

Figure D1. Diagrams of body-part tally systems from Lean's (1992) thesis.

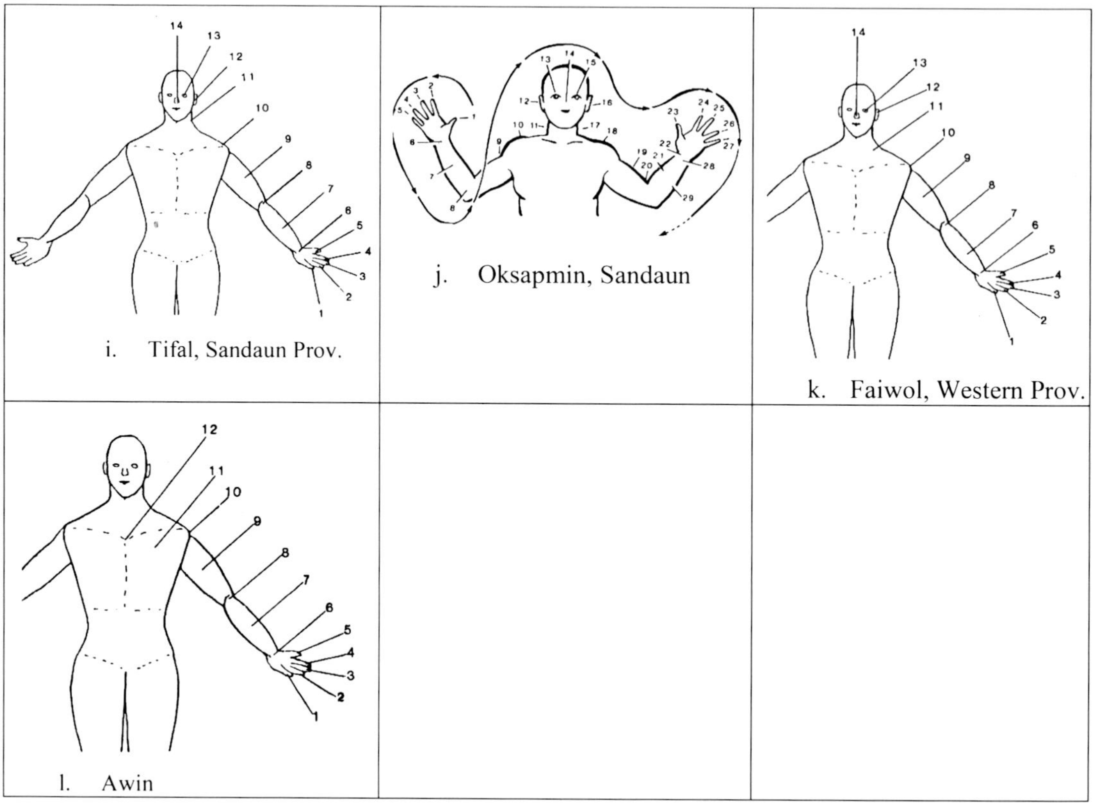

Figure D1. (continued)

Table D2
Summary of Lean's collated data on body-part tally systems.

Classification	Language	Province	Cycles	Frame words	Comments
S-R,S,SH, Duna-Pogaia Family	Duna (Yuna)	Hela	14, (15?)	1–14	14 cycle asymmetric body part tally system moving up one side and across to the eye and ear of the other side
S-R,S,SH, Sanio Family	Hewa	Hela	27	1–14	27 part body-part tally system
S-R,S,SH, Sanio Family	Piame	East Sepik	?	1–10, ?	From thumb to shoulder used in body-part tally system but not sufficient data to determine cycle
	Sanio	East Sepik	27, 2	1–14	Body-part tally with 27 cycle points with nose as centre symmetry point. A separate 2 cycle numeral system too
S-R,S,SH, Alamblak Family	Alamblak	East Sepik	29	1–15	Two systems coexist with a body-part tally system. Men and women have different 29 cycle body-part tally systems using left arm to nose (men) and down right arm
Arai (Left May) Phylum	Ama	East Sepik	12?	1–6, 7, 9, 11	Body-part tally is unusual in including the navel and switching from left and right breast and shoulders. Unusual if only a system up to 12 but no more data suggested, only 'many'
TNGP, East New Guinea Highlands Stock, West Central Family, Enga Sub-Family	Ipili	Enga	28	1–28	27 body-parts are used with some parts between 6 and 14 and 15–22 having the names of the body parts (i.e. not the digits of the hands). The nose is the central number. 28 is man or "pondo" in both dialects. A second set of words could be ordinal numbers rather than cardinal numbers in both dialects
TNGP, Eleman Family, Western Eleman Sub-Family	Orokolo	Gulf	27	1–14	Body part tally of 27 parts with modification to accommodate introduced decimal system, abandoning at 10
TNGP, Purari Family-Isolate	Purari	Gulf	23, 2′, 3?	1–12	Body tally, modified 2 cycle and 7=3+3+1 etc indicating a 3 cycle or part of a pairs Motu-type system
TNGP, Inland Gulf Sub-Phylum Level Family	Minanibai	Gulf	?	?	Insufficient data to confirm a body-part tally system
TNGP, Turama-Kikorian, Turama-Omati Family	Omati	Gulf	?	1–10	Insufficient data to confirm body-part tally system
Pawaian Stock-Level Isolate	Pawaia	Gulf	?	1–5, 10	Possible body-part tally sysem
TNGP, East New Guinea Highlands Stock, Kalam Family	Kalam	Madang	23	1–12	A body-part tally system with a 23- cycle, the symmetrical mid-point of the cycle being tallied at 12 at a point at a base of the throat
TNGP, Mabuso Stock, Hanseman Family	Murupi	Madang	23	1–12	23- cycle body-part tally one with sternum as midpoint
TNGP, East New Guinea Highlands Stock, East Central Family	Gende	Madang	31	1–16	31 points with nose as midpoint; also 2 cycle system

(continued)

Table D2
(continued)

Classification	Language	Province	Cycles	Frame words	Comments
TNGP, Brahman Stock-level Family	Biyom	Madang	?	1–5?	Insufficient data to confirm a body part tally system
TNGP, Isumrud Stock, Kowan Family	Waskia	Madang	?, 2, 5	1, 2, 5	2, 5 cycle system but insufficient data to confirm digit tally as some indication of body-part tally system
Nagatman Phylum-Level Isolate	Nagatman	Sandaun	74?	1–5, 10 recorded	Left side presumably from little finger, side of body, left leg to big toe and then revese order on right
Sepik-Ramu Phylum, Leonhard Schultze Stock, Family	Tuwari	Sandaun	?	1–14?	Body parts system with words at least to 10
Senagi Stock, Family	Anggor	Sandaun	2′, 18, 23	1–12, 20–23	Asymmetric to 18, or symmetric 23 using sternum as midpoint; 1–4 are numerals from binary system
Papuan	Mianmin	Sandaun	2, 27	1, 2, 7–14	Body part tally system with 27 tally points and a 2 cycle system, the words for the first 6 numbers are the first words of the body-part tally system in one source
Phylum Level Isolate	Yuri	Sandaun	23	1–12	23 part body part system with sternum as midpoint
TNGP, Oksapmin Sub-Phylum-Level Isolate	Oksapmin	Sandaun	27	1–14	Body parts counting system of 27 parts with an additional 2 cycle counting system; truncated to 20 (or taken to 30) with money influence
TNGP Ok-Awyu, Ok Family	Telefol	Sandaun	27	1–14	Body parts system of 27 parts
	Tifal	Sandaun	27	1–14	Body parts tally system with 27 parts
Kwomtari Stock, Baibai Family	Baibai	Sandaun	?	1–5	Finger names for 1–4 and 5 is not "hand"
Kwomtari Stock, Kwomtari Family	Kwomtari	Sandaun	?	1–5	CSQ suggested body-part tally with body parts & both hands, 5 is not "hand"; used in a 10-cycle system (not binary)
TNGP, East New Guinea Highlands Stock, Wiru Family-Level Isolate	Wiru	Southern Highlands	23?, 45?, 4	1–23?, 45?; 1–4	Body part tally, probably of 23 parts and a 4 cycle system, 5 = 4 + 1, 8 = 2 × 4
TNGP, East New Guinea Highlands Stock, West-Central Family	Kewa	Southern Highlands	18, 33, 35, 47 (68)	1–18, 22, 24	47 body-part system and others, 4 cycle system used as well as the body-part tally; later some truncated to 10, 18

(continued)

Table D2
(continued)

Classification	Language	Province	Cycles	Frame words	Comments
TNGP, East New Guinea Highlands Stock, West-Central	Mendi	Southern Highlands	20 or 39?	1–20	West Kewa similar to South Mendi; some variation between informants of body parts; also counted in pairs and quartets
Family Australian	Mendi	Southern Highlands	4, 48	1–4, 16, to 20, 24, 28	4 cycle unit system within a 48 cycle systems—Angal Heneng d.; counted in sets of 20 and also 16 or 24 or 28
	Mabuiag	Western	18	1–10	Australian language. Small counting system data suggest a digit tally but also seems to be a body part tally system to cycle 18 or 19
Trans-Fly Sub-Phylum-Level Stock, Eastern Trans-Fly Family	Gidra	Western	19	1–10	Body part replaced by digit tally systems apparently
Trans-Fly Sub-Phylum-Level Stock, Eastern Trans-Fly Family South Central Papuan, Pahoturi	Bine	Western	19, 2, 3, 5	1, 2, 3, 5 or up to 10	Older data suggest a body-part system with 10 as midpoint and with 2 cycle numerals for 1–3; possible 3 cycle data with 4=3+1, 3+2, 3+3, 3+3+1, 3+3+2, 3+3+3, 3+3+3+1 or hybrid of 5 and 3 cycles
River Family	Miriam	Western	29? or 25?	1–16	Body-part tally system and a 2 cycle system. 4=2+2, 5=2+2+1
	Agob	Western	18	1–9	Numerals for 1, with 1–4 as combinations. 5=2+2+1 and not a known hand morphene. Also a body part tally system. By Lean's time only numerals were used in various combinations
TNGP, Awin-Pare Family	Awin	Western	2′, 23	1, 2, 4, 5–12	2 cycle with 3=2+1 but from 5 on body part tally of 23 cycle matches part words
TNGP, Inland Gulf Sub-Phylum Level Family, Minanibai	Tao-Suamato	Western	?	1–5, 10	5 is "thumb" and 10 is "upper arm"
TNGP, Lowland Ok Sub-	Ninggirum	Western	2′, 30	1, 2, 4 to 15	Body part tally system with 30 cycle crossing directly across from left eye to right. Numerals for 1, 2 3=2+1, 4, 5
TNGP, Lowland Ok Sub- TNGP, Mountain Ok Sub-Family	Yonggom	Family Western	2′, ?	1, 2, 4, to ?	2 cycle modified and probably body part system of 27–29
	Faiwol	Western	2, 27	1, 2, 5 to 14	Body part tally with numerals for 2, 3, 4 from a 2-cycle
TNGP, Bosavi Family	Edolo (Etolo)	Southern Highlands	17	1–17	17 parts on right side of body only to nose, multiples of 17 for larger numbers
TNGP, Bosavi Family TNGP, Bosavi Family	Onabasulu	Southern Highlands	33	1–17	Symmetrical from right side to left, nose as midpoint
	Kaluli	Southern Highlands	35	1–18	Lean considered body-part tally and gave 1–5, 10. 5 and 10 do not include a hand morph, symmetrical with ridge of nose as midpoint, usually from left, right little finger is complete count, use 1, 2, 3 as noun modifiers and higher for multiples of 35

(continued)

Table D2
(continued)

Classification	Language	Province	Cycles	Frame words	Comments
TNGP, Bosavi Family TNGP, East Strickland Family	Kasua	Southern Highlands	35	1–18	Lean noted 1–5; 5 has no hand morph. symmetrical with ridge of nose as midpoint, uses sides of nose too
	Eibela	Western & Southern Highlands	35	1–18	symmetrical from left side, ridge of nose as midpoint, uses sides of nose too
	Bedamuni	Western	33	1–17	symmetrical from right side to left, nose as midpoint
	Konai (Kalai)	Western	13	1–13	one side of body from left little finger to nose; Lean recorded 1–5 with 5 as thumb; also numerals from a 2 cycle and possible 5 cycle; Årsjö & Årsjö (2000) provides detailed body-tally system
TNGP, East Strickland Family	Kubo	Western	27, 2′	1–14	Lean included as Nomad dialect with 2′ cycle, right little finger start
	Odoodee	Western	12	1–12	One-side of body only from left hand little finger to join of neck and shoulder
	Febi	Western	28	1–14	Symmetrical from left hand little finger with two central points of nose then mouth
	Samo		20, 2′	1 to 16 and 20	Lean included as Nomad dialect with 2′ cycle with 10 as "many" (same word as Honibu dialect for 5). Shaw, long term resident, (2009) cited in Dwyer gave 16 cycle points with 11 as top of head, fewer points than most, Shaw did not see 16, 17, 19 but 10 was "two hands" (fist together), 20 was fist together "many hands"

Note: TNGP is Trans New Guinea Phylum. S-R,S,SH is Sepik-Ramu Phylum, Sepik Sub-Phylum, Sepik Hill Stock.

Appendix E
Maps of Languages and Diagram of Proto Oceania

Glen Lean and Kay Owens

Abstract In this Appendix, a selection of maps prepared by Lean are provided to make it easier to locate the languages referred to in Appendices B and C and in the chapters. These maps were based on Wurm and Hattori's (1981) atlas. They are not all the maps used by Lean to locate languages as he combined them with geographical topological maps often made by the Australian army during World War II to locate villages mentioned by students and field trips. The Proto Austronesian Proto Oceania diagram assists with the text on Oceania languages.

Keywords Language maps of New Guinea • language maps of Melanesia • Non-Austronesian languages • Oceania languages.

Maps

The maps in this Appendix are a selection from the maps that Lean (1992) used in his Appendices. They are based on Wurm and Hattori's (1981) language atlas. The list of names given with each map is that used by Lean. Readers should consult the Ethnologue, recent papers or other on-line resources for the latest language names. In Wurm and Hattori's time some current languages were considered dialects but they are now considered languages in their own right. A glance at the tables in Appendices B and C will soon show the diversity of names for languages partly as a result of the person recording a particular name and partly because the language itself does not have a name for itself other than a generic term. An example is Tinatatuna "proper language" for the Tolai as Paraide uses although Lean used the word Tolai for their language in preference to using the commonly used word Kuanua. It is also noted that the Southern Highlands Province PNG is now Southern Highlands Province and Hela Province PNG, and Western Highlands Province PNG is now Western Highlands Province and Jiwaka Province. Irian Jaya, now known as two Provinces of Indonesia, we refer to as West Papua being the name that refugees from this country gave for their land. In Chapter 8, there are another four maps of languages referred to in that chapter.

K. Owens et al., *History of Number*, History of Mathematics Education,
DOI 10.1007/978-3-319-45483-2

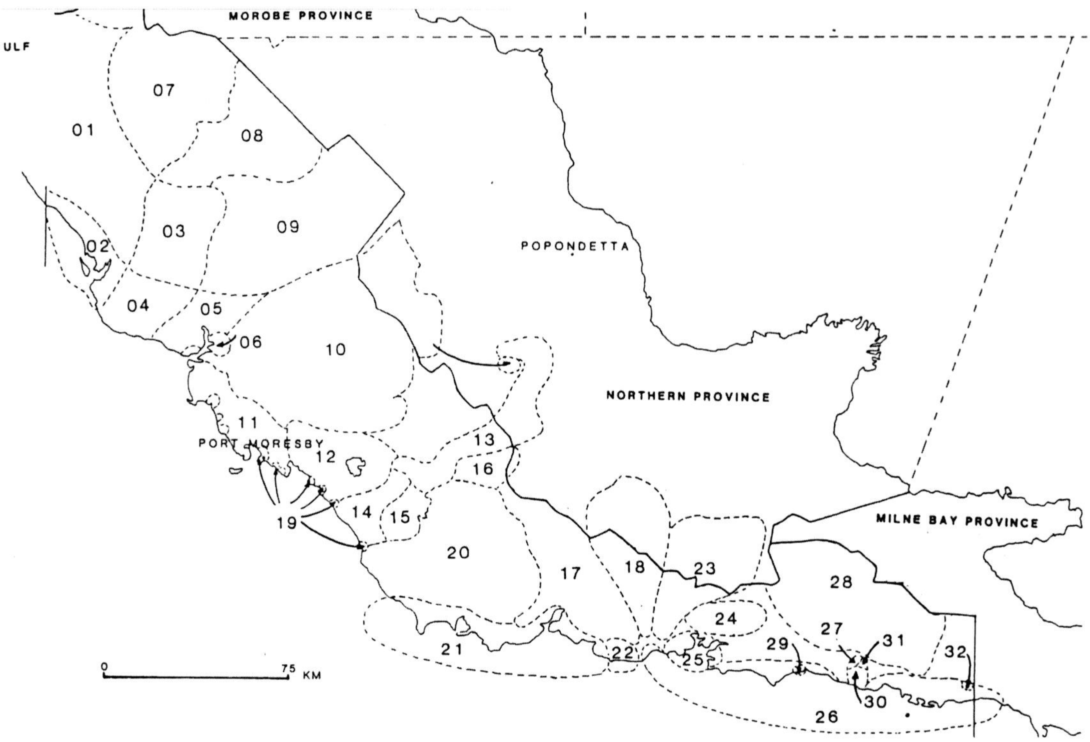

1	Mekeo	9	Fuyuge	17	Maria	25	Morawa
2	Roro	10	Mountain Koiari	18	Abia	26	Magi
3	Kuni	11	Koita	19	Motu	27	Laua
4	Nara	12	Koiari	20	Sinagoro	28	Daga
5	Gabadi	13	Barai	21	Keapara	29	Ouma
6	Doura	14	Humene	22	Domu	30	Magori
7	Kunimaipa	15	Kwale	23	Bauwaki	31	Yoba
8	Tauade	16	Doromu	24	Binahari	32	Bina

Figure E1. Map of languages in Central Province, PNG.

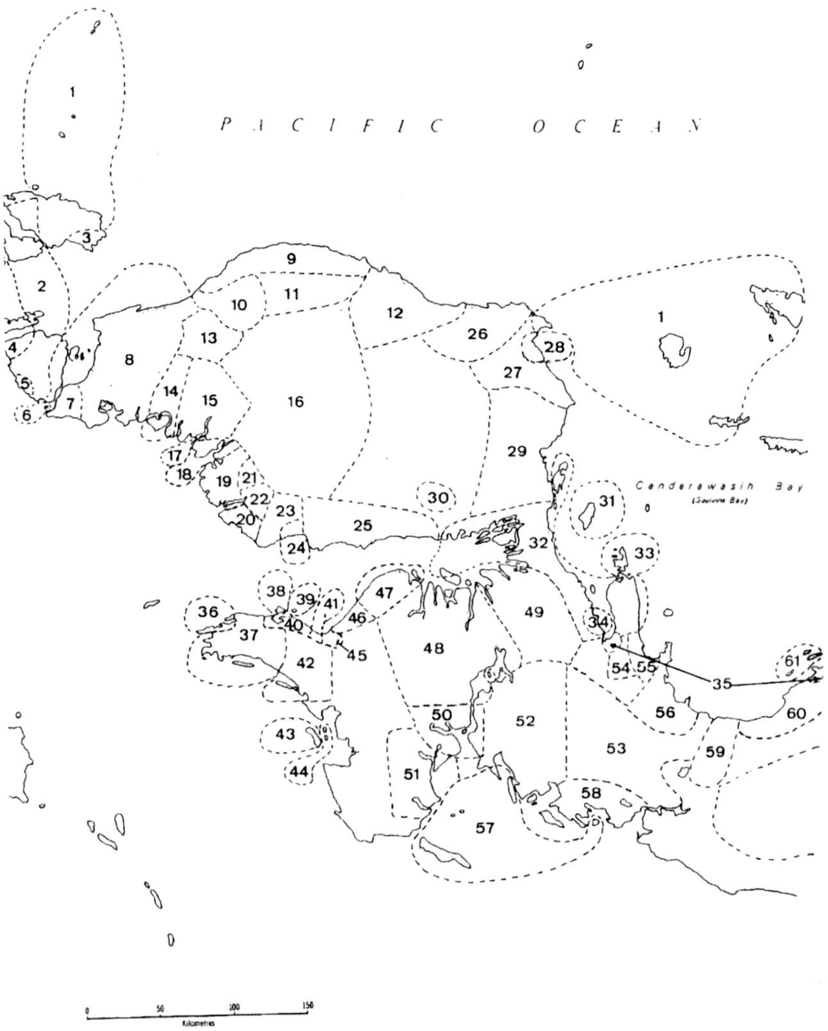

1	Biak	9	Karon Pantai	17	Kuwani	25	Arandai	33	Ron	41	Erokwanas	49	Nabi	57	Kaiwai
2	Maya	10	Madik	18	Konda	26	Meax	34	Dusner	42	Baham	50	Iria	58	Semimi
3	Laganyan	11	Karon Dori	19	Yahadian	27	Hattam	35	Waropen	43	Karas	51	Asienara	59	Iresim
4	Maden	12	Amberbaken	20	Inanwatan	28	Borai	36	Onin	44	Uruangnirin	52	Mairasi	60	Tarunggare
5	Palamul	13	Moraid	21	Kampong Baru	29	Mantion	37	Iha	45	Mor	53	NE Mairasi	61	Môr
6	Durianker	14	Kalabra	22	Puragi	30	Meninggo	38	Sekar	46	Barau	54	Tandia		
7	Seget	15	Tehit	23	Kasuweri	31	Moeswar	39	Arguni	47	Tanahmerah	55	Yeretuar		
8	Moi	16	Brat	24	Tarof	32	Wandamen	40	Bedoanas	48	Irahutu	56	Yaur		

Figure E2. Languages of western West Papua.

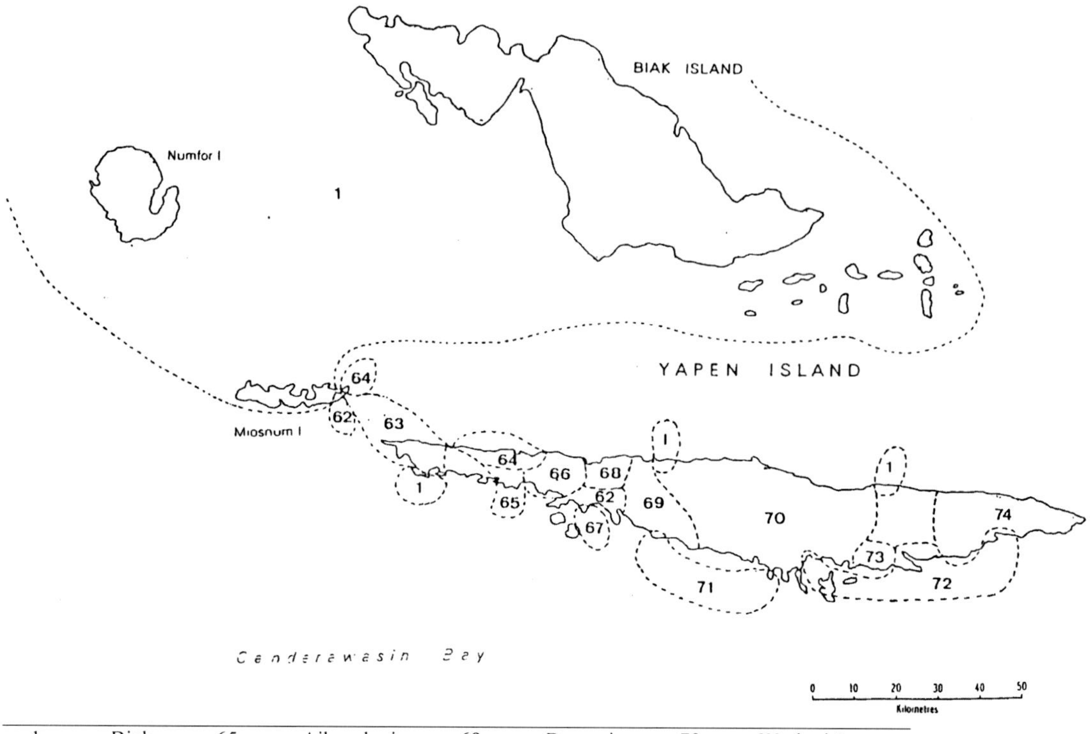

1	Biak	65	Aibondeni	69	Busami	73	Wadapi-Laut
62	Ansus	66	Marau	70	Yava	74	Wabo
63	Woi	67	Papuma	71	Serui-Laut		
64	Pom	68	Munggui	72	Ambai		

Figure E3. North central section of West Papua.

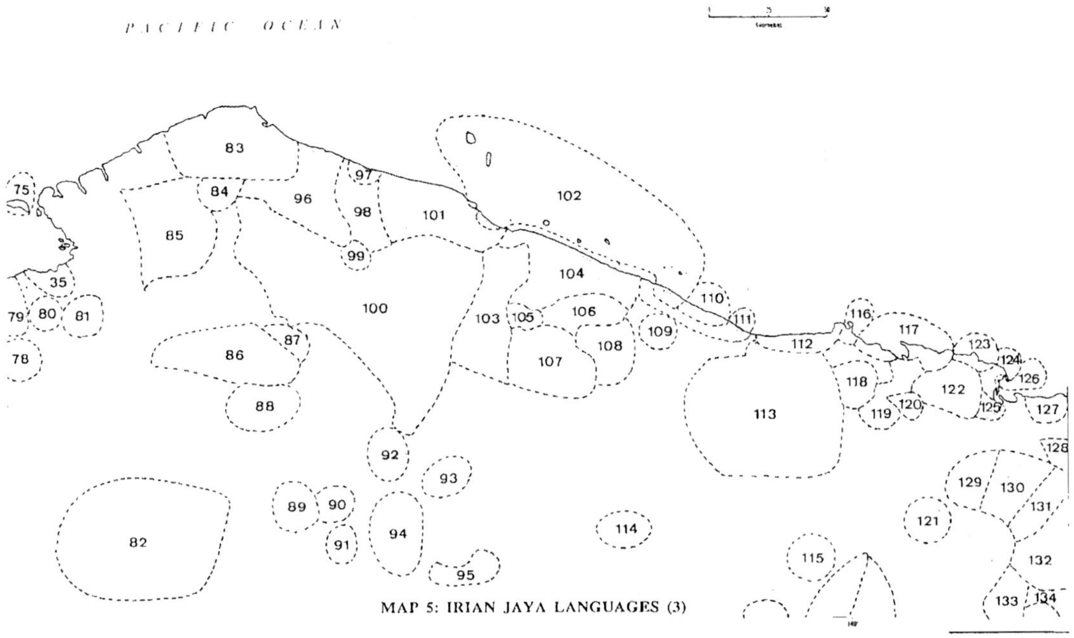

75	Kurudu	85	Bapu	95	Taworta	105	Bonerif	115	Sause	125	Nafri
76	Sauri	86	Bauzi	96	Airoran	106	Itik	116	Demta	126	Kayupulau
77	Kofei	87	Taurap	97	Massep	107	Mander	117	Tanahmerah	127	Sko
78	Siromi	88	Baburiwa	98	Samarokena	108	Wares	118	Nimboran	128	Sangke
79	Baropsi	89	Taogwe	99	Sasawa	109	Maremgi	119	Mekwei-Gresi-Kwansu	129	Awyi
80	Bonefa	90	Taori-Kei	100	Kwerba	110	Bongo	120	Kemtuk	130	Taikat
81	Nisa	91	Taori-So	101	Saberi	111	Mawes	121	Morwap	131	Manem
82	Turu	92	Aikwai	102	Sobei	112	Tarpia	122	Sentani	132	Waris
83	Warenbori	93	Papasena	103	Berik	113	Uria	123	Orma	133	Senggi
84	Pauwi	94	Weretai	104	Kwesten	114	Foau	124	Yotafa	134	Waina

Figure E4. North east coast and inland languages of West Papua.

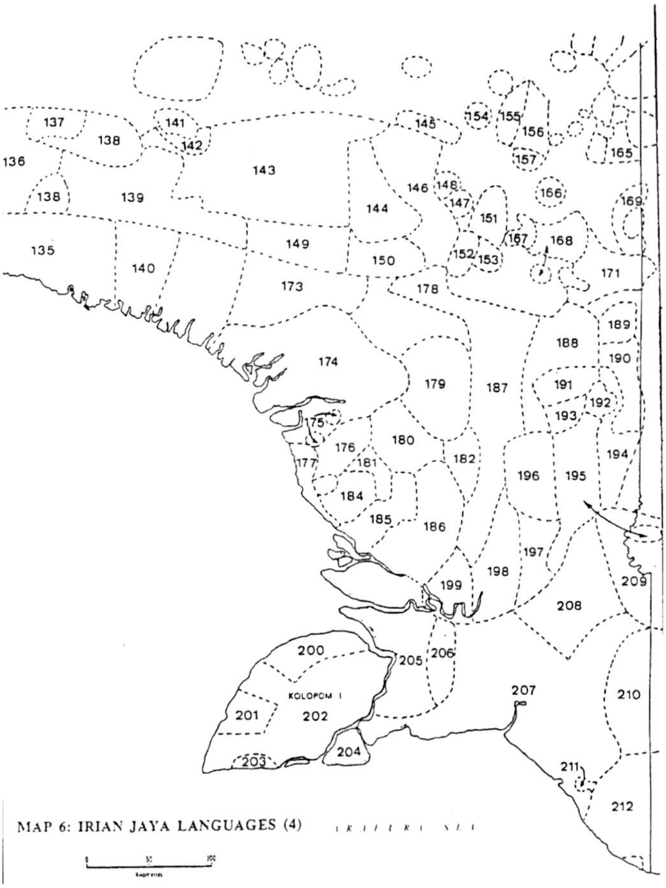

135	Kamoro	145	Dabra	155	Narau	165	Emumu	175	Warkay-Bipim	185	Tamagario	195	Kaeti	205	Yelmek
136	Egaki	146	North Ngalik	156	Kaure	166	Sirkai	176	Sawuy	186	Yaqay	196	Aghu	206	Maklew
137	Wodani	147	Nipsan	157	Kosare	167	Eipo-Mek	177	Casuarina Coast Asmat	187	Mapi River	197	Kia River	207	Marind
138	Moni	148	Kosarek	158	Molof	168	Ketengban	178	Somahai	188	Upper Digul	198	Ederah River	208	Bian Marind
139	Uhunduni	149	Nduga	159	Usku	169	Pyu	179	Citak Asmat	189	Iwur	199	Siagha-Yenimu	209	**Boazi**
140	Sempan	150	South Ngalik	160	Tofamna	170	Biksi	180	Pisa	190	North Kati	200	**Riantana**	210	Yey
141	Dem	151	Nalca	161	Yafi	171	Ngalum	181	Airo-Sumaghaghe	191	Kotogüt	201	**Ndom**	211	Moraori
142	Wano	152	Korapun	162	Dera	172	Kinome	182	Upper Kaene River	192	Wambon	202	Kimaghama	212	**Kanum**
143	Western Dani	153	Goliath	163	Dubu	173	North Asmat	183	Kaugat	193	Wanggom	203	Koneraw		
144	Grand Valley Dani	154	Kapori	164	Towei	174	Central Asmat	184	Kaygir	194	South Kati	204	Mombum		

Figure E5. Southern coast of West Papua, Kolopom Island and nearby to Western Province, Papua New Guinea.

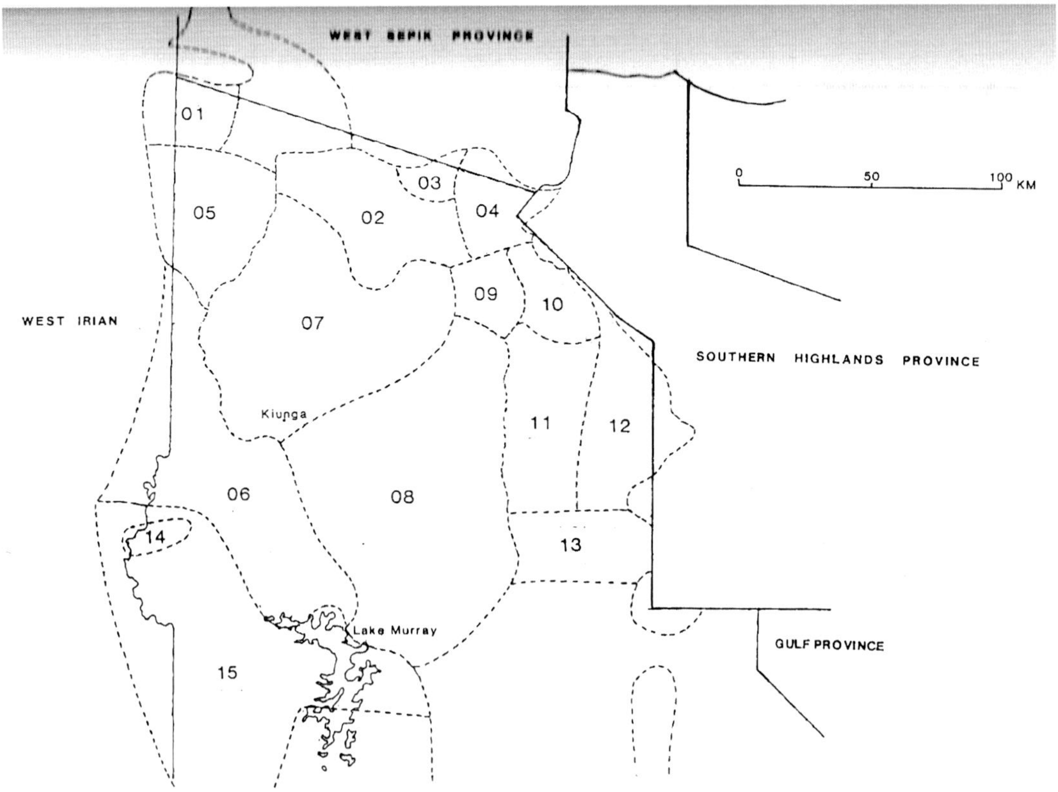

1	Kauwol	5	Ninggirum	9	Konai	13	Tomu
2	Faiwol	6	Yonggom	10	Agala	14	Kaeti
3	Setaman	7	Awin	11	Nomad	15	Boazi
4	Bimin	8	Pa	12	Beami		

Figure E6. Languages of west Western Province, Papua New Guinea.

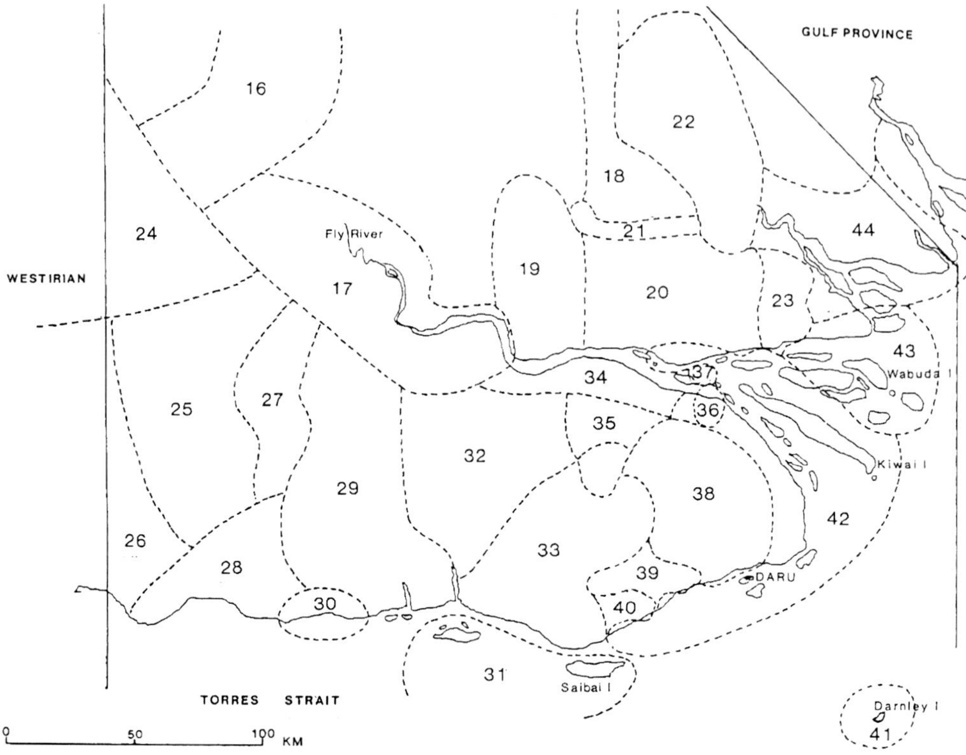

16	Zimakani	22	Tao-Suamato	27	Upper Morehead	33	Agöb	39	Bine
17	Suki	23	Waia	28	Lower Morehead	34	Lewada	40	Gizra
18	Kamula	24	Yey	29	Nambu	35	Mutum	41	Miriam
19	Ari-Waruna	25	Tonda	30	Dorro	36	Tirio	42	Southern Kiwai
20	Gogodala	26	Kanum	31	Mabuiag	37	Aturu	43	Wabuda
21	Bainapi			32	Idi	38	Gidra	44	Bamu Kiwai

Figure E7. Languages of south west Western Province PNG.

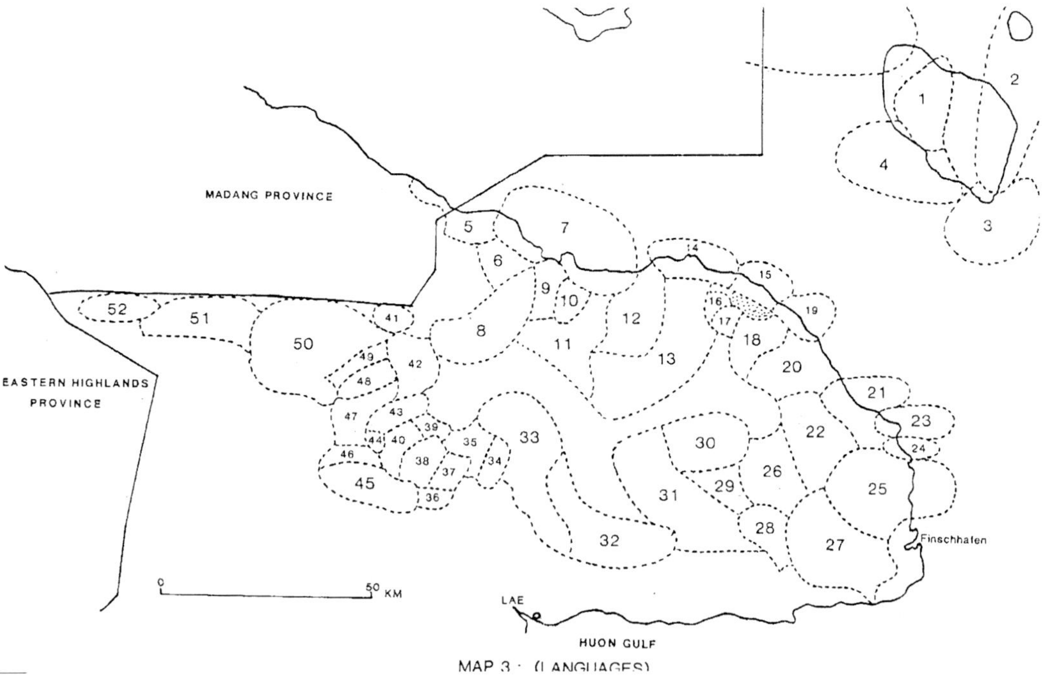

1	Kovai	11	Timbe	21	Migabac	31	Burum	41	Som	51	Ufim
2	Mangap	12	Selepet	22	Dedua	32	Mesem	42	Sakam	52	Wasembo
3	Mutu	13	Komba	23	Momare	33	Nabak	43	Gusan		
4	Barim	14	Sio	24	Sene	34	Nuk	44	Finungwa		
5	Roinji	15	Gitua	25	Kâte	35	Nek	45	Urii		
6	Nenaya	16	Kumukio	26	Kube	36	Munkip	46	Mamaa		
7	Malasanga	17	Kinalakna	27	Mape	37	Nakama	47	Irumu		
8	Yau	18	Nomu	28	Kosorong	38	Numanggang	48	Yagawak		
9	Komutu	19	Sialum	29	Yaknge	39	Sauk	49	Bam		
10	Weliki	20	Ono	30	Tobo	40	Nimi	50	Wantoat		

Figure E8. Languages of Siassi and Huon Peninsula of Morobe Province PNG.

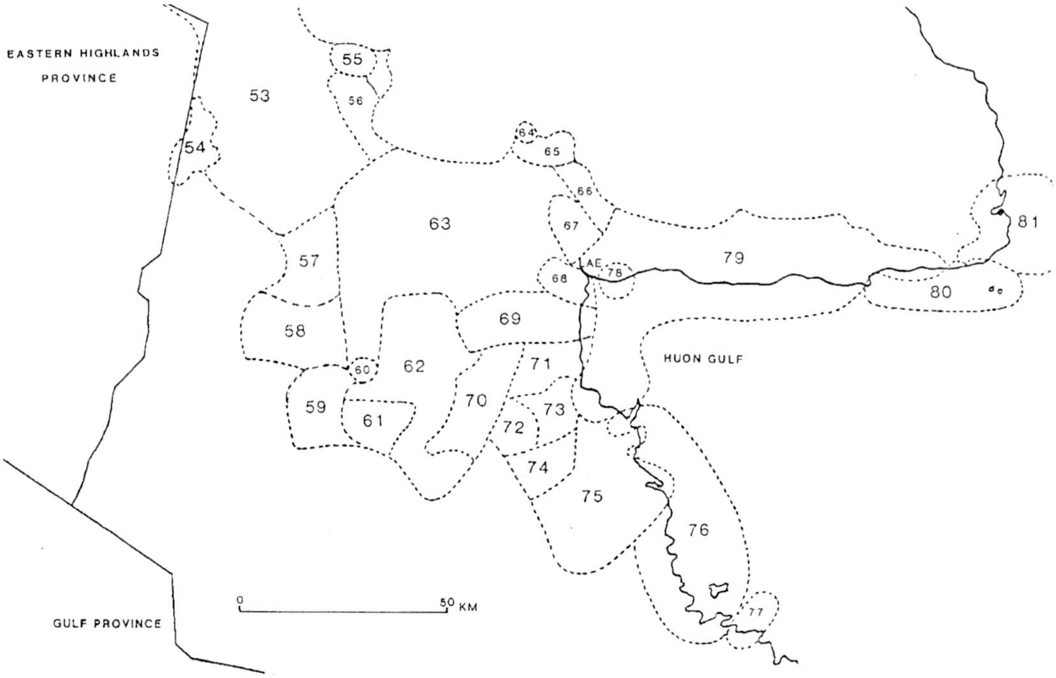

53	Adzera	58	Middle Watut	63	Wampur	68	Labu	73	Hote
54	Wampur	59	South Watut	64	Sirak (Nafi)	69	Mapos Buang	74	Misim
55	Sukurum	60	Piu	65	Duwet	70	Manga Buang	75	Kaiwa
56	Sirasira	61	Kapin	66	Musom	71	Vehes	76	Kela
57	North Watut	62	Mumeng	67	Yalu	72	Yamap	77	Numbami

Figure E9. Languages in central Morobe Province, PNG.

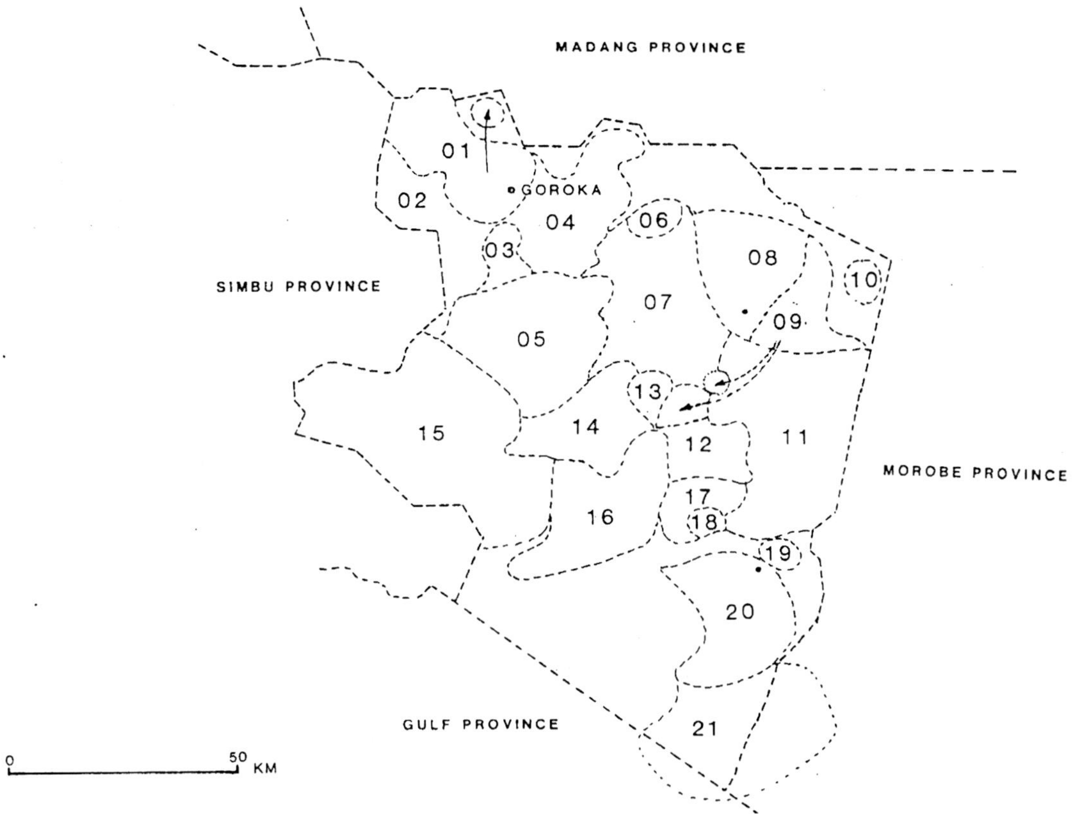

1	Gahuku-Asaro	5	Yagaria	9	Gadsup	13	Usarufa	17	Awa
2	Siane	6	Abaga	10	Binumarien	14	Yate	18	Owena
3	Yabiyufa	7	Kamano	11	Tairora	15	Gimi	19	Kenati
4	Benabena	8	Agarabi	12	Auyana	16	Fore	20	Baruya
								21	Yagwoia

Figure E10. Languages of Eastern Highlands Province, PNG.

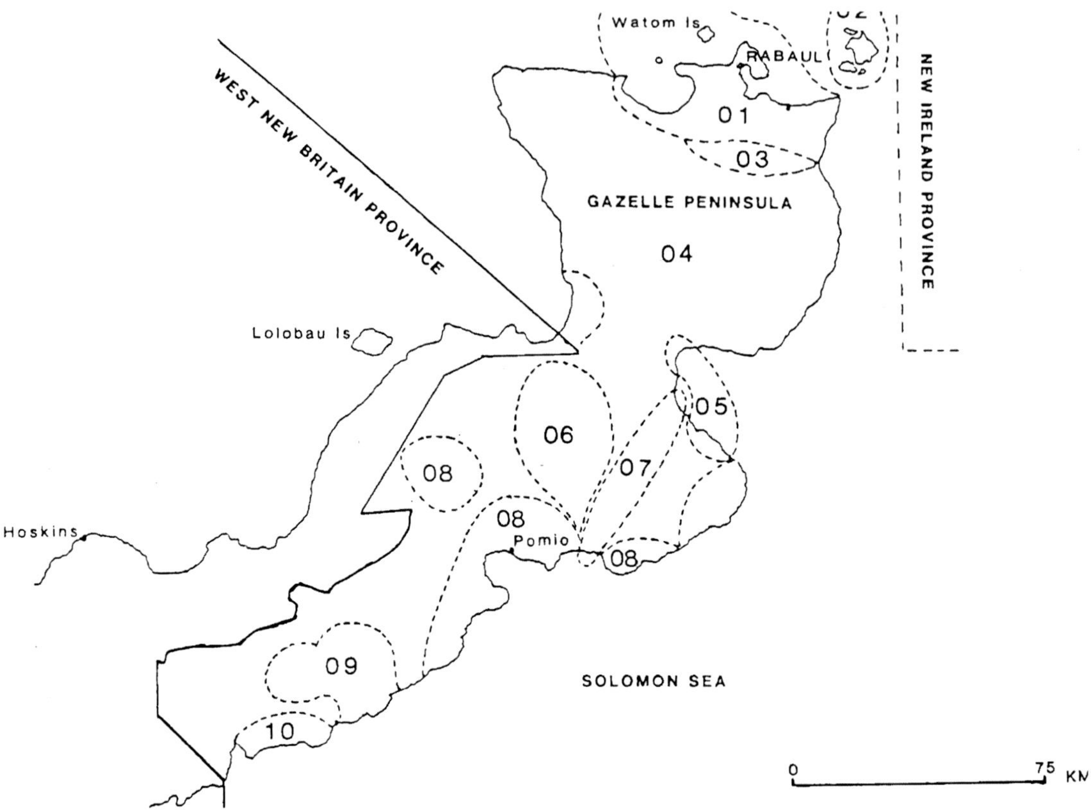

1	Tolai or Tinatatuna	3	Taulil	5	Sulka	7	Tomoip	9	Mamusi-Kakuna
2	Duke of York	4	Baining	6	Kol-Sui	8	Mengen	10	Uvol

Figure E11. Languages of East New Britain, PNG.

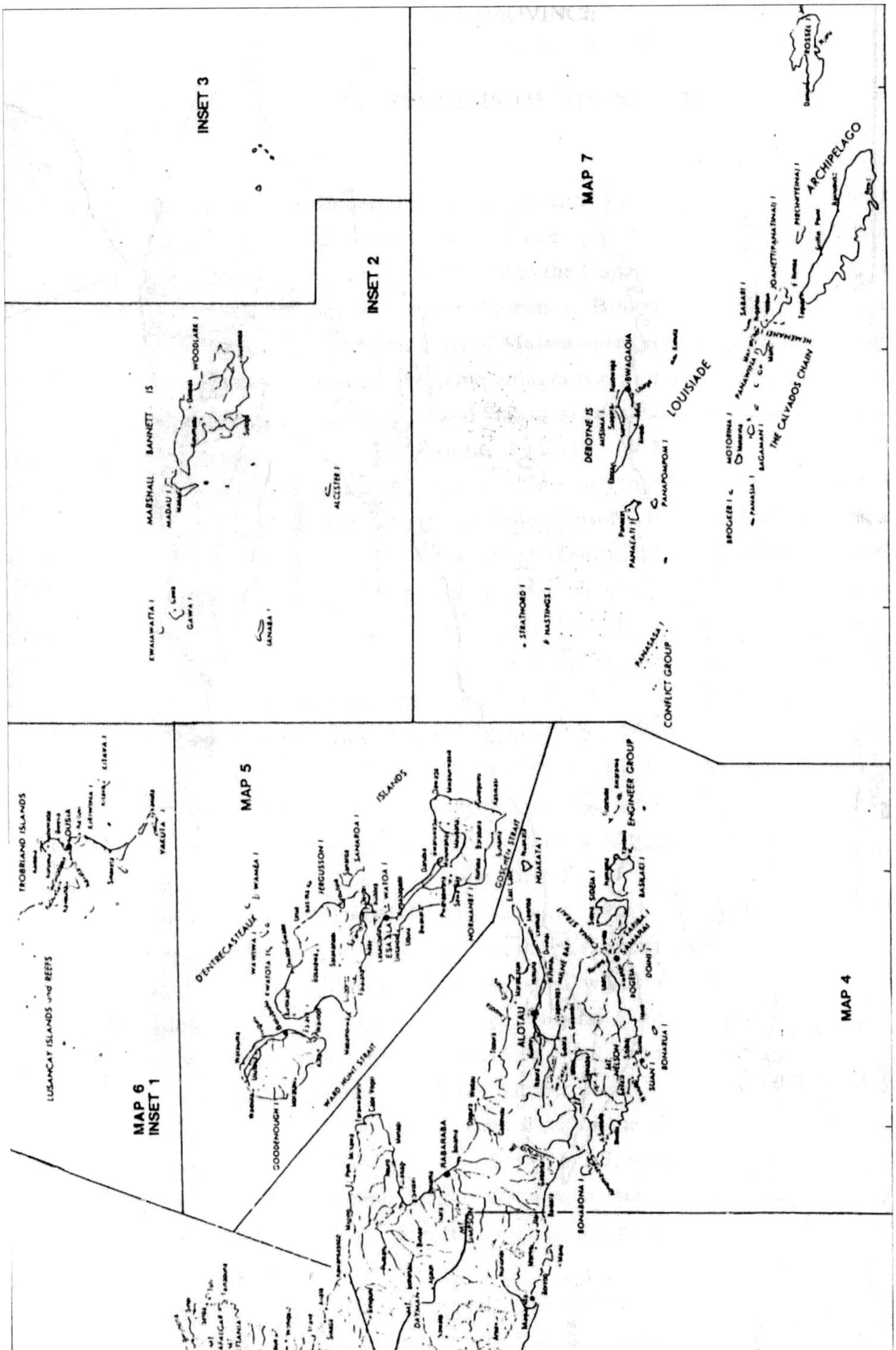

Figure E12. Mainland and island regions of Milne Bay Province, PNG.

Note. Dobu is spoken in parts of the D'Encastreau Islands, Yele on Rossel Island in the Louisiade Archepelago, Kilivila in the Trobriand Islands.

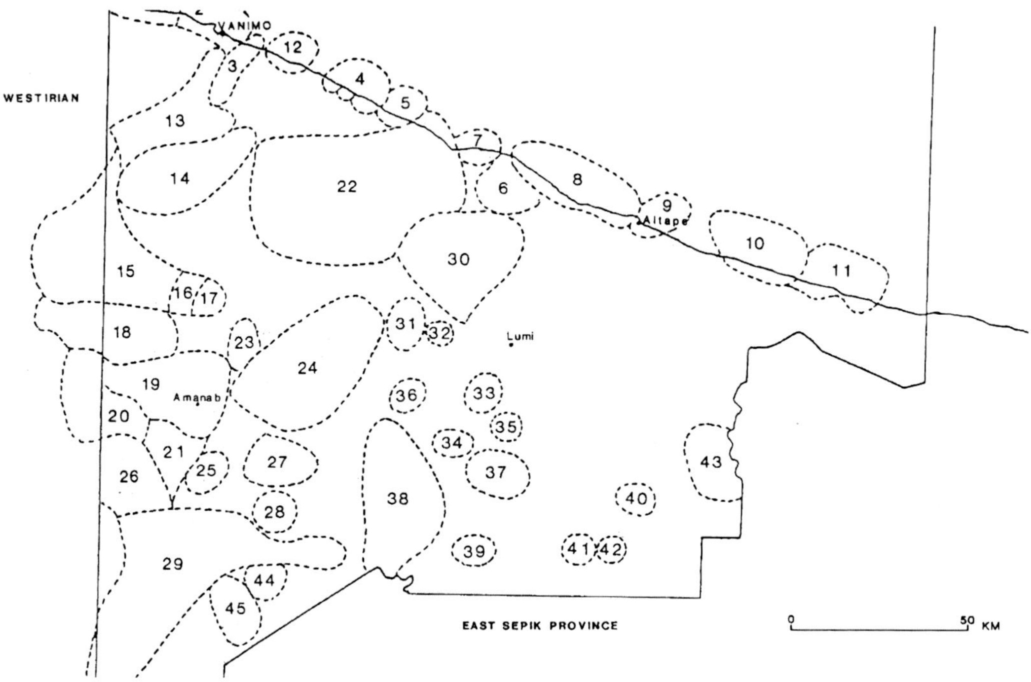

1	Wutung	9	Tomleo	17	Simog	25	Biaka	33	Autu	41	Kalou
2	Vanimo	10	Ali	18	Waina	26	Yuri	34	Bouye	42	Pasi
3	Krisa	11	Ulu-Suain	19	Amanab	27	Nagataman	35	Karawa	43	Mehek
4	Rawo	12	Ningera	20	Dera	28	Busa	36	Ak	44	Musian
5	Puari	13	Kilmeri	21	Anggor	29	Abau	37	Awun	45	Amto
6	Warapu	14	Pagi	22	Fas	30	One	38	Namie		
7	Sera	15	Waris	23	Baibai	31	Seta	39	Amal		
8	Sissano	16	Daonda	24	Kwomtari	32	Seti	40	Pahi		

Figure E13. Languages of the northern region of Sandaun (West Sepik) Province PNG.

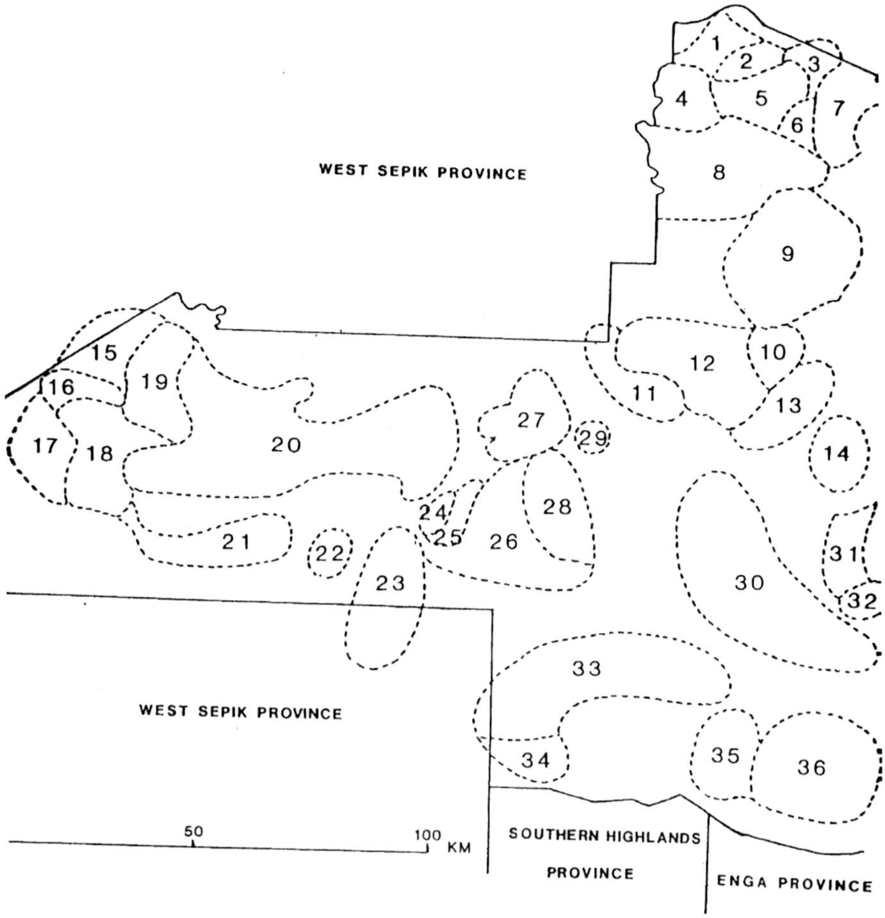

1	Kombio	9	Kwasengan	17	Bo	25	Pai	33	Paka
2	Yambes	10	Kaunga	18	Nemo	26	Sanio	34	Gabiano
3	Wom	11	Mayo	19	Ama	27	Wogamusin	35	Piame
4	Urim	12	Kwoma	20	Iwam	28	Bitara	36	Bikaru
5	Urat	13	Manambu	21	Owiniga	29	Ngala		
6	Bumbita	14	Yerakai	22	Papi	30	Bahinemo		
7	Southern Arapesh	15	Rocky Peak	23	Yabio	31	Mari		
8	Kwanga	16	Iteri	24	Walio	32	Watakataui		

Figure E14. Languages of upper Sepik River area of East Sepik Province.

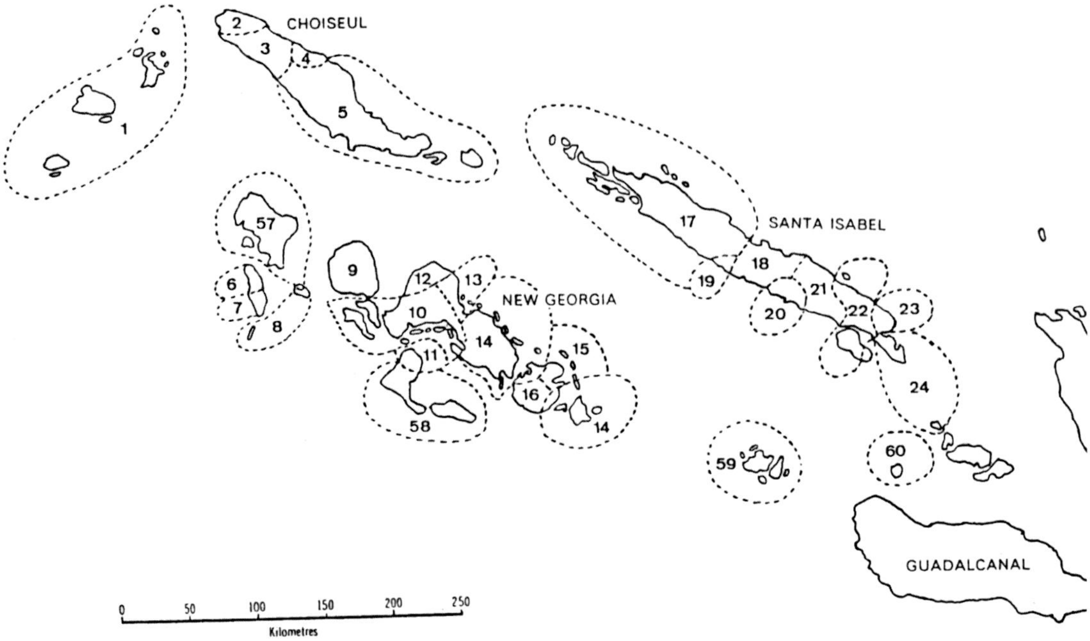

1	Mono-Alu	9	Nduke	17	Kia	57	Kazukuru		
2	Vaghua	10	Roviana	18	Kokota	58	Mbilua		
3	Varisi	11	Ughele	19	Laghu	59	Mbaniata		
4	Ririo	12	Kusaghe	20	Zazao	60	lavukaleve		
5	Central-East Choiseul	13	Hoava	21	Blablanga				
6	Ghanongga	14	Marovo	22	Maringe				
7	Lungga	15	Mbareke	23	Nggao				
8	Simbo	16	Vangunu	24	Mbughotu				

Figure E15. Languages of western Solomon Islands.

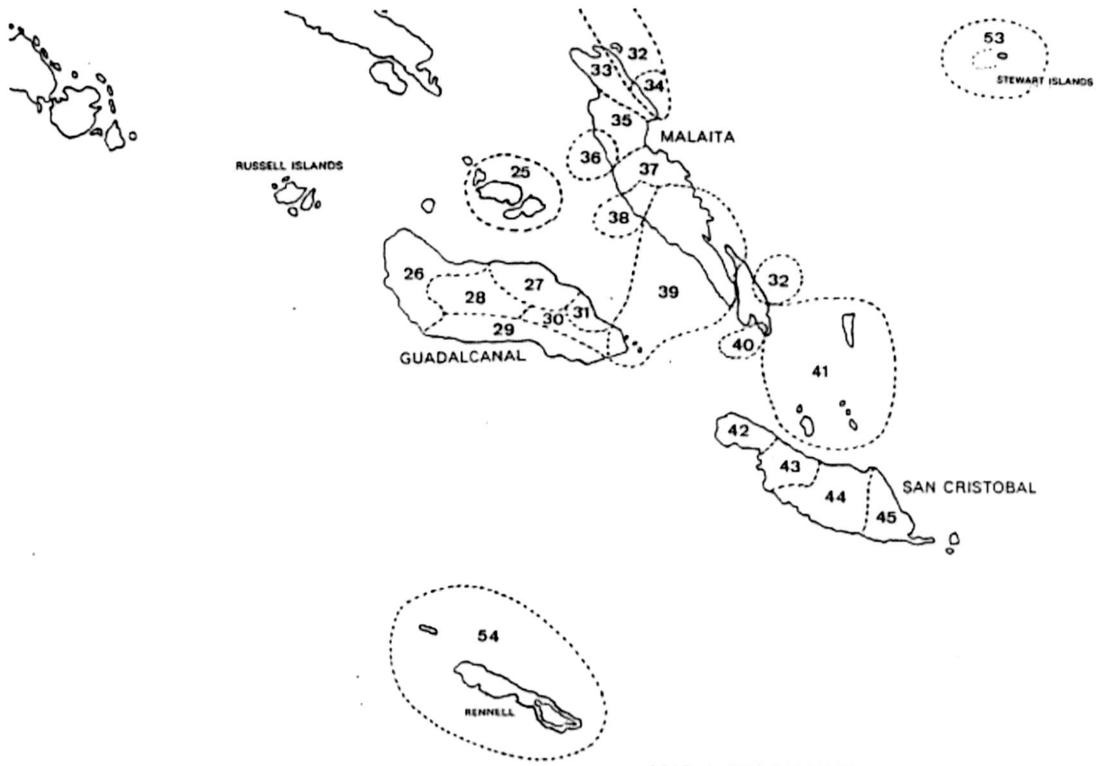

25	Nggela	33	North Malaita	41	South Malaita
26	West Guadalcanal	34	Kwai	42	Arosi
27	Lengo	35	Kwara'ae	43	Faghani
28	Malango	36	Langalanga	44	Bauro
29	Talise	37	Kwaio	45	Kahua
30	Mbirao	38	Dori'o	53	Sikaiana
31	Longgu	39	'Are'are	54	Rennellese
32	Lau	40	Oroha		

Figure E16. Languages of eastern Solomon Islands.

Diagram of Proto-Oceanic

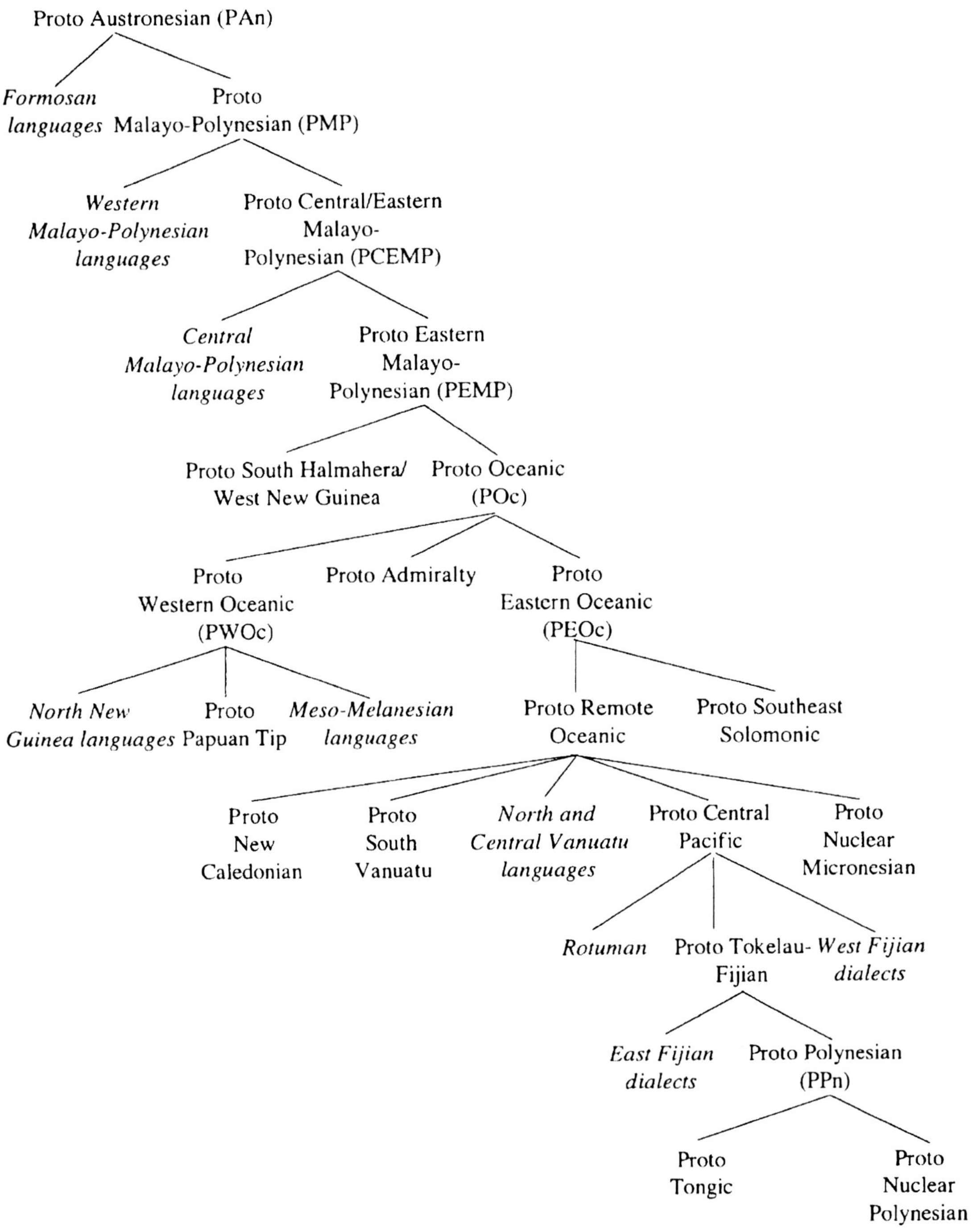

Figure E17. Tree Diagram Developed by Ross for Proto-Austronesia and Proto-Oceanic Languages.
Source. Ross, Pawley, and Osmond (2003).

References

Allen, J., & Hurd, C. (1963). *Standard survey word lists, Bougainville District*. Ukarumpa, Eastern Highlands Province, Papua New Guinea: Summer Institute of Linguistics Library.

Anceaux, J. (1958). Languages of the Bomberai Peninsula. *Nieuw Guinea Studien, 2*.

Anonymous. (1913). Wörterverzeichnisse von Papua Sprachen aus Hollandisch Neuguinea. *Anthropos, 8*, 254–260.

Armstrong, W. (1923). Report on anthropology of the South-Eastern Division. *Territory of Papua, Anthropology Report, 2*, 1–31.

Armstrong, W. (1924). Rossel Island money: A unique monetary system. *The Economic Journal, 34*, 423–429.

Armstrong, W. (1928). *Rossel Island: An ethnological study*. Cambridge, UK: Cambridge University Press.

Arnold, J. (1928). *A grammar of the language of Edugaura*. Port Moresby, Papua New Guinea: Government Printer.

Arnold, J. (1931). *A grammar of the language of Edugaura*. Port Moresby, Papua New Guinea: Government Printer.

Årsjö, S., & Årsjö, B. (2000). Konai. In J. Brownie (Ed.), *Sociolinguistic and literacy studies: South-West, Sepik and Morobe: Data papers on Papua New Guinea languages* (Vol. 46, pp. 26–81). Ukarumpa, Papua New Guinea: SIL.

Årsjö, S., & Årsjö, B. (2005). Organised phonology data Konai (Kal̠ai) language. *Summer Institute of Linguistics*. Available at http://wwwĐ01.sil.org/pacific/png/pubs/928474531159/Konai_OPD.pdf. Accessed 29 December 2015.

Bastard, E. (1919). Vocabularies: Abau Station, Eastern Division: *Annual Report for Papua 1917–1918* (pp. 88–92, Appendix E).

Bastard, E. (1920). Vocabularies Abau Station, Eastern Division: *Annual Report for Papua 1918–1919* (pp. 109–113, Appendix D).

Beaglehole, E., & Beaglehole, P. (1938). Ethnology of Pukapuka. *Bernice P. Bishop Museum Bulletin, 150*.

Bellamy, R. (1917). *Vocabularies: Losuia Trobriand Station, South Eastern Division (Bwaioa Tribe): Annual Report for Papua 1914–1915* (pp. 178, Appendix 174p).

Blackwood, B. (1931–1932). A report on field work in Buka and Bougainville. *Oceania, 2*, 199–219.

Bradshaw, J. (1994). Topics in the description of Kiriwina (Book Review). *Oceanic Linguistics, 33*(2), 584–586. doi:10.2307/3623142.

Briley, J. (1977). Some counting systems of Irian Jaya. *Irian: Bulletin of Irian Jaya Development, 6*(3), 28–32.

British New Guinea. (1890a). *Aboriginal vocabularies of Koiari Goto and Koita Ga: Annual Report for 1889–1890* (pp. 131–140, Appendix X4).

British New Guinea. (1890b). *Aboriginal vocabulary of Murua: Annual Report for 1889–1890* (pp. 148–152, Appendix X7).

British New Guinea. (1890). Collection of words of Rossel Island dialect *Annual Report for 1889–90* (p. 157, Appendix X10).

British New Guinea. (1892). *Vocabulary of dialect spoken by natives on the south coast of British New Guinea from Domara to Mairu: Annual Report for 1890–1891* (pp. 115–128, Appendix GG112).

British New Guinea. (1893). *Vocabulary of the Kiriwina dialect, British New Guinea: Annual Report for 1891–1892* (pp. 102–106, Appendix U).

British New Guinea. (1895). *Aboriginal vocabulary of Yela (Rossel Island): Annual Report for 1893–1894* (pp. 116–120, Appendix GG112).

Bromilow, W. (1893). *Aboriginal Vocabulary of Dobu: Annual Report on British New Guinea from 1891 to 1892* (pp. 106–110, Appendix U).

Bus, G. (1951). The *te* festival or gift exchange in Enga (Central Highlands of New Guinea). *Anthropos, 46*(5-6), 813–824.

Capell, A. (1962). *A linguistic survey of the south-western Pacific* (Technical Paper 136). Noumea: South Pacific Commission.

Capell, A. (1969). *A survey of New Guinea languages*. Sydney: Sydney University Press.

Capell, A. (1971). The Austronesian languages of Australian New Guinea. In T. Sebeok (Ed.), *Current trends in linguistics* (Linguistics in Oceania, Vol. 8, pp. 240–340). The Hague, The Netherlands: Mouton.

Carrier, A. (1981). Counting and calculation on Ponam Island. *Journal of the Polynesian Society, 90*(4), 465–479.

Chinnery, E. (1928). *Certain natives in south New Britain and Dampier Straits, Territory of New Guinea* (Anthropological Report No 3).

Chinnery, E. (1931). *The natives of South Bougainville and Mortlocks (Taku)* (Territory of New Guinea Anthropological Report No. 5, pp. 69–156).

Chowning, A. (1989). *The "Papuan Tip" languages reconsidered.* Paper presented at the VICAL 1: Papers from the Fifth International Conference on Austronesian languages, Auckland, New Zealand.

Chowning, A. (1976). Austronesian languages: New Britain. In S. Wurm (Ed.), *New Guinea area languages and language study, vol. 2: Austronesian languages* (Vol. C-39, pp. 365–386). Canberra, Australia: Pacific Linguistics.

Codrington, R. (1885). *The Melanesian languages*. Oxford, UK: Clarendon Press.

Conrad, R. (1978). *A survey of the Arapesh Language Family of Papua New Guinea* (Workpapers in Papua New Guinea Languages 25, pp. 57–78).

Conrad, R., & Dye, W. (1975). *Some language relationships in the upper Sepik region of Papua New Guinea* (Vol. A-10, pp. 1–35). Canberra, Australia: Pacific Linguistics.

Cowan, H. (1953). *Voorloopige resultaten van een ambtelijk taalonderzoek in Nieuw-Guinea*. The Hague, The Netherlands: M. Nijhoff.

Crotty, J. (1951). First dictionary of the Tchaga language, Central Highlands, New Guinea. *Anthropos, 46*, 933–963.

d'Armandville, W. (1903). Woordenlijst der taal die gesproken wordt in het gebergte van Kapaur tot aan Sekar. *Tijdschrift voor Indische Taal-, Land-, en Volkenkunde, 46*, 1–70.

Dempwolff, O. (1905). Beiträge zur Kenntnis der Sprachen von Deutsch-Neuguinea. *Mitteilungen des Seminars für orientalische Sprachen zu Berlin, 8*, 182–254.

Die deutsche marine-expedition 1907–1909 (1908). Marine Rundschau (pp. 19, 424–425; 675–476; 816–417; 951–452; 1069–1070; 1209–1011; 1474–1075).

Dutton, T. (1969). *The peopling of Central Papua: Some preliminary observations* (Vol. B-9). Canberra, Australia: Pacific Linguistics.

Dutton, T. (1973). *A checklist of languages and present-day villages of central and South-East mainland Papua* (Vol. B-24). Canberra, Australia: Pacific Linguistics.

Dutton, T. (1970). Notes on the languages of the Rigo area of the central district of Papua. In S. Wurm & D. Laycock (Eds.), *Pacific linguistics studies in honour of Arthur Capell* (Vol. C-13, pp. 879–983). Canberra, Australia: Pacific Linguistics.

Dutton, T. (1975). A Koita grammar sketch and vocabulary. In T. Dutton (Ed.), *Studies in languages of central and South-East Papua* (Vol. C-29, pp. 281–412). Canberra, Australia: Pacific Linguistics.

Dwyer, P., & Minnegal, M. (2016). Counting systems of the Strickland-Bosavi languages, Papua New Guinea. *Language and Linguistics in Melanesia, 34*(1), 1–36.

Fellowes, S. (1902). *Grammar and dictionary of the Kiriwina dialect: Annual Report on British New Guinea from 1900–1901* (pp. 171–196, Appendix N172).

Fisher, J. (~2010). *Enriching students' learning through ethnomathematics in Kuruti elementary schools in Papua New Guinea*. Lae, Papua New Guinea: ResearchGate.

Fortune, R. (1942). Arapesh. *American Ethnological Society, 19*, 1–60.

Fox, C. (1931). Arosi numerals and numeration. *Journal of the Polynesian Society, 40*, 236–243.

Friederici, G. (1913). Wissenschaftliche ergebnisse einer amtlichen forschungsreise nach dem Bismarck-Archipel im Jahre 1908. 3: Untersuchungen über eine melanesische wanderstrasse. *Mitteilung aus den Deutschen Schutzgebieten, Ergänzungsheft 7.*

Frizzi, E. (1914). *Ein Beitrag zur ethnologie von Bougainville und Buka mit spezieller berucksichtigung der Nasioi Archiv, Baesslet* (Vol. 6). Leipzig: Trübner.

Galis, K. (1960). Telsystemen in Nederlands-Nieuw-Guinea. *Nieuw Guinea Studien, 4*(2), 131–150.

Gerstner, A. (1963). Grammatik der Alubänsprache. *Micro-Bibliotheca Anthropos, 37.*

Girard, F. (1968–1969). Les notions de nobre et de temps chez les Buang de Nouvelle-Guinée (District du Morobe). *L'Ethnographie*, 160–178.

Glasgow, D., & Loving, R. (1964). *Languages of the Maprik sub-district* (p. 11). Port Moresby, PNG: Department of Information and Extension Services.

Grisward, P. (1910). Notes grammaticales sur la langue des Teleï, Bougainville, Iles Salomones. *Antropos, 5*(82–94), 381–406.

Guise, R. (1892). *Vocabulary of Bula'a dialect, spoken in the Central District of British New Guinea, by some coast tribes* (Annual report on British New Guinea from 1890 to 1891) (pp. 108–114, Appendix GG101).

Healey, A. (1964). *The Ok Language Family of New Guinea* (PhD). Australian National University, Canberra, Australia.

Henderson, J. (1975). Yeletnye, the language of Rossel Island. In T. Dutton (Ed.), *Studies in languages of central and south-east Papua* (Vol. C-29, pp. 817–834). Canberra, Australia: Pacific Linguistics.

Holzknecht, S. (1988). *The Markham languages of Papua New Guinea. A history of the Austronesian languages of the Markham and Ramu valleys, Papua New Guinea* (PhD). Australian National University, Canberra, Australia.

Holzknecht, S. (1989). *The Markham languages of Papua New Guinea* (Vol. C-115). Canberra, Australia: Pacific Linguistics.

Hooley, B. (1971). Austronesian languages of the Morobe District, Papua New Guinea. *Oceanic Linguistics, 10*, 79–151.

Hooley, B. (1978). Number and time in Central Buang. *Kivang: Journal of the Linguistic Society of Papua New Guinea, 11*(2), 152–170.

Hurd, C., & Hurd, P. (1966). *Nasioi language course*. Port Moresby, PNG: Department of Information and Extension Services.

Kirschbaum, F. (1938). Über Zahlensysteme im Zentralgebirge von Neuguinea. *Anthropos, 33*, 278–279.

Klaffl, J., & Vormann, F. (1905). Die Sprachen des Berlinhafen-Bezirks in Deutsch-Neuguinea. *Mitteilungen des Seminar für orientalische Sprachen zu Berlin, 8*, 1–138.

Kluge, T. (1938). *Die Zahlbegriffe der Australier, Papua und Bantuneger nebst einer Einleitung ueber die Zahl, ein Beitrag zur Geistesgeschichte des Menschen*. Berlin, Germany.

Kluge, T. (1941). *Die Zahlenbegriffe der Sprachen Central- und Südostasiens, Indonesiens, Micronesiens, Melanesiens und Polynesiens mit Nachträgen zu den Bänden 2–4*. Berlin, Germany. Ein fünfter Beitrag zur Geistesgeschichte des Menschen nebst einer principiellen Untersuchung über die Tonsprachen. Original unpublished manuscript held at Yale University, microfilm held at Australian National University, Canberra, Australia.

Kolia, J. (1975). A Balawaia grammar sketch and vocabulary. In T. Dutton (Ed.), *Studies in languages of central and South-East Papua* (Vol. C-29, pp. 107–226). Canberra, Australia: Pacific Linguistics.

Kopamu, S. (2005). Mathematical concepts in Enga Tokples. *Papua New Guinea Journal of Mathematics, Computing & Education, 7*(1), 13–18.

Lancy, D. (1978). Indigenous mathematics systems: An introduction. *Papua New Guinea Journal of Education, 14*(Special Issue), 6–15.

Lancy, D. (1983). *Cross-cultural studies in cognition and mathematics*. New York, NY: Academic Press.

Lang, A. (1973). *Enga dictionary* (Vol. C-20). Canberra, Australia: Pacific Linguistics.

Laufer, B. (1946–1949). Rigenmucha, das Höchste Wesen der Baining (neubritannien). *Anthropos, 41/44*, 497–560.

Lawes, F. (1894). *Vocabulary of words spoken by the Keakalo tribe, from Vererupu to Paramana Point* (British New Guinea Annual Report 1892–1893) (pp. 103–106).

Lawes, W. (1885, revised 1896). *Grammar and vocabulary of the language spoken by the Motu tribe*. Sydney, Australia: C. Potter, Government Printer.

Lawes, W. (1890). *Comparative view of New Guinea dialects: Annual Report on British New Guinea from 1889 to 1890* (pp. 158–167, Appendix X111).

Lawton, R. (1977). Missionary Lingue Franche: Dobu. In S. Wurm (Ed.), *New Guinea area languages and language study, vol. 3: Language, culture, society, and the modern world*, 2 fascicles (Vol. C-40, pp. 907–946). Canberra, Australia: Pacific Linguistics.

Lawton, R. (1991). *Topics in the descriptions of Kiriwina* (Vol. D-84). Canberra, Australia: Pacific Linguistics.

Laycock, D. (1973). *Sepik languages: Checklist and preliminary classification* (Vol. B-25). Canberra, Australia: Pacific Linguistics.

Laycock, D. (1975a). Observations on number systems and semantics. In S. Wurm (Ed.), *New Guinea area languages and language study. Vol. 1: Papuan languages and the New Guinea linguistic scene* (pp. 219–233). Canberra, Australia: Pacific Linguistics.

Laycock, D. (1975b). The Torricelli Phylum. In *New Guinea area languages and language study, 1: Papuan languages and the New Guinea linguistic scene* (Vol. C-38, pp. 767–780). Canberra, Australia: Pacific Linguistics.

Laycock, D. (1976). Austronesian languages: Sepik Provinces. In S. Wurm (Ed.), *New Guinea area languages and language study, vol. 2: Austronesian languages, Pacific Linguistics* (Vol. C-39, pp. 399–418). Canberra, Australia: Australian National University.

Le Roux, C. (1950). *De Bergpapoea's van Nieuw-Guinea en hun Woongebied* (2 Vols.). Leiden: E. J. Brill.

Lean, G. (1992). *Counting systems of Papua New Guinea and Oceania* (Unpublished PhD thesis). PNG University of Technology, Lae, Papua New Guinea. Retrieved from http://www.uog.ac.pg/glec/

Leverd, A. (1922). Polynesian language of Uvea, Loyalty Islands. *Journal of the Polynesian Society, 31*, 94–103.

Lincoln, P. (1976). Rai Coast word lists. Unpublished manuscript.

Lindrud, S. (1980). The languages and communities of the Open Bay, Wide Bay and southern Gazelle Peninsula region. In R. Johnston (Ed.), *Language, communication and development in New Britain*. Ukarumpa: Summer Institute of Linguistics.

Lister-Turner, R., Clark, J., & Chatterton, P. (1931). *A dictionary of the Motu language of Papua* (2nd ed.). Sydney, Australia: Government Printer.

Lithgow, D. (1976). Austronesian languages: Milne Bay and adjacent islands (Milne Bay Province). In S. Wurm (Ed.), *New Guinea area languages and language study, Vol. 2: Austronesian languages* (Vol. C-39, pp. 157–170; 441–523). Canberra, Australia: Pacific Linguistics.

Loving, R., & Bass, J. (1964). *Languages of the Amanab sub-district*. Port Moresby, PNG: Department of Information and Extension Services.

Malinowski, B. (1917–1920). Classificatory particles in the language of the Kiriwina. *Bulletin of the School of Oriental Studies, London Institution, 1*, 33–78.

Malinowski, B. (1920). Kula: The circulating exchange of valuables in the archipelagoes of eastern New Guinea. *MAN, 20*, 97–105.

Malinowski, B. (1922). *Argonauts of the Western Pacific*. London, UK: Routledge and Kegan Paul.

Mark, K. (2003). *Ethno-mathematics: Is there any mathematics in my culture? Deriving mathematics from culture*. Goroka, Papua New Guinea: University of Goroka.

McAdam, T. (1926). *Vocabulary of native languages: Annual Report to the Council of the League of Nations on the Administration of the Territory of New Guinea for 1924–1925* (pp. 81–93, Appendix B).

McElhanon, K. (1984). *A linguistic field guide to the Morobe Province, Papua New Guinea* (Vol. D-57). Canberra, Australia: Pacific Linguistics.

Meggitt, M. (1957/1958). The Enga of the New Guinea Highlands: Some preliminary observations. *Oceania, 28*, 253–330.

Meggitt, M. (1958). Mae Enga time-reckoning and calendar New Guinea. *MAN, 58*, 74–77.

Müller, H. (1907). Grammatik der Mengen-sprache. *Anthropos, 2*(80–90), 241–254.

National Statistical Office (1982). *1980 National population census. Final figures: Census Unit population* (19 vols.). Port Moresby, PNG: National Statistical Office.

National Statistical Office. (2014a). *2011 National population and housing census - Momase Region*. Retrieved from Port Moresby, PNG: https://www.nso.gov.pg/

National Statistical Office. (2014b). *2011 National population and housing census - Southern Region*. Retrieved from Port Moresby, PNG: https://www.nso.gov.pg/

Neuhaus, K. (1954). Grammatik der Lir-Sprache in Melanesien. *Micro-Bibliotheca Anthropos, 20*.

New Guinea Report. (1924). *Vocabulary of native languages, Sepik River, Aitape District, and portions of Talasea District: Annual Report to the Council of the League of Nations on the administration of the Territory of New Guinea for 1922–1923* (pp. 128–134, Appendix C).

Oliver, D. (1949). Studies in the anthropology of Bougainville, Solomon Islands. *Papers of the Peabody Museum of American Archeology and Ethnology* (Vol. 49, pp. 6–27). Cambridge, MA: Harvard University.

Oliver, D. (1967). *A Solomon Island society: Kinship and leadership among the Siuai of Bougainville*. Boston, MA: Beacon Press.

Onagi, G. (~2008). *Let's speak Magi* (A dictionary and grammar of Magi). Goroka: University of Goroka.

Owens, K. (2001). The work of Glendon Lean on the counting systems of Papua New Guinea and Oceania. *Mathematics Education Research Journal, 13*(1), 47–71.

Panoff, M. (1969). The notion of time among the Maenge people of New Britain. *Ethnology, 8*, 153–166.

Panoff, M. (1970). Father arithmetic: Numeration and counting in New Britain. *Ethnology, 9*(4), 358–365.

Parkinson, R. (1907). *Dreissig jahre in der Südsee*. Stuttgart, Germany: Strecker und Schröder.

Patrick, H. (1981). *Bibliography of the Summer Institute of Linguistics, Papua New Guinea Branch, 1956–1980*. Ukarumpa, PNG: SIL.

Pawley, A. (1972). On the internal relationships of Eastern Oceanic languages (from 1969 MS). In R. Green & M. Kelly (Eds.), *Studies in Oceanic culture history 3: Pacific anthropological records 13* (pp. 1–142). Honolulu, HI: Bishop Museum.

Pawley, A. (1976). Austronesian languages: Western parts of south-eastern mainland Papua. In S. Wurm (Ed.), *New Guinea area languages and language study, vol. 2: Austronesian languages* (Vol. C-39, pp. 301–320). Canberra, Australia: Pacific Linguistics.

Peekel, G. (1909). Grammatik der Neu-Mecklenburgischen sprache, speziell der Pala-Sprache. *Archiv für das Studium Deutscher Kolonialsprachen, 9*.

Rath, D. (1980). The languages and communities of the Mengen region. In R. Johnston (Ed.), *Language, communication and development in New Britain* (pp. 197–224). Ukarumpa, Eastern Highlands Province, Papua New Guinea: Summer Institute of Linguistics.

Rausch, P. (1912). Die Sprache von Südost-Bougainville, deutsche Salomoninseln. *Anthropos, 7*, 105–134; 585–616; 964–994.

Ray, S. (1895). The languages of British New Guinea. *The Journal of the Anthropological Institute of Great Britain and Ireland, 24*, 15–39. doi:10.2307/2842475.

Ray, S. (1907). *Reports of the Cambridge anthropological expedition to Torres Straits* (Vol. 3). Cambridge: Cambridge University Press.

Ray, S. (1912a). Notes on the languages in the east of Netherlands New Guinea. In A. Wollaston (Ed.), *Pygmies and Papuans: The stone age today In Dutch New Guinea* (pp. 322–345). London, UK: Smith Elder.

Ray, S. (1912b). Polynesian linguistics. *Journal of the Polynesian Society, 21*(65–76), 164–172.

Ray, S. (1917). Polynesian languages of the Solomon Islands. *Journal of the Polynesian Society, 26*, 36–43; 99–105; 170–179.

Ray, S. (1919a). The languages of northern Papua. *Journal of the Royal Anthropological Institute of Great Britain and Ireland, 49*, 317–341.

Ray, S. (1919b). Polynesian linguistics. *Journal of the Polynesian Society, 28*, 168–177.

Ray, S. (1919c). The Polynesian languages of Melanesia. *Anthropos, 14*, 46–96.

Ray, S. (1927). The Papuan languages. In *Festschrift meinhof, sprachwissenschaftliche und andere studien* (pp. 377–385). Hamburg, Germany: L. Friederichsen & Co.

Ray, S. (1929). The languages of the Central Division of Papua. *Journal of the Royal Anthropological Institute of Great Britain and Ireland, 59*, 65–96.

Ray, S. (1938a). The languages of the Eastern and South-Eastern Divisions of Papua. *Journal of the Royal Anthropological Institute of Great Britain and Ireland, 68*, 153–208.

Ray, S. (1938b). The languages of the Eastern Louisiade Archipelago. *Bulletin of the School of Oriental Studies, London Institution, 9*(2), 363–384.

Ross, M. (1992). The position of Gumawana in the Papuan Tip Cluster. *Languages and Linguistics in Melanesia, 23*, 139–165.

Ross, M., Pawley, A., & Osmond, M. (2003). *The lexicon of Proto Oceanic: The culture and environment of ancestral Oceanic society. 2: The physical environment*. Canberra, Australia: Pacific Linguistics, Research School of Pacific and Asian Studies ANU.

Salzmann, Z. (1950). A method for analyzing numerical systems. *Word, 6*, 78–83.

Saville, W. (1912). A grammar of the Mailu language, Papua. *Journal of the Royal Anthropological Institute of Great Britain and Ireland, 42*, 397–436.

Saxe, G. (1981). Body parts as numerals: A developmental analysis of numeration among the Oksapmin in Papua New Guinea. *Child Development, 52*, 306–316.

Saxe, G. (2012). *Cultural development of mathematical ideas: Papua New Guinea studies*. New York, NY: Cambridge University Press.

Schmidt, M. (1900). Die sprachlichen Verhältnisse von Deutsch Neu-Guinea. *Zeitschrift für Afrikanische und Ozeanische Sprachen, 5*, 345–384.

Schmidt, W. (1909). Neuendeckte Papuasprachen von den Salomoninseln (Bougainville). *Globus, 95*, 206–207.

Seidenberg, A. (1960). The diffusion of counting practices. *University of California Publications in Mathematics, 3*, 215–300.

Seligman, C. (1910). *The Melanesians of British New Guinea*. Cambridge: Cambridge University Press.

Seligmann, C. (1912–1913). Five Melanesian vocabularies from British New Guinea. *Zeitschrift fur Kolonialsprachen, 3*, 182–200.

Senft, G. (1986). *Kilivila, the language of the Trobriand Islanders*. Berlin, Germany: Mouton de Gruyter.

Shaw, D. (2009*). Numeral systems of the world's languages: Samo, Papua New Guinea*. Available at https://mpiĐlingweb.shh.mpg.de/numeral/Samo.htm. Accessed 19 January 2016.

SIL. (n.d.). *Ethnologue: Languages of the world*. Retrieved from http://www.ethnologue.com/

Smith, G. (1984). *Morobe counting systems: An investigation into the numerals of the Morobe Province, Papua New Guinea* (MPhil). Papua New Guinea University of Technology, Lae, Papua New Guinea.

Smith, J., & Weston, P. (1975). *Notes on Mianmin grammar* (Workpapers in Papua New Guinea Languages 7, pp. 35–142).

Smythe, W. E. (n.d*.). Admiralty Islands linguistics, Part 7: Gele' vocabulary*. Ukarumpa, Eastern Highlands Province, Papua New Guinea: Unpublished type script held at SIL library.

Stokhof, W. A. L. (1983). *Holle lists: Vocabularies in languages of Indonesia, Irian Jaya: Papuan languages, northern languages, central highlands languages* (Vol. 5, No. 2). Canberra, Australia: Pacific Linguistics.

Strathern, A. (1971). *The rope of moka: big-men and ceremonial exchange in Mount Hagen, New Guinea*. Cambridge, UK: University Press.

Strathern, A. (1977). Mathematics in the *moka*. *Papua New Guinea Journal of Education, Special Edition The Indigenous Mathematics Project, 13*(1), 16–20.

Strong, W. (1911). *Notes on the languages of the north-eastern and adjoining Divisions: Annual Report for Papua 1910–1911* (pp. 203–217).

Symons, A. (1917). *Vocabulary - Kulumadau Station, South-Eastern Division (Tribe: Laughlin Islands): Annual Report for 1914–1915 Papua* (pp. 179, Appendix 174q).

Symons, A. (1919). *Vocabulary - Kulumadau Station, South-Eastern Division (Tribe: Marshall Group): Annual Report for 1917–1918, Papua* (pp. 87, Appendix E).

Taylor, A. (1970). *Syntax and phonology of Motu (Papua): A transformational approach*. PhD, Australian National University, Canberra, Australia.

Thomson, N. (1975a). *The dialects of Magi* (Vol. A-40, pp. 37–90). Canberra, Australia: Pacific Linguistics.

Thomson, N. (1975b). Magi phonology and grammar—Fifty years afterwards. In T. Dutton (Ed.), *Studies in languages of central and south-east Papua* (Vol. C-29, pp. 599–667). Canberra, Australia: Pacific Linguistics, Australian National University.

Thune, C. (1978). Numbers and counting in Loboda: An example of a non-numerically oriented culture. *Papua New Guinea Journal of Education, 14*, 69–80.

Thurnwald, R. (1909). Reisebericht aus Buin und Kieta. *Zeitschrift für Ethnologie, 41*, 512–532.

Tranel, W. (1952). Völkerkundliche und sprachliche aufzeichmungen aus dem moando-sprachgebiet in nordost-Neuguinea. *Anthropos, 47*, 447–473.

Tryon, D. (1976). *New Hebrides languages: An internal classification* (Vol. C-50). Canberra, Australia: Pacific Linguistics.

Tryon, D. (2006). Proto-Austronesian and the major subgroups of Austronesian. In P. Bellwood, J. Fox, & D. Tryon (Eds.), *The Austronesians: Historical and comparative perspectives* (pp. 19–42). Canberra, Australia: ANU Press.

Tryon, D., & Hackman, B. (1983). *Solomon Islands languages: An internal classification* (Vol. C-72). Canberra, Australia: Pacific Linguistics.

van der Stap, P. (1966). Outline of Dani morphology. In *Verhandelingen van het Koninklijk Instituut voor Taal-, Land- en Volkenkunde* (Vol. 48). The Hague, The Netherlands: M. Nijhoff.

Vicedom, G., & Tischner, H. (1943–1948). *Die Mbowamb: die Kultur der Hagenberg-Stämme im östlichen Zentral-Neuguinea*. (3 vols. (1943 volume published by de Gruyter)). Museum für Völkerkunde und Vorgeschichte, Hamburg, Germany.

Voorhoeve, C. (1975). *Languages of Irian-Jaya: Checklist, preliminary classification, language maps, wordlists* (Pacific Linguistics, Vol. B-31). Canberra, Australia: Pacific Linguistics.

Werner, E. (1909). Im westlichen Finisterregebirge und an der Nordküste von Deutsch-Neuguinea. *Globus, 55*(73–82), 107–112.

Wheeler, G. (1910–1911). A note on the Telei speech of South Bougainville, Solomon Islands. *Zeitschrift für kolonial-sprachen*, 4, 290–304.

Wirz, P. (1923). Worterverzeichnis der Snrache des Stammes Mansibaber (von Hollandisch-Nord-Neuguinea). *Tijdschrift voor Indische taal-, land- en volkenkunde, 62*, 189–208.

Wirz, P. (1924). Anthropologische und ethnologische Ergebnisse der Central Neu-Guinea Expedition 1921–1922. In *Nova-Guinea* (Vol. 16, Book 1). Leiden: E. J. Brill.

Woodford, C. (1907). Notes on Rennell Island. *MAN, 7*(24), 33–37.

Wurm, S. (1975). The East Papuan Family in general. In S. Wurm (Ed.), *New Guinea area languages and language study. Vol. 1: Papuan languages and the New Guinea linguistic scene* (Vol. C-38, pp. 783–804). Canberra, Australia: Pacific Linguistics.

Wurm, S. (1982). Papuan languages. In R. May & H. Nelson (Eds.), *Melanesia: Beyond diversity* (pp. 225–240). Canberra, Australia: Research School of Pacific Studies, Australian National University.

Wurm, S., & Hattori, S. (Eds.). (1981). *Language atlas of the Pacific area. Part 1: New Guinea area, Oceania, Australia*. Canberra, Australia: Australian Academy of the Humanities.

Z'Graggen, J. (1975a). *Comparative wordlists of the Admiralty Islands languages collected by W. E. Smythe* (Workpapers in Papua New Guinea Languages 14, pp. 117–216).

Z'Graggen, J. (1975b). *The languages of the Madang District, Papua New Guinea* (Vol. B-41). Canberra, Australia: Pacific Linguistics.

Zöller, H. (1890). Untersuchungen über 24 Sprachen aus dem Schutzgebiet der Neuguinea-Kompanie. *Petermanns Geographische Mitteilungen, 36*, 122–128; 145–152; 181.

About the Authors

Glendon Angove Lean lectured at the Papua New Guinea University of Technology and Deakin University, Geelong, Victoria, Australia. He completed his doctoral research on the *Counting Systems of Papua New Guinea and Oceania* with the PNG University of Technology.

Patricia Paraide was a researcher with the PNG National Research Institute and is Associate Professor at Divine Word University, Madang, Papua New Guinea.

Charly Muke was a lecturer at the University of Goroka and is currently a mathematics teacher at St Theresa's College, Abergowrie, Townsville, Queensland, Australia.

Kay Owens taught at Balob Teachers College and the PNG University of Technology, during the same period as Glen Lean both in Lae, Morobe Province, PNG, and at the University of Western Sydney (now Western Sydney University) and is currently at Charles Sturt University, Dubbo, NSW, Australia.

Top row: Kay Owens (photographer – Janet Henderson); Glen Lean in PNG and at graduation (photographer – Jane Jones);
Bottom row: Patricia Paraide (photographer - Stephanie Paraide); Charly Muke in red shirt delivering books to a village school Jiwaka Province, PNG (photographer – Kay Owens)

K. Owens et al., *History of Number*, History of Mathematics Education,
DOI 10.1007/978-3-319-45483-2

Author Index

K. Owens et al., *History of Number*, History of Mathematics Education,
DOI 10.1007/978-3-319-45483-2

Subject Index

K. Owens et al., *History of Number*, History of Mathematics Education,
DOI 10.1007/978-3-319-45483-2

D

E